Evolution and Development
of Neural Circuits

A subject collection from *Cold Spring Harbor Perspectives in Biology*

Evolution and Development of Neural Circuits

A subject collection from *Cold Spring Harbor Perspectives in Biology*

EDITED BY

Laura C. Andreae
King's College London

Justus M. Kebschull
Johns Hopkins University

Anthony M. Zador
Cold Spring Harbor Laboratory

COLD SPRING HARBOR LABORATORY PRESS
Cold Spring Harbor, New York • www.cshlpress.org

Evolution and Development of Neural Circuits

A subject collection from *Cold Spring Harbor Perspectives in Biology*
Articles online at www.cshperspectives.org

Executive Editor	Richard Sever
Project Supervisor	Barbara Acosta
Editorial Assistant	Danett Gil
Permissions Administrator	Carol Brown
Production Editor	Diane Schubach
Production Manager/Cover Designer	Denise Weiss
Publisher	John Inglis

Front cover artwork: Image of a futuristic brain from the Lumiere Collective © Shutterstock (ID 2521846167).

Library of Congress Cataloging-in-Publication Data

Names: Andreae, Laura C., 1971- editor.
Title: Evolution and development of neural circuits / edited by Laura C. Andreae, King's College London, Justus M. Kebschull, Johns Hopkins University, and Anthony M. Zador, Cold Spring Harbor Laboratory.
Description: Cold Spring Harbor, New York : Cold Spring Harbor Laboratory Press, [2025] | "A subject collection from Cold Spring Harbor Perspectives in Biology." | Includes bibliographical references and index.
Summary: "The organization of many neurons into complex circuits is critical for brain function. This volume explores how these circuits evolved, develop and function in the nervous system"-- Provided by publisher.
Identifiers: LCCN 2024027527 (print) | LCCN 2024027528 (ebook) | ISBN 9781621824848 (hardcover) | ISBN 9781621824855 (epub)
Subjects: LCSH: Neural circuitry. | Neural circuitry--Evolution. | Neural circuitry--Adaptation.
Classification: LCC QP363.3 .E86 2025 (print) | LCC QP363.3 (ebook) | DDC 612.8/2--dc23/eng/20240826
LC record available at https://lccn.loc.gov/2024027527
LC ebook record available at https://lccn.loc.gov/2024027528

All World Wide Web addresses are accurate to the best of our knowledge at the time of printing.

For a complete catalog of all Cold Spring Harbor Laboratory Press publications, visit our website at www.cshlpress.org.

Contents

Contents

Preface

DECIPHERING THE EVOLUTIONARY AND DEVELOPMENTAL MECHANISMS that shape the architecture of neuronal circuits promises to reveal the principles by which brains enable behavior and cognition. In recent years, rapid technical and conceptual advances in neuroanatomy, molecular genetics, physiology and theory have enabled deep insights into these mechanisms across model systems ranging from flies to humans.

This volume brings together perspectives from investigators who have driven these discoveries. A key theme across articles is the leveraging of evolutionary change, both divergent and convergent, across species to provide a larger framework for neuronal circuit design and function. This evolutionary theme is integrated with developmental perspectives, illustrating how conserved molecular building blocks and selection pressures have given rise to both shared and divergent circuit motifs across species, and leads to a series of articles covering the mechanisms and principles of circuit assembly. This major theme dissects how molecular, cellular, functional, and theoretical approaches have all been used to understand the development of diverse cell types and circuits in the brain, to offer a tantalizing view on a rapidly expanding field of neuroscience.

Together, the chapters collected here offer a view of how neural circuits develop, diversify, and function. While no single volume can capture the full breadth of this field, we hope that the contributions assembled here will serve both as a resource for specialists and as an entry point for students and investigators from neighboring disciplines.

We would like to thank the authors for their generosity in sharing their expertise, and the staff of Cold Spring Harbor Laboratory Press for their support in bringing this project to completion.

LAURA C. ANDREAE
JUSTUS M. KEBSCHULL
ANTHONY M. ZADOR

Neuronal Circuit Evolution: From Development to Structure and Adaptive Significance

Nikolaos Konstantinides[1] and Claude Desplan[2]

[1]Université Paris Cité, CNRS, Institut Jacques Monod, F-75013 Paris, France

[2]Department of Biology, New York University, New York, New York 10003, USA

Correspondence: nikos.konstantinides@ijm.fr; cd38@nyu.edu

Neuronal circuits represent the functional units of the brain. Understanding how the circuits are generated to perform computations will help us understand how the brain functions. Nevertheless, neuronal circuits are not engineered, but have formed through millions of years of animal evolution. We posit that it is necessary to study neuronal circuit evolution to comprehensively understand circuit function. Here, we review our current knowledge regarding the mechanisms that underlie circuit evolution. First, we describe the possible genetic and developmental mechanisms that have contributed to circuit evolution. Then, we discuss the structural changes of circuits during evolution and how these changes affected circuit function. Finally, we try to put circuit evolution in an ecological context and assess the adaptive significance of specific examples. We argue that, thanks to the advent of new tools and technologies, evolutionary neurobiology now allows us to address questions regarding the evolution of circuitry and behavior that were unimaginable until very recently.

A neuronal circuit is formed when a group of neurons are connected to each other by synapses and participate in the same function. Historically, the term was first used to explain the amplification through feedback of neuronal circuit activity that could lead to neurological problems by the psychiatrist Lawrence Kubie in 1930 (Kubie 1930). It was then appropriated and immortalized by Walter Pitts who, in the beginning of the 1940s, studied patterns of excitation and inhibition in neuronal circuits (Pitts 1942). Since then, the notion of neuronal circuits has been used as a functional unit of larger brain networks, as it can perform basic logical and/or arithmetic functions, which is why it has inspired the design of artificial neural networks in computing. While neuronal circuits can become extremely complex, our basic understanding of them comes from studies of relatively simple circuits.

The sea slug *Aplysia californica* has been used historically as a neuroscience research model because of the relative simplicity of its neuronal circuitry involved in learning, as well as the large size of its neurons. The simple underlying circuitry of the *Aplysia* gill and siphon withdrawal reflex allowed Eric Kandel and others to study nonassociative learning (habituation, dishabituation, and sensitization) and start uncovering the cellular and circuit basis of a complex behavior. Similarly, the relatively small crustacean stomatogastric nervous system (composed of about 30 neurons) that controls the motion of the gut has been used as a model circuit to understand motor pattern generation (Marder and Calabrese 1996) and how central pattern generators (CPGs) can be activated and modified by neuromodulators. Despite being relatively simple, these models are still very difficult to completely understand, and the question arises as to how did the first (and presumably simpler) neural circuits evolve?

Before studying how neuronal circuits evolved, it is necessary to understand the origins and evolution of neurons. The evolutionary origin of neurons is still unresolved; in fact, recent phylogenetic studies that place ctenophores as sister to all other animals (Schultz et al. 2023) suggest that nervous systems evolved twice independently in the Metazoa, once in Ctenophora, and once in the common ancestor of Bilateria and Cnidaria. This is also supported by fundamental differences in the architecture of their respective neuronal systems (Burkhardt et al. 2023) and the molecular components of their synapses (Arendt 2020). In any case, it is likely that neurons evolved in an early metazoan from existing epithelial cells (neurosecretory or mechanosensory cells) either to coordinate a bodily response to environmental challenges or to control cilia beating (discussed in more detail later). These early cells would probably qualify as sensory neurons that transmitted information to motor cells, via electrical or chemical protosynapses. These protoneurons then acquired their current sophisticated synaptic machinery in a stepwise manner (Arendt 2020).

While these protoneurons were probably sufficient in small animals for sensory-to-motor transformation, the evolution of neuronal circuits allowed animals to integrate information from multiple sensory cells in an efficient manner (Jékely 2011). This generated an intermediate cell (a motor neuron) that could integrate neuronal input in a sensory modality-specific manner and allow for a reduction of wiring length, which decreases the required energy to build axons and signal through them.

Building and maintaining a nervous system is a very costly investment. Indeed, whether one looks at a whole brain, circuits, or cells, their physical location is decided by the effort to minimize unnecessary wiring (Cherniak 1994, 1995).

Therefore, the selective advantages of an increasingly complex circuit should outweigh the metabolic, wiring, and signaling costs. By reducing axonal length to minimize these costs, neuronal cells were concentrated in specific locations, which probably led to the centralization of the nervous system soon after the evolution of the first circuits, as elements of centralization can be found in Cnidaria. This centralization is more prominent in Bilateria, which led to the gradual formation of a central processing unit, the brain, which allowed also for an increase in cognitive capacity (Martinez and Sprecher 2020), which in turn let bilaterians expand, differentiate, invade different environments, and generate an impressive organismal diversity.

It comes as no surprise that our understanding of the evolution of neurons, neuronal circuits, and central nervous systems comes from comparative approaches.

- Comparisons at the level of gene expression have been instrumental in the identification of homologous neuronal structures and circuits across different taxa.

- Comparisons at the level of neuronal type composition have allowed us to discover instances and mechanisms of neuronal type evolution.

- Comparisons at the level of circuitry (i.e., the evolution of new connections, changes in synaptic strength, or sign—excitatory or inhibitory—revealed the evolution of circuit structure).

- Comparisons at the level of brain regions uncovered cases of circuit or region duplications and eliminations.

Altogether, comparative studies have painted a picture of the mechanisms that allowed circuits to evolve, how this affects neuronal circuit structure, and the selective pressure that led to these changes. In this review, we first focus on the genetic and developmental mechanisms that underlie the generation of neuronal circuitry. Then, we delve into the structural and functional changes that have occurred in these neuronal circuits throughout evolution. We further examine the adaptive significance of evolutionary changes

Cite this article as *Cold Spring Harb Perspect Biol* doi: 10.1101/cshperspect.a041493

in neuronal circuitry in their ecological context. Finally, we conclude by touching upon emerging research areas and future directions.

GENETIC AND DEVELOPMENTAL MECHANISMS

How can a neuronal circuit evolve? To answer this question (reviewed by Tosches 2017), we need to consider the position circuits acquire in the evolutionary process. Obviously, natural selection acts upon animal behavior: Evolutionary pressure and natural selection will allow individ-

uals with certain behaviors to survive. For these behaviors to be transmitted to the next generation, they must result from genetic differences that occurred through random mutation in certain genes. As multiple genes are involved in a given behavior at different levels of neurogenesis and/or neuronal function, they might affect neuronal identity or circuit structure. These genes can be divided into four broad categories (Fig. 1) that are discussed below.

1. Genes that affect the effective size of the neuronal progenitor pool, such as members of signaling pathways (Hh, Notch, insulin, and oth-

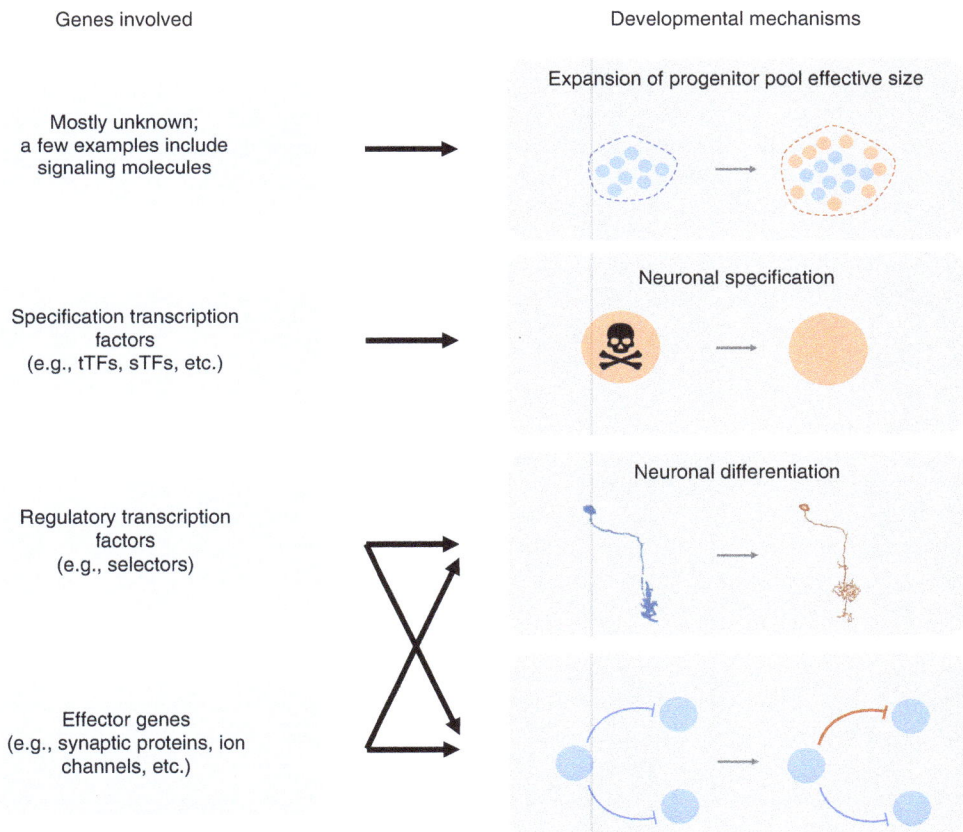

Figure 1. Genetic and developmental mechanisms of circuit evolution. We can distinguish genes that are involved in neuronal circuit evolution into four categories. Alterations in the expression pattern of these genes can differentially affect neuronal developmental mechanisms and lead to the evolution of different neuronal circuits. Genes involved in the expansion of the progenitor pool effective size, as well as specification factors, act at the level of the progenitors and can lead to the evolution and development of new neuronal types. On the other hand, regulatory transcription factors and their downstream effectors act at the level of the postmitotic neurons and can affect specific neuronal characters (such as arborization locations or synaptic strength).

ers) (Homem et al. 2015). The number, as well as the timing of divisions of the progenitors, could have profound effects on the evolution of new circuitry. An increase in the effective size of neuronal progenitors relieves the newly evolved neurons from selective pressure (as their ancestral counterparts continue to play their ancestral roles) and allows them to diversify freely (Oakley 2003; Chakraborty and Jarvis 2015; Grillner and Robertson 2016). The modular organization of animal brains from insects to vertebrates suggests that duplication or multiplication of existing pools of progenitors is at the basis of neuronal circuit evolution. One such example is the evolution of cerebellar nuclei; while the cerebellum itself is conserved in vertebrates, the numbers of its nuclei vary from none (jawless vertebrates) to three in mammals. The cellular composition and architecture of each nucleus in mice (three nuclei) and chickens (two nuclei) suggest that there exists an archetypal nucleus that has repeatedly duplicated during vertebrate evolution (Kebschul et al. 2020). Similarly, the vertebrate cerebral cortex has been suggested to evolve through the expansion of radial glial progenitors and/or transit amplifying progenitors that has allowed the expansion of the three-layered cerebral cortex in turtles to the six-layered one in mammals (Briscoe and Ragsdale 2018; Tosches et al. 2018; Lin et al. 2021). Importantly, all radial glial progenitors undergo a fairly fixed set of divisions to output a specific number of neurons that occupy progressively the more superficial layers of the cortex, which supports their common evolutionary origin. Interestingly, sister neurons (neurons that originate from the same division of a radial glial progenitor) preferentially form synapses with each other (Lin et al. 2023). This raises an appealing model where whole circuits can evolve through the expansion of the radial glial progenitor pool. This modular organization is not restricted to vertebrates. The insect brain is also clonally organized (i.e., distinct neuronal lineages of stem cells—called neuroblasts in insects—occupy specific parts of the insect brain and are involved in the same function) (Ito et al. 2013). Because of this lineage-based architecture, it has been proposed that neural networks can evolve and diversify in a modular

manner in different species (Kandimalla et al. 2023).

Similarly, the timing and duration of neuronal progenitor proliferation also play an important role (reviewed in Fenlon 2022). For example, differences in the timing (heterochronies) in the switch from proliferative to neurogenic divisions of radial glial progenitors and in the duration of the neurogenesis period can explain, to a large extent, the variation in the number of cortical neurons (Picco et al. 2018) and the expansion of the upper cortical layer (Caviness et al. 1995; Rakic 1995; Kriegstein et al. 2006). Genes involved in signaling pathways, such as Notch (NOTCH2NL; Fiddes et al. 2018; Suzuki et al. 2018) and Wnt pathways (FZD8; Boyd et al. 2015), are also implicated in the dynamics of progenitor proliferation (Suzuki 2020; Pinson and Huttner 2021).

2. Neuronal specification genes are regulators of gene expression and secreted molecules that act usually in neuronal progenitors and are responsible for specifying the identity of their neuronal progeny. Temporal and spatial transcription factors, for example, pattern neuronal progenitors based on their age and location in the tissue and allow them to generate diverse neurons. Altering the expression of any of these genes during development has an immediate impact on neuronal identity. Nevertheless, it is still unknown how the expression of these genes has changed during evolution.

One of the potential neuronal fates that is specified by these genes is cell death; neurons are preprogrammed to undergo programmed cell death, which occurs immediately after their specification. Preventing cell death in different neuronal lineages in *Drosophila* allows neurons to acquire a different fate when mature, and incorporate into circuits. Interestingly, this new circuitry can often be found to be present in other insects (Pop et al. 2020; Prieto-Godino et al. 2020), suggesting that there is a reserve of regulatory networks that can be recruited in different species.

3. Regulatory effector genes control gene expression and act cell autonomously downstream from neuronal specification genes to implement the specified identity of the neuron. They regulate

(directly or indirectly) effector gene expression. Altering the expression of these genes can change some or all aspects of neuronal identity. For example, *Fezf2* is expressed in mouse corticospinal neurons and is able to regulate both neurotransmitter identity by activating glutamatergic genes (such as *VGlut1*) and inhibiting GABAergic ones (such as *Gad1*), and connectivity by regulating the expression of *EphB1* that controls the ipsilateral extension to the corticospinal tract (Lodato et al. 2014).

Regulatory effector genes that are expressed throughout development and persist into adulthood have been termed terminal selectors. The concept of terminal selector genes that define neuronal identity has been defined in worms where they act in a concerted manner to regulate all neuronal type-specific effector genes. Misexpression or down-regulation of terminal selectors leads to a complete change of neuronal identity; they have, therefore, been hypothesized to operate as a main evolutionary driver to generate diverse cell types (Arlotta and Hobert 2015; Cros and Hobert 2022). Such terminal selectors have been found in many species, including *Drosophila*, where targeted modifications of terminal selector expression in the visual brain allow the transdetermination of a neuron into another type (Özel et al. 2022).

4. Finally, effector genes actually implement identity. These are ion channels, genes involved in the synthesis of neurotransmitters and their receptors, synaptic genes, etc. While changes in these genes do not impact neuronal circuitry per se, they can have immediate effects on animal behavior. Changes in effector genes have been found in closely related species. A very prominent example is the genetic cause of the behavioral differences between the prairie vole (*Microtus ochrogaster*) and the montane vole (*Microtus montanus*): The former forms monogamous pairs while the latter is solitary and does not associate with former mates. This behavioral distinction is caused by a difference in the expression distribution of the receptors for the neuromodulatory neuropeptides oxytocin and vasopressin, which are responsible for the pair-bonding behavior in prairie voles (Insel et al. 1995; Young et al. 1997, 1999; Katz and Harris-Warrick 1999). Notably, this has been challenged recently, as it was shown that prairie voles can form long-term pairs in the absence of an oxytocin receptor (Berendzen et al. 2023). Similarly, the cosmopolitan drosophilids, *Drosophila melanogaster* and *Drosophila simulans*, are reproductively isolated from each other partly by a difference in their response to a pheromone that is produced by the *D. melanogaster* females, 7,11-heptacosadiene, which attracts *D. melanogaster* males but repulses *D. simulans* males. This differential response is due to a change in the balance between excitation and repression in response to the pheromone from the presynaptic neurons (vAB3 and mAL neurons, respectively) onto the courtship-promoting P1 neurons (Seeholzer et al. 2018).

It is clear that genetic changes can affect neuronal development and, thus, neuronal identity and circuitry at different levels and to different extents. Neurodevelopmental genes and their regulatory networks that are active at the progenitor state (i.e., the first two categories) (Fig. 1) are largely more conserved and, hence, less often involved, probably due to their limited number (progenitor pool size is regulated by a handful of signaling pathways and neuronal specification genes tend to be members of very conserved transcription factor families). Moreover, most of these genes are pleiotropic and involved in many different processes; therefore, changes in these genes are rarer. On the other hand, neuronal type-specific genes, such as regulatory effectors and effector genes are likely evolutionary tinkerers (Jacob 1977) that can change neuronal type identity or parts of it between closely related species (Fig. 1).

STRUCTURAL AND FUNCTIONAL EVOLUTION OF NEURONAL CIRCUITS

The evolutionary origin of neurons is still unresolved (Jékely 2011; Arendt 2021), although it has been proposed that neurons evolved at least twice in Metazoa. Recent single-cell mRNA sequencing data from Placozoa that lack clearly identifiable neuronal cells identified key neuronal components (such as presynaptic scaffold) in peptidergic secretory cells that come from progenitors with neurogenic ontogenetic modules

(Najle et al. 2023). These data support the idea that neuronal cells evolved from neurosecretory cells and acted as "sensorimotor neurons" to coordinate potentially ciliary beating (Fig. 2A).

How did the first circuits evolve? It is likely that the first circuits originated from these sensorimotor neurons through a "division of labor" (Mackie 1970; Arendt 2008) where the sensorimotor neuron duplicated and generated two neurons (a sensory neuron and a motor neuron) that were synaptically connected and formed the first basal neuronal circuit (Fig. 2B). An alternative scenario for the evolution of the first circuit also relies on the "division of labor" of an ancestral myoepithelial sensory cell, whose duplication led to a sensory neuron in the surface and a motor neuron. Finally, it has also been proposed that the first neuronal circuits evolved to coordinate body parts of an increasingly complex system (Jékely et al. 2015).

While less likely, we cannot dismiss the possibility that neuronal circuits evolved more than once: they may have evolved (1) from secretory cells, as supported by molecular evidence from Placozoa (Najle et al. 2023) and vertebrates that show a close relationship between neuronal and secretory cells, and (2) from myocytes, as suggested by the close relation of ectodermal neurons and myocytes in cnidarians (Arendt 2021).

After basal connectivity was acquired, circuits slowly became more complex. It is possible that the first interneurons arose to integrate and process information from multiple sensory protoneurons (Fig. 2C; Jékely 2011). Moreover, sensory information might have had to be differently processed before forwarding it to the motor cells. For example, an interneuron might serve to send both excitatory signals to one motor cell as well as inhibitory information to a different motor cell to achieve more elaborate motor output (Fig. 2C). This interneuron may have duplicated during evolution and divided its labor to increase the complexity of the circuit. Similarly, during eye evolution, it has been hypothesized that with the increase in the number of photoreceptors, some of them assumed a "processing role" and became interneurons. With increased processing needs, the interneurons likely divided their labor and generated first- and second-order interneu-

rons that were likely organized into what we recognize today as optic ganglia (neuropils) (Arendt et al. 2009).

As animals became more complex with diverse cell types, the neuronal circuits had to adapt. Several mechanisms can collectively account for the generation of more complex circuitry (Fig. 2D):

- As mentioned in the previous section, neural circuits may evolve by the duplication of progenitor regions that generate entire circuit modules and their subsequent divergence. This mode of evolution is very prevalent and efficient as it relies on functioning preexisting modules being co-opted for a different function. Besides the evolution of cerebellar nuclei mentioned earlier (Kebschul et al. 2020), the evolution of vertebrate basal ganglia has been proposed to rely on modular evolution, where each module controls a specific behavior, such as locomotion, eye movements, posture, and chewing. Each of these modular circuits contains a pathway of striatal projection neurons that disinhibit the brainstem motor centers and control a motor program, and a pathway of a different type of striatal projection neurons and intrinsic basal ganglia nuclei that inhibit competing motor programs (Grillner and Robertson 2016). These modules could be theoretically gained or lost independently during evolution. A similar mode of evolution can also be found in the hypothalamus (Xie and Dorsky 2017).

- A subcategory of the above mechanism is the exaptation (co-option of a trait for a function that is different from the one for which it originally evolved) in different contexts of self-organizing neuronal mini-circuits, such as CPGs. In the vertebrate spinal cord, CPG circuits that are responsible for limb locomotion originated in vertebrates lacking limbs (Grillner 2006; Grillner and Jessell 2009; Dasen 2017). In particular, the limb flexor CPGs may have been co-opted from the CPGs that were used to activate axial muscles during undulatory locomotion, while the extensor CPG might have evolved from premotor circuits

Cite this article as *Cold Spring Harb Perspect Biol* doi: 10.1101/cshperspect.a041493

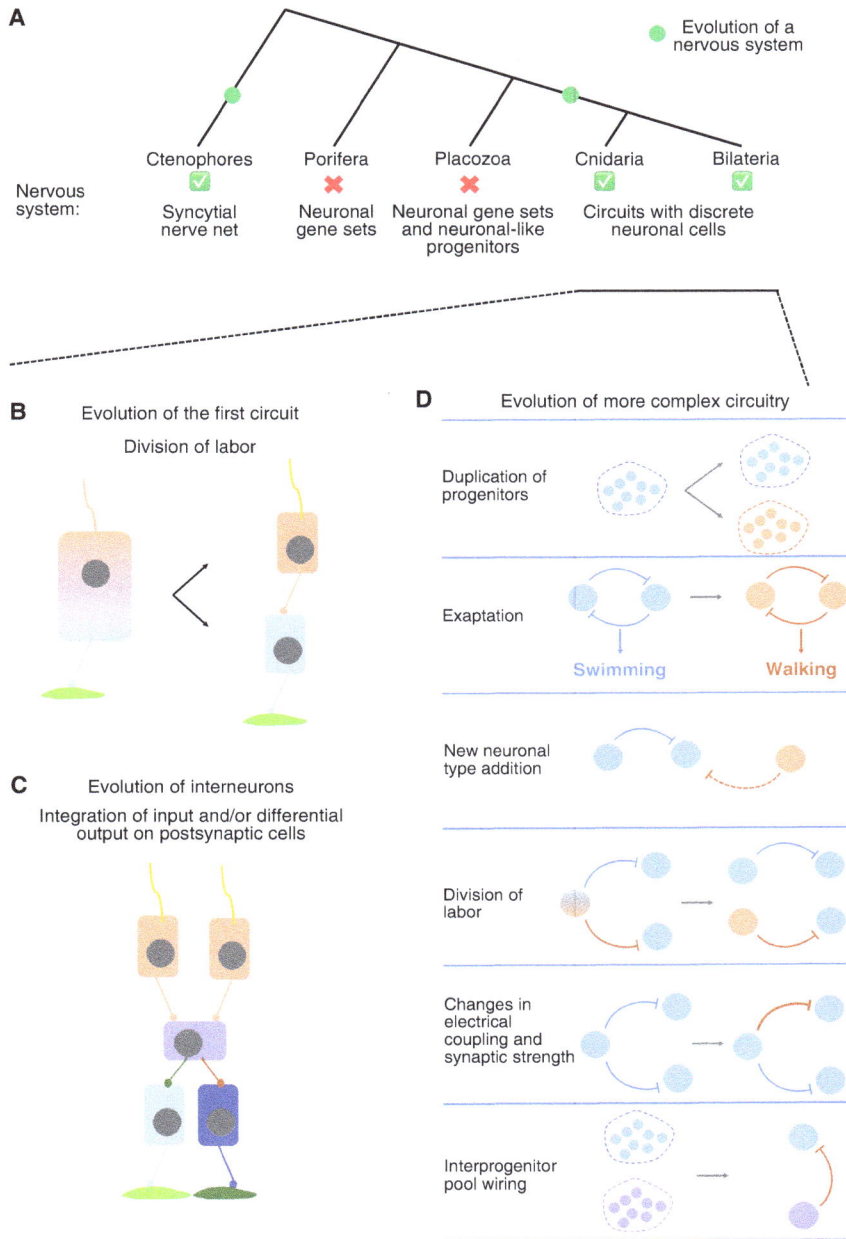

Figure 2. Structural evolution of neuronal circuits. (*A*) Based on current knowledge, the most parsimonious scenario for the evolution of neurons includes their independent emergence in ctenophores and the common ancestor of cnidarian and bilaterian animals. On the other hand, while Porifera and Placozoa lack bona fide neurons, they have neuronal gene sets and neuronal-like progenitors, respectively, which argues for a stepwise evolution of neurons. (*B*) Focusing on the common ancestor of cnidarians and bilaterians, the evolution of the first neuronal circuit involved probably the division of labor of an ancestral sensorimotor cell (orange–blue cell) and the evolution of synaptically connected sensory (orange) and motor (blue) cells. (*C*) Further elaboration of neuronal circuits and the evolution of interneurons (purple) was probably triggered by the need for integration of presynaptic input (orange cells) and/or the need for differential output onto postsynaptic partners (blue cells) (e.g., inhibitory [red] and excitatory [green]). (*D*) Subsequently, a number of different mechanisms were involved in the further complexification of neuronal circuitry, including duplication of progenitors, exaptation of neuronal circuits from one function (e.g., swimming) to another (e.g., walking), the addition of new neuronal types in existing circuitry, division of labor of one cell type into two cell types, changes in electrical coupling or synaptic strength, and interprogenitor pool wiring.

used for postural correction (Bagnall and McLean 2014; Dasen 2017).

- Another obvious way to complexify circuits is adding new neuronal types into preexisting circuits. For example, based on transcriptomic comparisons of neurons in mammals and reptiles, it appears that the neocortical circuits are evolutionary mosaics of deeply conserved GABAergic interneurons and very divergent glutamatergic pyramidal neurons (Tosches et al. 2018; Tosches 2021; Hain et al. 2022). Similarly, in the spinal cord, although ipsilateral V2a-type excitatory neurons, and commissural V0-type inhibitory neurons can be found in all vertebrates (lamprey, tadpole, zebrafish, and mouse), the V3, V1, V2b, and dorsal dI6 interneurons have only evolved in jawed vertebrates (Wilson and Sweeney 2023).

- The "division of labor" model also accounts for adding complexity to neuronal circuits. In the same example of the spinal cord, the complexity in movement patterns and gaits is correlated with the division of cardinal classes of interneurons into distinct subtypes.

- Circuits can also evolve by changes in electrical coupling and synaptic strength (Katz 2007), as exemplified by the species-specific mate preferences between *D. melanogaster* and *D. simulans* that depend on alterations in the synaptic strength of vAB3 and mAL pathways to P1 neurons (Seeholzer et al. 2018).

- Circuits can also evolve and become more complex by interprogenitor pool wiring (Suzuki and Sato 2017) (i.e., by combining neurons that come from different progenitor pools). An obvious example of this is the mammalian cerebral cortex that is formed by the GABAergic cells that migrate from the eminences and the glutamatergic cells that are born in the cortex.

- Finally, circuits can evolve by the addition (Edwards et al. 1999) and elimination of synapses (Ebbesson 1980; Tosches 2021), which can alter the functional output of the neuronal circuit.

The above mechanisms can account, more or less, for the evolution of the complex neuronal circuits that we observe today. As a rule, as animal bodies became more complex with more cell types, they had to be controlled by more complex nervous systems; therefore, most of the mechanisms did increase the number and synaptic complexity of neuronal cell types. A few exceptions of circuit simplifications serve to prove the rule.

ADAPTIVE SIGNIFICANCE AND ECOLOGICAL CONTEXT

Before delving into behavioral evolution, it is important to note that hotspots of genetic variation that allow neuronal systems to adapt to evolutionary challenges are often found in genes that can alter sensory perception; sensory genes, such as receptors for chemosensation, evolve very rapidly (Cande et al. 2013). They can also affect metabolic processes outside of the nervous system; for example, genetic variation that affects metabolism influences eating and drinking behavior without necessarily affecting circuitry. These types of behavioral evolution have been reviewed elsewhere (Bendesky and Bargmann 2011; Niepoth and Bendesky 2020); we will focus on how neuronal circuit evolution can change behavioral output in response to different environments and lifestyles.

While the evolution of genetic and developmental mechanisms, as well as that of neuronal circuitry, has been explored in the past, this has not been done in a comprehensive manner for the evolution of behaviors. For this reason, we explore different examples of behavioral evolution that are caused or accompanied by circuitry changes in an effort to identify some general rules.

The Case of Nudibranchs

Nudibranch (which are marine gastropod mollusks) swimming has been a fruitful example to compare similar behaviors and their respective circuits. Nudibranchs undulate their bodies either from left to right or dorsally to ventrally. These types of swimming have evolved independently in different species. Interestingly, different circuitries can give rise to the same behaviors.

Cite this article as *Cold Spring Harb Perspect Biol* doi: 10.1101/cshperspect.a041493

Such an example is the circuitry underlying the same left–right body flexion of two nudibranchs, *Melibe leonina* and *Dendronotus iris*: in this circuitry, while the behavioral rhythm and neuronal composition are the same between the two species, the connectivity of homologous neurons differs. Moreover, while neurons that are part of the dorsoventral swim CPG have homologs in species that undulate their bodies from left to right, these homologous neurons are not part of the left–right swim circuitry. These examples indicate that multiple circuits that can serve the exact same (rhythmic) behavior (Fig. 3A; Newcomb et al. 2012; Sakurai and Katz 2015; Katz 2016; Jourjine and Hoekstra 2021). A general rule that arises from these studies is that one cannot homologize behaviors on the basis of how similar or dissimilar their underlying circuits are and, vice versa, the circuitry cannot be predicted by behavior, although it can be a good approximation (Newcomb et al. 2012).

Circuit Modularity as a Means to Evolve New Behaviors

Circuits can evolve by whole duplication of progenitor regions that lead to duplicated circuits. But how does this influence behaviors? In birds, whole circuit duplication has been suggested to contribute to song learning (Feenders et al. 2008; Chakraborty and Jarvis 2015). In particular, it was hypothesized that, in vocal learners, the forebrain motor learning circuit that connects to the brainstem circuits that control vocalization is a product of duplication of other motor learning circuits that receive auditory, somatosensory, or other sensory input (Fig. 3B).

Circuit modularity allows also the modular tinkering of neuronal pathways. In insect courtship, *fruitless*-expressing neurons in the central brain are important for the execution of courtship behavior per se in males. Species-specific differences are usually impacting either the sensory pathways that feed into the *fruitless* neurons, or the downstream motor pathways (Ding et al. 2019; Sato et al. 2020). For example, while the structural, electrical, and neurochemical properties of the *fruitless*-expressing pIP10 neurons of *D. melanogaster* and *Drosophila yakuba* are the same, their activation leads to songs that differ in structure and frequency; this difference has to be attributed to the downstream motor pathway.

Finally, circuit modularity allows for the mixing and matching of different circuits. *Drosophila pseudoobscura* males have incorporated a regurgitation behavior in their courtship, whereby they offer a nuptial proboscis-to-proboscis gift to the female before attempting copulation. An evolutionary scenario that could account for this is that the neuronal circuit that was responsible for regurgitation as a feeding behavior became at some point during evolution postsynaptic to the courtship command system.

Collectively, the above examples argue for the importance of circuit modularity for the rapid evolution of new behaviors.

Neuronal Circuits and Social Interactions

Neuronal circuits can also affect and be influenced by the evolution of social interactions. In humans, FOXP2 is the only gene that is currently linked to central aspects of speech and language (Fisher and Scharff 2009), as the haploinsufficiency of FOXP2 impairs both of them. It does so by causing alterations in corticostriatal and corticocerebellar circuits (Vargha-Khadem et al. 2005). Moreover, expression of humanized FOXP2 in mice leads to a decrease in dopamine levels, as well as an increase in total dendrite length and synaptic plasticity of medium spiny neurons (Fig. 3C; Enard et al. 2009). Similarly, in worms, a single gene can affect their "social interaction" in terms of aggregation. The laboratory strain N2 shows low aggregation, as opposed to wild-type strains. This has been attributed to a single mutation in the *npr-1* gene that leads to a single amino acid difference. This mutation causes RMG neurons to reduce the number of electrical synapses they form with sensory neurons that stimulate aggregation.

While the above examples offer a number of basic rules, it is obvious that our knowledge regarding the neuronal circuitry underpinnings of behavioral evolution is very fragmented. With the increasing ability of producing connectomes of differently sized brains, we will soon acquire a more holistic view on this subject.

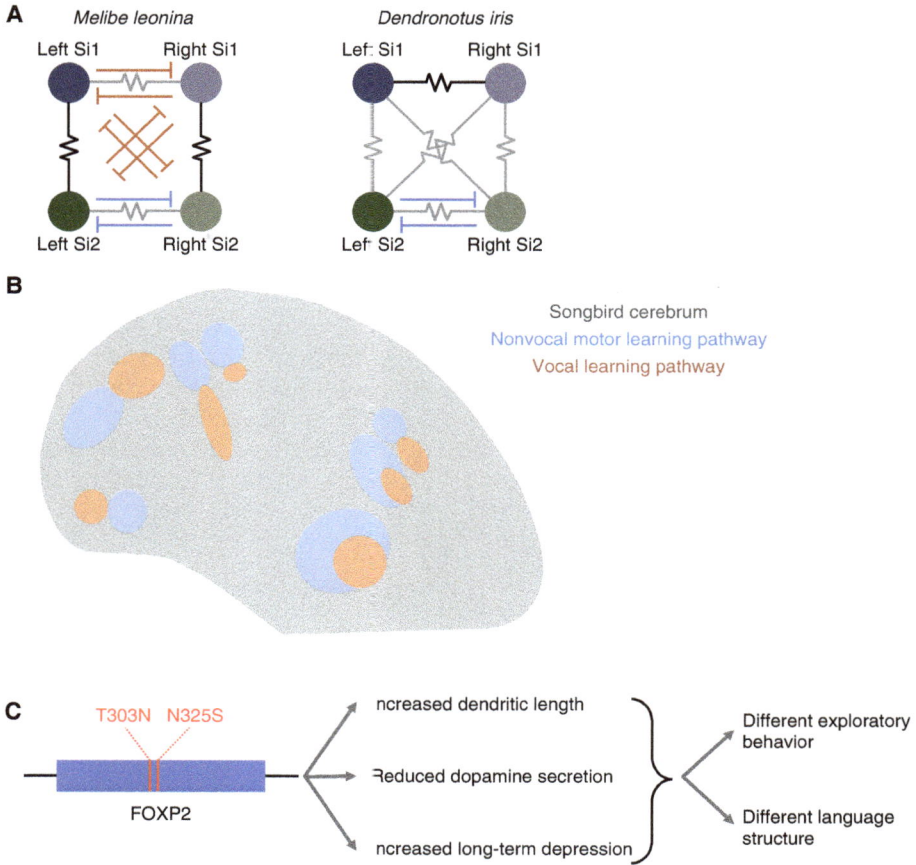

Figure 3. Neuronal circuits and behavior. (*A*) The central pattern generator that controls left–right swimming in both the nudibranchs *Melibe leonina* and *Dendronotus iris* consists of two bilaterally paired neurons, Si1 and Si2. Notably, despite the fact that this similar behavior is driven by homologous neurons in the two species, the underlying synaptic and electrical connectivity is substantially different. Differences in synapses are shown in orange. Darker lines show stronger coupling. (Panel *A* based on data in Sakurai and Katz 2015.) (*B*) In the cerebrum of vocal learners (such as songbirds), a duplication of the nonvocal motor learning nuclei (blue) led to the evolution of the adjacent vocal learning pathway (orange), which is necessary for the production of the learned vocalizations. (Panel *B* based on data in Feenders et al. 2008.) (*C*) Two amino acid substitutions that occurred in FOXP2 in the human lineage (after the divergence from that of the chimpanzees) led to a number of neuronal differences in the corticostriatal and corticocerebellar circuits: First, they led to an increased dendritic length and long-term neuronal depression, as well as a decrease in the secretion of dopamine. These neuronal changes caused differences in language structure and exploratory behavior that are believed to be important for the human ability to articulate speech and develop language.

CONCLUDING REMARKS

We presented the current knowledge of the evolution of neuronal circuitry, from genes to circuits to behavior: We described how genetic and developmental mechanisms that underlie circuit differentiation suggest that different genes can affect neuronal circuitry at different developmental levels and, hence, to a different extent. We then reviewed current ideas regarding the evolution of the first circuits and different ways by which that circuits can evolve increasingly complex structures. Finally, we examined the adaptive significance of evolutionary changes in neuronal circuitry (i.e., how circuits evolved to perform various functions to control different behaviors).

One of the main limitations to understand neuronal evolution has been the difficulty in identifying and comparing homologous circuits between different animals outside the traditional model organisms. Given the fairly recent technological developments that give us access to almost all animals, the future of this field is bright. On the one hand, the advent of single-cell sequencing allows the identification and comparison of orthologous cell types in different species. On the other hand, the increasing feasibility of obtaining new connectomes and the constant development of tools to analyze connectomic data will allow us to compare the circuits in which these orthologous cell types participate. Moreover, not being constrained by model animals will facilitate the investigation of the phylogenetic tree of animals more or less uniformly.

Circuits evolve under an evolutionary pressure to "perform" differently. Small-scale behavioral differences might not be qualitatively detectable. Therefore, it is essential to measure and compare behaviors quantitatively. The development of deep learning methods for pose estimation (i.e., detection of the position and orientation of an animal), such as SLEAP (Pereira et al. 2022) and DeepLabCut (Mathis et al. 2018), allows for the detailed and accurate characterization of different behaviors that could then be quantitatively compared between different species. Importantly, these methods perform multianimal pose estimation that should enable the quantitation of social behaviors and the concurrent quantitation of behaviors of many animals at once, which can increase statistical power in the analysis of subtle behavioral differences. In combination with connectomics, this will allow us to address how differences in neuronal circuitry are translated into behavioral differences.

This toolkit can also be applied to understand how the human brain evolved (reviewed recently in Sousa et al. 2017; Vanderhaeghen and Polleux 2023). The human brain represents without doubt the most complex tissue in any animal. While the development of the human brain is not exceptionally different from that of nonhu-

man primates, the differences in cognitive capacities are extreme. A number of human-specific genomic changes have been described that might differentiate human brain development from that of nonhuman primates and cause increased size, neuronal number, as well as a rewired and more complex neuronal circuitry; however, it is likely that we are still missing a very large part of the story. Very recently, transcriptional and epigenetic atlases of the cellular composition of adult and developing humans, as well as nonhuman primates, marmosets, and macaques brains, were made available (Ament et al. 2023; Braun et al. 2023; Chiou et al. 2023; Jorstad et al. 2023a, b; Komiyama 2023; Krienen et al. 2023; Li et al. 2023; Maroso 2023; Micali et al. 2023; Siletti et al. 2023; Tian et al. 2023; Velmeshev et al. 2023; Zhu et al. 2023). These and other efforts will allow us to understand which genetic and developmental mechanisms enabled the human brain to support higher cognitive functions.

While the above techniques can generate unprecedented knowledge regarding the development of the human brain, different experimental tools are needed to probe the function of the identified genes and circuits. Since experimentation on human brains is restricted for ethical and practical reasons, the development of protocols for the generation of human brain organoids ex vivo as well as from different animals (e.g., different vertebrates and primates) (Lázaro et al. 2023) will prove invaluable to test candidates generated from high-throughput sequencing studies, assess their role in the development of human-specific neuronal circuits, and, ultimately, understand the genetic and developmental mechanisms that are responsible for the generation of this complex structure and will also allow for the comparisons of these mechanisms in brain organoids coming from animals that span the phylogenetic tree (Paşca 2018; Tambalo and Lodato 2020; Velasco et al. 2020; Lin et al. 2021; Uzquiano and Arlotta 2022) and might also aid significantly in the understanding of human neurodevelopmental disorders.

ACKNOWLEDGMENTS

We thank Gáspár Jékely and Simon Sprecher for feedback on the manuscript as well as the

Konstantinides and Desplan laboratories for discussions that helped shape our understanding of circuit evolution. Research in the N.K. laboratory on the evolution of neuronal types has received funding from the European Research Council (ERC) under the European Union's Horizon 2020 research and innovation programme (grant agreement no. 949500). C.D. is supported by the National Institutes of Health (NIH) (R01 EY13010 and R01EY13012).

REFERENCES

Ament SA, Cortes-Gutierrez M, Herb BR, Mocci E, Colantuoni C, McCarthy MM. 2023. A single-cell genomic atlas for maturation of the human cerebellum during early childhood. *Sci Transl Med* **15:** eade1283. doi:10.1126/scitranslmed.ade1283

Arendt D. 2008. The evolution of cell types in animals: emerging principles from molecular studies. *Nat Rev Genet* **9:** 868–882. doi:10.1038/nrg2416

Arendt D. 2020. The evolutionary assembly of neuronal machinery. *Curr Biol* **30:** R603–R616. doi:10.1016/j.cub.2020.04.008

Arendt D. 2021. Elementary nervous systems. *Philos Trans R Soc Lond B Biol Sci* **376:** 20200347. doi:10.1098/rstb.2020.0347

Arendt D, Hausen H, Purschke G. 2009. The "division of labour" model of eye evolution. *Philos Trans R Soc Lond B Biol Sci* **364:** 2809–2817. doi:10.1098/rstb.2009.0104

Arlotta P, Hobert O. 2015. Homeotic transformations of neuronal cell identities. *Trends Neurosci* **38:** 751–762. doi:10.1016/j.tins.2015.10.005

Bagnall MW, McLean DL. 2014. Modular organization of axial microcircuits in zebrafish. *Science* **343:** 197–200. doi:10.1126/science.1245629

Bendesky A, Bargmann CI. 2011. Genetic contributions to behavioural diversity at the gene–environment interface. *Nat Rev Genet* **12:** 809–820. doi:10.1038/nrg3065

Berendzen KM, Sharma R, Mandujano MA, Wei Y, Rogers FD, Simmons TC, Seelke AMH, Bond JM, Larios R, Goodwin NL, et al. 2023. Oxytocin receptor is not required for social attachment in prairie voles. *Neuron* **111:** 787–796.e4. doi:10.1016/j.neuron.2022.12.011

Boyd JL, Skove SL, Rouanet JP, Pilaz LJ, Bepler T, Gordân R, Wray GA, Silver DL. 2015. Human-chimpanzee differences in a FZD8 enhancer alter cell-cycle dynamics in the developing neocortex. *Curr Biol* **25:** 772–779. doi:10.1016/j.cub.2015.01.041

Braun E, Danan-Gotthold M, Borm LE, Lee KW, Vinsland E, Lönnerberg P, Hu L, Li X, He X, Andrusivová Ž, et al. 2023. Comprehensive cell atlas of the first-trimester developing human brain. *Science* **382:** eadf1226. doi:10.1126/science.adf1226

Briscoe SD, Ragsdale CW. 2018. Homology, neocortex, and the evolution of developmental mechanisms. *Science* **362:** 190–193. doi:10.1126/science.aau3711

Burkhardt P, Colgren J, Medhus A, Digel L, Naumann B, Soto-Angel JJ, Nordmann EL, Sachkova MY, Kittelmann M. 2023. Syncytial nerve net in a ctenophore adds insights on the evolution of nervous systems. *Science* **380:** 293–297. doi:10.1126/science.ade5645

Cande J, Prud'homme B, Gompel N. 2013. Smells like evolution: the role of chemoreceptor evolution in behavioral change. *Curr Opin Neurobiol* **23:** 152–158. doi:10.1016/j.conb.2012.07.008

Caviness VS, Takahashi T, Nowakowski RS. 1995. Numbers, time and neocortical neuronogenesis: a general developmental and evolutionary model. *Trends Neurosci* **18:** 379–383. doi:10.1016/0166-2236(95)93933-O

Chakraborty M, Jarvis ED. 2015. Brain evolution by brain pathway duplication. *Philos Trans R Soc Lond B Biol Sci* **370:** 20150056. doi:10.1098/rstb.2015.0056

Cherniak C. 1994. Component placement optimization in the brain. *J Neurosci* **14:** 2418–2427. doi:10.1523/JNEUROSCI.14-04-02418.1994

Cherniak C. 1995. Neural component placement. *Trends Neurosci* **18:** 522–527. doi:10.1016/0166-2236(95)98373-7

Chiou KL, Huang X, Bohlen MO, Tremblay S, DeCasien AR, O'Day DR, Spurrell CH, Gogate AA, Zintel TM, Unit CBR, et al. 2023. A single-cell multi-omic atlas spanning the adult rhesus macaque brain. *Sci Adv* **9:** eadh1914. doi:10.1126/sciadv.adh1914

Cros C, Hobert O. 2022. *Caenorhabditis elegans* sine oculis/SIX-type homeobox genes act as homeotic switches to define neuronal subtype identities. *Proc Natl Acad Sci* **119:** e2206817119. doi:10.1073/pnas.2206817119

Dasen JS. 2017. Master or servant? Emerging roles for motor neuron subtypes in the construction and evolution of locomotor circuits. *Curr Opin Neurobiol* **42:** 25–32. doi:10.1016/j.conb.2016.11.005

Ding Y, Lillvis JL, Cande J, Berman GJ, Arthur BJ, Long X, Xu M, Dickson BJ, Stern DL. 2019. Neural evolution of context-dependent fly song. *Curr Biol* **29:** 1089–1099.e7. doi:10.1016/j.cub.2019.02.019

Ebbesson SOE. 1980. The parcellation theory and its relation to interspecific variability in brain organization, evolutionary and ontogenetic development, and neuronal plasticity. *Cell Tissue Res* **213:** 179–212.

Edwards DH, Heitler WJ, Krasne FB. 1999. Fifty years of a command neuron: the neurobiology of escape behavior in the crayfish. *Trends Neurosci* **22:** 153–161. doi:10.1016/S0166-2236(98)01340-X

Enard W, Gehre S, Hammerschmidt K, Hölter SM, Blass T, Somel M, Brückner MK, Schreiweis C, Winter C, Sohr R, et al. 2009. A humanized version of Foxp2 affects cortico-basal ganglia circuits in mice. *Cell* **137:** 961–971. doi:10.1016/j.cell.2009.03.041

Feenders G, Liedvogel M, Rivas M, Zapka M, Horita H, Hara E, Wada K, Mouritsen H, Jarvis ED. 2008. Molecular mapping of movement-associated areas in the avian brain: a motor theory for vocal learning origin. *PLoS ONE* **3:** e1768. doi:10.1371/journal.pone.0001768

Fenlon LR. 2022. Timing as a mechanism of development and evolution in the cerebral cortex. *Brain Behav Evol* **97:** 8–32. doi:10.1159/000521678

Fiddes IT, Lodewijk GA, Mooring M, Bosworth CM, Ewing AD, Mantalas GL, Novak AM, van den Bout A, Bishara A,

Rosenkrantz JL, et al. 2018. Human-specific NOTCH2NL genes affect Notch signaling and cortical neurogenesis. *Cell* 173: 1356–1369.e22. doi:10.1016/j.cell.2018.03.051

Fisher SE, Scharff C. 2009. FOXP2 as a molecular window into speech and language. *Trends Genet* 25: 166–177. doi:10.1016/j.tig.2009.03.002

Grillner S. 2006. Biological pattern generation: the cellular and computational logic of networks in motion. *Neuron* 52: 751–766. doi:10.1016/j.neuron.2006.11.008

Grillner S, Jessell TM. 2009. Measured motion: searching for simplicity in spinal locomotor networks. *Curr Opin Neurobiol* 19: 572–586. doi:10.1016/j.conb.2009.10.011

Grillner S, Robertson B. 2016. The basal ganglia over 500 million years. *Curr Biol* 26: R1088–R1100. doi:10.1016/j.cub.2016.06.041

Hain D, Gallego-Flores T, Klinkmann M, Macias A, Ciirdaeva E, Arends A, Thum C, Tushev G, Kretschmer F, Tosches MA, et al. 2022. Molecular diversity and evolution of neuron types in the amniote brain. *Science* 377: eabp8202. doi:10.1126/science.abp8202

Homem CCF, Repic M, Knoblich JA. 2015. Proliferation control in neural stem and progenitor cells. *Nat Rev Neurosci* 16: 647–659. doi:10.1038/nrn4021

Insel T, Winslow J, Wang Z, Young L, Hulihan TJ. 1995. Oxytocin and the molecular basis of monogamy. *Adv Exp Med Biol* 395: 227–234.

Ito M, Masuda N, Shinomiya K, Endo K, Ito K. 2013. Systematic analysis of neural projections reveals clonal composition of the *Drosophila* brain. *Curr Biol* 23: 644–655. doi:10.1016/j.cub.2013.03.015

Jacob F. 1977. Evolution and tinkering. *Science* 196: 1161–1166. doi:10.1126/science.860134

Jékely G. 2011. Origin and early evolution of neural circuits for the control of ciliary locomotion. *Proc Biol Sci* 278: 914–922.

Jékely G, Keijzer F, Godfrey-Smith P. 2015. An option space for early neural evolution. *Philos Trans R Soc Lond B Biol Sci* 370: 20150181. doi:10.1098/rstb.2015.0181

Jorstad NL, Close J, Johansen N, Yanny AM, Barkan ER, Travaglini KJ, Bertagnolli D, Campos J, Casper T, Crichton K, et al. 2023a. Transcriptomic cytoarchitecture reveals principles of human neocortex organization. *Science* 382: eadf6812. doi:10.1126/science.adf6812

Jorstad NL, Song JHT, Exposito-Alonso D, Suresh H, Castro-Pacheco N, Krienen FM, Yanny AM, Close J, Gelfand E, Long B, et al. 2023b. Comparative transcriptomics reveals human-specific cortical features. *Science* 382: eade9516. doi:10.1126/science.ade9516

Jourjine N, Hoekstra HE. 2021. Expanding evolutionary neuroscience: insights from comparing variation in behavior. *Neuron* 109: 1084–1099. doi:10.1016/j.neuron.2021.02.002

Kandimalla P, Omoto JJ, Hong EJ, Hartenstein V. 2023. Lineages to circuits: the developmental and evolutionary architecture of information channels into the central complex. *J Comp Physiol A Neuroethol Sens Neural Behav Physiol* 209: 679–720. doi:10.1007/s00359-023-01616-y

Katz PS. 2007. Evolution and development of neural circuits in invertebrates. *Curr Opin Neurobiol* 17: 59–64. doi:10.1016/j.conb.2006.12.003

Katz PS. 2016. Evolution of central pattern generators and rhythmic behaviours. *Philos Trans R Soc Lond B Biol Sci* 371: 20150057. doi:10.1098/rstb.2015.0057

Katz PS, Harris-Warrick RM. 1999. The evolution of neuronal circuits underlying species-specific behavior. *Curr Opin Neurobiol* 9: 628–633. doi:10.1016/S0959-4388(99)00012-4

Kebschul JM, Richman EB, Ringach N, Friedmann D, Albarran E, Kolluru SS, Jones RC, Allen WE, Wang Y, Cho SW, et al. 2020. Cerebellar nuclei evolved by repeatedly duplicating a conserved cell-type set. *Science* 370: eabd5059. doi:10.1126/science.abd5059

Komiyama T. 2023. Diversity of primate brain cells unraveled. *Sci Adv* 9: eadl0650. doi:10.1126/sciadv.adl0650

Kriegstein A, Noctor S, Martínez-Cerdeño V. 2006. Patterns of neural stem and progenitor cell division may underlie evolutionary cortical expansion. *Nat Rev Neurosci* 7: 883–890. doi:10.1038/nrn2008

Krienen FM, Levandowski KM, Zaniewski H, del Rosario RCH, Schroeder ME, Goldman M, Wienisch M, Lutservitz A, Beja-Glasser VF, Chen C, et al. 2023. A marmoset brain cell census reveals regional specialization of cellular identities. *Sci Adv* 9: eadk3986. doi:10.1126/sciadv.adk3986

Kubie LS. 1930. A theoretical application to some neurological problems of the properties of excitation waves which move in closed circuits. *Brain* 53: 166–177. doi:10.1093/brain/53.2.166

Lázaro J, Costanzo M, Sanaki-Matsumiya M, Girardot C, Hayashi M, Hayashi K, Diecke S, Hildebrandt TB, Lazzari G, Wu J, et al. 2023. A stem cell zoo uncovers intracellular scaling of developmental tempo across mammals. *Cell Stem Cell* 30: 938–949.e7. doi:10.1016/j.stem.2023.05.014

Li YE, Preissl S, Miller M, Johnson ND, Wang Z, Jiao H, Zhu C, Wang Z, Xie Y, Poirion O, et al. 2023. A comparative atlas of single-cell chromatin accessibility in the human brain. *Science* 382: eadf7044. doi:10.1126/science.adf7044

Lin Y, Yang J, Shen Z, Ma J, Simons BD, Shi SH. 2021. Behavior and lineage progression of neural progenitors in the mammalian cortex. *Curr Opin Neurobiol* 66: 144–157. doi:10.1016/j.conb.2020.10.017

Lin Y, Zhang XJ, Yang J, Li S, Li L, Lv X, Ma J, Shi SH. 2023. Developmental neuronal origin regulates neocortical map formation. *Cell Rep* 42: 112170. doi:10.1016/j.celrep.2023.112170

Lodato S, Molyneaux BJ, Zuccaro E, Goff LA, Chen HH, Yuan W, Meleski A, Takahashi E, Mahony S, Rinn JL, et al. 2014. Gene co-regulation by Fezf2 selects neurotransmitter identity and connectivity of corticospinal neurons. *Nat Neurosci* 17: 1046–1054. doi:10.1038/nn.3757

Mackie GO. 1970. Neuroid conduction and the evolution of conducting tissues. *Q Rev Biol* 45: 319–332. doi:10.1086/406645

Marder E, Calabrese RL. 1996. Principles of rhythmic motor pattern generation. *Physiol Rev* 76: 687–717. doi:10.1152/physrev.1996.76.3.687

Maroso M. 2023. A quest into the human brain. *Science* 382: 166–167. doi:10.1126/science.adl0913

Martinez P, Sprecher SG. 2020. Of circuits and brains: the origin and diversification of neural architectures. *Front Ecol Evol* 8: 82. doi:10.3389/fevo.2020.00082

Mathis A, Mamidanna P, Cury KM, Abe T, Murthy VN, Mathis MW, Bethge M. 2018. Deeplabcut: markerless pose estimation of user-defined body parts with deep learning. *Nat Neurosci* **21**: 1281–1289. doi:10.1038/s41593-018-0209-y

Micali N, Ma S, Li M, Kim SK, Mato-Blanco X, Sindhu SK, Arellano JI, Gao T, Shibata M, Gobeske KT, et al. 2023. Molecular programs of regional specification and neural stem cell fate progression in macaque telencephalon. *Science* **382**: eadf3786. doi:10.1126/science.adf3786

Najle SR, Grau-Bové X, Elek A, Navarrete C, Cianferoni D, Chiva C, Cañas-Armenteros D, Mallabiabarrena A, Kamm K, Sabidó E, et al. 2023. Stepwise emergence of the neuronal gene expression program in early animal evolution. *Cell* **186**: 4676–4693.e29. doi:10.1016/j.cell.2023.08.027

Newcomb JM, Sakurai A, Lillvis JL, Gunaratne CA, Katz PS. 2012. Homology and homoplasy of swimming behaviors and neural circuits in the Nudipleura (Mollusca, Gastropoda, Opisthobranchia). *Proc Natl Acad Sci* **109**: 10669–10676. doi:10.1073/pnas.1201877109

Niepoth N, Bendesky A. 2020. How natural genetic variation shapes behavior. *Annu Rev Genom Hum Genet* **21**: 437–463. doi:10.1146/annurev-genom-111219-080427

Oakley TH. 2003. The eye as a replicating and diverging, modular developmental unit. *Trends Ecol Evol* **18**: 623–627. doi:10.1016/j.tree.2003.09.005

Özel MN, Gibbs CS, Holguera I, Soliman M, Bonneau R, Desplan C. 2022. Coordinated control of neuronal differentiation and wiring by sustained transcription factors. *Science* **378**: eadd1884. doi:10.1126/science.add1884

Paşca SP. 2018. The rise of three-dimensional human brain cultures. *Nature* **553**: 437–445. doi:10.1038/nature25032

Pereira TD, Tabris N, Matsliah A, Turner DM, Li J, Ravindranath S, Papadoyannis ES, Normand E, Deutsch DS, Wang ZY, et al. 2022. SLEAP: a deep learning system for multi-animal pose tracking. *Nat Methods* **19**: 486–495. doi:10.1038/s41592-022-01426-1

Picco N, García-Moreno F, Maini PK, Woolley TE, Molnár Z. 2018. Mathematical modeling of cortical neurogenesis reveals that the founder population does not necessarily scale with neurogenic output. *Cereb Cortex* **28**: 2540–2550. doi:10.1093/cercor/bhy068

Pinson A, Huttner WB. 2021. Neocortex expansion in development and evolution—from genes to progenitor cell biology. *Curr Opin Cell Biol* **73**: 9–18. doi:10.1016/j.ceb.2021.04.008

Pitts W. 1942. The linear theory of neuron networks: the static problem. *Bull Math Biophys* **4**: 169–175. doi:10.1007/BF02478112

Pop S, Chen CL, Sproston CJ, Kondo S, Ramdya P, Williams DW. 2020. Extensive and diverse patterns of cell death sculpt neural networks in insects. *eLife* **9**: 1–31.

Prieto-Godino LL, Silbering AF, Khallaf MA, Cruchet S, Bojkowska K, Pradervand S, Hansson BS, Knaden M, Benton R. 2020. Functional integration of "undead" neurons in the olfactory system. *Sci Adv* **6**: eaaz7238. doi:10.1126/sciadv.aaz7238

Rakic P. 1995. A small step for the cell, a giant leap for mankind: a hypothesis of neocortical expansion during evolution. *Trends Neurosci* **18**: 383–388. doi:10.1016/0166-2236(95)93934-P

Sakurai A, Katz PS. 2015. Phylogenetic and individual variation in gastropod central pattern generators. *J Comp Physiol A Neuroethol Sens Neural Behav Physiol* **201**: 829–839. doi:10.1007/s00359-015-1007-6

Sato K, Tanaka R, Ishikawa Y, Yamamoto D. 2020. Behavioral evolution of *Drosophila*: unraveling the circuit basis. *Genes (Basel)* **11**: 157. doi:10.3390/genes11020157

Schultz DT, Haddock SHD, Bredeson JV, Green RE, Simakov O, Rokhsar DS. 2023. Ancient gene linkages support ctenophores as sister to other animals. *Nature* **618**: 110–117. doi:10.1038/s41586-023-05936-6

Seeholzer LF, Seppo M, Stern DL, Ruta V. 2018. Evolution of a central neural circuit underlies *Drosophila* mate preferences. *Nature* **559**: 564–569. doi:10.1038/s41586-018-0322-9

Siletti K, Hodge R, Albiach AM, Lee KW, Ding S-L, Hu L, Lönnerberg P, Bakken T, Casper T, Clark M, et al. 2023. Transcriptomic diversity of cell types across the adult human brain. *Science* **382**: eadd7046. doi:10.1126/science.add7046

Sousa AMM, Meyer KA, Santpere G, Gulden FO, Sestan N. 2017. Evolution of the human nervous system function, structure, and development. *Cell* **170**: 226–247. doi:10.1016/j.cell.2017.06.036

Suzuki IK. 2020. Molecular drivers of human cerebral cortical evolution. *Neurosci Res* **151**: 1–14. doi:10.1016/j.neures.2019.05.007

Suzuki T, Sato M. 2017. Inter-progenitor pool wiring: an evolutionarily conserved strategy that expands neural circuit diversity. *Dev Biol* **431**: 101–110. doi:10.1016/j.ydbio.2017.09.029

Suzuki IK, Gacquer D, Van Heurck R, Kumar D, Wojno M, Bilheu A, Herpoel A, Lambert N, Cheron J, Polleux F, et al. 2018. Human-specific NOTCH2NL genes expand cortical neurogenesis through Delta/Notch regulation. *Cell* **173**: 1370–1384.e16. doi:10.1016/j.cell.2018.03.067

Tambalo M, Lodato S. 2020. Brain organoids: human 3D models to investigate neuronal circuits assembly, function and dysfunction. *Brain Res* **1746**: 147028. doi:10.1016/j.brainres.2020.147028

Tian W, Zhou J, Bartlett A, Zeng Q, Liu H, Castanon RG, Kenworthy M, Altshul J, Valadon C, Aldridge A, et al. 2023. Single-cell DNA methylation and 3D genome architecture in the human brain. *Science* **382**: eadf5357. doi:10.1126/science.adf5357

Tosches MA. 2017. Developmental and genetic mechanisms of neural circuit evolution. *Dev Biol* **431**: 16–25. doi:10.1016/j.ydbio.2017.06.016

Tosches MA. 2021. From cell types to an integrated understanding of brain evolution: the case of the cerebral cortex. *Annu Rev Cell Dev Biol* **37**: 495–517. doi:10.1146/annurev-cellbio-120319-112654

Tosches MA, Yamawaki TM, Naumann RK, Jacobi AA, Tushev G, Laurent G. 2018. Evolution of pallium, hippocampus, and cortical cell types revealed by single-cell transcriptomics in reptiles. *Science* **360**: 881–888. doi:10.1126/science.aar4237

Uzquiano A, Arlotta P. 2022. Brain organoids: the quest to decipher human-specific features of brain development. *Curr Opin Genet Dev* **75**: 101955. doi:10.1016/j.gde.2022.101955

Vanderhaeghen P, Polleux F. 2023. Developmental mechanisms underlying the evolution of human cortical circuits. *Nat Rev Neurosci* **24:** 213–232. doi:10.1038/s41583-023-00675-z

Vargha-Khadem F, Gadian DG, Copp A, Mishkin M. 2005. FOXP2 and the neuroanatomy of speech and language. *Nat Rev Neurosci* **6:** 131–138. doi:10.1038/nrn1605

Velasco S, Paulsen B, Arlotta P. 2020. 3D brain organoids: studying brain development and disease outside the embryo. *Annu Rev Neurosci* **43:** 375–389. doi:10.1146/annurev-neuro-070918-050154

Velmeshev D, Perez Y, Yan Z, Valencia JE, Castaneda-Castellanos DR, Wang L, Schirmer L, Mayer S, Wick B, Wang S, et al. 2023. Single-cell analysis of prenatal and postnatal human cortical development. *Science* **382:** eadf0834. doi:10.1126/science.adf0834

Wilson AC, Sweeney LB. 2023. Spinal cords: symphonies of interneurons across species. *Front Neural Circuits* **17:** 1146449. doi:10.3389/fncir.2023.1146449

Xie Y, Dorsky RI. 2017. Development of the hypothalamus: conservation, modification and innovation. *Development* **144:** 1588–1599. doi:10.1242/dev.139055

Young LJ, Winslow JT, Nilsen R, Insel TR. 1997. Species differences in V1a receptor gene expression in monogamous and nonmonogamous voles: behavioral consequences. *Behav Neurosci* **111:** 599–605. doi:10.1037/0735-7044.111.3.599

Young LJ, Nilsen R, Waymire KG, MacGregor GR, Insel TR. 1999. Increased affiliative response to vasopressin in mice expressing the V1a receptor from a monogamous vole. *Nature* **400:** 766–768. doi:10.1038/23475

Zhu K, Bendl J, Rahman S, Vicari JM, Coleman C, Clarence T, Latouche O, Tsankova NM, Li A, Brennand KJ, et al. 2023. Multi-omic profiling of the developing human cerebral cortex at the single-cell level. *Sci Adv* **9:** eadg3754. doi:10.1126/sciadv.adg3754

Cell Adhesion Molecule Signaling at the Synapse: Beyond the Scaffold

Ben Verpoort[1,2] and Joris de Wit[1,2]

[1]VIB-KU Leuven Center for Brain and Disease Research, 3000 Leuven, Belgium

[2]KU Leuven, Department of Neurosciences, Leuven Brain Institute, 3000 Leuven, Belgium

Correspondence: joris.dewit@kuleuven.be

Synapses are specialized intercellular junctions connecting pre- and postsynaptic neurons into functional neural circuits. Synaptic cell adhesion molecules (CAMs) constitute key players in synapse development that engage in homo- or heterophilic interactions across the synaptic cleft. Decades of research have identified numerous synaptic CAMs, mapped their *trans*-synaptic interactions, and determined their role in orchestrating synaptic connectivity. However, surprisingly little is known about the molecular mechanisms that translate *trans*-synaptic adhesion into the assembly of pre- and postsynaptic compartments. Here, we provide an overview of the intracellular signaling pathways that are engaged by synaptic CAMs and highlight outstanding issues to be addressed in future work.

Synapses are highly organized and specialized intercellular junctions that mediate connectivity between neurons and provide the basic structural units for neuronal communication. The appropriate formation of chemical synapses requires the coordinated assembly of the synaptic vesicle release machinery on the presynaptic side and the neurotransmitter receptor machinery on the postsynaptic side into precisely aligned nanocolumns across the synaptic cleft (Biederer et al. 2017; Chen et al. 2018; Nozawa et al. 2022). Both the pre- and postsynaptic compartment are composed of large and distinct protein complexes, which are built from the neuron's repertoire of scaffold proteins, neurotransmitter receptors, ion channels, and cell adhesion molecules (CAMs) (Emes and Grant 2012; Grant 2019). Synaptic CAMs engage in homo- or heterophilic *trans*-synaptic interactions to form cleft-spanning complexes. Since pioneering work showing that the postsynaptic CAM neuroligin-1 (NLGN1) presented on the surface of heterologous cells can induce presynaptic differentiation in axons of cocultured neurons, a process that can be blocked by interfering with the presynaptic CAM neurexin (NRXN) (Scheifele et al. 2000; Dean et al. 2003), numerous *trans*-synaptic adhesive complexes with the ability to induce pre- or postsynaptic differentiation have been identified. We refer to these proteins as synaptogenic CAMs herein.

A picture has emerged in which NRXNs and LAR-type receptor protein tyrosine phosphatases (RPTPs) act as presynaptic adhesive platforms that engage in extracellular interactions with a wide variety of nonoverlapping postsynaptic CAMs, such as neuroligins (NLGNs), secreted cerebellins (CBLNs), and multiple families of leucine-rich repeat (LRR)-containing cell-surface proteins (Takahashi and Craig 2013; Südhof 2017; Gomez et al. 2021). In addition, many more adhesive cleft-spanning complexes with a critical role in synapse development have been identified, including interactions between ephrins and Eph receptors, between SynCAMs, and between teneurins or fibronectin leucine-rich repeat transmembrane proteins (FLRTs) and latrophilins (LPHNs) (de Wit and Ghosh 2016; Südhof 2018; Sanes and Zipursky 2020; Kim et al. 2021). This molecular diversity is substantially amplified through alternative splicing (Um and Ko 2013; Boucard et al. 2014; Lu et al. 2018; Gomez et al. 2021) and posttranslational modifications (Jeong et al. 2017; Noborn and Sterky 2023).

Despite intense research on synaptic CAMs, surprisingly little is known about how *trans*-synaptic adhesion couples to downstream intracellular signaling pathways to orchestrate synapse development. Synaptic CAMs are thought to regulate pre- and postsynaptic development via their association with key scaffold proteins that organize neurotransmitter release sites and postsynaptic density assembly, respectively (Jang et al. 2017; Han et al. 2019b). However, few studies have focused on the physiological relevance of these interactions. Moreover, emerging data indicate the existence of additional mechanisms. Rather than providing an overview of *trans*-synaptic mechanisms that have been extensively covered elsewhere (Kim et al. 2021, 2022), we summarize here recent progress in our understanding of the intracellular signal transduction cascades employed by synaptic CAMs. Because of space constraints, we limit our discussion to synaptogenic CAMs at excitatory synapses and focus on mammalian work.

SYNAPTIC CAMs ORGANIZE SYNAPSES VIA SCAFFOLD-MEDIATED INTERACTIONS

Via their intracellular PSD95/Discs large/ZO-1 (PDZ)-binding domain, many synaptic CAMs bind to various members of the membrane-associated guanylate kinase (MAGUK) family of scaffold proteins (Han and Kim 2008). Through this association with MAGUKs, synaptic CAMs have been postulated to modulate the organization of postsynaptic receptors and regulate synaptic strength (Fig. 1A; Jang et al. 2017). Presynaptically, synaptic CAMs are thought to aid in the assembly of the neurotransmitter release machinery (Fig. 1A; Han et al. 2019b). However, the functional importance of these PDZ-mediated interactions and the underlying molecular mechanisms remain poorly understood.

PRESYNAPTIC NRXNs AND LAR-RPTPs COUPLE TO ACTIVE ZONE ASSEMBLY

NRXNs

The NRXN family consists of three genes (*Nrxn1–3*), each transcribed from two independent promoters, generating a long (α) and short (β) isoform that differ in their extracellular domain but share identical intracellular sequences (Reissner et al. 2013). In addition, the *Nrxn1* gene encodes a third isoform, NRXN1γ, that lacks all extracellular domains apart from a few membrane-proximal sequences (Sterky et al. 2017). Furthermore, all *Nrxn* genes undergo extensive alternative splicing at six canonical sites generating thousands of theoretically possible isoforms, hundreds of which are differentially expressed across various brain regions (Schreiner et al. 2014; Treutlein et al. 2014). Finally, NRXNs are posttranslationally modified and carry an extracellular heparan sulfate (HS) polysaccharide side chain (Zhang et al. 2018). A series of genetic studies using conditional deletions of *Nrxns* in different circuits produced divergent phenotypes, ranging from synapse loss to impairments in basal synaptic transmission and reduced action potential–evoked presynaptic Ca^{2+} influx (Anderson et al. 2015; Aoto et al. 2015; Chen et al. 2017; Boxer et al. 2021). Subsequent work demonstrated that NRXNs are responsible for the tight spatial coupling of Ca^{2+} channels to neurotransmitter release sites in the active zone (Luo et al. 2020), confirming their essential role for presynaptic Ca^{2+} influx (Missler et al. 2003).

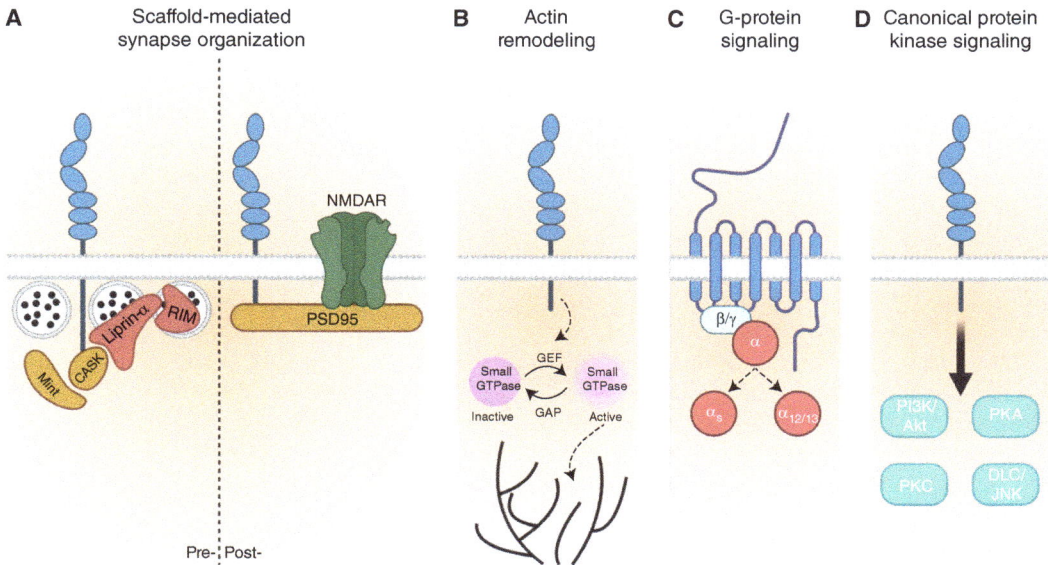

Figure 1. Summary of synaptic cell adhesion molecule (CAM)-mediated intracellular signaling pathways. (*A*) Synaptic CAMs mediate synapse development via binding to pre- or postsynaptic scaffold proteins such as calmodulin-dependent serine protein kinase (CASK) and PSD95, respectively. These scaffold proteins are involved in the assembly of the neurotransmitter release machinery and the organization of postsynaptic receptor complexes. Additional pathways include (*B*) signaling via guanine nucleotide exchange factors (GEFs) and GTPase activating proteins (GAPs) to control actin cytoskeletal dynamics, (*C*) signaling via G proteins downstream of adhesive G protein–coupled receptor (aGPCRs), and (*D*) signaling via canonical protein kinases.

Intracellularly, NRXNs bind the presynaptic scaffold proteins calcium/calmodulin-dependent serine protein kinase (CASK) (Hata et al. 1996) and Mints (Biederer and Südhof 2000), the FERM-domain containing protein 4.1 (Biederer and Südhof 2001) and synaptotagmin (Hata et al. 1993). In turn, CASK binds to α-liprins that are part of a large core active zone protein complex regulating neurotransmitter release, including synaptic vesicle docking and priming, and the recruitment of Ca²⁺ channels (Südhof 2012). Thus, via CASK, NRXNs could mediate the precise assembly of presynaptic Ca²⁺ channels in the active zone. However, a detailed mechanistic understanding of such coupling is still lacking.

In addition to identifying numerous *trans*-synaptic CAM pairs, the heterologous synapse formation assay or coculture assay (Biederer and Scheiffele 2007) has provided mechanistic insight into our understanding of how synaptic CAMs might couple to intracellular signaling

pathways to mediate synapse development. In this assay, a synaptic CAM of interest is expressed in heterologous cells, such as HEK239T or COS7 cells, that will induce pre- or postsynaptic specializations in cocultured neurons. When presynaptic NRXNs are expressed in heterologous cells, postsynaptic specializations are formed in contacting dendrites of cocultured neurons (Graf et al. 2004). Conversely, the postsynaptic NRXN ligands NLGNs or LRRTMs will induce the formation of presynaptic specializations in axons of cocultured neurons (Scheiffele et al. 2000; de Wit et al. 2009; Ko et al. 2009; Linhoff et al. 2009).

The cytoplasmic domain of NRXNs is dispensable for the induction of presynaptic specializations onto NLGN1- or LRRTM2-expressing COS7 cells (Gokce and Südhof 2013; Roppongi et al. 2020). These findings suggest that *cis* interactions with a coreceptor are required for NRXN-mediated presynaptic differentiation in response to NLGN1 or LRRTM2 binding (Fig. 2). The existence of endogenous glycosyl-phosphatidylinositol (GPI)-anchored NRXN isoforms (Hauser et al. 2022) is consistent with this notion. Indeed, LAR-RPTPs were found to interact with the extracellular HS sidechains of NRXNs (Han et al. 2020a). Short hairpin RNA (shRNA)-mediated knockdown (KD) of PTPσ in cultured hippocampal neurons inhibits excitatory presynaptic assembly induced by NLGN1-expressing COS7 cells. In addition, a large number of intracellular PTPσ-binding partners, including α-liprins, were shown to be necessary for the assembly of presynaptic specializations induced by NLGN1-expressing COS7 cells (Han et al. 2020a). Since NLGN1 and PTPσ do not interact in *trans*, these findings suggest that NRXNs couple to PTPσ's intracellular signaling machinery to induce the formation of presynaptic specializations in response to NLGN1 binding (Fig. 2A).

In contrast to heterologous synapses formed onto NGLN1-expressing COS7 cells, loss of PTPσ in cultured hippocampal neurons does not influence their ability to induce presynaptic specializations onto LRRTM2-expressing heterologous cells (Ko et al. 2015; Condomitti et al. 2018; Roppongi et al. 2020). This indicates the existence of an additional, as-yet unidentified NRXN coreceptor that is engaged by LRRTM2 (Fig. 2B). PTPρ, an RPTP different from the LAR-RPTP subfamily that also interacts with synaptic adhesion molecules (Lim et al. 2009), is one possible candidate that could interact with NRXNs in *cis* to transduce an intracellular signal in response to LRRTM2 binding.

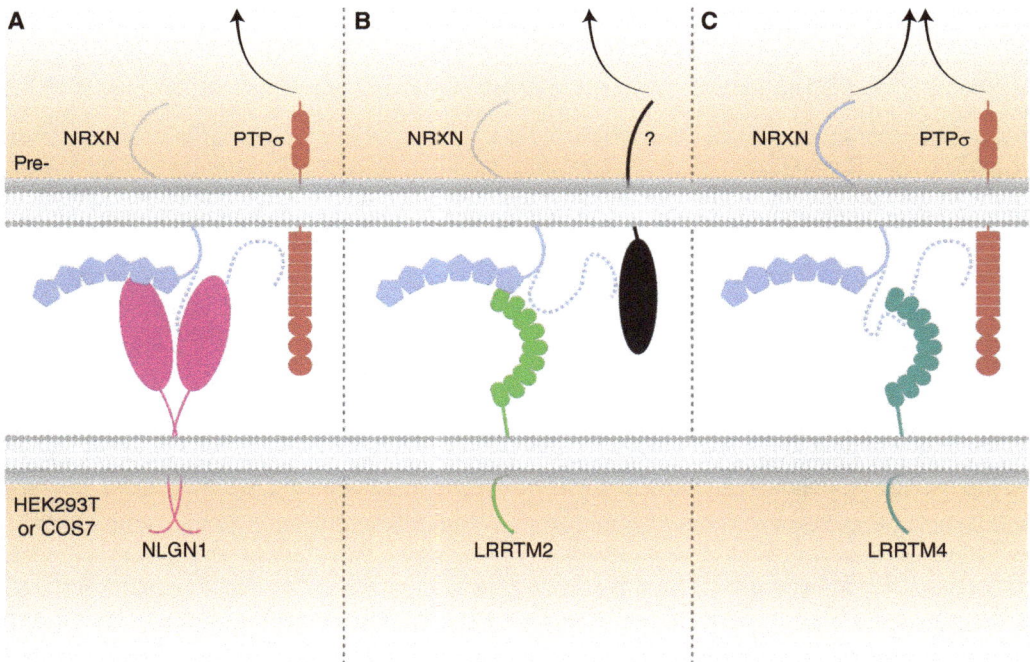

Figure 2. Induction of presynaptic specializations onto heterologous cells expressing different postsynaptic neurexin (NRXN) ligands. (*A*) The intracellular tail of NRXN is dispensable for the formation of NLGN1-induced presynaptic specializations. Instead, NRXN engages PTPσ's intracellular signaling machinery to induce presynaptic development. (*B*) Similarly, the formation of LRRTM2-induced presynaptic specializations does not require the intracellular tail of NRXNs. However, PTPσ is not involved during this process, indicating the existence of an as-yet unidentified NRXN coreceptor. (*C*) LRRTM4-mediated induction of presynaptic specializations requires the intracellular tail of NRXN. Here, NRXN and PTPσ cooperate to allow full-strength synaptogenesis induced by LRRTM4.

Cite this article as *Cold Spring Harb Perspect Biol* doi: 10.1101/cshperspect.a041501

Surprisingly, presynaptic assembly induced by LRRTM3- and LRRTM4-expressing COS7 cells does require the cytoplasmic domain of NRXNs (Roppongi et al. 2020). Whereas shRNA-mediated KD of NRXNs fully blocks LRRTM4-induced presynaptic differentiation, shRNA-mediated KD of PTPσ reduces LRRTM4's synaptogenic potency. Together, these results indicate that NRXN does not engage PTPσ as a coreceptor to transduce an intracellular signal, but that PTPσ and NRXN cooperate to allow full-strength synaptogenesis induced by LRRTM4 (Fig. 2C). Given that NRXNs' cytoplasmic domain is necessary for the formation of LRRTM3- and LRRTM4-induced presynaptic specializations, it is likely that a different set of intracellular proteins is recruited compared to NLGN1- or LRRTM2-induced presynaptic specializations.

Interestingly, an engineered pair of CAMs consisting of tightly interacting bacterial proteins with no eukaryotic homologs requires coupling to the intracellular sequence of NRXNs to induce presynaptic specializations, consistent with the notion that their cytoplasmic tail can couple to intracellular signaling pathways (Hale et al. 2023).

Together, these findings raise the intriguing possibility that various postsynaptic ligands differentially engage the cytoplasmic domain of NRXNs to enable unique signaling pathways.

LAR-RPTPs

Like NRXNs, members of the presynaptic LAR-RPTP family (LAR, PTPσ, and PTPδ) exist in multiple isoforms that perform distinct functions at various synapses (Han et al. 2020b,c; Kim et al. 2020; Park et al. 2020; Sclip and Südhof 2020). At hippocampal CA1-subicular synapses, PTPσ maintains excitatory synapse density and presynaptic neurotransmitter release (Han et al. 2020b).

Whereas NRXNs are required for the induction of presynaptic specializations onto NLGN- and LRRTM-expressing heterologous cells, LAR-RPTPs are responsible for the presynaptic differentiation induced by a different set of postsynaptic CAMs, including Slit- and Trk-like

proteins (SLITRKs). LAR-RPTPs contain a cytosolic catalytically inactive D2 domain and a catalytically active D1 domain that confers tyrosine phosphatase activity (Han et al. 2016a). Via its D2 domain, LAR-RPTPs bind to an array of presynaptic scaffolds and components of the core active zone protein complex, including CASK and α-liprins, respectively (Han et al. 2018). Substantiating these interactions, shRNA-mediated KD of liprin-α2, liprin-α3, caskins, rab3-interacting molecules (RIMs), and RIM-binding proteins (RIM-BPs) impairs the formation of presynaptic specializations onto SLITRK6-expressing HEK239T cells (Han et al. 2018, 2020a; Bomkamp et al. 2019). Accordingly, a PTPσ deletion mutant lacking the D2 domain fails to rescue heterologous synapse formation, excitatory synapse density, and proper synaptic vesicle localization in PTPσ-deficient cultured hippocampal neurons (Han et al. 2018, 2020b).

Thus, coculture assay-based experiments indicate that presynaptic NRXNs and LAR-RPTPs can mediate the assembly of presynaptic terminals via their cytoplasmic interactions with various scaffolds and components of the core active zone protein complex. Many of these proteins indirectly associate with each other through multiple interactions, forming a large intracellular network. It remains to be determined to what extent these intracellular interactions also occur in intact circuits.

POSTSYNAPTIC CAMs MODULATE NEUROTRANSMITTER RECEPTOR FUNCTION

NLGNs

The NLGN family, arguably the most extensively characterized NRXN ligands, comprise four members (NLGN1–4) that localize to different types of synapses and govern distinct properties. Whereas NLGN1 and NLGN2 localize to excitatory and inhibitory synapses, respectively (Song et al. 1999; Poulopoulos et al. 2009), NLGN3 is present at both synapse types (Budreck and Scheiffele 2007) and NLGN4 at glycinergic synapses (Hoon et al. 2011). NLGN1–3 contain two alternative splice sites located in their extracellular domain. Intracellularly, all NLGNs harbor a PDZ- and gephyrin-binding domain (Bemben et al. 2015).

Experiments using acute hippocampal slices from *Nlgn1–3* triple conditional knockout (KO) mice with a constitutive *Nlgn4* deletion (*Nlgn1–4* quadruple KO mice), revealed that deletion of all *Nlgns* does not affect AMPAR-mediated synaptic transmission, but strongly decreases NMDAR-mediated synaptic transmission and long-term potentiation (LTP) in CA1 pyramidal neurons (Wu et al. 2019). These decreases were entirely attributed to NLGN1 function, since *Nlgn1* conditional KO mice phenocopied the removal of all NLGNs and NLGN1 was able to fully rescue these synaptic changes in *Nlgn1–4* quadruple KO mice (Wu et al. 2019). A GPI-anchored NLGN1 failed to rescue NMDAR-mediated synaptic transmission in *Nlgn1–4* quadruple KO mice, highlighting the importance of its intracellular tail (Wu et al. 2019). A full-length NRXN-binding-deficient NLGN1 mutant completely rescued NMDAR-mediated synaptic transmission, indicating that *trans*-synaptic binding to NRXN is dispensable in this process. On the other hand, LTP was fully rescued using GPI-anchored NLGN1 and required binding to presynaptic NRXNs (Wu et al. 2019). Thus, NLGN1 requires its intracellular tail to maintain NMDAR-mediated synaptic transmission, but this domain is dispensable for its role in LTP, demonstrating a striking uncoupling of the NLGN1 domains required for these two processes. Although not tested here, binding of NLGN1 via its intracellular PDZ-binding domain to the postsynaptic scaffold protein PSD95 provides a plausible mechanism to regulate NMDAR-mediated synaptic transmission (Irie et al. 1997). Conditional genetic deletion constitutes an attractive genetic strategy for acute removal of *Nlgn* expression, but achieving complete KO of all four NLGNs in a given cell type using this approach remains challenging.

In contrast to the findings mentioned above, others have demonstrated a role for NLGN1 in controlling AMPAR-mediated synaptic transmission. In organotypic hippocampal slices from constitutive *Nlgn1* KO mice, overexpression of NLGN1 wild-type (WT), but not a mu-

tant lacking its PDZ-binding domain, augments AMPAR-mediated synaptic responses in CA1 pyramidal neurons. NMDAR-mediated synaptic responses on the other hand were unaffected following this manipulation (Letellier et al. 2018). NLGNs undergo extensive posttranslational modifications that regulate their folding, surface trafficking, and interaction with other proteins (Jeong et al. 2017). NLGN1 phosphorylation at intracellular tyrosine residue Y782 favors interaction with PSD95 and enables synaptic recruitment of AMPARs (Giannone et al. 2013; Letellier et al. 2018). Also contradicting the findings of Wu et al. who demonstrated that NLGN1's intracellular tail is dispensable for LTP, phosphorylation at intracellular tyrosine residue Y782 was shown to be a key event dictating the LTP response (Letellier et al. 2018). Bypassing potential confounds introduced by manipulating NLGN expression levels, a follow-up study acutely altered endogenous NLGN1 phosphorylation using a photo-activatable fibroblast growth factor receptor 1 (FGFR1) and confirmed the importance of intracellular signaling via NLGN1 tyrosine residue Y782 for LTP (Letellier et al. 2020). Organotypic cultures offer experimental accessibility, but cell density and synaptic connectivity change over time, which may affect results obtained using this approach.

In another study, NLGN3 was found to enhance AMPAR-mediated synaptic transmission independent of its association with PSD95. Removal of the PDZ-binding domain, or the region encompassing tyrosine residue Y782 in NLGN3, did not impair its ability to enhance AMPAR-mediated synaptic transmission following overexpression in organotypic hippocampal slices in which *Nlgn1–3* were knocked down using microRNAs (Shipman et al. 2011). Instead, a small region closer to the transmembrane domain encompassing three amino acids was found to be responsible for increasing AMPAR-mediated synaptic responses. Similar results were obtained for NLGN1 (Shipman et al. 2011). Which binding partner(s) might be involved in this process is not known.

In conclusion, considerable discrepancy exists in the literature regarding the role of NLGN1

 Cite this article as *Cold Spring Harb Perspect Biol* doi: 10.1101/cshperspect.a041501

in controlling NMDAR- or AMPAR-mediated synaptic transmission and LTP. Furthermore, whether interactions with postsynaptic scaffolds via the NLGN PDZ-binding domain are required for these processes remains controversial. These differences might arise from variations in the use of certain experimental preparations (acute vs. organotypic slices), the approach used to manipulate NLGN1 levels (KD vs. KO) and the time at which these perturbations take place (constitutive vs. conditional). Given that NLGNs form heterodimers (Poulopoulos et al. 2012), a genetic background in which all NLGNs are removed is necessary to exclude the possibility that rescue constructs couple to intracellular signaling pathways of remaining endogenous NLGNs.

LRRTMs

LRRTMs form a family of four members, LRRTM1–4, that localize to excitatory synapses. LRRTM1 and LRRTM2 are important for AMPAR-mediated synaptic transmission and LTP in the mature hippocampus, and are involved in synapse formation during development (Bhouri et al. 2018; Schroeder et al. 2018; Dhume et al. 2022). LRRTM3 and LRRTM4 are selectively required in hippocampal granule cells to maintain excitatory synapse density and support LTP-driven synaptic surface expression of AMPARs (de Wit et al. 2013; Siddiqui et al. 2013; Um et al. 2016). Even though all LRRTMs can bind PSD95 through their carboxy-terminal PDZ-binding domain (de Wit et al. 2009; Linhoff et al. 2009; Siddiqui et al. 2013; Um et al. 2016; Karimi et al. 2021), the functional relevance of this interaction remains unclear.

Experiments using acute hippocampal slices from *Lrrtm1/2* double conditional KO mice demonstrated that LRRTM1 and LRRTM2 maintain AMPAR-mediated, but not NMDA-mediated synaptic responses, and are required for LTP in CA1 pyramidal neurons (Bhouri et al. 2018). Which domains in LRRTM1 and LRRTM2 are responsible to maintain AMPAR-mediated synaptic responses was not assessed in the study (Bhouri et al. 2018). LTP can be rescued using a GPI-linked LRRTM2, but not when key

extracellular residues that are thought to mediate NRXN binding (but see Yamagata et al. 2018) are mutated and introduced into this construct. Thus, LRRTM2 binding to NRXNs is required for LTP, but an interaction with PSD95 is dispensable in this process (Bhouri et al. 2018).

In contrast to LRRTM2, both LRRTM3 and LRRTM4 exist as short and long isoforms that are developmentally regulated and show distinct expression patterns. The long isoforms constitute variants containing an intracellular tail ending in a non-PDZ-binding domain. Only the short isoforms can interact with PSD95 (Um et al. 2016). While both isoforms reach the synaptic surface equally well, only the short LRRTM3 isoform can rescue chemical LTP-induced AMPAR surface expression in cultured hippocampal neurons from *Lrrtm3* KO mice (Um et al. 2016). These findings underscore the importance of the intracellular PDZ-binding domain of LRRTM3. Activity-dependent AMPAR insertion in cultured neurons from *Lrrtm4* KO hippocampal granule cells is also impaired (Siddiqui et al. 2013), suggesting that LRRTM4 might operate in the same manner. Whether these results can be translated to a more physiologically relevant setting, such as acute hippocampal slices from *Lrrtm3* or *Lrrtm4* conditional KO mice, and whether activity-dependent AMPAR insertion during LTP might require the PDZ-binding domains of LRRTM3 or LRRTM4 in this context, remain to be explored.

Other postsynaptic CAMs such as the LAR-RPTP-binding partners SLITRK2 (Han et al. 2019a) and netrin-G ligand 3 (NGL3) (Kim et al. 2006), as well as synaptic adhesion-like molecules 1–3 (SALM1–3) (Lie et al. 2018) also bind to PSD95. However, to what extent these interactions are directly involved in modulating neurotransmitter receptor function in vivo requires further investigation.

SIGNALING BEYOND SYNAPTIC SCAFFOLDS

In addition to MAGUKs, synaptic CAMs can interact with multiple other proteins residing in the PSD, linking cell adhesion with dynamic cytoskeletal remodeling (Fig. 1B) or G protein signaling (Fig 1C). Canonical signal transduction pathways involving various protein kinases have also been demonstrated to act downstream of synaptic CAMs (Fig. 1D).

MODULATION OF THE ACTIN CYTOSKELETON

Dynamic changes in the actin cytoskeleton support the growth and remodeling of dendritic spines during synapse development. In addition, reorganization of the actin cytoskeleton plays a key role in the assembly of the presynaptic active zone machinery (Spence and Soderling 2015; Gentile et al. 2022). The activity of proteins that regulate actin assembly and disassembly are tightly controlled by Rho and Ras family GTPases such as Rho, Rac, and Rap. In turn, these small GTPases are themselves regulated by various guanine nucleotide exchange factors (GEFs) and GTPase-activating proteins (GAPs) that turn GTPase activity on or off, respectively (Tolias et al. 2011).

Recent evidence indicates that NLGN1 has the ability to couple to actin-related pathways via interactions with both GEFs and GAPs. First, the PDZ-binding domain of NLGN1 has been demonstrated to pull down spine-associated Rap-GAP (SPAR) from brain lysates, a GAP that tunes down activity of the small GTPase Rap1 (Liu et al. 2016). Activity-induced proteolytic cleavage of NLGN1's intracellular domain inhibits SPAR and allows Rap1-mediated phosphorylation of cofilin to promote actin assembly. NLGN1-mediated inhibition of SPAR was shown to promote dendritic spine formation and synaptic strength in CA1 pyramidal neurons in vivo. Additionally, SPAR inhibition by NLGN1 attenuates long-term depression (LTD) while it facilitates LTP in CA1 pyramidal neurons from acute hippocampal slices (Liu et al. 2016).

Second, using quantitative proteomic analysis, two Rac1 GEFs with key roles in excitatory synapse development, Trio and Kalirin (Paskus et al. 2020), were identified as intracellular NLGN1-binding partners (Paskus et al. 2019; Tian et al. 2021). ShRNA-mediated KD of Kalirin or Trio in cultured hippocampal neurons

prevented the ability of NLGN1 to augment dendritic spine number, a phenotype that was aggravated by the simultaneous KD of both Kalirin and Trio. Furthermore, the observed increase in NMDAR- and AMPAR-mediated synaptic transmission following overexpression of NLGN1 in CA1 pyramidal neurons from acute hippocampal slices depended on its interaction with Kalirin or Trio (Paskus et al. 2019; Tian et al. 2021). The intracellular NLGN1 domain responsible for binding to Kalirin or Trio is not known.

SynCAM1, one of four SynCAM (SynCAM1–4) family members with a key role in regulating excitatory synapse development in various circuits, has been demonstrated to interact with Farp1, a GEF for the small GTPase Rho (Cheadle and Biederer 2012). ShRNA-mediated KD of Farp1 in cultured hippocampal neurons prevented the ability of SynCAM1 to promote dendritic spine formation. Mechanistically, SynCAM1 maintains synaptic Farp1 levels, allowing activation of Rho to promote actin polymerization in dendritic spines (Cheadle and Biederer 2012). The FERM-binding domain in SynCAM1's cytosolic tail mediates binding to Farp1. SynCAMs also contain a PDZ-binding domain that interacts with several MAGUKs (Frei and Stoeckli 2017). However, the functional significance of these interactions remains unknown.

G PROTEIN SIGNALING

LPHNs (LPHN1–3), members of the adhesive G protein–coupled receptor (aGPCR) family, interact with teneurins and FLRTs to control the development of specific synapses in CA1 pyramidal neurons and cerebellar Purkinje cells. In the hippocampus, LPHN2 is postsynaptically required for the development of perforant path synapses, while postsynaptic LPHN3 maintains Schaffer-collateral synaptic connections in the same CA1 pyramidal neuron (Sando et al. 2019; Zhang et al. 2020). These trans-synaptic interactions are mediated by LPHNs' unusually large extracellular region harboring various adhesion domains. Their intracellular tail contains multiple highly conserved sequences and a PDZ-binding domain that interacts with Shank scaffold proteins (Lala and Hall 2022). Acting as canonical GPCRs, LPHNs activate diverse G protein–dependent signaling pathways via their seven-pass transmembrane region. However, until recently, it was unclear whether signaling via G proteins is physiologically relevant and required for LPHN-mediated synapse development.

LPHN2 and LPHN3 were shown to act as constitutive GPCRs up-regulating cAMP levels (Sando and Südhof 2021). Abolishing their G protein coupling ability and thus restricting cAMP production rendered them unable to direct the formation of Schaffer-collateral and perforant path synapses onto CA1 hippocampal neurons (Sando and Südhof 2021). Together with the observation that local postsynaptic inhibition of cAMP production was similarly shown to be required for the formation of the aforementioned synapses (Sando et al. 2022), these findings indicate that compartmentalized cAMP-mediated LPHN signaling assembles postsynaptic specializations. Increased levels of cAMP upon overexpression of LPHN2 and LPHN3 indicate coupling to $G\alpha_s$. This has been confirmed, at least for LPHN3 (Nazarko et al. 2018; Moreno-Salinas et al. 2022). However, others found that LPHN2 and LPHN3 couple to $G\alpha_{12}$ and $G\alpha_{13}$, but not $G\alpha_s$ (Mathiasen et al. 2020; Pederick et al. 2023). A detailed molecular understanding of how such coupling to $G\alpha_s$ or $G\alpha_{12/13}$ mediates synapse development is still lacking, but likely involves activation of PKA and the small GTPase Rho, respectively.

Brain-specific angiogenesis inhibitors (BAIs) constitute another class of postsynaptically localized aGPCRs and comprise three family members (BAI1–3) (Lala and Hall 2022). Reducing BAI1 levels impairs dendritic spine formation in cultured hippocampal neurons as well as in CA1 pyramidal neurons in vivo (Duman et al. 2013; Tu et al. 2018). BAI1 has been shown to stimulate the small GTPase Rho via its coupling to $G\alpha_{12/13}$ (Stephenson et al. 2013; Kishore et al. 2016). However, whether this signaling pathway is involved in dendritic spine development remains to be determined. A noncanonical, G protein-independent signaling pathway has been proposed that mediates BAI1's ability to promote dendritic spine formation. Here, BAI1 interacts with and recruits the GEF Tiam1 to postsynaptic sites allowing for the activation of the small GTPase Rac1 and dendritic spine formation to proceed (Duman et al. 2013; Tu et al. 2018).

CONVERGENCE ONTO CANONICAL SIGNAL TRANSDUCTION PATHWAYS

Once activated, membrane-bound receptor tyrosine kinases (RTKs) and RPTPs have the ability to modulate numerous downstream canonical signal transduction pathways (Schlessinger 2014; Lee et al. 2015). At synapses, these pathways can be modulated by a limited number of synaptic CAMs of which the cytoplasmic tail exhibits enzymatic activity, including members of the LAR-RPTP family, receptor tyrosine kinase C (TrkC), and Eph receptors (Dabrowski and Umemori 2011). Signaling via ephrin–Eph receptor interactions has recently been reviewed elsewhere and will not be discussed here (Washburn et al. 2023).

In addition to the catalytically inactive D2 domain that interacts with presynaptic scaffolds, LAR-RPTPs also contain a D1 domain that confers tyrosine phosphatase activity (Han et al. 2016a). Many of the identified substrates of the LAR-RPTPs D1 domain, including N-cadherin, β-catenin, Abelson kinase (Abl), Enabled (Ena), and p250GAP, play a role during actin polymerization (Coles et al. 2015). ShRNA-mediated KD of these substrates in cultured hippocampal neurons inhibited their ability to induce presynaptic specializations onto SLITRK6-expressing HEK293T cells (Han et al. 2020a). Moreover, a PTPσ point mutant that abolishes its phosphatase activity fails to rescue heterologous synapse formation, excitatory synapse density, and proper synaptic vesicle localization in PTPσ-deficient cultured hippocampal neurons (Han et al. 2020a,b). These findings suggest that LAR-RPTP signaling mechanisms converge on the organization of the presynaptic active zone via its D2 domain as well as on the modulation of the presynaptic actin cytoskeletal network via its D1 domain. Additional signaling modules are likely to be involved given that unbiased phosphoproteomic analyses from *Ptpσ* conditional KO mice demonstrated widespread changes in the phosphotyrosine levels of many presynaptic as well as postsynaptic proteins (Kim et al. 2020).

TrkC, one of the postsynaptic PTPσ-binding partners (Takahashi et al. 2011), contains an intracellular tyrosine kinase domain shown to be important for its synapse-promoting activity in cultured hippocampal neurons (Han et al. 2016b). WT TrkC, but not a point mutant that abolishes its tyrosine kinase activity, is able to rescue the loss of postsynaptic specializations in TrkC-deficient cultured hippocampal neurons. TrkC downstream signaling can result in the activation of extracellular signal-regulated kinase (ERK), phosphatidylinositol 3-kinase (PI3K)/Akt (protein kinase B), and protein kinase C (PKC) pathways (Han et al. 2016b). However, which substrates in these signaling cascades are engaged by TrkC to promote its synapse-inducing activity remains to be determined.

A recent study examined whether any of these and other canonical protein kinase pathways might be required during the formation of heterologous synapses induced by various other synaptic CAMs. Extensive pharmacological screening demonstrated that the dual leucine zipper kinase (DLK)/c-Jun amino-terminal kinase (JNK) and protein kinase A (PKA) signaling pathways are involved in the formation of presynaptic specializations onto NLGN1- and LRRTM2-expressing HEK293T cells (Jiang et al. 2021). Both pathways were also shown to be responsible for the induction of postsynaptic specializations onto NRXN1-expressing HEK293T cells. Interestingly, the PI3K/Akt pathway was required for the induction of postsynaptic, but not presynaptic specializations. To circumvent issues regarding drug specificity and a lack of subcellular control when bath-applying drugs, PTEN was targeted to postsynaptic sites by fusing it to the PSD protein Homer1. PTEN is a phosphatase that hydrolyses PIP3 to PIP2 and thus prevents Akt activation. Overexpression of PTEN-Homer1 in cultured hippocampal neurons prevented the formation of postsynaptic specializations onto NRXN1-expressing HEK239T cells. In addition, this manipulation reduced synapse density and resulted in a strong suppression of spontaneous excitatory synaptic transmission (Jiang et al. 2021).

It would be interesting to investigate to what extent the formation of postsynaptic specializations induced by other presynaptic CAMs, such as members of the LAR-RPTP family, requires the same signaling pathways. PI3K is likely to be an important signaling node given that removal of NGL-3, a postsynaptic PTPσ ligand (Kwon et al. 2010), leads to aberrant Akt/glycogen synthase kinase 3β (GSK3β) signaling (Lee et al. 2019). Additionally, earlier work demonstrated that interleukin (IL)-1 receptor accessory protein-like 1 (IL1RAPL1), a postsynaptic PTPδ ligand, can activate JNK (Pavlowsky et al. 2010). How these signaling pathways are coupled to synaptic CAMs warrants further investigation but might include the involvement of scaffolds that bridge components of the signal transduction cascade with synaptic CAMs (Xu et al. 2019).

OUTLOOK

Considerable progress has been made in identifying synaptic CAMs, their *trans*-synaptic interactors, and their role in the development of synaptic connectivity, but the mechanisms translating extracellular adhesion into intracellular signaling cascades are largely unknown. While most synaptic CAMs bind to scaffold proteins, the physiological relevance of these interactions remains poorly understood. Presynaptically, evidence largely derived from in vitro studies indicates that synaptic CAMs associate with scaffold proteins and components of the core active zone protein complex to instruct the assembly of presynaptic terminals. On the other hand, conflicting results have been obtained on the role of postsynaptically localized CAMs and their interactions with scaffold proteins, depending on the experimental strategy used.

Beyond scaffold proteins, synaptic CAMs can interact with various other cytoplasmic interactors to couple *trans*-synaptic adhesion to downstream signaling pathways. Synaptic CAMs associate with several proteins that mediate actin remodeling, a key process involved in the development of both pre- and postsynaptic specializations. aGPCRs may act as constitutive GPCRs to instruct synapse development via their coupling to $G\alpha_s$ or $G\alpha_{12/13}$. A role for canonical protein kinase signaling has also been suggested.

While these studies provide an initial characterization of the possible signaling pathways involved, an important aspect to be addressed in future studies is to determine what cellular processes these pathways impinge on to instruct synapse development, in the context of intact neural circuits. For example, PI3K/Akt-related signaling may impact on the regulation of local protein synthesis at synapses via the mechanistic target of rapamycin (mTOR). Indeed, the synaptic CAM ErbB4 mediates excitatory synapse development in the mouse neocortex by enabling mTOR activity, ultimately resulting in the local synaptic translation of several AMPAR-related proteins and other synaptic CAMs (Bernard et al. 2022).

Several reports indicate that the same synaptic CAM can perform divergent functions at different synapses, including NRXNs (Chen et al. 2017), LAR-RPTPs (Han et al. 2020b; Kim et al. 2020), and LPHNs (Zhang et al. 2020; Sando and Südhof 2021), in some cases at different inputs onto the same neuron (Schroeder et al. 2018). This raises the important question of the extent to which signaling pathways that have been identified in culture systems are activated in various circuits.

Additionally, it remains unclear how CAM-mediated signaling pathways confer synapse specificity. For example, even though both LPHN2 and LPHN3 may instruct synapse development via cAMP-mediated signaling (Sando and Südhof 2021), they cannot functionally substitute for one another in directing perforant path and Schaffer-collateral synapses, respectively (Sando et al. 2019). Many synaptic CAMs cooperate to promote synapse development (Stan et al. 2010; Li et al. 2017; Molumby et al. 2017; Roppongi et al. 2020; Steffen et al. 2021; de Arce et al. 2023). Given that different synapses may express distinct sets of synaptic CAMs (Condomitti et al. 2018; Schroeder et al. 2018; Apóstolo et al. 2020; Marcassa et al. 2023), their concerted action could possibly engage unique signaling pathways to enable the development of specific synaptic connections.

Finally, given their strong involvement in the etiology of many neurodevelopmental disorders, a better understanding of how synaptic CAMs signal at the synapse may provide invaluable insight toward the development of therapeutic strategies.

ACKNOWLEDGMENTS

We thank De Wit laboratory members Dan Dascenco and Marinka Brouwer for critical reading of the manuscript, and Stephanie Chanda for input on the initial stage of the manuscript. Figures were created using BioRender. B.V. is supported by an FWO PhD Fellowship Fundamental Research (11A0421N). J.d.W. is supported by FWO Research Projects G0C4518N, G0A8320N, and G0A8720N; FWO EOS Project G0H2818N; SAO Grant 2019/0013, and a Methusalem Grant from the KU Leuven/Flemish Government.

REFERENCES

Anderson GR, Aoto J, Tabuchi K, Földy C, Covy J, Yee AX, Wu D, Lee S-J, Chen L, Malenka RC, et al. 2015. β-Neurexins control neural circuits by regulating synaptic endocannabinoid signaling. *Cell* **162:** 593–606. doi:10.1016/j.cell.2015.06.056

Aoto J, Földy C, Ilcus SMC, Tabuchi K, Südhof TC. 2015. Distinct circuit-dependent functions of presynaptic neurexin-3 at GABAergic and glutamatergic synapses. *Nat Neurosci* **18:** 997–1007. doi:10.1038/nn.4037

Apóstolo N, Smukowski SN, Vanderlinden J, Condomitti G, Rybakin V, Ten Bos J, Trobiani L, Portegies S, Vennekens KM, Gounko NV, et al. 2020. Synapse type-specific proteomic dissection identifies IgSF8 as a hippocampal CA3 microcircuit organizer. *Nat Commun* **11:** 5171. doi:10.1038/s41467-020-18956-x

Bemben MA, Shipman SL, Nicoll RA, Roche KW. 2015. The cellular and molecular landscape of neuroligins. *Trends Neurosci* **38:** 496–505. doi:10.1016/j.tins.2015.06.004

Bernard C, Exposito-Alonso D, Selten M, Sanalidou S, Hanusz-Godoy A, Aguilera A, Hamid F, Oozeer F, Maeso P, Allison L, et al. 2022. Cortical wiring by synapse type-specific control of local protein synthesis. *Science* **378:** eabm7466. doi:10.1126/science.abm7466

Bhouri M, Morishita W, Temkin P, Goswami D, Kawabe H, Brose N, Südhof TC, Craig AM, Siddiqui TJ, Malenka R. 2018. Deletion of LRRTM1 and LRRTM2 in adult mice impairs basal AMPA receptor transmission and LTP in hippocampal CA1 pyramidal neurons. *Proc Natl Acad Sci* **115:** E5382–E5389. doi:10.1073/pnas.1803280115

Biederer T, Südhof TC. 2000. Mints as adaptors. Direct binding to neurexins and recruitment of munc18. *J Biol Chem* **275:** 39803–39806. doi:10.1074/jbc.C000656200

Biederer T, Südhof TC. 2001. CASK and protein 4.1 support F-actin nucleation on neurexins. *J Biol Chem* **276:** 47869–47876. doi:10.1074/jbc.M105287200

Biederer T, Scheiffele P. 2007. Mixed-culture assays for analyzing neuronal synapse formation. *Nat Protoc* **2:** 670–676. doi:10.1038/nprot.2007.92

Biederer T, Kaeser PS, Blanpied TA. 2017. Transcellular nanoalignment of synaptic function. *Neuron* **96:** 680–696. doi:10.1016/j.neuron.2017.10.006

Bomkamp C, Padmanabhan N, Karimi B, Ge Y, Chao JT, Loewen CJR, Siddiqui TJ, Craig AM. 2019. Mechanisms of PTPσ-mediated presynaptic differentiation. *Front Synaptic Neurosci* **11:** 17. doi:10.3389/fnsyn.2019.00017

Boucard AA, Maxeiner S, Südhof TC. 2014. Latrophilins function as heterophilic cell-adhesion molecules by binding to teneurins: regulation by alternative splicing. *J Biol Chem* **289:** 387–402. doi:10.1074/jbc.M113.504779

Boxer EE, Seng C, Lukacsovich D, Kim J, Schwartz S, Kennedy MJ, Földy C, Aoto J. 2021. Neurexin-3 defines synapse- and sex-dependent diversity of GABAergic inhibition in ventral subiculum. *Cell Rep* **37:** 110098. doi:10.1016/j.celrep.2021.110098

Budreck EC, Scheiffele P. 2007. Neuroligin-3 is a neuronal adhesion protein at GABAergic and glutamatergic synapses. *Eur J Neurosci* **26:** 1738–1748. doi:10.1111/j.1460-9568.2007.05842.x

Cheadle L, Biederer T. 2012. The novel synaptogenic protein Farp1 links postsynaptic cytoskeletal dynamics and transsynaptic organization. *J Cell Biol* **199:** 985–1001. doi:10.1083/jcb.201205041

Chen LY, Jiang M, Zhang B, Gokce O, Südhof TC. 2017. Conditional deletion of all neurexins defines diversity of essential synaptic organizer functions for neurexins. *Neuron* **94:** 611–625.e4. doi:10.1016/j.neuron.2017.04.011

Chen H, Tang A-H, Blanpied TA. 2018. Subsynaptic spatial organization as a regulator of synaptic strength and plasticity. *Curr Opin Neurobiol* **51:** 147–153. doi:10.1016/j.conb.2018.05.004

Coles CH, Jones EY, Aricescu AR. 2015. Extracellular regulation of type IIa receptor protein tyrosine phosphatases: mechanistic insights from structural analyses. *Semin Cell Dev Biol* **37:** 98–107. doi:10.1016/j.semcdb.2014.09.007

Condomitti G, Wierda KD, Schroeder A, Rubio SE, Vennekens KM, Orlandi C, Martemyanov KA, Gounko NV, Savas JN, de Wit J. 2018. An input-specific orphan receptor GPR158-HSPG interaction organizes hippocampal mossy fiber-CA3 synapses. *Neuron* **100:** 201–215.e9. doi:10.1016/j.neuron.2018.08.038

Dabrowski A, Umemori H. 2011. Orchestrating the synaptic network by tyrosine phosphorylation signalling. *J Biochem* **149:** 641–653. doi:10.1093/jb/mvr047

Dean C, Scholl FG, Choih J, DeMaria S, Berger J, Isacoff E, Scheiffele P. 2003. Neurexin mediates the assembly of presynaptic terminals. *Nat Neurosci* **6:** 708–716. doi:10.1038/nn1074

de Arce KP, Ribic A, Chowdhury D, Watters K, Thompson GJ, Sanganahalli BG, Lippard ETC, Rohlmann A, Strittmatter SM, Missler M, et al. 2023. Concerted roles of LRRTM1 and SynCAM 1 in organizing prefrontal cortex synapses and cognitive functions. *Nat Commun* **14:** 459. doi:10.1038/s41467-023-36042-w

de Wit J, Ghosh A. 2016. Specification of synaptic connectivity by cell surface interactions. *Nat Rev Neurosci* **17:** 22–35. doi:10.1038/nrn.2015.3

de Wit J, Sylwestrak E, O'Sullivan ML, Otto S, Tiglio K, Savas JN, Yates JR 3rd, Comoletti D, Taylor P, Ghosh A. 2009. LRRTM2 interacts with Neurexin1 and regulates excitatory synapse formation. *Neuron* **64:** 799–806. doi:10.1016/j.neuron.2009.12.019

de Wit J, O'Sullivan ML, Savas JN, Condomitti G, Caccese MC, Vennekens KM, Yates JR 3rd, Ghosh A. 2013. Unbiased discovery of glypican as a receptor for LRRTM4 in regulating excitatory synapse development. *Neuron* **79:** 696–711. doi:10.1016/j.neuron.2013.06.049

Dhume SH, Connor SA, Mills F, Tari PK, Au-Yeung SHM, Karimi B, Oku S, Roppongi RT, Kawabe H, Bamji SX, et al. 2022. Distinct but overlapping roles of LRRTM1 and LRRTM2 in developing and mature hippocampal circuits. *eLife* **11:** e64742 doi:10.7554/eLife.64742

Duman JG, Tzeng CP, Tu Y-K, Munjal T, Schwechter B, Ho TS-Y, Tolias KF. 2013. The adhesion-GPCR BAI1 regulates synaptogenesis by controlling the recruitment of the Par3/Tiam1 polarity complex to synaptic sites. *J Neurosci* **33:** 6964–6978. doi:10.1523/JNEUROSCI.3978-12.2013

Emes RD, Grant SGN. 2012. Evolution of synapse complexity and diversity. *Annu Rev Neurosci* **35:** 111–131. doi:10.1146/annurev-neuro-062111-150433

Frei JA, Stoeckli ET. 2017. SynCAMs—from axon guidance to neurodevelopmental disorders. *Mol Cell Neurosci* **81:** 41–48. doi:10.1016/j.mcn.2016.08.012

Gentile JE, Carrizales MG, Koleske AJ. 2022. Control of synapse structure and function by actin and its regulators. *Cells* **11:** 603. doi:10.3390/cells11040603

Giannone G, Mondin M, Grillo-Bosch D, Tessier B, Saint-Michel E, Czöndör K, Sainlos M, Choquet D, Thoumine O. 2013. Neurexin-1β binding to neuroligin-1 triggers the preferential recruitment of PSD-95 versus gephyrin through tyrosine phosphorylation of neuroligin-1. *Cell Rep* **3:** 1996–2007. doi:10.1016/j.celrep.2013.05.013

Gokce O, Südhof TC. 2013. Membrane-Tethered monomeric neurexin LNS-domain triggers synapse formation. *J Neurosci* **33:** 14617–14628. doi:10.1523/JNEUROSCI.1232-13.2013

Gomez AM, Traunmüller L, Scheiffele P. 2021. Neurexins: molecular codes for shaping neuronal synapses. *Nat Rev Neurosci* **22:** 137–151. doi:10.1038/s41583-020-00415-7

Graf ER, Zhang X, Jin S-X, Linhoff MW, Craig AM. 2004. Neurexins induce differentiation of GABA and glutamate postsynaptic specializations via neuroligins. *Cell* **119:** 1013–1026. doi:10.1016/j.cell.2004.11.035

Grant SGN. 2019. Synapse diversity and synaptome architecture in human genetic disorders. *Hum Mol Genet* **28:** R219–R225. doi:10.1093/hmg/ddz178

Hale WD, Südhof TC, Huganir RL. 2023. Engineered adhesion molecules drive synapse organization. *Proc Natl Acad Sci* **120:** e2215905120. doi:10.1073/pnas.2215905120

Han K, Kim E. 2008. Synaptic adhesion molecules and PSD-95. *Prog Neurobiol* **84:** 263–283. doi:10.1016/j.pneurobio.2007.10.011

Cite this article as *Cold Spring Harb Perspect Biol* doi: 10.1101/cshperspect.a041501

Han KA, Jeon S, Um JW, Ko J. 2016a. Emergent synapse organizers: LAR-RPTPs and their companions. *Int Rev Cell Mol Biol* **324:** 39–65. doi:10.1016/bs.ircmb.2016.01.002

Han KA, Woo D, Kim S, Choii G, Jeon S, Won SY, Kim HM, Heo WD, Um JW, Ko J. 2016b. Neurotrophin-3 regulates synapse development by modulating TrkC-PTPσ synaptic adhesion and intracellular signaling pathways. *J Neurosci* **36:** 4816–4831. doi:10.1523/JNEUROSCI.4024-15.2016

Han KA, Ko JS, Pramanik G, Kim JY, Tabuchi K, Um JW, Ko J. 2018. PTPσ drives excitatory presynaptic assembly via various extracellular and intracellular mechanisms. *J Neurosci* **38:** 6700–6721. doi:10.1523/JNEUROSCI.0672-18.2018

Han KA, Kim J, Kim H, Kim D, Lim D, Ko J, Um JW. 2019a. Slitrk2 controls excitatory synapse development via PDZ-mediated protein interactions. *Sci Rep* **9:** 17094. doi:10.1038/s41598-019-53519-1

Han KA, Um JW, Ko J. 2019b. Intracellular protein complexes involved in synapse assembly in presynaptic neurons. *Adv Protein Chem Struct Biol* **116:** 347–373. doi:10.1016/bs.apcsb.2018.11.008

Han KA, Kim Y-J, Yoon TH, Kim H, Bae S, Um JW, Choi S-Y, Ko J. 2020a. LAR-RPTPs directly interact with neurexins to coordinate bidirectional assembly of molecular machineries. *J Neurosci* **40:** 8438–8462. doi:10.1523/JNEUROSCI.1091-20.2020

Han KA, Lee H-Y, Lim D, Shin J, Yoon TH, Lee C, Rhee J-S, Liu X, Um JW, Choi S-Y, et al. 2020b. PTPσ controls presynaptic organization of neurotransmitter release machinery at excitatory synapses. *iScience* **23:** 101203. doi:10.1016/j.isci.2020.101203

Han KA, Lee H-Y, Lim D, Shin J, Yoon TH, Liu X, Um JW, Choi S-Y, Ko J. 2020c. Receptor protein tyrosine phosphatase delta is not essential for synapse maintenance or transmission at hippocampal synapses. *Mol Brain* **13:** 94. doi:10.1186/s13041-020-00629-x

Hata Y, Davletov B, Petrenko AG, Jahn R, Südhof TC. 1993. Interaction of synaptotagmin with the cytoplasmic domains of neurexins. *Neuron* **10:** 307–315. doi:10.1016/0896-6273(93)90320-Q

Hata Y, Butz S, Südhof TC. 1996. CASK: a novel dlg/PSD95 homolog with an N-terminal calmodulin-dependent protein kinase domain identified by interaction with neurexins. *J Neurosci* **16:** 2488–2494. doi:10.1523/JNEUROSCI.16-08-02488.1996

Hauser D, Behr K, Konno K, Schreiner D, Schmidt A, Watanabe M, Bischofberger J, Scheiffele P. 2022. Targeted proteoform mapping uncovers specific Neurexin-3 variants required for dendritic inhibition. *Neuron* **110:** 2094–2109.e10. doi:10.1016/j.neuron.2022.04.017

Hoon M, Soykan T, Falkenburger B, Hammer M, Patrizi A, Schmidt K-F, Sassoè-Pognetto M, Löwel S, Moser T, Taschenberger H, et al. 2011. Neuroligin-4 is localized to glycinergic postsynapses and regulates inhibition in the retina. *Proc Natl Acad Sci* **108:** 3053–3058. doi:10.1073/pnas.1006946108

Irie M, Hata Y, Takeuchi M, Ichtchenko K, Toyoda A, Hirao K, Takai Y, Rosahl TW, Südhof TC. 1997. Binding of neuroligins to PSD-95. *Science* **277:** 1511–1515. doi:10.1126/science.277.5331.1511

Jang S, Lee H, Kim E. 2017. Synaptic adhesion molecules and excitatory synaptic transmission. *Curr Opin Neurobiol* **45:** 45–50. doi:10.1016/j.conb.2017.03.005

Jeong J, Paskus JD, Roche KW. 2017. Posttranslational modifications of neuroligins regulate neuronal and glial signaling. *Curr Opin Neurobiol* **45:** 130–138. doi:10.1016/j.conb.2017.05.017

Jiang X, Sando R, Südhof TC. 2021. Multiple signaling pathways are essential for synapse formation induced by synaptic adhesion molecules. *Proc Natl Acad Sci* **118:** e2000173118. doi:10.1073/pnas.2000173118

Karimi B, Silwal P, Booth S, Padmanabhan N, Dhume SH, Zhang D, Zahra N, Jackson MF, Kirouac GJ, Ko JH, et al. 2021. Schizophrenia-associated LRRTM1 regulates cognitive behavior through controlling synaptic function in the mediodorsal thalamus. *Mol Psychiatry* **26:** 6912–6925. doi:10.1038/s41380-021-01146-6

Kim S, Burette A, Chung HS, Kwon S-K, Woo J, Lee HW, Kim K, Kim H, Weinberg RJ, Kim E. 2006. NGL family PSD-95-interacting adhesion molecules regulate excitatory synapse formation. *Nat Neurosci* **9:** 1294–1301. doi:10.1038/nn1763

Kim K, Shin W, Kang M, Lee S, Kim D, Kang R, Jung Y, Cho Y, Yang E, Kim H, et al. 2020. Presynaptic PTPσ regulates postsynaptic NMDA receptor function through direct adhesion-independent mechanisms. *eLife* **9:** e54224. doi:10.7554/eLife.54224

Kim HY, Um JW, Ko J. 2021. Proper synaptic adhesion signaling in the control of neural circuit architecture and brain function. *Prog Neurobiol* **200:** 101983. doi:10.1016/j.pneurobio.2020.101983

Kim J, Wulschner LEG, Oh WC, Ko J. 2022. *Trans*-synaptic mechanisms orchestrated by mammalian synaptic cell adhesion molecules. *Bioessays* **44:** e2200134. doi:10.1002/bies.202200134

Kishore A, Purcell RH, Nassiri-Toosi Z, Hall RA. 2016. Stalk-dependent and stalk-independent signaling by the adhesion G protein-coupled receptors GPR56 (ADGRG1) and BAI1 (ADGRB1). *J Biol Chem* **291:** 3385–3394. doi:10.1074/jbc.M115.689349

Ko J, Fuccillo MV, Malenka RC, Südhof TC. 2009. LRRTM2 functions as a neurexin ligand in promoting excitatory synapse formation. *Neuron* **64:** 791–798. doi:10.1016/j.neuron.2009.12.012

Ko JS, Pramanik G, Um JW, Shim JS, Lee D, Kim KH, Chung G-Y, Condomitti G, Kim HM, Kim H, et al. 2015. PTPσ functions as a presynaptic receptor for the glypican-4/LRRTM4 complex and is essential for excitatory synaptic transmission. *Proc Natl Acad Sci* **112:** 1874–1879. doi:10.1073/pnas.1410138112

Kwon S-K, Woo J, Kim S-Y, Kim H, Kim E. 2010. *Trans*-synaptic adhesions between netrin-G ligand-3 (NGL-3) and receptor tyrosine phosphatases LAR, protein-tyrosine phosphatase δ (PTPδ), and PTPσ via specific domains regulate excitatory synapse formation. *J Biol Chem* **285:** 13966–13978. doi:10.1074/jbc.M109.061127

Lala T, Hall RA. 2022. Adhesion G protein-coupled receptors: structure, signaling, physiology, and pathophysiology. *Physiol Rev* **102:** 1587–1624. doi:10.1152/physrev.00027.2021

Lee H, Yi J-S, Lawan A, Min K, Bennett AM. 2015. Mining the function of protein tyrosine phosphatases in health and disease. *Semin Cell Dev Biol* **37**: 66–72. doi:10.1016/j.semcdb.2014.09.021

Lee H, Shin W, Kim K, Lee S, Lee E-J, Kim J, Kweon H, Lee E, Park H, Kang M, et al. 2019. NGL-3 in the regulation of brain development, Akt/GSK3b signaling, long-term depression, and locomotive and cognitive behaviors. *PLoS Biol* **17**: e2005326. doi:10.1371/journal.pbio.2005326

Letellier M, Szíber Z, Chamma I, Saphy C, Papasideri I, Tessier B, Sainlos M, Czöndör K, Thoumine O. 2018. A unique intracellular tyrosine in neuroligin-1 regulates AMPA receptor recruitment during synapse differentiation and potentiation. *Nat Commun* **9**: 3979. doi:10.1038/s41467-018-06220-2

Letellier M, Lagardère M, Tessier B, Janovjak H, Thoumine O. 2020. Optogenetic control of excitatory post-synaptic differentiation through neuroligin-1 tyrosine phosphorylation. *eLife* **9**: e52027. doi:10.7554/eLife.52027

Li J, Han W, Pelkey KA, Duan J, Mao X, Wang Y-X, Craig MT, Dong L, Petralia RS, McBain CJ, et al. 2017. Molecular dissection of Neuroligin 2 and Slitrk3 reveals an essential framework for GABAergic synapse development. *Neuron* **96**: 808–826.e8. doi:10.1016/j.neuron.2017.10.003

Li J, Shalev-Benami M, Sando R, Jiang X, Kibrom A, Wang J, Leon K, Katanski C, Nazarko O, Lu YC, et al. 2018. Structural basis for teneurin function in circuit-wiring: a toxin motif at the synapse. *Cell* **173**: 735–748.e15. doi:10.1016/j.cell.2018.03.036

Lie E, Li Y, Kim R, Kim E. 2018. SALM/lrfn family synaptic adhesion molecules. *Front Mol Neurosci* **11**: 105. doi:10.3389/fnmol.2018.00105

Lim S-H, Kwon S-K, Lee MK, Moon J, Jeong DG, Park E, Kim SJ, Park BC, Lee SC, Ryu SE, et al. 2009. Synapse formation regulated by protein tyrosine phosphatase receptor T through interaction with cell adhesion molecules and Fyn. *EMBO J* **28**: 3564–3578. doi:10.1038/emboj.2009.289

Linhoff MW, Laurén J, Cassidy RM, Dobie FA, Takahashi H, Nygaard HB, Airaksinen MS, Strittmatter SM, Craig AM. 2009. An unbiased expression screen for synaptogenic proteins identifies the LRRTM protein family as synaptic organizers. *Neuron* **61**: 734–749. doi:10.1016/j.neuron.2009.01.017

Liu A, Zhou Z, Dang R, Zhu Y, Qi J, He G, Leung C, Pak D, Jia Z, Xie W. 2016. Neuroligin 1 regulates spines and synaptic plasticity via LIMK1/cofilin-mediated actin reorganization. *J Cell Biol* **212**: 449–463. doi:10.1083/jcb.201509023

Luo F, Sclip A, Jiang M, Südhof TC. 2020. Neurexins cluster Ca^{2+} channels within the presynaptic active zone. *EMBO J* **39**: e103208. doi:10.15252/embj.2019103208

Marcassa G, Dascenco D, de Wit J. 2023. Proteomics-based synapse characterization: from proteins to circuits. *Curr Opin Neurobiol* **79**: 102690. doi:10.1016/j.conb.2023.102690

Mathiasen S, Palmisano T, Perry NA, Stoveken HM, Vizurraga A, McEwen DP, Okashah N, Langenhan T, Inoue A, Lambert NA, et al. 2020. G12/13 is activated by acute tethered agonist exposure in the adhesion GPCR ADGRL3. *Nat Chem Biol* **16**: 1343–1350. doi:10.1038/s41589-020-0617-7

Missler M, Zhang W, Rohlmann A, Kattenstroth G, Hammer RE, Gottmann K, Südhof TC. 2003. α-Neurexins couple Ca^{2+} channels to synaptic vesicle exocytosis. *Nature* **423**: 939–948. doi:10.1038/nature01755

Molumby MJ, Anderson RM, Newbold DJ, Koblesky NK, Garrett AM, Schreiner D, Radley JJ, Weiner JA. 2017. γ-Protocadherins interact with neuroligin-1 and negatively regulate dendritic spine morphogenesis. *Cell Rep* **18**: 2702–2714. doi:10.1016/j.celrep.2017.02.060

Moreno-Salinas AL, Holleran BJ, Ojeda-Muñiz EY, Correoso-Braña KG, Ribalta-Mena S, Ovando-Zambrano J-C, Leduc R, Boucard AA. 2022. Convergent selective signaling impairment exposes the pathogenicity of latrophilin-3 missense variants linked to inheritable ADHD susceptibility. *Mol Psychiatry* **27**: 2425–2438. doi:10.1038/s41380-022-01537-3

Nazarko O, Kibrom A, Winkler J, Leon K, Stoveken H, Salzman G, Merdas K, Lu Y, Narkhede P, Tall G, et al. 2018. A comprehensive mutagenesis screen of the adhesion GPCR latrophilin-1/ADGRL1. *iScience* **3**: 264–278. doi:10.1016/j.isci.2018.04.019

Noborn F, Sterky FH. 2023. Role of neurexin heparan sulfate in the molecular assembly of synapses—expanding the neurexin code? *FEBS J* **290**: 252–265. doi:10.1111/febs.16251

Nozawa K, Sogabe T, Hayashi A, Motohashi J, Miura E, Arai I, Yuzaki M. 2022. In vivo nanoscopic landscape of neurexin ligands underlying anterograde synapse specification. *Neuron* **110**: 3168–3185.e8. doi:10.1016/j.neuron.2022.07.027

Park H, Choi Y, Jung H, Kim S, Lee S, Han H, Kweon H, Kang S, Sim WS, Koopmans F, et al. 2020. Splice-dependent *trans*-synaptic PTPδ-IL1RAPL1 interaction regulates synapse formation and non-REM sleep. *EMBO J* **39**: e104150. doi:10.15252/embj.2019104150

Paskus JD, Tian C, Fingleton E, Shen C, Chen X, Li Y, Myers SA, Badger JD, Bemben MA, Herring BE, et al. 2019. Synaptic Kalirin-7 and Trio interactomes reveal a GEF protein-dependent Neuroligin-1 mechanism of action. *Cell Rep* **29**: 2944–2952.e5. doi:10.1016/j.celrep.2019.10.115

Paskus JD, Herring BE, Roche KW. 2020. Kalirin and Trio: RhoGEFs in synaptic transmission, plasticity, and complex brain disorders. *Trends Neurosci* **43**: 505–518. doi:10.1016/j.tins.2020.05.002

Pavlowsky A, Gianfelice A, Pallotto M, Zanchi A, Vara H, Khelfaoui M, Valnegri P, Rezai X, Bassani S, Brambilla D, et al. 2010. A postsynaptic signaling pathway that may account for the cognitive defect due to IL1RAPL1 mutation. *Curr Biol* **20**: 103–115. doi:10.1016/j.cub.2009.12.030

Pederick DT, Perry-Hauser NA, Meng H, He Z, Javitch JA, Luo L. 2023. Context-dependent requirement of G protein coupling for Latrophilin-2 in target selection of hippocampal axons. *eLife* **12**: e83529. doi:10.7554/eLife.83529

Poulopoulos A, Aramuni G, Meyer G, Soykan T, Hoon M, Papadopoulos T, Zhang M, Paarmann I, Fuchs C, Harvey K, et al. 2009. Neuroligin 2 drives postsynaptic assembly at perisomatic inhibitory synapses through gephyrin and collybistin. *Neuron* **63**: 628–642. doi:10.1016/j.neuron.2009.08.023

Poulopoulos A, Soykan T, Tuffy LP, Hammer M, Varoqueaux F, Brose N. 2012. Homodimerization and isoform-specific heterodimerization of neuroligins. *Biochem J* **446**: 321–330. doi:10.1042/BJ20120808

Reissner C, Runkel F, Missler M. 2013. Neurexins. *Genome Biol* **14**: 213. doi:10.1186/gb-2013-14-9-213

Roppongi RT, Dhume SH, Padmanabhan N, Silwal P, Zahra N, Karimi B, Bomkamp C, Patil CS, Champagne-Jorgensen K, Twilley RE, et al. 2020. LRRTMs organize synapses through differential engagement of neurexin and PTPσ. *Neuron* **106:** 108–125.e12. doi:10.1016/j.neuron.2020.01.003

Sando R, Südhof TC. 2021. Latrophilin GPCR signaling mediates synapse formation. *eLife* **10:** e65717. doi:10.7554/eLife.65717

Sando R, Jiang X, Südhof TC. 2019. Latrophilin GPCRs direct synapse specificity by coincident binding of FLRTs and teneurins. *Science* **363:** eaav7969. doi:10.1126/science.aav7969

Sando R, Ho ML, Liu X, Südhof TC. 2022. Engineered synaptic tools reveal localized cAMP signaling in synapse assembly. *J Cell Biol* **221.** doi:10.1083/jcb.202109111

Sanes JR, Zipursky SL. 2020. Synaptic specificity, recognition molecules, and assembly of neural circuits. *Cell* **181:** 536–556. doi:10.1016/j.cell.2020.04.008

Scheiffele P, Fan J, Choih J, Fetter R, Serafini T. 2000. Neuroligin expressed in nonneuronal cells triggers presynaptic development in contacting axons. *Cell* **101:** 657–669. doi:10.1016/S0092-8674(00)80877-6

Schlessinger J. 2014. Receptor tyrosine kinases: legacy of the first two decades. *Cold Spring Harb Perspect Biol* **6:** a008912. doi: 10.1101/cshperspect.a008912

Schreiner D, Nguyen T-M, Russo G, Heber S, Patrignani A, Ahrné E, Scheiffele P. 2014. Targeted combinatorial alternative splicing generates brain region-specific repertoires of neurexins. *Neuron* **84:** 386–398. doi:10.1016/j.neuron.2014.09.011

Schroeder A, Vanderlinden J, Vints K, Ribeiro LF, Vennekens KM, Gounko NV, Wierda KD, de Wit J. 2018. A modular organization of LRR protein-mediated synaptic adhesion defines synapse identity. *Neuron* **99:** 329–344.e7. doi:10.1016/j.neuron.2018.06.026

Sclip A, Südhof TC. 2020. LAR receptor phospho-tyrosine phosphatases regulate NMDA-receptor responses. *eLife* **9:** e53406. doi:10.7554/eLife.53406

Shipman SL, Schnell E, Hirai T, Chen B-S, Roche KW, Nicoll RA. 2011. Functional dependence of neuroligin on a new non-PDZ intracellular domain. *Nat Neurosci* **14:** 718–726. doi:10.1038/nn.2825

Siddiqui TJ, Tari PK, Connor SA, Zhang P, Dobie FA, She K, Kawabe H, Wang YT, Brose N, Craig AM. 2013. An LRRTM4-HSPG complex mediates excitatory synapse development on dentate gyrus granule cells. *Neuron* **79:** 680–695. doi:10.1016/j.neuron.2013.06.029

Song J-Y, Ichtchenko K, Südhof TC, Brose N. 1999. Neuroligin 1 is a postsynaptic cell-adhesion molecule of excitatory synapses. *Proc Natl Acad Sci* **96:** 1100–1105. doi:10.1073/pnas.96.3.1100

Spence EF, Soderling SH. 2015. Actin out: regulation of the synaptic cytoskeleton. *J Biol Chem* **290:** 28613–28622. doi:10.1074/jbc.R115.655118

Stan A, Pielarski KN, Brigadski T, Wittenmayer N, Fedorchenko O, Gohla A, Lessmann V, Dresbach T, Gottmann K. 2010. Essential cooperation of N-cadherin and neuroligin-1 in the transsynaptic control of vesicle accumulation. *Proc Natl Acad Sci* **107:** 11116–11121. doi:10.1073/pnas.0914233107

Steffen DM, Ferri SL, Marcucci CG, Blocklinger KL, Molumby MJ, Abel T, Weiner JA. 2021. The γ-protocadherins interact physically and functionally with Neuroligin-2 to negatively regulate inhibitory synapse density and are required for normal social interaction. *Mol Neurobiol* **58:** 2574–2589. doi:10.1007/s12035-020-02263-z

Stephenson JR, Paavola KJ, Schaefer SA, Kaur B, Van Meir EG, Hall RA. 2013. Brain-specific angiogenesis inhibitor-1 signaling, regulation, and enrichment in the postsynaptic density. *J Biol Chem* **288:** 22248–22256. doi:10.1074/jbc.M113.489757

Sterky FH, Trotter JH, Lee S-J, Recktenwald CV, Du X, Zhou B, Zhou P, Schwenk J, Fakler B, Südhof TC. 2017. Carbonic anhydrase-related protein CA10 is an evolutionarily conserved pan-neurexin ligand. *Proc Natl Acad Sci* **114:** E1253–E1262. doi:10.1073/pnas.1621321114

Südhof TC. 2012. The presynaptic active zone. *Neuron* **75:** 11–25. doi:10.1016/j.neuron.2012.06.012

Südhof TC. 2017. Synaptic neurexin complexes: a molecular code for the logic of neural circuits. *Cell* **171:** 745–769. doi:10.1016/j.cell.2017.10.024

Südhof TC. 2018. Towards an understanding of synapse formation. *Neuron* **100:** 276–293. doi:10.1016/j.neuron.2018.09.040

Takahashi H, Craig AM. 2013. Protein tyrosine phosphatases PTPδ, PTPσ, and LAR: presynaptic hubs for synapse organization. *Trends Neurosci* **36:** 522–534. doi:10.1016/j.tins.2013.06.002

Takahashi H, Arstikaitis P, Prasad T, Bartlett TE, Wang YT, Murphy TH, Craig AM. 2011. Postsynaptic TrkC and presynaptic PTPσ function as a bidirectional excitatory synaptic organizing complex. *Neuron* **69:** 287–303. doi:10.1016/j.neuron.2010.12.024

Tian C, Paskus JD, Fingleton E, Roche KW, Herring BE. 2021. Autism spectrum disorder/intellectual disability-associated mutations in Trio disrupt Neuroligin 1-mediated synaptogenesis. *J Neurosci* **41:** 7768–7778. doi:10.1523/JNEUROSCI.3148-20.2021

Tolias KF, Duman JG, Um K. 2011. Control of synapse development and plasticity by Rho GTPase regulatory proteins. *Prog Neurobiol* **94:** 133–148. doi:10.1016/j.pneurobio.2011.04.011

Treutlein B, Gokce O, Quake SR, Südhof TC. 2014. Cartography of neurexin alternative splicing mapped by single-molecule long-read mRNA sequencing. *Proc Natl Acad Sci* **111:** E1291–E1299. doi:10.1073/pnas.1403244111

Tu Y-K, Duman JG, Tolias KF. 2018. The adhesion-GPCR BAI1 promotes excitatory synaptogenesis by coordinating bidirectional *trans*-synaptic signaling. *J Neurosci* **38:** 8388–8406. doi:10.1523/JNEUROSCI.3461-17.2018

Um JW, Ko J. 2013. LAR-RPTPs: synaptic adhesion molecules that shape synapse development. *Trends Cell Biol* **23:** 465–475. doi:10.1016/j.tcb.2013.07.004

Um JW, Choi T-Y, Kang H, Cho YS, Choii G, Uvarov P, Park D, Jeong D, Jeon S, Lee D, et al. 2016. LRRTM3 regulates excitatory synapse development through alternative splicing and neurexin binding. *Cell Rep* **14:** 808–822. doi:10.1016/j.celrep.2015.12.081

Washburn HR, Chander P, Srikanth KD, Dalva MB. 2023. Transsynaptic signaling of Ephs in synaptic development, plasticity, and disease. *Neuroscience* **508:** 137–152. doi:10.1016/j.neuroscience.2022.11.030

Wu X, Morishita WK, Riley AM, Hale WD, Südhof TC, Malenka RC. 2019. Neuroligin-1 signaling controls LTP and NMDA receptors by distinct molecular pathways. *Neuron* **102**: 621–635.e3. doi:10.1016/j.neuron.2019.02.013

Xu J, Du Y-L, Xu J-W, Hu X-G, Gu L-F, Li X-M, Hu P-H, Liao T-L, Xia Q-Q, Sun Q, et al. 2019. Neuroligin 3 regulates dendritic outgrowth by modulating Akt/mTOR signaling. *Front Cell Neurosci* **13**: 518. doi:10.3389/fncel.2019.00518

Yamagata A, Goto-Ito S, Sato Y, Shiroshima T, Maeda A, Watanabe M, Saitoh T, Maenaka K, Terada T, Yoshida T, et al. 2018. Structural insights into modulation and selectivity of transsynaptic neurexin-LRRTM interaction. *Nat Commun* **9**: 3964. doi:10.1038/s41467-018-06333-8

Zhang P, Lu H, Peixoto RT, Pines MK, Ge Y, Oku S, Siddiqui TJ, Xie Y, Wu W, Archer-Hartmann S, et al. 2018. Heparan sulfate organizes neuronal synapses through neurexin partnerships. *Cell* **174**: 1450–1464.e23. doi:10.1016/j.cell.2018.07.002

Zhang RS, Liakath-Ali K, Südhof TC. 2020. Latrophilin-2 and latrophilin-3 are redundantly essential for parallel-fiber synapse function in cerebellum. *eLife* **9**: e54443. doi:10.7554/eLife.54443

Cite this article as *Cold Spring Harb Perspect Biol* doi: 10.1101/cshperspect.a041501

Development of Prefrontal Circuits and Cognitive Abilities

Jastyn A. Pöpplau and Ileana L. Hanganu-Opatz

Institute of Developmental Neurophysiology, Center for Molecular Neurobiology, Hamburg Center of Neuroscience (HCNS), University Medical Center Hamburg-Eppendorf, Hamburg 20246, Germany

Correspondence: jastyn.poepplau@zmnh.uni-hamburg.de; hangop@zmnh.uni-hamburg.de

The prefrontal cortex is considered as the site of multifaceted higher-order cognitive abilities. These abilities emerge late in life long after full sensorimotor maturation, in line with the protracted development of prefrontal circuits that has been identified on molecular, structural, and functional levels. Only recently, as a result of the impressive methodological progress of the last several decades, the mechanisms and clinical implications of prefrontal development have begun to be elucidated, yet major knowledge gaps still persist. Here, we provide an overview on how prefrontal circuits develop to enable multifaceted cognitive processing at adulthood. First, we review recent insights into the mechanisms of prefrontal circuit assembly, with a focus on the contribution of early electrical activity. Second, we highlight the major reorganization of prefrontal circuits during adolescence. Finally, we link the prefrontal plasticity during specific developmental time windows to mental health disorders and discuss potential approaches for therapeutic interventions.

The prefrontal cortex (PFC), the brain region located in the anterior pole of the neocortex, has been the focus of research since the railroad worker Phineas Gage experienced, in 1848, a dramatic personality change after a rod penetrated and seriously injured his PFC (Harlow 1849). Since then, a wealth of studies documented that the PFC is the hub of various aspects of cognitive flexibility, such as attention, working memory, decision making, emotions, and social behavior. While these abilities exist in a basic form in all mammalian species, they reached the highest degree of complexity and sophistication in humans, since to adapt and develop flexible strategies is crucial for evolutionary success. Consequently, it is not surprising that the PFC is the latest developed region of the neocortex, phylogenetically as well as ontogenetically, representing the highest level of the cortical hierarchy (Fuster 2001; Miller and Cohen 2001; Chini and Hanganu-Opatz 2021).

The uniqueness of human PFC is boon and bane at the same time. The PFC is largest in humans, covering one-third of the entire cortical mantle and containing almost twice as much cortical gray matter as in macaques (Donahue et al. 2018). Initiated more than a century ago as a result of a tragic work accident, the examination of the PFC using the outcome of lesions led to the first coarse "mapping" and definition of the PFC and its subregions (Fuster 2001). Traditionally, the PFC is considered to be a granular

Cite this article as *Cold Spring Harb Perspect Biol* doi: 10.1101/cshperspect.a041502

brain region (i.e., possessing a granular layer 4) that receives dense projections from the medio-dorsal nucleus (MD) of the thalamus. However, this can be applied as a definition only for the PFC of primates and even their PFC has been shown to contain agranular or poorly developed granular regions. Moreover, it was revealed that the connectivity-based definition of the PFC lacks precision, since the MD projections do not exclusively target frontal regions and the MD is not the only thalamic nucleus tightly connected with the PFC (Preuss and Wise 2022). The lack of a rigorous definition of the PFC is mainly due to ethical and technical considerations that severely limit direct access to and manipulation of single neurons and circuits in humans. Therefore, fundamental research with animal models, including rodents and nonhuman primates, is of paramount importance (Carlén 2017). Even if the connectivity patterns, parcellation, and layered structure are different in rodents, and to a lesser extent in nonhuman primates, cross-species comparisons based on the "functional homology" of prefrontal regions, which enable similar cognitive abilities (although not necessarily evolutionarily or structural homologous), offer a promising research direction (Preuss 1995; Ährlund-Richter et al. 2019).

Rodents possess fewer frontal areas than primates and all lack a granular layer 4. Thalamic MD projections target several cortical regions that are considered subdivisions of the rodent medial PFC (mPFC): cingulate (Cg), prelimbic (PL), and infralimbic (IL) cortices. Based on a functional and less structural homology, these regions are considered to correspond to primate areas 24, 32, and 25, respectively (Carlén 2017; Laubach et al. 2018; van Heukelum et al. 2020; Preuss and Wise 2022). There is abundant literature documenting the involvement of rodent mPFC in cognitive processing (Carlén 2017; Chini and Hanganu-Opatz 2021; Le Merre et al. 2021). Similar to the primate PFC, the rodent mPFC shows a dorsoventral gradient in cognitive functioning, with more dorsal regions underlying decision making and attention and more ventral regions supporting emotions and motivation (Fig. 1; Anastasiades and Carter 2021).

The unique prefrontal rules of local and long-range wiring as well as microcircuit structures accounting for these complex behaviors have been extensively investigated in adult rats and mice. Briefly, the adult mPFC receives projections from nearly all other neocortical regions as well as from a large number of subcortical areas, including the thalamus, basolateral amygdala, hippocampus (HP), and claustrum. As most of these monosynaptic connections are reciprocal, with the exception of HP–mPFC communication lacking direct prefrontal projections to HP, the mPFC is well positioned to compute diverse information to generate a suited output via top-down control. The mPFC shows a layered hierarchy. Dense projections from the thalamic MD and from other subcortical and cortical regions target the strongly interconnected layer 2/3

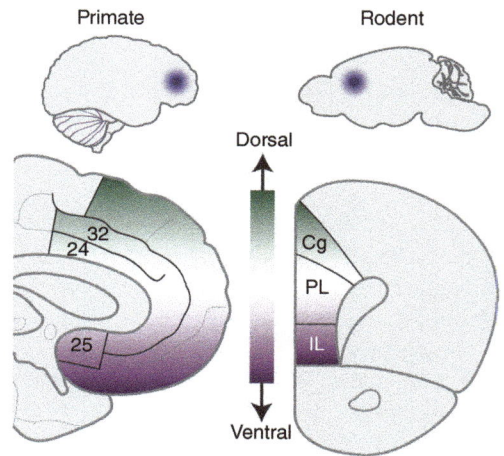

Figure 1. Functional and anatomical divisions of primate and rodent frontal cortex. (*Left*) Section of the primate frontal cortex (*bottom*). Location of the section in the primate brain is indicated by a blue circle (*top*). (*Right*) Section of the rodent frontal cortex (*bottom*). Location of the section in the rodent brain is indicated by a blue circle (*top*). The purple to green colors correspond to the dorsoventral gradient. More dorsal frontal regions are involved in decision making and attention, whereas more ventral regions support emotions and motivation. Brodmann areas 24, 32, and 25 of the primate prefrontal cortex (PFC) correspond to the rodent medial PFC (mPFC) areas, cingulate (Cg), prelimbic (PL), and infralimbic (IL) cortices, respectively.

that forward amplified inputs toward layer 5/6, which represents the main output layer of mPFC (Anastasiades and Carter 2021).

In contrast, the development of prefrontal circuits is much less well understood. Only recently, following unprecedented methodological progress to monitor and interfere with developmental events, key questions regarding the (1) mechanisms that underlie the formation and dynamic refinement of prefrontal circuits along development, (2) the role of early electrical activity for these processes, and (3) the contribution of developing circuits for the maturation of cognitive processing have begun to be addressed. Here, we summarize this recent knowledge and highlight the milestones of PFC development in relation to cognitive abilities. While the vast majority of data has been obtained from rats and mice, we will discuss how they might be instrumental for inferring the (mis)developmental of prefrontal circuits in humans (Fig. 2).

MILESTONES OF STRUCTURAL PREFRONTAL DEVELOPMENT

Despite its privileged function as a hub of cognition, the PFC generally follows the general principles of neocortical development that are conserved across mammalian species. However, the processes occurring from neurulation until complete circuit refinement are considered to expand over a longer time window than those in sensory or motor cortical regions (Schubert et al. 2015; Lim et al. 2018; Chini and Hanganu-Opatz 2021; Kolk and Rakic 2022). This "delayed" maturation of prefrontal circuits has profound implications for behavioral abilities in health and disease.

The sequence of developmental processes in the PFC has been largely investigated. After the neural tube has formed, progenitor cells in the dorsal neuroepithelium start to proliferate, initiating cortical surface expansion. In the PFC, this expansion depends on growth factors of the Fgf family, which enable a differential and independent growth rate of the PFC in relation to other cortical areas (Kolk and Rakic 2022). Initially, stem cells in the ventricular zone proliferate by symmetric cell division (for more details, see Geschwind and Rakic 2013). As soon as asym-

metric cell division of early arising radial glia cells begins, the number of immature excitatory neurons and additional radial glia cells greatly increases. The earliest generation of neurons migrates along a scaffold, which is built by radial glia cells toward the outer pial surface to build the preplate. Subsequent generations of excitatory neurons, which mainly comprise pyramidal neurons, migrate radially to form the cortical plate by splitting the preplate into the outer marginal zone (later layer 1) and subjacent subplate (Marín and Rubenstein 2003; Kirischuk et al. 2017). The subplate is a transient layer that contains diverse populations of pioneer neurons with remarkably mature morphological and transcriptomic profiles already at embryonic stages. These neurons assist the cortical circuit construction by facilitating the growth of prefrontal afferents and efferents. In the PFC, the subplate size is much larger in comparison to other cortical layers and reaches its maximum in the human brain (Hoerder-Suabedissen and Molnár 2015; Chini and Hanganu-Opatz 2021; Kolk and Rakic 2022). The subplate is considered to facilitate the ingrowth and synapse formation for numerous modulatory projections, yet it does not assist the thalamic MD innervation, which, as mentioned above, has been traditionally used to define the PFC and is already present in rodents near birth (Chini and Hanganu-Opatz 2021).

Modulatory serotonergic, noradrenergic, dopaminergic, and cholinergic projections as well as glutamatergic and γ-aminobutyric acid (GABA)ergic axons enter the rodent mPFC during the prenatal stage not only via the subplate but also along the marginal zone, where they are most likely released via volume transmission in close proximity to the transient population of Cajal–Retzius neurons. Neurotransmitter receptors are already expressed by progenitor cells before the first incoming projections arrive. Thus, external neurotransmitter sources (i.e., placenta) might contribute to the regulation of cell division, neuronal migration, differentiation, and synaptogenesis (Kolk and Rakic 2022).

In the next step, migrating pyramidal neurons build the cortical layers in an inside-out manner, with a birthdate-dependent laminar gradient from layer 6 to layer 2. The diversity of

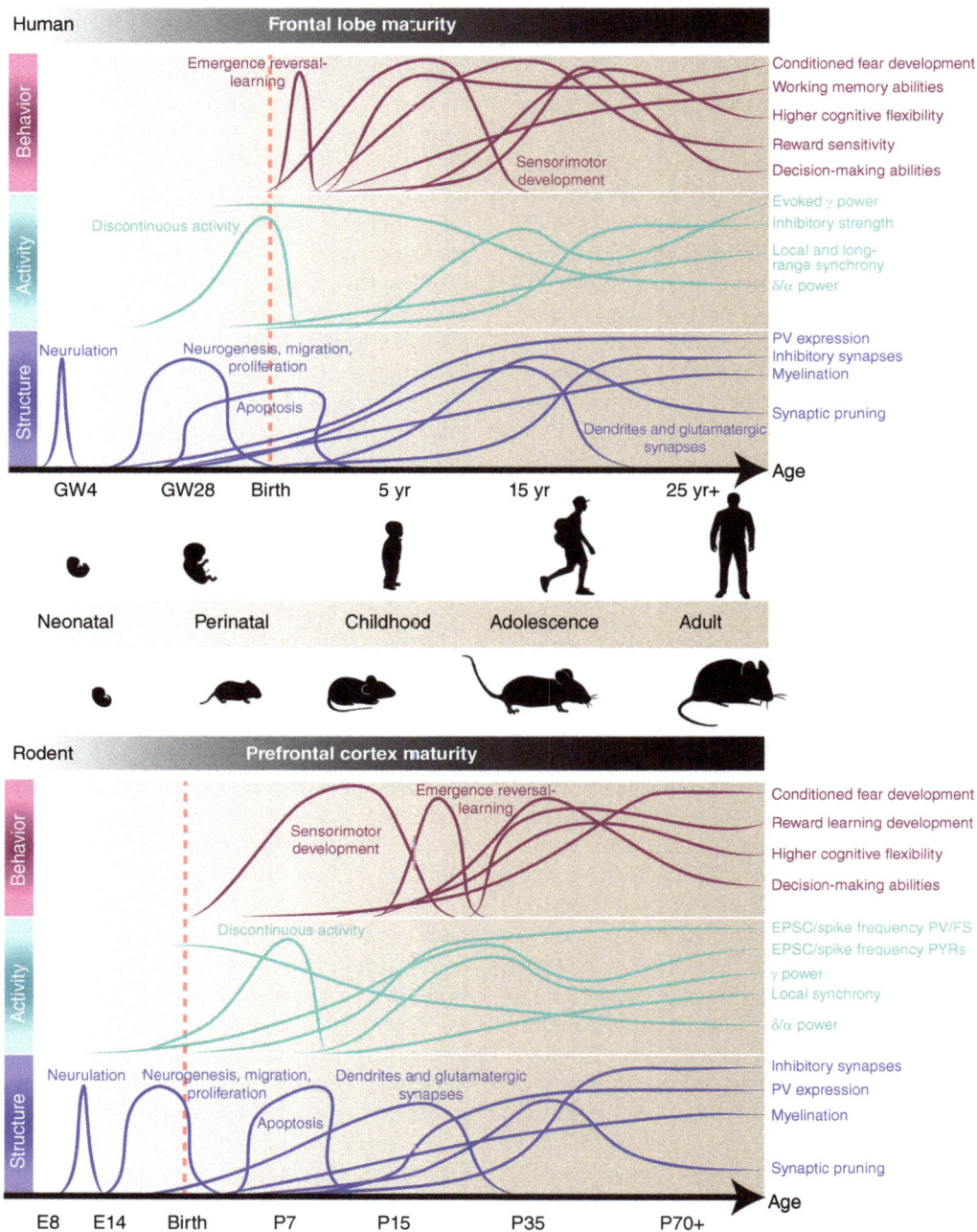

Figure 2. Timeline of major processes during human and rodent prefrontal development. The structural (blue) development of the frontal cortex is largely established during early development. Around birth, electrical activity (turquoise) emerges in the frontal cortex and contributes to the activity-dependent refinement of initial circuits and behavioral abilities (pink). Human frontal lobe maturity (*top*) and rodent prefrontal cortex maturity (*bottom*) follow the same developmental milestones with a less protracted temporal profile in rodents. (GW) Gestational week, (yr) years, (E) embryonic day, (P) postnatal day, (PYR) pyramidal neurons, (FS) fast-spiking, (PV) parvalbumin, (EPSC) excitatory postsynaptic potential, (δ) delta, (α) alpha, (γ) gamma.

pyramidal neurons arises most likely postmitotically during lamina allocation and is less determined by the dividing progenitor pool (Di Bella et al. 2021). In contrast to pyramidal neurons, interneurons are generated in the subpallium with major progenitor pools in the medial and caudal ganglionic eminence. The temporal and spatial specification of progenitor cells within the subpallium mainly accounts for the diversity of interneurons. However, postmitotic cues, such as Dlx and Gad1, are also required for unfolding interneuronal identity (Lim et al. 2018). Postmitotic interneurons migrate tangentially to the developing cortical layers either above or below the marginal zone or ventricular zone, respectively. While interneurons start to migrate earlier than pyramidal neurons, they reach the cortical targets long after the pyramidal neurons have initiated the building of a layered cortical structure. When they have arrived at their destination, interneurons switch from tangential to radial migration to allocate into specific layers. This process of interneuronal sorting is guided by cues released from pyramidal neurons. During this process, medial ganglionic eminence-derived interneurons display a comparable birthdate-dependent, inside-out organization as identified for pyramidal neurons. In contrast, caudal ganglionic eminence-derived interneurons start to proliferate later than interneurons from the medial ganglionic eminence, and the vast majority (~75%) are present in layers 1–3 regardless of their birthdate (Miyoshi and Fishell 2011; Lim et al. 2018; Chini and Hanganu-Opatz 2021). Besides neurons, glial cells populate the developing PFC. After the peak of neurogenesis, radial glia give rise to astrocytes. On the other hand, microglia already originate from the mesoderm and populate the brain during neurogenesis. Overall, in the primate PFC, glia cells outnumber neurons, suggesting an evolutionary advantage of glia cell function for the development of cognitive abilities (Hattori 2022; Kolk and Rakic 2022).

Arriving at their final destination, prefrontal neurons undergo a profound morphological reorganization that serves the establishment of initial synaptic contacts. The extension of the axon and formation of an arborized dendritic tree is controlled by molecular cues. Whereas the growth cone of axons toward their targets is guided by semaphorins, Reln and adhesion molecules are critical for synaptogenesis and synaptic contact formation. During early development, many of these established circuits are transient, and are the subject of profound age-dependent remodeling. Even the number of initially generated neurons substantially decreases during rodent postnatal development, with an apoptotic reduction of 12% for pyramidal neurons and 30% for interneurons (Wong et al. 2018; Hanganu-Opatz et al. 2021). Whether neurons persist during development depends on their circuit embedding. Increased excitatory input from pyramidal neurons to interneurons increases the interneuronal survival rate (Wong et al. 2018; Duan et al. 2020). This indicates that early electrical activity might have a key role in shaping the structure and function of the developing PFC.

EARLY PATTERNS OF ELECTRICAL ACTIVITY IN THE DEVELOPING PREFRONTAL CORTEX

As highlighted above, the PFC initially consists of a patchwork of immature neurons, which first build functional synapses under the influence of genetic factors. This early developmental stage is mainly activity-independent. With ongoing maturation, this immature circuitry undergoes major refinement processes that are mediated by electrical activity (Hanganu-Opatz 2010).

The developing PFC follows the general scheme of neocortical development. In rodents, the first patterns of electrical activity have been reported to emerge endogenously (i.e., in the absence of environmental triggers) at late prenatal stages (Corlew et al. 2004). Around birth, these highly discontinuous events (i.e., periods of electrical silence interrupted by oscillatory discharges with frequencies ranging from 4 to 20 Hz) occur in most neocortical regions at a low rate (1–2 events per minute). These early events originate in the posterior cerebral cortex and spread out to the whole cortex toward the anterior pole (Cirelli and Tononi 2015; Kirischuk et al. 2017). They share similar dynamics and spatial distribution across phylogenetically distant species, ranging from fishes, rodents (defined as spindle bursts), up to humans (defined as delta brushes) (Kirischuk et al. 2017; Chini and Hanganu-Opatz 2021). With ongoing development these low-frequency, high-amplitude events occur more frequently, while the periods of electrical "silence" shorten until electrical activity switches from discontinuous to continuous patterns. At the same time, the electrical activity becomes more rhythmic and is interspersed with higher frequency bursts that accelerate with age (Kirischuk et al. 2017; Chini and Hanganu-Opatz 2021). The emergence of continuous activity goes hand in hand with the unfolding of an active sleep–wake cycle. Early patterns of activity do not relate to a specific behavioral state, being equally present during wakefulness as well as during active and quiet sleep. This dissociation between electrical activity and early behavioral state is shared by all investigated mammalian species and might be relevant for the age-specific effects of anesthetics on brain activity and consciousness (Cirelli and Tononi 2015; Chini et al. 2019).

Overall, the dynamics of prefrontal activity patterns resemble the ones characterized in detail for sensory cortices and the archicortex, yet the onset of discontinuous events and the switch to continuous activity are slightly delayed (Brockmann et al. 2011; Cirelli and Tononi 2015; Shen and Colonnese 2016; Chini et al. 2022). In rodents, a few days after birth, the first discontinuous events within 4–10 Hz can be detected, although they are sparse (~10% of total recording time) and have low amplitude and power (Chini et al. 2022). Toward the beginning of the second postnatal week, the occurrence and power of these events augment, while the interspersed γ band events accelerate from 16 to 50 Hz until the end of the fourth postnatal week (Fig. 3; Brockmann et al. 2011; Bitzenhofer et al. 2020). Shortly before eye-opening and appearance of an active sleep–wake cycle at the end of the second postnatal week, the prefrontal activity switches from discontinuous to continuous patterns that entrain prefrontal neurons, shaping and stabilizing the developing circuits (Chini et al. 2022).

The emergence and age-dependent dynamics of early patterns of prefrontal activity are controlled by afferents from several cortical and subcortical areas. Sparse experimental data document

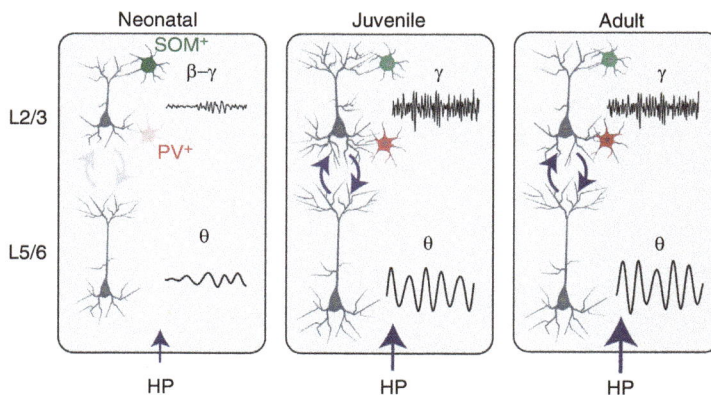

Figure 3. Patterns of electrical activity in the rodent medial prefrontal cortex (mPFC) at different developmental stages and their underlying mechanism and connectivity. During the whole of development, excitatory hippocampal inputs (arrows) target the prefrontal L5/6, boosting θ-band activity in the mPFC. The hippocampal drive and the resulting prefrontal–hippocampal synchrony strengthen with age. Fast oscillatory activity critically depends on the activity of pyramidal neurons in prefrontal L2/3. This locally generated activity augments with age and evolves along development from initially β-low γ-band (~16 Hz) to γ-band (50–80 Hz). While the contribution of distinct interneuronal populations is less well elucidated, it is hypothesized that somatostatin-positive (SOM+) interneurons (green) mediate the generation of β–γ rhythms during neonatal age and parvalbumin-positive (PV+) interneurons (red) control the emergence of mature γ rhythms. (HP) Hippocampus, (L) layer, (θ) theta, (β) beta, (γ) gamma.

the ability of cholinergic projections from the basal forebrain nuclei to modulate the amplitude and frequency of prefrontal oscillations during neonatal development (Janiesch et al. 2011). Other subcortical areas, such as the striatum and thalamus might equally modulate the excitability of prefrontal circuits (G Man, M Chini, and IL Hanganu-Opatz, unpubl.). However, the best-documented input driving the oscillatory events in the developing mPFC comes from the intermediate and ventral HP (Brockmann et al. 2011; Ahlbeck et al. 2018; Song et al. 2022). During the first postnatal week, hippocampal axons, mainly but not exclusively, target prefrontal layers 5/6 and the excitatory monosynaptic inputs facilitate the emergence and maturation of oscillatory events in mPFC. Discontinuous hippocampal θ bursts drive time-locked bursts of nested β/low-γ activity (Brockmann et al. 2011; Ahlbeck et al. 2018). Moreover, the early hippocampal sharp wave ripples provide a strong excitatory drive that leads to neuronal burst firing in mPFC (Fig. 3; Pochinok et al. 2024).

Recently, the development of opto- and chemogenetic tools enabled the first insights into the cellular mechanisms underlying the generation of activity patterns in mPFC development. While the θ activity (4–10 Hz) requires an external boosting drive (e.g., from HP), the fast activity that intersperses the θ oscillations seems to be locally generated. For this, the synchronized firing of layer 2/3, but not of layer 5/6 pyramidal neurons is of critical relevance (Bitzenhofer et al. 2017). The age-dependent acceleration of fast activity is driven by the layer 2/3 pyramidal neuron (Bitzenhofer et al. 2020). The features of these early, fast oscillations markedly differ from their adult counterpart, the γ rhythm, in several aspects such as rhythmicity, frequency, duration, and occurrence. In the adult brain, γ oscillations have been proposed to increase information processing and are correlated to a number of operations ranging from perception to cognition. Mature γ rhythmogenesis is commonly associated with precisely timed and coordinated interactions of excitatory and inhibitory neurons. In particular, inhibition of the perisomatic region of pyramidal neurons, which is mainly targeted by parvalbumin-positive (PV$^+$) interneurons, is essential for the generation of γ

oscillations (Buzsáki and Wang 2012; Cardin 2016; Sohal 2016). During the first postnatal week, the bouts of β/low-γ oscillations in the mPFC already rely on the glutamatergic drive of pyramidal neurons onto interneurons (Bitzenhofer et al. 2015), indicating common principles for γ generation across development. Indeed, the timeline of PV$^+$ interneuron maturation follows a course similar to prefrontal γ development. PV expression can be detected in the rodent mPFC from the end of the second postnatal week on, and reaches mature levels only toward the end of the juvenile period (Bitzenhofer et al. 2020; Kalemaki et al. 2022). Moreover, the typical fast-spiking phenotype of PV$^+$ interneurons has first been detected at juvenile age. It is accompanied by a continuous increase of excitatory and inhibitory postsynaptic potentials in fast-spiking interneurons, reflecting their progressive embedding into prefrontal circuits (Fig. 3; Okaty et al. 2009; Miyamae et al. 2017).

The protracted development of fast-spiking PV$^+$ interneurons aligns with the delayed onset of mature prefrontal γ oscillations. Whereas PV$^+$ interneurons underlie the generation of γ oscillations, dendrite-targeting somatostatin-positive (SOM$^+$) interneurons have been found to contribute to β/low-γ activity (Chen et al. 2017; Hakim et al. 2018). Consequently, the acceleration of prefrontal γ frequency might relate to a transition of interneuronal dominance from SOM$^+$ to late-maturing PV$^+$ interneurons. SOM$^+$ interneurons originate at an earlier time point than PV$^+$ interneurons and show a premature embedding into cortical circuits with transient connections (Lim et al. 2018). During the first three postnatal weeks, the strength of cortical synapses between SOM$^+$ interneurons and pyramidal neurons gradually decreases, whereas reciprocal connections between PV$^+$ interneurons and pyramidal neurons progressively augment in their density and efficiency (Anastasiades et al. 2016; Lim et al. 2018; Pan et al. 2019). Moreover, SOM$^+$ interneurons seem to mediate the integration of PV$^+$ interneurons into SOM$^+$-dominated circuits. Within sensory cortical areas, early silencing of SOM$^+$ interneurons prevents the emergence of perisomatic motifs and PV$^+$-mediated feedforward inhibition (Tuncdemir et al. 2016; Dard et al. 2022). In the mPFC, the shift from SOM to PV dominance might also mediate the sparsification of activity as network events become less globally organized and transform toward more local circuit events, leading to an increase in the overall inhibitory tone (Fig. 3; Chini et al. 2022).

In addition to the intralaminar, the cross-laminar connectivity is also comprehensively refined during postnatal development. While at the end of the second postnatal week, layer 2/3 but not layer 5/6 pyramidal neurons are already functionally interconnected, the translaminar connectivity is functionally silent. This is due to the lack of functional α-amino-3-hydroxy-5-methyl-4-isoxazolepropionic acid receptors. The cross-laminar connections become active shortly after the switch from discontinuous to continuous activity during the third postnatal week in rodents (Fig. 3; Anastasiades and Butt 2012; Anastasiades et al. 2016).

Overall, the developing milestones in mPFC resemble the ones reported for primary sensory cortices, yet have a delayed timeline. This prolonged time window of plasticity might be required for the emergence and calibration of the highly complex functions that the PFC accounts for. Thus, it is not surprising that profound changes continue to occur within prefrontal circuits until the end of adolescence, the time point when mature cognitive functions emerge (Chini and Hanganu-Opatz 2021; Klune et al. 2021).

LATE DEVELOPMENT OF PREFRONTAL CIRCUITS AND EMERGENCE OF COGNITIVE ABILITIES

Complex neural circuits distributed throughout the brain account for cognitive behaviors, such as planning actions, solving problems, and adapting to new situations according to external information and internal states. However, it is accepted that the PFC acts as a hub region within these circuits, coordinating the tradeoff between the stability and flexibility of neural representations. For this function, prefrontal synchrony plays an important role (Hanganu-Opatz et al. 2023). In all mammalian species, cognitive skills still continue to improve long after the emergence and fine-tuning of sensorimotor abilities. This delayed matu-

ration aligns with the dynamics of structural and functional maturation of prefrontal circuits.

Most cognitive abilities emerge toward the end of development. At this late stage of maturation, structural changes are rather modest when compared to the ones occurring during early development. In contrast, a considerable functional remodeling of initial circuits takes place, which is considered a marker of the adolescent period. The extended process of prefrontal development is not linear. Instead, it is paced in an age-dependent manner with experience-sensitive phases of rapid changes and augmented plasticity that align with nonlinear changes in cognitive skill achievement (Caballero et al. 2016; Chini and Hanganu-Opatz 2021; Kolk and Rakic 2022). On an anatomical level, the synaptogenesis, which is initiated during early development, reaches its maximum during adolescence and is followed by synaptic pruning, a process that extends until adulthood (Kolk and Rakic 2022). In humans, this process of gray matter thinning and an increase in white matter continues until the third decade of life. Postmortem studies revealed the most prominent overproduction of spines in the dorsolateral PFC, particularly, in layer 3. The process of pruning is not restricted to dendritic spines but also affects axonal interhemispheric connections of the corpus callosum that undergo substantial sparsification at this age (Wahlstrom et al. 2010; Petanjek et al. 2011; Sturman and Moghaddam 2011; Kolk and Rakic 2022).

The reorganization through pruning of prefrontal circuits is present in rodents as well. During the fifth postnatal week, the volume of the mPFC peaks due to an exorbitant overproduction of spines. This process is accompanied by a turnover of synapses that reaches the highest magnitude during adolescence and continues until adulthood. The pruning includes compartments of the dendritic tree in addition to synaptic contacts (Koss et al. 2014; Delevich et al. 2018). Microglia, which undergo major changes in their morphology and function throughout adolescence, are key mediators of synaptic pruning. Whereas their number postnatally decreases, microglia show the most rounded and least ramified phenotype during adolescence, when they engulf synaptic elements, especially in layer 2/3 of the mPFC (Mallya et al. 2019; Zimmermann et al. 2019; Pöpplau et al. 2024).

The structural reorganization identified during adolescence underlies substantial functional changes. Imaging and electroencephalogram studies revealed, at this late developmental stage, unique patterns of activity during cognitive tasks in humans. Similar prefrontal regions as in adults are activated during these tasks, yet adolescents show a different magnitude, temporal or spatial activation patterns, and interconnectivity as well as a less focal activation and a lower signal-to-noise ratio. Similarly, stimulus-induced γ activity has been found to broadly peak during adolescence, yet neuronal synchrony and cognitive performance continuously increase until adulthood, suggesting that cognitive performance relies on precise local processing in the PFC (Uhlhaas et al. 2010; Sturman and Moghaddam 2011).

The underlying cellular and synaptic mechanisms accounting for adolescence-specific activity have been mainly elucidated in rodents, capitalizing on electrophysiological recordings in vivo and in vitro. During adolescence, increased firing rates and frequency of postsynaptic currents of excitatory but not inhibitory neurons, particularly in prefrontal layer 2/3, might augment the broadband γ activity in the mPFC. As a result, the establishment of long-range connectivity and circuit reorganization via activity-dependent pruning might be boosted. The subsequent decrease in broadband γ power and increase in prefrontal synchrony most likely emerge from the progressive shift toward inhibition (Caballero and Tseng 2016; Nabel et al. 2020; Pöpplau et al. 2024). In addition to the shift in interneuronal dominance of SOM$^+$ and calretinin-positive interneurons toward PV$^+$ interneurons, the changes in the composition of GABA$_A$ and N-methyl-D-aspartate receptors promote the acceleration and potentiation of synaptic inhibition and transmission (Caballero and Tseng 2016; Konstantoudaki et al. 2018). Moreover, adult-like PV-mediated hyperpolarizing GABAergic inhibition of the axon initial segment of prefrontal layer 2/3 pyramidal neurons has first been detected at adolescence (Rinetti-Vargas et al. 2017). Whereas glutamatergic synapses peak during adolescence, the number of

inhibitory synapses progressively increases after adolescence (Wahlstrom et al. 2010; Caballero and Tseng 2016). In agreement with this, excitatory neurons undergo maximal transcriptional changes during adolescence, whereas inhibitory neurons appear to be more evolutionarily conserved (Bhattacherjee et al. 2019).

During adolescence, several neuromodulatory systems additionally contribute to the reorganization of circuits. Among the monoaminergic neurotransmitters, dopamine undergoes the most protracted development and substantially contributes to the strengthening of prefrontal inhibition. Whereas the density of dopaminergic projections targeting the mPFC continuously increases until adulthood, the density of D1- and D2-type dopaminergic receptors peaks during adolescence. Accordingly, the excitatory control by dopamine of GABAergic transmission first emerges toward the end of development (Caballero and Tseng 2016; Caballero et al. 2016; Walker et al. 2017).

As highlighted above, microglia contribute to the structural reorganization of prefrontal circuits during late development, yet the interactions between neuronal and microglial activity are poorly elucidated. Active neurons release adenosine triphosphate that causes microglia-induced synapse remodeling. Activated microglia have been shown to release brain-derived neurotrophic factor (BDNF), which increases network excitability (Ferrini and De Koninck 2013; Ball et al. 2022). Notably, prefrontal BDNF levels augment between adolescence and adulthood (Larsen and Luna 2018). Inactive synapses, on the other hand, are tagged by so-called "eat-me" factors that initiate phagocytic activity of microglia cells. During adolescence, "eat-me" factors are enriched, promoting long-term depression (LTD) and microglia-mediated pruning (Tseng and O'Donnell 2007; Zhang et al. 2021; Schalbetter et al. 2022). Correspondingly, HP-initiated LTD is not present in the PFC before adulthood (Zimmermann et al. 2019). LTD induction through weakening of synaptic strength is promoted by microglia-driven trogocytosis (i.e., incomplete engulfment of synapses) (Weinhard et al. 2018). These data provide the first insights into the tight mutual interactions between neuronal and microglial activity as the substrate of prefrontal reorganization during adolescence.

The consequence of circuit reorganization during adolescence is the progressive maturation of higher-order cognitive functions. Most executive abilities, such as working memory, decision making, switching, as well as the control of emotions, evolve during adolescence to reach adult performance. The protracted PFC development when compared to cortical and subcortical areas accounting for emotional states, most likely leads to adolescence-characteristic behaviors, such as heightened risk-taking, reduced response inhibition, fear extinction, and increased reward sensitivity, which are present in most mammalian species (Klune et al. 2021; Uhlhaas et al. 2023). From an evolutionary perspective, such risk-seeking behavior might appear detrimental for species survival. However, it supports the transition to autonomy and increased sociability toward strangers (Wahlstrom et al. 2010; Klune et al. 2021). The distinct trajectories of structural and functional development of PFC underlie the behavioral dynamics during late development. While decision making and working memory emerge during childhood, they show specific trajectories throughout adolescence (Klune et al. 2021). In line with the continuous increase in γ activity and prefrontal synchrony, working memory abilities increase almost linearly until adulthood (Conklin et al. 2007; Uhlhaas et al. 2010). On the other hand, despite their characteristic risk-seeking, adolescents are flexible and creative in problem solving, having improved decision-making abilities when compared to adults (Crone and Dahl 2012; Hauser et al. 2015).

Thus, the development of cognitive abilities mirrors the dynamics of structural and functional changes in the PFC.

DISEASE-RELEVANT MISWIRING DURING PREFRONTAL DEVELOPMENT

Since the prefrontal function accounts for cognitive processing and a wealth of behavioral abilities, prefrontal dysfunctions might be the source of severe cognitive deficits. Such deficits have been reported to represent the core symptom of several devastating neuropsychiatric disorders, such as schizophrenia, bipolar disorders, or autism spectrum disorders. Most patients do not respond to common treatment approaches and even responders suffer from a noncurative, mainly symptom-based, treatment with many off-target effects (Millan et al. 2012, 2016; Greenberg et al. 2015; Uhlhaas et al. 2023). The individual and societal burden of these disorders highlight the pressing need for novel therapeutic strategies. They can emerge only through a better understanding of the etiology and mechanisms of disease. While the classification of mental disorders according to symptomatology and clinical representation is highly debated and varies between different diagnostic guides, such as ICD-11 and DSM-5 (Stein et al. 2020; Uhlhaas et al. 2023), it is generally accepted that many mental disorders share a common etiology of combined genetic and environmental factors (such as food, drugs, viral infection, radiation, hypoxia, cannabis use) (Kolk and Rakic 2022) that perturb normal brain development. These factors cumulate in their output and create a domino effect, leading to an adverse influence on subsequent developmental milestones (Horváth and Mirnics 2009; Feigenson et al. 2014; Uhlhaas et al. 2023).

Around 75% of patients experience the onset of diseases during childhood or adolescence (i.e., before the age of 25) (Uhlhaas et al. 2023). More than 900 genes have been linked with increased susceptibility for mental disorders and are potential prognostic biomarkers. However, only around one-third of genetic high-risk patients develop disease symptoms throughout life, indicating a multifactorial nature. Moreover, risk genes across mental diseases show a high degree of genetic overlap, leading to substantial comorbidity among diseases. Most of these risk genes are strongly expressed during development and trigger a common cascade of lasting alterations with shared behavioral endophenotypes, especially within the cognitive domain. Thus, identifying risk genes might not be sufficient for therapeutic approaches (Spencer-Smith and Anderson 2009; Marín 2016; The Brainstorm Consortium et al. 2018; Parenti et al. 2020).

Considering the alterations in prefrontal circuits that are common across species and relate to abnormal cognitive behavior, the investigation

Cite this article as *Cold Spring Harb Perspect Biol* doi: 10.1101/cshperspect.a041502

of rodent models for disease might be a promising strategy toward therapeutic approaches. Time windows of high prefrontal plasticity, the perinatal and adolescent period of development, have been associated with an increased vulnerability for mental diseases. Targeting circuit dysfunction during these phases of development might open new therapeutic opportunities (Selemon and Zecevic 2015; Chini and Hanganu-Opatz 2021). Studies in mice revealed that circuit abnormalities at these specific time points of prefrontal development lead to long-lasting cognitive impairments (Shao et al. 2013; Rubino et al. 2015; Yang et al. 2015; Xu et al. 2019; Bitzenhofer et al. 2021). For example, the decreased excitability of spine-lacking pyramidal neurons in the prefrontal layer 2/3 of a dual genetic-environmental mouse model of schizophrenia leads to weaker local connectivity during neonatal development that persists until adulthood and causes poor cognitive abilities (Xu et al. 2019; Chini et al. 2020). Not only local but also long-range miswiring during development has been identified to lead to cognitive impairment in mental disorders. At adulthood, the PFC function strongly depends on inputs to generate an appropriate output that, in turn, needs to be adequately processed in downstream regions. Studies of the last decades uncovered the contribution of different circuits for specific cognitive abilities. For example, whereas the prefrontal–amygdala circuit is primarily associated with altered social cognition and emotional regulation (Rasetti et al. 2009; Gangopadhyay et al. 2021), the prefrontal–hippocampal circuit affects working and recognition memory abilities (Godsil et al. 2013; Adams et al. 2020). However, the disease-characteristic development of long-range circuits involving the PFC is still poorly understood. The first insights reveal that despite the multifactorial etiology and broad spectrum of symptoms, some functional deficits and miswiring patterns occurring during distinct stages of development are common across diseases. For example, the early prefrontal–hippocampal miswiring is present in both mouse models of schizophrenia and autism spectrum disorders (Richter et al. 2019; Chini et al. 2020; Scharrenberg et al. 2022; Song et al. 2022).

Identification of common trajectories of dysfunction and miswiring might facilitate the identification of targets amenable for future therapies.

CONCLUDING REMARKS

The findings summarized above provide first insights into the milestones, mechanisms, and relevance of prefrontal development. While this substantial knowledge gain of recent years highlights the important role of circuit formation and refinement for later cognitive performance, the causes and mechanisms of altered prefrontal development and especially the rescue strategies with therapeutic potential for mental diseases are still largely unclear. For this, future studies should capitalize on the plastic potential of the adolescent period of PFC development (Klune et al. 2023). Due to ethical and technical limitations, the early stage of prominent circuit refinement occurring at the neo-/perinatal age has poor translational potential. Thus, the second window of opportunity, taking place during adolescence with the nonlinear reorganization of circuits and behavior, might be instrumental for the development of therapeutic approaches in prodromal high-risk subjects. Interventions during specific time windows of high plasticity might be able to block the domino effect of accumulating risks. To develop such interventions, it is critical to understand how risk factors on different levels (temporal, genetic, environmental, spatial) interact to result in asymptomatic pathologies or in disease onset. Achieving this goal has proven to be notoriously difficult and remains a challenge for future studies.

The development of PFC requires time to bring complex cognitive abilities to fruition. It is, therefore, not surprising that even after centuries of research, the PFC is still a "black box" of developmental neuroscience.

REFERENCES

Adams RA, Bush D, Zheng F, Meyer SS, Kaplan R, Orfanos S, Marques TR, Howes OD, Burgess N. 2020. Impaired θ phase coupling underlies frontotemporal dysconnectivity in schizophrenia. *Brain* **143:** 1261–1277. doi:10.1093/brain/awaa035

Ahlbeck J, Song L, Chini M, Bitzenhofer SH, Hanganu-Opatz IL. 2018. Glutamatergic drive along the septo-temporal axis of hippocampus boosts prelimbic oscillations in the neonatal mouse. *eLife* **7**: e33158. doi:10.7554/eLife.33158

Ährlund-Richter S, Xuan Y, van Lunteren JA, Kim H, Ortiz C, Pollak Dorocic I, Meletis K, Carlén M. 2019. A whole-brain atlas of monosynaptic input targeting four different cell types in the medial prefrontal cortex of the mouse. *Nat Neurosci* **22**: 657–668. doi:10.1038/s41593-019-0354-y

Anastasiades PG, Butt SJ. 2012. A role for silent synapses in the development of the pathway from layer 2/3 to 5 pyramidal cells in the neocortex. *J Neurosci* **32**: 13085–13099. doi:10.1523/JNEUROSCI.1262-12.2012

Anastasiades PG, Carter AG. 2021. Circuit organization of the rodent medial prefrontal cortex. *Trends Neurosci* **44**: 550–563. doi:10.1016/j.tins.2021.03.006

Anastasiades PG, Marques-Smith A, Lyngholm D, Lickiss T, Raffiq S, Kätzel D, Miesenböck G, Butt SJ. 2016. GABAergic interneurons form transient layer-specific circuits in early postnatal neocortex. *Nat Commun* **7**: 10584. doi:10.1038/ncomms10584

Ball JB, Green-Fulgham SM, Watkins LR. 2022. Mechanisms of microglia-mediated synapse turnover and synaptogenesis. *Prog Neurobiol* **218**: 102336. doi:10.1016/j.pneurobio.2022.102336

Bhattacherjee A, Djekidel MN, Chen R, Chen W, Tuesta LM, Zhang Y. 2019. Cell type-specific transcriptional programs in mouse prefrontal cortex during adolescence and addiction. *Nat Commun* **10**: 4169. doi:10.1038/s41467-019-12054-3

Bitzenhofer SH, Sieben K, Siebert KD, Spehr M, Hanganu-Opatz IL. 2015. Oscillatory activity in developing prefrontal networks results from θ-γ-modulated synaptic inputs. *Cell Rep* **11**: 486–497. doi:10.1016/j.celrep.2015.03.031

Bitzenhofer SH, Ahlbeck J, Wolff A, Wiegert JS, Gee CE, Oertner TG, Hanganu-Opatz IL. 2017. Layer-specific optogenetic activation of pyramidal neurons causes β-γ entrainment of neonatal networks. *Nat Commun* **8**: 14563. doi:10.1038/ncomms14563

Bitzenhofer SH, Pöpplau JA, Hanganu-Opatz I. 2020. γ Activity accelerates during prefrontal development. *eLife* **9**: e56795. doi:10.7554/eLife.56795

Bitzenhofer SH, Pöpplau JA, Chini M, Marquardt A, Hanganu-Opatz IL. 2021. A transient developmental increase in prefrontal activity alters network maturation and causes cognitive dysfunction in adult mice. *Neuron* **109**: 1350–1364.e6. doi:10.1016/j.neuron.2021.02.011

Brockmann MD, Pöschel B, Cichon N, Hanganu-Opatz IL. 2011. Coupled oscillations mediate directed interactions between prefrontal cortex and hippocampus of the neonatal rat. *Neuron* **71**: 332–347. doi:10.1016/j.neuron.2011.05.041

Buzsáki G, Wang XJ. 2012. Mechanisms of γ oscillations. *Annu Rev Neurosci* **35**: 203–225. doi:10.1146/annurev-neuro-062111-150444

Caballero A, Tseng KY. 2016. GABAergic function as a limiting factor for prefrontal maturation during adolescence. *Trends Neurosci* **39**: 441–448. doi:10.1016/j.tins.2016.04.010

Caballero A, Granberg R, Tseng KY. 2016. Mechanisms contributing to prefrontal cortex maturation during adolescence. *Neurosci Biobehav Rev* **70**: 4–12. doi:10.1016/j.neubiorev.2016.05.013

Cardin JA. 2016. Snapshots of the brain in action: local circuit operations through the lens of γ oscillations. *J Neurosci* **36**: 10496–10504. doi:10.1523/JNEUROSCI.1021-16.2016

Carlén M. 2017. What constitutes the prefrontal cortex? *Science* **358**: 478–482. doi:10.1126/science.aan8868

Chen G, Zhang Y, Li X, Zhao X, Ye Q, Lin Y, Tao HW, Rasch MJ, Zhang X. 2017. Distinct inhibitory circuits orchestrate cortical β and γ band oscillations. *Neuron* **96**: 1403–1418.e6. doi:10.1016/j.neuron.2017.11.033

Chini M, Hanganu-Opatz IL. 2021. Prefrontal cortex development in health and disease: lessons from rodents and humans. *Trends Neurosci* **44**: 227–240. doi:10.1016/j.tins.2020.10.017

Chini M, Gretenkord S, Kostka JK, Pöpplau JA, Cornelissen L, Berde CB, Hanganu-Opatz IL, Bitzenhofer SH. 2019. Neural correlates of anesthesia in newborn mice and humans. *Front Neural Circuits* **13**: 38. doi:10.3389/fncir.2019.00038

Chini M, Pöpplau JA, Lindemann C, Carol-Perdiguer L, Hnida M, Oberländer V, Xu X, Ahlbeck J, Bitzenhofer SH, Mulert C, et al. 2020. Resolving and rescuing developmental miswiring in a mouse model of cognitive impairment. *Neuron* **105**: 60–74.e7. doi:10.1016/j.neuron.2019.09.042

Chini M, Pfeffer T, Hanganu-Opatz I. 2022. An increase of inhibition drives the developmental decorrelation of neural activity. *eLife* **11**: e78811. doi:10.7554/eLife.78811

Cirelli C, Tononi G. 2015. Cortical development, electroencephalogram rhythms, and the sleep/wake cycle. *Biol Psychiatry* **77**: 1071–1078. doi:10.1016/j.biopsych.2014.12.017

Conklin HM, Luciana M, Hooper CJ, Yarger RS. 2007. Working memory performance in typically developing children and adolescents: behavioral evidence of protracted frontal lobe development. *Dev Neuropsychol* **31**: 103–128. doi:10.1207/s15326942dn3101_6

Corlew R, Bosma MM, Moody WJ. 2004. Spontaneous, synchronous electrical activity in neonatal mouse cortical neurones. *J Physiol* **560**: 377–390. doi:10.1113/jphysiol.2004.071621

Crone EA, Dahl RE. 2012. Understanding adolescence as a period of social-affective engagement and goal flexibility. *Nat Rev Neurosci* **13**: 636–650. doi:10.1038/nrn3313

Dard RF, Leprince E, Denis J, Rao-Balappa S, Suchkov D, Boyce R, Lopez C, Giorgi-Kurz M, Szwagier T, Dumont T, et al. 2022. The rapid developmental rise of somatic inhibition disengages hippocampal dynamics from self-motion. *eLife* **11**: e78116. doi:10.7554/eLife.78116

Delevich K, Thomas AW, Wilbrecht L. 2018. Adolescence and "late blooming" synapses of the prefrontal cortex. *Cold Spring Harb Symp Quant Biol* **83**: 37–43. doi:10.1101/sqb.2018.83.037507

Di Bella DJ, Habibi E, Stickels RR, Scalia G, Brown J, Yadollahpour P, Yang SM, Abbate C, Biancalani T, Macosko EZ, et al. 2021. Molecular logic of cellular diversification in the mouse cerebral cortex. *Nature* **595**: 554–559. doi:10.1038/s41586-021-03670-5

Donahue CJ, Glasser MF, Preuss TM, Rilling JK, Van Essen DC. 2018. Quantitative assessment of prefrontal cortex in humans relative to nonhuman primates. *Proc Natl Acad Sci* 115: E5183–E5192. doi:10.1073/pnas.1721653115

Duan ZRS, Che A, Chu P, Modol L, Bollmann Y, Babij R, Fetcho RN, Otsuka T, Fuccillo MV, Liston C, et al. 2020. GABAergic restriction of network dynamics regulates interneuron survival in the developing cortex. *Neuron* 105: 75–92.e5. doi:10.1016/j.neuron.2019.10.008

Feigenson KA, Kusnecov AW, Silverstein SM. 2014. Inflammation and the two-hit hypothesis of schizophrenia. *Neurosci Biobehav Rev* 38: 72–93. doi:10.1016/j.neubiorev.2013.11.006

Ferrini F, De Koninck Y. 2013. Microglia control neuronal network excitability via BDNF signalling. *Neural Plast* 2013: 429815. doi:10.1155/2013/429815

Fuster JM. 2001. The prefrontal cortex—an update: time is of the essence. *Neuron* 30: 319–333. doi:10.1016/S0896-6273(01)00285-9

Gangopadhyay P, Chawla M, Dal Monte O, Chang SWC. 2021. Prefrontal–amygdala circuits in social decision-making. *Nat Neurosci* 24: 5–18. doi:10.1038/s41593-020-00738-9

Geschwind DH, Rakic P. 2013. Cortical evolution: judge the brain by its cover. *Neuron* 80: 633–647. doi:10.1016/j.neuron.2013.10.045

Godsil BP, Kiss JP, Spedding M, Jay TM. 2013. The hippocampal-prefrontal pathway: the weak link in psychiatric disorders? *Eur Neuropsychopharmacol* 23: 1165–1181. doi:10.1016/j.euroneuro.2012.10.018

Greenberg PE, Fournier AA, Sisitsky T, Pike CT, Kessler RC. 2015. The economic burden of adults with major depressive disorder in the United States (2005 and 2010). *J Clin Psychiatry* 76: 155–162. doi:10.4088/JCP.14m09298

Hakim R, Shamardani K, Adesnik H. 2018. A neural circuit for γ-band coherence across the retinotopic map in mouse visual cortex. *eLife* 7: e28569. doi:10.7554/eLife.28569

Hanganu-Opatz IL. 2010. Between molecules and experience: role of early patterns of coordinated activity for the development of cortical maps and sensory abilities. *Brain Res Rev* 64: 160–176. doi:10.1016/j.brainresrev.2010.03.005

Hanganu-Opatz IL, Butt SJB, Hippenmeyer S, De Marco García NV, Cardin JA, Voytek B, Muotri AR. 2021. The logic of developing neocortical circuits in health and disease. *J Neurosci* 41: 813–822. doi:10.1523/JNEUROSCI.1655-20.2020

Hanganu-Opatz IL, Klausberger T, Sigurdsson T, Nieder A, Jacob SN, Bartos M, Sauer JF, Durstewitz D, Leibold C, Diester I. 2023. Resolving the prefrontal mechanisms of adaptive cognitive behaviors: a cross-species perspective. *Neuron* 111: 1020–1036. doi:10.1016/j.neuron.2023.03.017

Harlow JM. 1849. Passage of an iron rod through the head. *Northwest Med Surg J* 1: 513–518.

Hattori Y. 2022. The behavior and functions of embryonic microglia. *Anat Sci Int* 97: 1–14. doi:10.1007/s12565-021-00631-w

Hauser TU, Iannaccone R, Walitza S, Brandeis D, Brem S. 2015. Cognitive flexibility in adolescence: neural and behavioral mechanisms of reward prediction error processing in adaptive decision making during development.

Neuroimage 104: 347–354. doi:10.1016/j.neuroimage.2014.09.018

Hoerder-Suabedissen A, Molnár Z. 2015. Development, evolution and pathology of neocortical subplate neurons. *Nat Rev Neurosci* 16: 133–146. doi:10.1038/nrn3915

Horváth S, Mirnics K. 2009. Breaking the gene barrier in schizophrenia. *Nat Med* 15: 488–490. doi:10.1038/nm0509-488

Janiesch PC, Krüger HS, Pöschel B, Hanganu-Opatz IL. 2011. Cholinergic control in developing prefrontal-hippocampal networks. *J Neurosci* 31: 17955–17970. doi:10.1523/JNEUROSCI.2644-11.2011

Kalemaki K, Velli A, Christodoulou O, Denaxa M, Karagogeos D, Sidiropoulou K. 2022. The developmental changes in intrinsic and synaptic properties of prefrontal neurons enhance local network activity from the second to the third postnatal weeks in mice. *Cereb Cortex* 32: 3633–3650. doi:10.1093/cercor/bhab438

Kirischuk S, Sinning A, Blanquie O, Yang JW, Luhmann HJ, Kilb W. 2017. Modulation of neocortical development by early neuronal activity: physiology and pathophysiology. *Front Cell Neurosci* 11: 379. doi:10.3389/fncel.2017.00379

Klune CB, Jin B, DeNardo LA. 2021. Linking mPFC circuit maturation to the developmental regulation of emotional memory and cognitive flexibility. *eLife* 10: e64567. doi:10.7554/eLife.64567

Klune CB, Goodpaster CM, Gongwer MW, Gabriel CJ, Chen R, Jones NS, Schwarz LA, DeNardo LA. 2023. Developmentally distinct architectures in top-down circuits. bioRxiv doi:10.1101/2023.08.27.555010

Kolk SM, Rakic P. 2022. Development of prefrontal cortex. *Neuropsychopharmacology* 47: 41–57. doi:10.1038/s41386-021-01137-9

Konstantoudaki X, Chalkiadaki K, Vasileiou E, Kalemaki K, Karagogeos D, Sidiropoulou K. 2018. Prefrontal cortical-specific differences in behavior and synaptic plasticity between adolescent and adult mice. *J Neurophysiol* 119: 822–833. doi:10.1152/jn.00189.2017

Koss WA, Belden CE, Hristov AD, Juraska JM. 2014. Dendritic remodeling in the adolescent medial prefrontal cortex and the basolateral amygdala of male and female rats. *Synapse* 68: 61–72. doi:10.1002/syn.21716

Larsen B, Luna B. 2018. Adolescence as a neurobiological critical period for the development of higher-order cognition. *Neurosci Biobehav Rev* 94: 179–195. doi:10.1016/j.neubiorev.2018.09.005

Laubach M, Amarante LM, Swanson K, White SR. 2018. What, if anything, is rodent prefrontal cortex? *eNeuro* 5: ENEURO.0315-18.2018. doi:10.1523/ENEURO.0315-18.2018

Le Merre P, Ährlund-Richter S, Carlén M. 2021. The mouse prefrontal cortex: unity in diversity. *Neuron* 109: 1925–1944. doi:10.1016/j.neuron.2021.03.035

Lim L, Mi D, Llorca A, Marín O. 2018. Development and functional diversification of cortical interneurons. *Neuron* 100: 294–313. doi:10.1016/j.neuron.2018.10.009

Mallya AP, Wang HD, Lee HNR, Deutch AY. 2019. Microglial pruning of synapses in the prefrontal cortex during adolescence. *Cereb Cortex* 29: 1634–1643. doi:10.1093/cercor/bhy061

Marín O. 2016. Developmental timing and critical windows for the treatment of psychiatric disorders. *Nat Med* **22:** 1229–1238. doi:10.1038/nm.4225

Marín O, Rubenstein JL. 2003. Cell migration in the forebrain. *Annu Rev Neurosci* **26:** 441–483. doi:10.1146/annurev.neuro.26.041002.131058

Millan MJ, Agid Y, Brüne M, Bullmore ET, Carter CS, Clayton NS, Connor R, Davis S, Deakin B, DeRubeis RJ, et al. 2012. Cognitive dysfunction in psychiatric disorders: characteristics, causes and the quest for improved therapy. *Nat Rev Drug Discov* **11:** 141–168. doi:10.1038/nrd3628

Millan MJ, Andrieux A, Bartzokis G, Cadenhead K, Dazzan P, Fusar-Poli P, Gallinat J, Giedd J, Grayson DR, Heinrichs M, et al. 2016. Altering the course of schizophrenia: progress and perspectives. *Nat Rev Drug Discov* **15:** 485–515. doi:10.1038/nrd.2016.28

Miller EK, Cohen JD. 2001. An integrative theory of prefrontal cortex function. *Annu Rev Neurosci* **24:** 167–202. doi:10.1146/annurev.neuro.24.1.167

Miyamae T, Chen K, Lewis DA, Gonzalez-Burgos G. 2017. Distinct physiological maturation of parvalbumin-positive neuron subtypes in mouse prefrontal cortex. *J Neurosci* **37:** 4883–4902. doi:10.1523/JNEUROSCI.3325-16.2017

Miyoshi G, Fishell G. 2011. GABAergic interneuron lineages selectively sort into specific cortical layers during early postnatal development. *Cereb Cortex* **21:** 845–852. doi:10.1093/cercor/bhq155

Nabel EM, Garkun Y, Koike H, Sadahiro M, Liang A, Norman KJ, Taccheri G, Demars MP, Im S, Caro K, et al. 2020. Adolescent frontal top-down neurons receive heightened local drive to establish adult attentional behavior in mice. *Nat Commun* **11:** 3983. doi:10.1038/s41467-020-17787-0

Okaty BW, Miller MN, Sugino K, Hempel CM, Nelson SB. 2009. Transcriptional and electrophysiological maturation of neocortical fast-spiking GABAergic interneurons. *J Neurosci* **29:** 7040–7052. doi:10.1523/JNEUROSCI.0105-09.2009

Pan NC, Fang A, Shen C, Sun L, Wu Q, Wang X. 2019. Early excitatory activity-dependent maturation of somatostatin interneurons in cortical layer 2/3 of mice. *Cereb Cortex* **29:** 4107–4118. doi:10.1093/cercor/bhy293

Parenti I, Rabaneda LG, Schoen H, Novarino G. 2020. Neurodevelopmental disorders: from genetics to functional pathways. *Trends Neurosci* **43:** 608–621. doi:10.1016/j.tins.2020.05.004

Petanjek Z, Judaš M, Šimić G, Rašin MR, Uylings HB, Rakic P, Kostović I. 2011. Extraordinary neoteny of synaptic spines in the human prefrontal cortex. *Proc Natl Acad Sci* **108:** 13281–13286. doi:10.1073/pnas.1105108108

Pochinok I, Stöber TM, Triesch J, Chii M, Hanganu-Opatz I. 2024. A developmental increase of inhibition promotes the emergence of hippocampal ripples. *Nat Commun* **15:** 738. doi:10.1038/s41467-024-44983-z

Pöpplau JA, Schwarze T, Dorofeikova M, Pochinok I, Günther A, Marquardt A, Hanganu-Opatz IL. 2024. Reorganization of adolescent prefrontal cortex circuitry is required for mouse cognitive maturation. *Neuron* **112:** 421–440.e7. doi:10.1016/j.neuron.2023.10.024

Preuss TM. 1995. Do rats have prefrontal cortex? The Rose-Woolsey-Akert program reconsidered. *J Cogn Neurosci* **7:** 1–24. doi:10.1162/jocn.1995.7.1.1

Preuss TM, Wise SP. 2022. Evolution of prefrontal cortex. *Neuropsychopharmacology* **47:** 3–19. doi:10.1038/s41386-021-01076-5

Rasetti R, Mattay VS, Wiedholz LM, Kolachana BS, Hariri AR, Callicott JH, Meyer-Lindenberg A, Weinberger DR. 2009. Evidence that altered amygdala activity in schizophrenia is related to clinical state and not genetic risk. *Am J Psychiatry* **166:** 216–225. doi:10.1176/appi.ajp.2008.08020261

Richter M, Murtaza N, Scharrenberg R, White SH, Johanns O, Walker S, Yuen RKC, Schwanke B, Bedürftig B, Henis M, et al. 2019. Altered TAOK2 activity causes autism-related neurodevelopmental and cognitive abnormalities through RhoA signaling. *Mol Psychiatry* **24:** 1329–1350. doi:10.1038/s41380-018-0025-5

Rinetti-Vargas G, Phamluong K, Ron D, Bender KJ. 2017. Periadolescent maturation of GABAergic hyperpolarization at the axon initial segment. *Cell Rep* **20:** 21–29. doi:10.1016/j.celrep.2017.06.030

Rubino T, Prini P, Piscitelli F, Zamberletti E, Trusel M, Melis M, Sagheddu C, Ligresti A, Tonini R, Di Marzo V, et al. 2015. Adolescent exposure to THC in female rats disrupts developmental changes in the prefrontal cortex. *Neurobiol Dis* **73:** 60–69. doi:10.1016/j.nbd.2014.09.015

Schalbetter SM, von Arx AS, Cruz-Ochoa N, Dawson K, Ivanov A, Mueller FS, Lin HY, Amport R, Mildenberger W, Mattei D, et al. 2022. Adolescence is a sensitive period for prefrontal microglia to act on cognitive development. *Sci Adv* **8:** eabi6672. doi:10.1126/sciadv.abi6672

Scharrenberg R, Richter M, Johanns O, Meka DP, Rücker T, Murtaza N, Lindenmaier Z, Ellegood J, Naumann A, Zhao B, et al. 2022. TAOK2 rescues autism-linked developmental deficits in a 16p11.2 microdeletion mouse model. *Mol Psychiatry* **27:** 4707–4721. doi:10.1038/s41380-022-01785-3

Schubert D, Martens GJ, Kolk SM. 2015. Molecular underpinnings of prefrontal cortex development in rodents provide insights into the etiology of neurodevelopmental disorders. *Mol Psychiatry* **20:** 795–809. doi:10.1038/mp.2014.147

Selemon LD, Zecevic N. 2015. Schizophrenia: a tale of two critical periods for prefrontal cortical development. *Transl Psychiatry* **5:** e623. doi:10.1038/tp.2015.115

Shao F, Han X, Shao S, Wang W. 2013. Adolescent social isolation influences cognitive function in adult rats. *Neural Regen Res* **8:** 1025–1030. doi:10.3969/j.issn.1673-5374.2013.11.008

Shen J, Colonnese MT. 2016. Development of activity in the mouse visual cortex. *J Neurosci* **36:** 12259–12275. doi:10.1523/JNEUROSCI.1903-16.2016

Sohal VS. 2016. How close are we to understanding what (if anything) γ oscillations do in cortical circuits? *J Neurosci* **36:** 10489–10495. doi:10.1523/JNEUROSCI.0990-16.2016

Song L, Xu X, Putthoff P, Fleck D, Spehr M, Hanganu-Opatz IL. 2022. Sparser and less efficient hippocampal-prefrontal projections account for developmental network dysfunction in a model of psychiatric risk mediated by gene-environment interaction. *J Neurosci* **42:** 601–618. doi:10.1523/JNEUROSCI.1203-21.2021

Cite this article as *Cold Spring Harb Perspect Biol* doi: 10.1101/cshperspect.a041502

Spencer-Smith M, Anderson V. 2009. Healthy and abnormal development of the prefrontal cortex. *Dev Neurorehabil* **12:** 279–297. doi:10.3109/17518420903090701

Stein DJ, Szatmari P, Gaebel W, Berk M, Vieta E, Maj M, de Vries YA, Roest AM, de Jonge P, Maercker A, et al. 2020. Mental, behavioral and neurodevelopmental disorders in the ICD-11: an international perspective on key changes and controversies. *BMC Med* **18:** 21. doi:10.1186/s12916-020-1495-2

Sturman DA, Moghaddam B. 2011. The neurobiology of adolescence: changes in brain architecture, functional dynamics, and behavioral tendencies. *Neurosci Biobehav Rev* **35:** 1704–1712. doi:10.1016/j.neubiorev.2011.04.003

The Brainstorm Consortium; Anttila V, Bulik-Sullivan B, Finucane HK, Walters RK, Bras J, Duncan L, Escott-Price V, Falcone GJ, Gormley P, et al. 2018. Analysis of shared heritability in common disorders of the brain. *Science* **360:** eaap8757. doi:10.1126/science.aap8757

Tseng KY, O'Donnell P. 2007. Dopamine modulation of prefrontal cortical interneurons changes during adolescence. *Cereb Cortex* **17:** 1235–1240. doi:10.1093/cercor/bhl034

Tuncdemir SN, Wamsley B, Stam FJ, Osakada F, Goulding M, Callaway EM, Rudy B, Fishell G. 2016. Early somatostatin interneuron connectivity mediates the maturation of deep layer cortical circuits. *Neuron* **89:** 521–535. doi:10.1016/j.neuron.2015.11.020

Uhlhaas PJ, Roux F, Rodriguez E, Rotarska-Jagiela A, Singer W. 2010. Neural synchrony and the development of cortical networks. *Trends Cogn Sci* **14:** 72–80. doi:10.1016/j.tics.2009.12.002

Uhlhaas PJ, Davey CG, Mehta UM, Shah J, Torous J, Allen NB, Avenevoli S, Bella-Awusah T, Chanen A, Chen EYH, et al. 2023. Towards a youth mental health paradigm: a perspective and roadmap. *Mol Psychiatry* **28:** 3171–3181. doi:10.1038/s41380-023-02202-z

van Heukelum S, Mars RB, Guthrie M, Buitelaar JK, Beckmann CF, Tiesinga PHE, Vogt BA, Glennon JC, Havenith MN. 2020. Where is cingulate cortex? A cross-species view. *Trends Neurosci* **43:** 285–299. doi:10.1016/j.tins.2020.03.007

Wahlstrom D, Collins P, White T, Luciana M. 2010. Developmental changes in dopamine neurotransmission in adolescence: behavioral implications and issues in assessment. *Brain Cogn* **72:** 146–159. doi:10.1016/j.bandc.2009.10.013

Walker DM, Bell MR, Flores C, Gulley JM, Willing J, Paul MJ. 2017. Adolescence and reward: making sense of neural and behavioral changes amid the chaos. *J Neurosci* **37:** 10855–10866. doi:10.1523/JNEUROSCI.1834-17.2017

Weinhard L, di Bartolomei G, Bolasco G, Machado P, Schieber NL, Neniskyte U, Exiga M, Vadisiute A, Raggioli A, Schertel A, et al. 2018. Microglia remodel synapses by presynaptic trogocytosis and spine head filopodia induction. *Nat Commun* **9:** 1228. doi:10.1038/s41467-018-03566-5

Wong FK, Bercsenyi K, Sreenivasan V, Portalés A, Fernández-Otero M, Marín O. 2018. Pyramidal cell regulation of interneuron survival sculpts cortical networks. *Nature* **557:** 668–673. doi:10.1038/s41586-018-0139-6

Xu X, Chini M, Bitzenhofer SH, Hanganu-Opatz IL. 2019. Transient knock-down of prefrontal DISC1 in immune-challenged mice causes abnormal long-range coupling and cognitive dysfunction throughout development. *J Neurosci* **39:** 1222–1235. doi:10.1523/JNEUROSCI.2170-18.2018

Yang XD, Liao XM, Uribe-Mariño A, Liu R, Xie XM, Jia J, Su YA, Li JT, Schmidt MV, Wang XD, et al. 2015. Stress during a critical postnatal period induces region-specific structural abnormalities and dysfunction of the prefrontal cortex via CRF1. *Neuropsychopharmacology* **40:** 1203–1215. doi:10.1038/npp.2014.304

Zhang YQ, Lin WP, Huang LP, Zhao B, Zhang CC, Yin DM. 2021. Dopamine D2 receptor regulates cortical synaptic pruning in rodents. *Nat Commun* **12:** 6444. doi:10.1038/s41467-021-26769-9

Zimmermann KS, Richardson R, Baker KD. 2019. Maturational changes in prefrontal and amygdala circuits in adolescence: implications for understanding fear inhibition during a vulnerable period of development. *Brain Sci* **9:** 65. doi:10.3390/brainsci9030065

Development and Evolution of Thalamocortical Connectivity

Zoltán Molnár[1] and Kenneth Y. Kwan[2]

[1]Department of Physiology, Anatomy and Genetics, Sherrington Building, University of Oxford, Oxford OX1 3PT, United Kingdom

[2]Michigan Neuroscience Institute (MNI), Department of Human Genetics, University of Michigan, Ann Arbor, Michigan 48109, USA

Correspondence: zoltan.molnar@dpag.ox.ac.uk; kykwan@umich.edu

Conscious perception in mammals depends on precise circuit connectivity between cerebral cortex and thalamus; the evolution and development of these structures are closely linked. During the wiring of reciprocal thalamus–cortex connections, thalamocortical axons (TCAs) first navigate forebrain regions that had undergone substantial evolutionary modifications. In particular, the organization of the pallial–subpallial boundary (PSPB) diverged significantly between mammals, reptiles, and birds. In mammals, transient cell populations in internal capsule and early corticofugal projections from subplate neurons closely interact with TCAs to guide pathfinding through ventral forebrain and PSPB crossing. Prior to thalamocortical axon arrival, cortical areas are initially patterned by intrinsic genetic factors. Thalamocortical axons then innervate cortex in a topographically organized manner to enable sensory input to refine cortical arealization. Here, we review the mechanisms underlying the guidance of thalamocortical axons across forebrain boundaries, the implications of PSPB evolution for thalamocortical axon pathfinding, and the reciprocal influence between thalamus and cortex during development.

Thalamus and six-layered isocortex form functional circuits that process sensory input, regulate brain state, and perform higher cognitive functions (Halassa 2022). The development and evolution of thalamus and isocortex are intimately linked. All cortical areas receive thalamocortical projections from specific thalamic nuclei (Jones 2007). Except for olfaction, all modalities of sensory information (e.g., vision, hearing, taste, touch, pain) are relayed to the isocortex through the thalamus (Sherman and Guillery 2005). All isocortical areas, in turn, project axons to the thalamus, thus enabling reciprocal communications via thalamo-cortico-thalamic circuits. We shall refer to isocortex as cortex in our review according to the definition of Northcutt and Kaas (1995). During development, bidirectional interactions between thalamus and cortex establish these connections; modality specific thalamic inputs play a role in the specification of cortical areas, whereas cortical connections to thalamus are vital for refining thalamic nuclei and hierarchy. Understanding thalamocortical circuits and their abnormalities is important to comprehend sensory perception, brain state control, and sleep, and has key implications for neurological and psychiatric conditions, such as epilepsy, synesthesia, and schizophrenia.

THE GUIDANCE OF EARLY THALAMIC PROJECTIONS THROUGH THE DIENCEPHALIC–TELENCEPHALIC BOUNDARY AND INTERNAL CAPSULE

In mouse, thalamic and cortical neurons are generated around the same embryonic periods (thalamus, embryonic day [E]12–15; cortex, E11–17; Shi et al. 2017). In macaque, thalamic neurons are born relatively earlier (thalamus E30–45, Spadory et al. 2022; visual cortex E45–102, Rakic 1974). The guidance of early thalamocortical connections relies on multiple mechanisms at different sectors along their trajectory to the cortex (Fig. 1A–C). Thalamic projections first descend through prethalamus and exit diencephalon, then extend along the primitive internal capsule, cross the boundaries between the pallidum/ventral pallium and pallium–subpallium, and grow through the cortical subplate zone (Molnár et al. 1998a,b, 2012). Early connectivity and migrating transient cell populations shape the trajectory of thalamocortical axons and assist their crossing of embryonic developmental boundaries. The diencephalic-telencephalic boundary (DTB) and pallial–subpallial boundary (PSPB) are the most vulnerable sectors of the thalamocortical pathway, with various guidance defects being reported in mutant mice with transcription factor or axon guidance molecule defects (summarized in Fig. 2; Molnár and Hannan 2000; López-Bendito and Molnár 2003; Molnár et al. 2012; Bandiera and Molnár 2022). In the earliest stages of thalamocortical development, guidepost cell populations in prethalamus, thalamic reticular nucleus, and perireticular nucleus form precocious projections to the dorsal thalamus that precede the early outgrowth of thalamocortical axons (DeCarlos and O Leary 1992; Métin and Godement 1996; Molnár et al. 1998a; Molnár and Cordery 1999a,b; Tuttle et al. 1999). The early scaffolds of these guidepost cells have been proposed to assist thalamocortical axon extension and entry to the telencephalon, whereas tangentially migrating neurons from the lateral ganglionic eminence (LGE) have been implicated for guidance through the internal capsule (Fig. 1A; Hanashima et al. 2006; López-Bendito et al. 2006).

Many transcription factors important for forebrain development are expressed along the trajectory of thalamocortical axons during their pathfinding. Genetic deletions of some of these transcription factors have revealed crucial insights into the many molecules that contribute to thalamocortical development (Fig. 2A). *Emx2*, for example, is expressed in a gradient manner in the cortex (caudal high; rostral low and medial high; lateral low) and near the DTB. *Emx2* deletion leads to misrouting of both corticothalamic and thalamocortical axons, the latter of which coincides with the displacement of internal capsule cells and disruption of their early thalamic projections (Fig. 2A′). This and other studies summarized in Figure 2A highlight key factors that guide thalamocortical axons through the ventral forebrain and toward the PSPB.

THALAMOCORTICAL AXON GUIDANCE THROUGH THE PSPB REQUIRES INTERACTIONS WITH THE EARLIEST GENERATED CORTICAL NEURONS

The crossing of the PSPB presents a major challenge to ascending thalamocortical projections. The PSPB is first established as a gene expression boundary (Smith-Fernandez et al. 1998; Molnár and Butler 2002). By the age at which thalamic axons approach the pallium (E13 in mouse), the PSPB has developed a striking radial glial fascicle and a dense pack of cells of the lateral cortical stream that each extend across the path of ascending thalamocortical axons (Chapouton et al. 2001; Carney et al. 2006, 2009). These are also features of the human PSPB (González-Arnay et al. 2017). It has been suggested that these characteristics make this region relatively nonpermissive to the passage of thalamic axons and that descending corticofugal axons from the cortex interact with ascending thalamic axons and assist them across this region (Molnár et al. 1998a,b; Molnár and Butler 2002).

Corticothalamic and thalamocortical projections grow toward the PSPB at approximately the same developmental stage (Fig. 1; E14.5 and E15.5). In rodents, the pioneer corticofugal axons arrive at the PSPB shortly before the thalamic axons. These initial corticofugal axons originate from cortical subplate neurons, among the earliest neurons born from the cortical germinal zones. It has been proposed that these descending corticofugal axons and the ascending thalamic axons cofasciculate to cross the nonpermissive PSPB (Molnár et al. 1998a,b). According to the "handshake hypothesis" (Fig. 1A′) the two fiber systems guide each other through this region of the embryonic forebrain, and ascending thalamic axons navigate to their appropriate cortical targets with help from reciprocal descending cortical axons (Molnár and Blakemore 1995; Molnár et al. 1998a,b). The nonpermissive environment near the PSPB causes the axons to fasciculate on each other in tight bundles, rather than spread out to extend as individual fibers (Fig. 1A). The breakdown of these interactions explains thalamocortical axon defects in some strains of mutant mice (Figs. 2 and 3; Hevner et al. 2002; Jones et al. 2002; López-Bendito et al. 2002; López-Bendito and Molnár 2003; Dwyer et al. 2011; Doyle et al. 2021).

The "handshake" hypothesis was based on in vivo observations in mouse and rat that demonstrated an intimate anatomical relationship between developing thalamic and early cortical axons (DeCarlos and O'Lerary 1992; Molnár et al. 1998a,b). In thalamus–cortex cocultures in vitro, cortical explants from different regions can nonspecifically accept innervations from any region of the thalamus (Molnár and Blakemore 1991, 1999). Therefore, it has been suggested that guidance from descending axons, which is present in vivo but disrupted in explant culture, might be necessary to achieve area-specific patterns of thalamocortical connectivity (Molnár and Blakemore 1991, 1999). The original formulation of the handshake hypothesis stated that "the descending and ascending axons each pioneer the pathway through their specific segment of the brain and, after a "handshake" near the internal capsule, each may guide the growth of the other over the PSPB and subsequently through the distal part of its trajectory" (Molnár and Blakemore 1991).

The "handshake" hypothesis only accounted for the encounter of the earliest corticofugal and thalamic projections in the internal capsule at the time of PSPB crossing. On a permissive surface of the culture dish, it has been demonstrated that thalamic and cortical growth cones often extended along axons of their own kind and after contacts between cortical and thalamic fibers, growth cones often collapsed and retracted (Bagnard et al. 2001). However, the relationship between early thalamic and corticofugal projections has not been explored in whole brain thalamocortical organotypic cultures, where the pathways would be surrounded by their natural environment. In contrast to the in vitro results, the in vivo cofasciculation patterns, which have been demonstrated in mouse and rat (Molnár et al. 1998a,b; Doyle et al. 2021), suggest that the developing striatum and cortical plate are each providing an extracellular environment that forces the thalamic projections to form large bundles rather than extend as individual fibers.

The use of constitutive deletion models has elucidated many key mechanisms of thalamocortical development (Fig. 2). The use of conditional deletion mutants has in addition directly tested the "handshake" hypothesis by blocking early corticofugal axonal development without disrupting the thalamus, subpallium, or PSPB (Fig. 3). Following conditional deletion of *Apc* from the cortical excitatory lineage using *Emx1-Cre* (Chen et al. 2012), thalamic axons still traverse the subpallium in topographic order, but they do not cross the PSPB (Fig. 3A). Interestingly, *Apc* conditional knockout (cKO) and wild-type cortex stimulate thalamic axon growth equally, suggesting that the inability of thalamic axons to cross the PSPB is unlikely to be explained by a loss of long-range chemoattraction. By providing evidence against alternative explanations and by showing that replacement of mutant cortex with control cortex restores both corticofugal efferents and thalamic axon crossing of the PSPB, the Chen et al. (2012) work provides strong evidence for the notion that cortical efferents are required for guidance of thalamocortical axons across the PSPB. This possibility is consistent with the observation that the earliest cohort of corticofugal fibers from subplate neurons can cross the PSPB before thalamic axons.

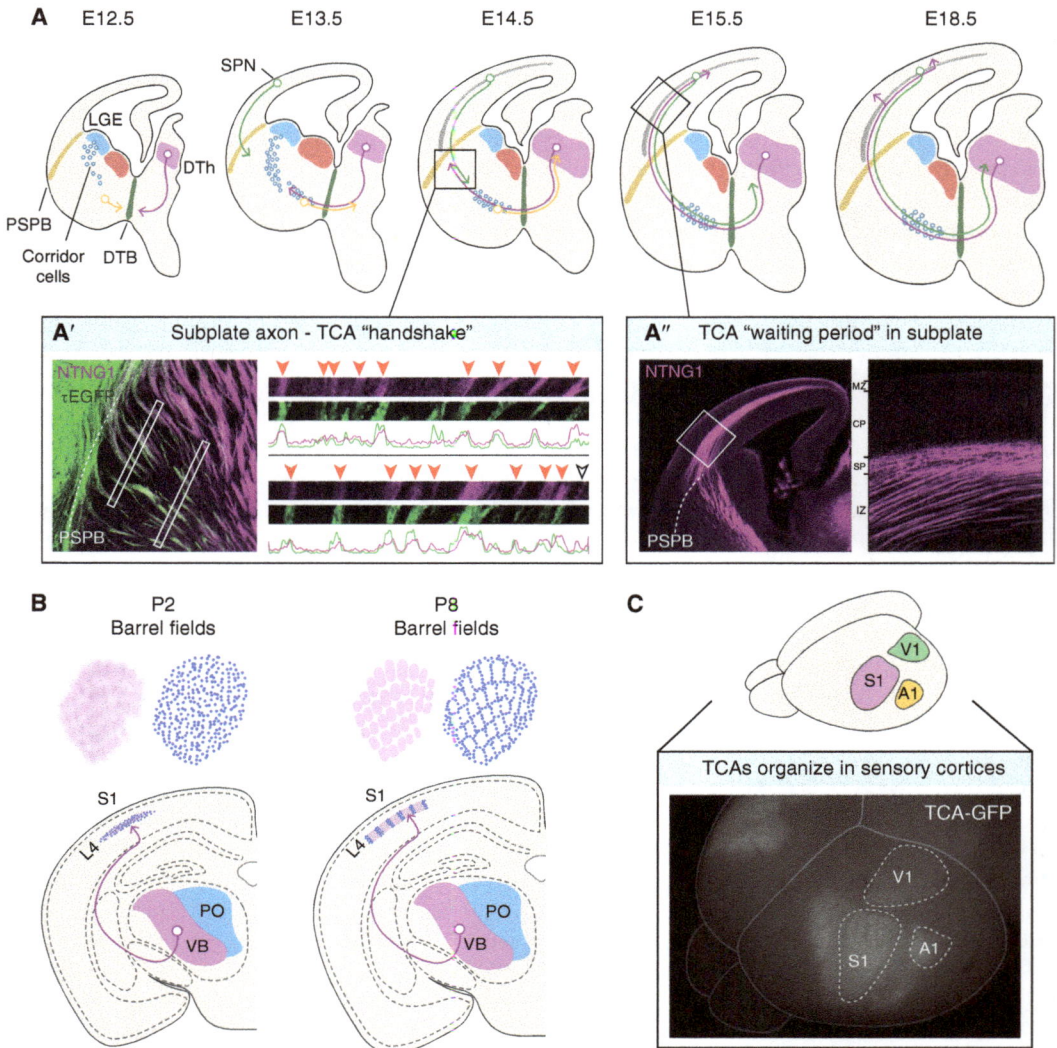

Figure 1. Development of thalamocortical connectivity in embryonic (*A*) and early postnatal mouse (*B,C*). (*A*) Schematic panels representing five stages (embryonic day [E]12.5–18.5) of the establishment of reciprocal connections between dorsal thalamus and cortex in mouse. At E12.5–14.5, guidepost cells (orange) in pre-thalamus and internal capsule have already developed projections to the dorsal thalamus (Molnár et al. 1998a; Braisted et al. 2000); these cells guide the thalamocortical axons (TCAs, magenta) through the diencephalic–telencephalic boundary (DTB, green) and toward internal capsule. Corridor cells (light blue) originate from the lateral ganglionic eminence (LGE) at E12.5 and migrate tangentially toward the diencephalon, where they form a permissive "corridor" for the thalamic projections to navigate through the internal capsule (López-Bendito et al. 2006). Perireticular cells have been proposed to regulate the entrance of TCAs into the subpallium (Métin and Godement 1996; Molnár et al. 1998a,b; Molnár and Cordery 1999a,b; Tuttle et al. 1999), whereas corridor cells orient the internal pathfinding of TCAs inside the medial ganglionic eminence (MGE) (López-Bendito et al. 2006). At E14.5 and E15.5, the early corticothalamic projections of subplate (SP) (layer 6b) neurons guide the thalamic projections across the pallial–subpallial boundary (PSPB, dark yellow) as originally proposed in the handshake hypothesis (Molnár and Blakemore 1995). (*Continued on following page.*)

In a more recent study, Doyle et al. (2021) first used a similar conditional strategy to delete the chromatin remodeler *Arid1a* from the *Emx1* lineage. In the *Emx1-Cre Arid1a* cKO, the ascending axons of thalamocortical neurons, which are not affected by cortical *Arid1a* deletion, are disrupted in their crossing of the PSPB and pathfinding into putative somatosensory cortex. Transcriptomic analysis revealed a selective disruption of subplate neuron gene expression following *Arid1a* deletion, suggesting a non-cell-autonomous contribution of subplate neurons to the axon guidance defects. Consistent with this, analysis of the PSPB reveals a reduction of corticofugal axons in the subpallium and a loss of subplate axon–thalamocortical axon cofasciculation (Fig. 3B). In the absence of "handshake" interaction from subplate neurons, thalamocortical axons become defasciculated, are unable to cross the PSPB along the normal trajectory and enter cortex via a narrow medial path. Together, these studies show that close cofasciculation with subplate neuron axons is required for thalamocortical axons to cross the PSPB, thus providing strong evidence supporting the "handshake" hypothesis (Molnár and Blakemore 1995).

THALAMOCORTICAL AXONS ACCUMULATE IN CORTICAL SUBPLATE PRIOR TO INVASION OF THE CORTICAL PLATE

Following PSPB crossing, thalamocortical axons enter the pallium at the level of the cortical subplate zone. Within this zone, subplate neurons play important guidance roles in addition to providing the earliest corticofugal axons for the "handshake" interactions. Upon entering the cortex, thalamocortical projections extend along the subplate, which is positioned at the boundary of cortical gray matter and developing white matter and has an impact on the targeting of thalamocortical projections. In the subplate, the ascending thalamocortical bundles break up into individual axons, suggesting that the environment in this compartment is more permissive compared to the PSPB. In the *Reeler* mouse and *Shaking Rat Kawasaki* mutants, however, the large fascicles continue through to the cortical plate (Molnár et al. 1998b; Higashi et al. 2005). The explanation lies in the relative position of subplate neurons. In wild-type mouse cortex, subplate cells form the deepest layer and reelin-expressing Cajal–Retzius cells form the most superficial marginal zone (Fig. 2B). Following preplate splitting, the cortical plate is formed in an inside-first, outside-last gradient, with earlier born cells located more deeply, and later born cells positioned more superficially (Kwan et al. 2012). In *Reeler* mouse and *Shaking Rat Kawasaki*, Cajal–Retzius cells do not express reelin, and the preplate fails to split into marginal zone and subplate. The cortical plate thus forms underneath this "superplate" in an outside-first, inside-last reverse gradient. In these mutants, thalamocortical projections target the superplate before they follow the same

---

Figure 1. (*Continued*) (*A′*) Cofasciculation of early corticofugal projections from SP in a Golli-t-eGFP mouse with thalamocortical projections. SP axons (green) and thalamocortical axons (NTNG1, magenta) cofasciculate while crossing the PSPB and the developing striatum. (*A″*) After entering cortex, thalamic projections do not immediately enter the dense cortical plate (CP) that contains still migrating cortical neurons; they "wait" or accumulate in the SP for several days prior to invading CP. *Upper* panels are modified from data in Molnár and Blakemore (1995) and Hanashima et al. (2006). *Lower* panels are based on data in Piñon et al. (2009) and Doyle et al. (2021). (*B*) At postnatal day (P)2, thalamocortical projections from ventrobasal (VB) complex of the dorsal thalamus (magenta) innervate the CP and begin to form periphery-related patterns in the whisker barrel field that is well delineated by P8. *Upper* schematic panels: Tangential sections through layer 4. Thalamocortical axons (magenta) and cortical neurons (blue dots) are distributed homogeneously in layer 4 at P2. By P8, the thalamocortical projections arrange into clear periphery-related patterns and most layer 4 neurons are in the septa of the cytoarchitectonic barrels. (*C*) By P8, genetically labeled TCA-GFP delineate primary somatosensory (S1), auditory (A1), and visual (V1) cortex on a wholemount brain. The periphery-related patterning of the thalamocortical axons within the barrel field of S1 is now established. S1 receives input from VB, A1 from medial geniculate nucleus (MGN), and V1 from dorsal lateral geniculate nucleus (dLGN). (SPN) Subplate neuron, (DTh), dorsal thalamus, (PO) posterior nucleus.

Cite this article as *Cold Spring Harb Perspect Biol* doi: 10.1101/cshperspect.a041503

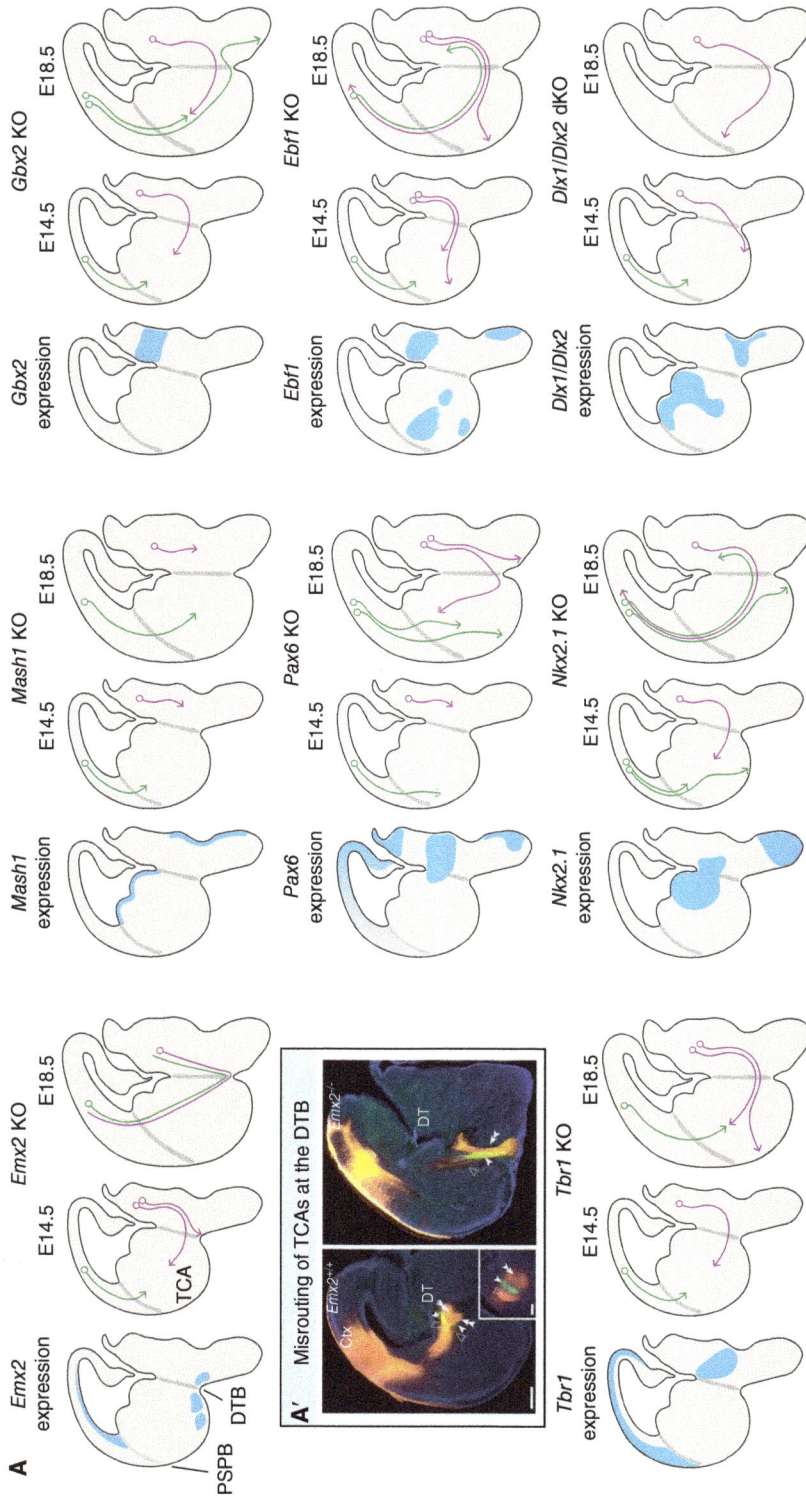

Figure 2. Schematic representations of the thalamocortical and corticothalamic guidance abnormalities that can be explained with the "handshake hypothesis" in selected mouse mutants (A) (figures are based on López-Bendito and Molnár 2003), and early generated subplate (SP) neurons govern the deployment of the thalamocortical projections to the cortical plate (B). In A, for each gene (*Emx2, Mash1, Gbx2, Pax6, Ebf1, Tbr1, Nkx2.1, Dlx1/Dlx2*) *left* panels depict the normal pattern of gene expression in wild-type (WT) mouse; *right* panels depict the thalamocortical and corticothalamic pathfinding abnormalities at E14.5 and E18.5, following gene deletion. The same schema and labeling conventions are used as in Figure 1. The original publications for these selected mutants: *Tbr1* (Hevner et al. 2002); *Gbx2* (Miyashita-Lin et al. 1999); *Mash1* (Tuttle et al. 1999); *Pax6* (Jones et al. 2002); *Nkx2.1* (Marín et al. 2002); *Ebf1* (Garel et al. 2002); *Dlx1/Dlx2* (Garel et al. 2002). (*Continued on following page.*)

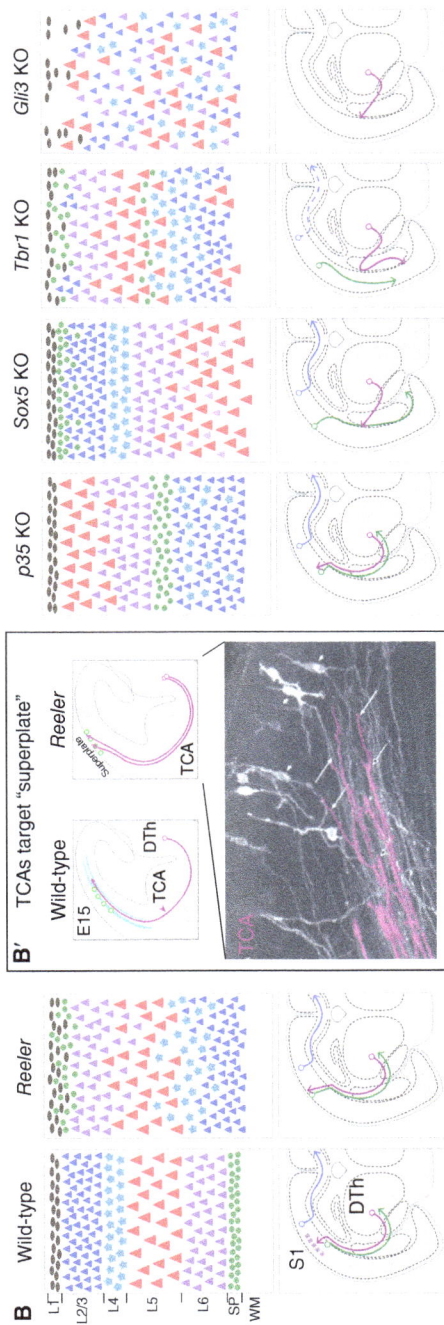

Figure 2. (*Continued*) (*A′*) The photomicrographs for *Emx2* depict the labeling pattern in the WT (*left*) and *Emx2*$^{-/-}$ (*right*) brains labeled from multiple points from the cerebral cortex. In WT, the anterogradely labeled corticothalamic and the retrogradely labeled thalamocortical projections crossed through the internal capsule and in the center of the diencephalic–telencephalic boundary, whereas in the *Emx2*$^{-/-}$ both sets of fibers descended into the ventral aspects of the diencephalic–telencephalic boundary, suggesting the dependence of both sets of fibers on the others. (DTh) Dorsal thalamus, (Ctx) cerebral cortex. (The original photomicrographs are reprinted from López-Bendito et al. 2002, with permission from John Wiley and Sons © 2002.) (*B*) Early generated SP neurons govern the deployment of the thalamocortical projections to the cortical plate. Changes in the positioning of the SP neurons (light green) in relation to the cortical plate neurons in various mutants (*Reeler*, *p35* knockout [KO], *Sox5* KO) and the effects of absent or reduced SP neurons on thalamocortical ingrowth (*Tbr1* KO, *Gli3* KO). The location of the SP neurons impacts the targeting of thalamocortical projections, and the outgrowth of cortical projections to the internal capsule and the corpus callosum (CC). Thalamocortical axons ([TCAs], magenta) target the SP, irrespective of its location within the cortex (*Reeler*, *p35* KO, *Sox5* KO). (*B′*) In *Reeler* mouse, the thalamocortical projections (magenta) cofasciculate with the early coticofugal projections as they cross the developing cortical plate and target superplate at E15. The example depicts anterogradely labeled thalamic axons (ending in growth cones, colored magenta) and back-labeled preplate neurons in the superplate (colored white). (Panels in *B′* are adapted from Molnár et al. 1998b.) In *p35* KO, the thalamocortical projections initially target the middle of the cortical plate, which is the location of the SP neurons in the mutant cortex. If SP neurons are reduced in numbers, then thalamocortical projections fail to enter the cortex (*Tbr1* and *Gli3* KO). (PSPB) Pallial–subpallial boundary, (DTB) diencephalic–telencephalic boundary. (Panels in *B* are modified from Hoerder-Suabedissen and Molnár 2015.)

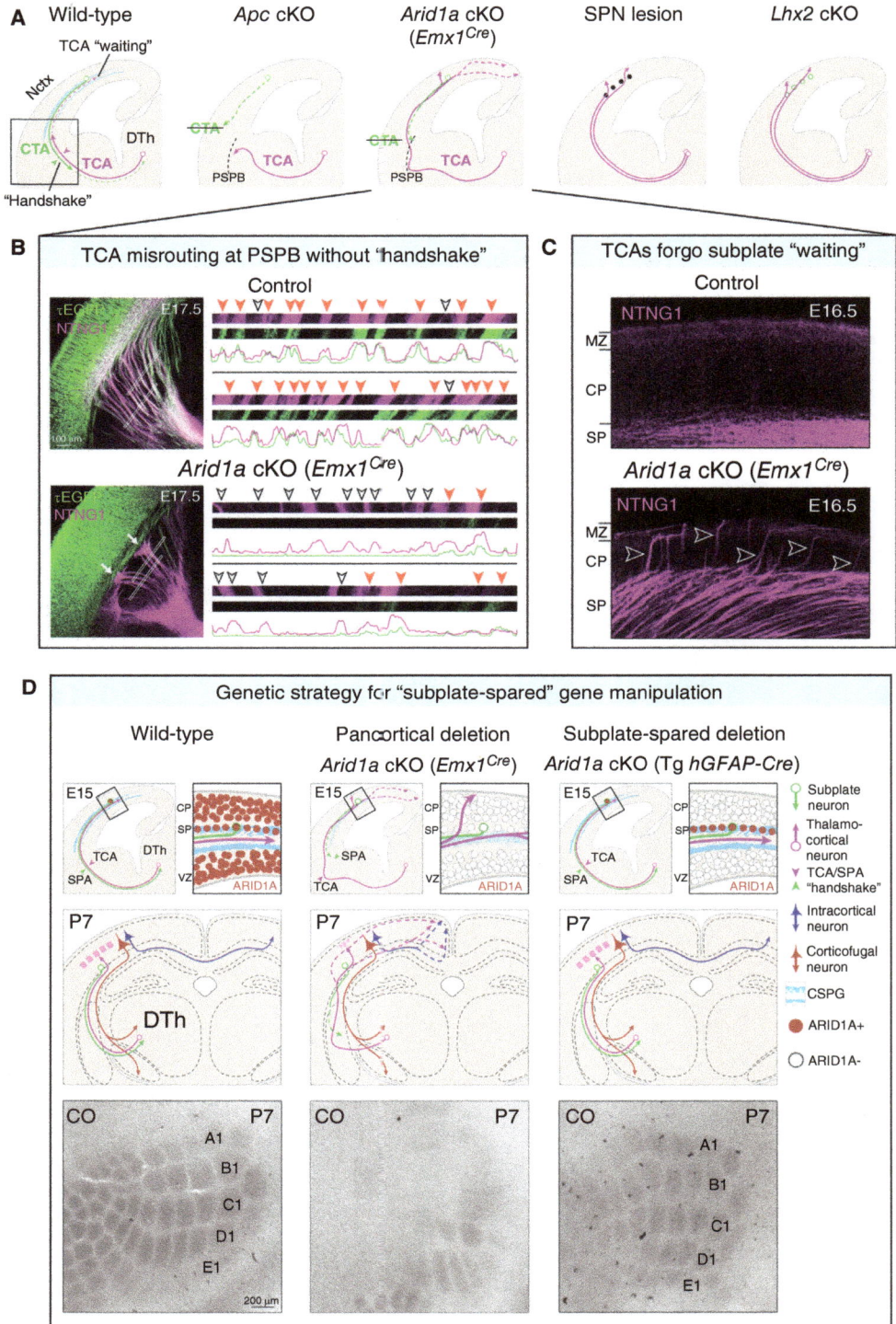

Figure 3. (*See following page for legend.*)

Cite this article as *Cold Spring Harb Perspect Biol* doi: 10.1101/cshperspect.a041503

developmental algorithm and invade the cortical plate (Fig. 2B).

In the cortex of *p35* knockout mice, the subplate is aberrantly positioned in the middle of the cortical plate, with deep-layer neurons being correctly positioned above the subplate and upper-layer neurons being mixed underneath (Rakić et al. 2006). In the *Sox5* knockout mouse cortex (Kwan et al. 2008), subplate cells are also located in the middle of the cortex, although they are more scattered (Fig. 2B). In *Reeler*, *p35* knockout, and *Sox5* knockout brains, subplate cells are displaced within the developing cortex but still form corticofugal projections that cross the PSPB. Thalamocortical axons are thus able to cross the PSPB and reach the cortex in *Reeler* and *p35* knockout brains, targeting the mispositioned subplate in these mutants. The absence or reduction of subplate neurons, in contrast, causes thalamocortical entry defects to the pallium. In the cortices of *Tbr1* knockout and *Gli3*$^{Xt/Pdn}$ mutant mice, the subplate is almost completely absent and subplate neuron markers are not detected. In both mutants, reelin-expressing Cajal–Retzius cells are also severely affected in their protein expression and cellular localization (Fig. 2B). With the loss of subplate neurons and the absence of their corticofugal fibers in the internal capsule, thalamocortical axons extend toward the PSPB but are unable to cross it (Fig. 2B).

These studies thus provide further support for the "handshake" hypothesis.

After extending along the subplate zone and nascent cortical gray matter, thalamocortical axons do not immediately invade the cortical plate; they accumulate in the subplate zone below, during what is known as the "waiting" period (Rakic 1974; Lund and Mustari 1977; Shatz and Luskin 1986). The initial deployment of thalamic projections and their accumulation in the subplate zone are orchestrated by molecular gradients and early activity patterns of subplate neurons, which form the earliest synapses in the cortex, including transient synapses with thalamocortical axons (Molnár et al. 2002, 2012). Subplate neurons have diverse origins (Hoerder-Suabedissen and Molnár 2015). Some are generated from apical radial glia progenitors through direct neurogenesis within the telencephalic germinal zone, but there is a considerable contribution from the Tbr2$^+$ intermediate progenitors (Vasistha et al. 2015). GABAergic subplate neurons can arrive by tangential migration from the rostral medial telencephalic wall similar to Cajal–Retzius neurons (Pedraza et al. 2014) or from the ganglionic eminences. The length of the waiting period differs among species; increased brain size is correlated with an extended waiting period (Kostovic and Rakic 1984, 1990; Catalano et al. 1991). In marsupials, the existence of the waiting period has been ques-

Figure 3. Subplate-dependent guidance of thalamocortical axons (TCAs). (*A*) Genetic ablation of early corticofugal projections from subplate neurons (*Apc* conditional knockout [cKO], *Arid1a* cKO, *Lhx2* cKO) or subplate lesion cause thalamocortical guidance defects. (*B*) TCAs (labeled by NTNG1 immunostaining, magenta) normally closely cofasciculate with subplate axons (labeled by the Golli-tEGFP transgene, green) at the pallial-subpallial boundary (PSPB). Following conditional deletion of *Arid1a* by *Emx1-Cre*, subplate axons are impaired in their innervation of the subpallium. In the absence of "handshake" interactions with subplate axons, TCAs fail to cross the PSPB (Doyle et al. 2021). (*C*) At E16.5, TCAs (magenta) normally accumulate in subplate during the "waiting" period. In *Emx1-Cre Arid1a* cKO, TCAs forgo the waiting period and prematurely invade the cortical plate (open arrowheads) (Doyle et al. 2021). (*D*) A genetic strategy for "subplate-spared" gene manipulation. Pioneered by Doyle et al. (2021), this approach compares *Emx1-Cre*, which mediates recombination from all cortical neural progenitor cells (including those that give rise to subplate neurons), with *hGFAP-Cre*, which mediates recombination from cortical progenitors about 2 d later (after subplate neurons have been generated). By comparing "pancortical deletion" (*Emx1-Cre*) with "subplate-spared deletion" (*hGFAP-Cre*), this method can interrogate gene necessity and sufficiency in subplate-mediated circuit development. Using this strategy, Doyle et al. (2021) found that *Emx1-Cre* deletion of *Arid1a* (*middle* panels) disrupts TCAs crossing of the PSPB and pathfinding into cortical barrels, whereas *hGFAP-Cre* "subplate-spared" deletion leads to none of these disruptions. Subplate neuron expression of *Arid1a* is therefore sufficient to support "handshake" with TCAs at the PSPB and the "waiting" period of TCA accumulation in the subplate. (SPN) Subplate neuron.

tioned, but it is indeed present; thalamic fibers accumulate in the location of subplate cells for a period proportional to the overall length of cortical development (Molnár et al. 1998c).

Early studies of fetal cat cortex have shown that excitotoxic ablation of subplate neurons leads thalamocortical axons to forgo the waiting period within the subplate zone, resulting in premature invasion of the cortical plate (Fig. 3A; Ghosh and Shatz 1993). Recent conditional deletion studies have further confirmed this role of subplate neurons. Following cortical deletion of *Arid1a* using *Emx1-Cre*, thalamocortical axons show a remarkable precocious penetration of the cortical plate as early as E15.5 (Fig. 3C; Doyle et al. 2021). Cortical deletion of *Lhx2* using *Emx1-Cre* causes a similar deficit (Fig. 3A; Pal et al. 2021). Together, lesion, constitutive deletion, and conditional deletion studies highlight the roles of subplate neurons not only in the guidance of thalamocortical axons across the PSPB, but also in the position, trajectory, and timing of thalamocortical axons during their innervation of the cortex.

GENETIC DISSECTION OF SUBPLATE NEURON FUNCTION IN TCA GUIDANCE USING SUBPLATE-SPARED GENE MANIPULATION

A key barrier to molecular study of subplate function, especially in the context of thalamocortical axon guidance, is the lack of specific genetic access to subplate neurons during critical embryonic stages of circuit formation. Although several published Cre lines have been shown to have varying degrees of subplate neuron specificity, Cre expression occurs too late in these lines for the study of circuit development in the fetal cortex (Hoerder-Suabedissen et al. 2013, 2018). In Doyle et al. (2021), the authors describe a novel genetic strategy to target subplate neurons They find that, whereas *Emx1-Cre* mediates recombination from all cortical neural progenitor cells, including those that give rise to subplate neurons, an alternative Cre line, *hGFAP-Cre*, mediates recombination from cortical progenitors about 2 days later; after subplate neurons have been gen-

erated. This approach thus leverages the comparison of "pancortical deletion" (*Emx1-Cre*) with "subplate-spared deletion" (*hGFAP-Cre*) to enable interrogation of gene necessity and sufficiency in subplate neuron-mediated circuit development (Fig. 3D). Using this strategy, they investigated the function of *Arid1a* in subplate neurons (Doyle et al. 2021). Whereas *Emx1-Cre* pancortical deletion of *Arid1a* disrupts thalamocortical axons in their crossing of the PSPB, pathfinding into cortex, and ultimately innervation of putative somatosensory cortex and later induce the formation of whisker barrels, *hGFAP-Cre* subplate-spared deletion of *Arid1a* leads to none of these disruptions. Thus, subplate neuron expression of *Arid1a* is sufficient to support "handshake" with thalamocortical axons at the PSPB, guidance of thalamocortical axons into the cortex, and the "waiting" period of thalamocortical axon accumulation in the subplate. This study suggests that these developmental functions are specific to subplate neurons; other cortical neurons are unable to compensate. Thus, subplate neurons have a special guidance role for the thalamocortical projections through the PSPB and into cortex, as originally suggested in the "handshake" hypothesis (Molnár and Blakemore 1995). In addition to thalamocortical axons, Doyle et al. (2021) find that callosal axon development, which is defective in the *Emx1-Cre* cKO, is also normal in the *hGFAP-Cre* cKO, suggesting a previously unappreciated role of subplate neurons in the guidance of callosal axons to the contralateral hemisphere.

THE DEVELOPMENT OF THE PALLIAL–SUBPALLIAL BOUNDARY IN MAMMALS REFLECT CHANGES THAT OCCURRED DURING EVOLUTION

The organization of the PSPB diverged significantly during evolution; these changes impacted the formation of reciprocal connectivity between thalamus and cortex. The PSPB has been correlated to the site of origin of the dorsal ventricular ridge (DVR), a structure that protrudes into the lateral ventricle in sauropsids (reptiles and birds) (Fig. 4; Fernandez et al. 1998; Puelles et al. 2016a, b,c; García-Moreno and Molnár 2020). In mam-

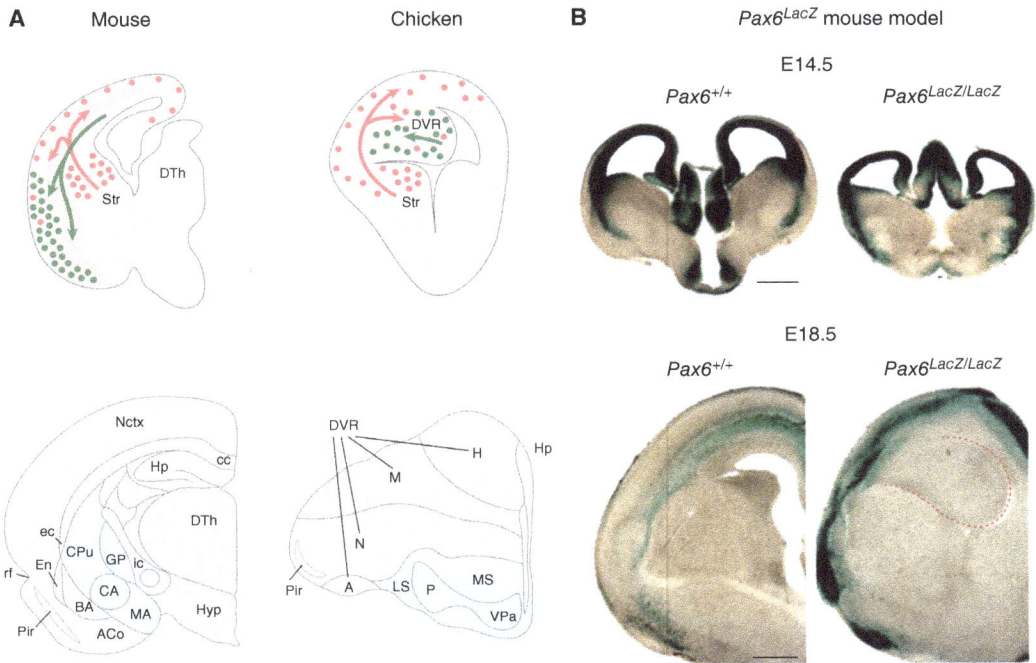

Figure 4. Comparisons of telencephalic organizations of developing (*upper* panels in *A*) and adult (*lower* panels in *A*) mouse and chicken forebrains and the Pax 6 conditional knockout (cKO) in *B*. (Panels in *A* based on Molnár and Butler 2002 and from Montiel et al. 2016. Panels in *B* reprinted from Jones et al. 2002, with permission from Company of Biologists © 2002.) (*A*) The mammalian brain organization is different, with an apparent dominance of the six layered dorsal cortex and the ventrolaterally migrated ventral pallial structures (claustrum, endopiriform nucleus, lateral amygdala, and lateral cortex (green arrows). The avian brain contains a ball-shaped structure that protrudes into the lateral ventricle, the wulst or dorsal ventricular ridge (DVR). The schematic diagrams depict the postulated homologies between avian and mammalian brains. Colors represent proposed homologies, based on current anatomical, developmental, and transcriptomic data. In mammals, the Pax6 territory is indicated in pink. Inhibitory, GABAergic neuronal precursors (red dots) originate from subpallial sources and migrate tangentially into the pallium in both mammals and sauropsids (red arrows). Excitatory, pyramidal-type neuronal precursors (green dots) of the lateral migratory stream traverse the Pax6 territory to reach lateral pallial regions in mammals but remain in situ within the DVR in birds. Despite its extensive target area, the lateral migratory stream is a subset of the radially migrating pallial neurons that is perpendicular to the trajectories of the growing thalamocortical connections during development. The tangentially migrating GABAergic cells have a similar origin from Dlx gene expression territories from the medial ganglionic eminence (origin of red arrows, anlage of striatum) and they migrate dorsal to the cortex in mammal and DVR and hyperpallium (dorsal cortex) in birds and reptiles. The panels demonstrate the alterations in the Pax6-lacZ KO mice that resembles the avian/reptilian morphology, with a large ball of cells protruding into the lateral ventricle. (*B*) Coronal sections through the forebrains and the Pax 6 cKO at E14.5 (*upper* panels) and E18.5 (*lower* panels). The LacZ expression (blue coloring) marks the stripe of transient Pax6-positive cells in the ventricular zone near the corticostriatal junction toward the ventrolateral telencephalon in Pax6–LacZ (mouse). The dark blue β-gal staining reflects Pax6 promoter activity. Pax6/LacZ$^{+/−}$ mouse (Jones et al. 2002) with normal lateral pallial sector. The dark b-gal staining reflects Pax6 promoter activity, which in +/− correlates with the Pax6 gene expression pattern. The Pax6 expression is gradually reduced but continues to be present in the ventral telencephalon. Pax6$^{−/−}$ mouse with a large, aberrant cell mass in the same position as the ADVR of sauropsids and severe malformation of lateral neocortex (LNC), the claustrum–endopiriform nucleus formation, and part of the amygdala (Molnár and Butler 2002). (DTh) Dorsal thalamus.

mals, the DVR is absent, but some components of this lineage migrated ventrolaterally to form the claustrum, endopiriform nucleus, and lateral amygdala along the PSPB (Bruguier et al. 2020). This ventrolateral migration thus adds to the complexity of the PSPB, presenting an especial challenge to the pathfinding of thalamocortical and corticothalamic projections in mammals. Interestingly, deletion of the *Pax6* gene in mouse can produce morphological alterations at the PSPB that result in an accumulation of cells aberrantly protruding into the lateral ventricle in a fashion that resembles the sauropsid DVR (Jones et al. 2002; Molnár and Butler 2002). This finding is thus consistent with the possibility that changes in genes that function in early forebrain development, such as *Pax6*, may have contributed to the evolutionary differences in these migratory streams.

The establishment of pallial subdomains near the PSPB is shaped by diffusible morphogenic factors and, importantly, by tangential migrations of cells from one brain compartment to another, a process that differs among taxa. Adult vertebrate brains reflect the differential impact of tangential migration during development (García-Moreno and Molnár 2020). In all vertebrate species that have been studied, including primates, tangentially migrating subpallial GABAergic neurons are present and their migration through the pallium is conserved throughout the vertebrate radiation (Cobos et al. 2001; Métin et al. 2007; Carrera et al. 2008; Moreno et al. 2008, 2010; García-Moreno et al. 2018). By contrast, glutamatergic intrapallial tangential migrations evolved independently (García-Moreno et al. 2018). We hypothesize that in the early evolution of mammals, genetic changes enabled pallial tangential migrations (Fig. 4) likely as a byproduct of earlier developing divergences such as pallial patterning, signaling, or proliferation dynamics (Molnár 2011; Garcia-Moreno and Molnár 2020); these migrations substantively impacted neocortical evolution.

The existence of tangential migratory streams has been investigated in embryonic chick brain (García-Moreno et al. 2018). In vivo full-lineage tracing of the chick cortical hem and ventral pallium reveals a complete lack of tangential migra-

tion of glutamatergic neurons toward the dorsal pallium; the cells originating in these subdomains do not migrate tangentially and stayed within the radial domains of their origins (i.e., cortical hem-generated cells remain in the hippocampal area and ventral pallium-generated cells remain in the nidopallium). Without tangential arrivals from other pallial sources, the avian dorsal pallium (also called hyperpallium) lacks the corticogenic instructions of mammalian Cajal–Retzius cells, which provide Reelin signaling, and subplate neurons, which guide axon pathfinding (García-Moreno et al. 2018). These crucial components of the preplate are the requisites of the radial inside-first, outside-last pattern of cortical development in mammals. Thus, this evolutionary difference in forebrain migratory streams may be a key contributor to the laminar and columnar organization of the mammalian neocortex versus the nuclear arrangement of the avian pallium. Gene expression and lineage data suggests a deep conservation of the radially exclusive development of lateral ventral pallia in amniotes (Puelles et al. 2016a,b,c; Garcia-Moreno and Molnár 2020). The mammalian novelty of these early and later migratory patterns of glutamatergic neurons is in contrast with the highly conserved tangential migratory patterns of GABAergic neurons in sauropsids (Fig. 4; Cobos et al. 2001; Métin et al. 2007; Carrera et al. 2008; Moreno et al. 2008; Rueda-Alaña et al. 2018; Garcia-Moreno and Molnár 2020).

Molnár and Butler (2002) postulated the collopallial field hypothesis, which examined the differences in the claustro-amygdalar formation between birds, reptiles, and mammals. The hypothesis held that in mammals, the thalamo-recipient collopallium differentiates into deep (claustro-amygdalar) and superficial (neocortical) components, whereas in sauropsids, and perhaps in platypus, this split may not occur. The original collopallial field hypothesis was based on a handful of gene expression patterns (Molnár and Butler 2002), and it was formulated before the modern tetrapartite pallium model that is now commonly used. The modern concept of VPall (olfactory), LPall (claustro-insular), and DPall (neocortical) pallial sectors (Puelles 2014, 2017) seems to have rendered obsolete these notions articulated in Puelles et al. (2000) and Molnár and Butler (2002) (Fig. 4). Altogether, the general presence of the claustrum in all extant mammalian taxa (excluding the uncertainty in monotremes), together with shared histochemical, molecular, cytoarchitectural, and hodological features, suggests that the claustrum was already present in the now extinct pan-mammalian common ancestor (Montiel et al. 2011; Puelles 2017; Suárez et al. 2018; Bruguier et al. 2020). We need much more detailed comparative lineage-tracing studies in this part of the brain to settle these issues. There has been huge progress in the understanding of the single-cell transcriptomic similarities (Tosches et al. 2018) and physiological properties (Norimoto et al. 2020; Fenk et al. 2023) of the "claustrum" of turtles and lizards, although relying on single-cell transcriptomic data and physiological similarities in the adult and in the absence of developmental cell lineage tracing might not be definitive to define homologies according to the criteria used by developmental evolutionary biologists. Species-dependent changes in neuroanatomical arrangements reflect evolutionary changes in development, including those of the PSPB at the time of thalamocortical pathfinding. The developing thalamocortical projections in the turtle embryo avoid the core of the developing DVR and negotiate their way to the dorsal cortex on an external path; whereas in mammals the thalamic projections arrive perpendicular to the lateral stream and the PSPB and they traverse through this region with the assistance of the early subplate projections (Cordery and Molnár 1999; Molnár and Cordery 1999a,b). It would be exciting to obtain data on the development of the thalamic nuclei that receive direct sensory input from the periphery (first-order thalamic nuclei) and contrast their development with the nuclei that only receive indirect sensory input via other structures such as superior colliculus or cortex (higher-order thalamic nuclei) in reptiles, birds, and mammals. The extensive evolutionary rearrangements within the PSPB explain why so many developmental defects are associated with abnormalities and thalamocortical and corticothalamic guidance defects (Fig. 2). The evolution of our brain is the evolution of its development (Horváth et al. 2022).

THALAMIC REGULATION OF CORTICAL DEVELOPMENT

Cortical areas are first patterned by intrinsic genetic factors prior to the arrival of thalamocortical axons (Rakic 1988). These factors include the morphogen FGF8 and transcription factors expressed in gradients across the cortex (PAX6, COUP-TFI, EMX2, and SP8) (Garel et al. 2003; Rash and Grove 2006; O'Leary and Sahara 2008). The basic pattern of neurogenesis, neuronal migration, and maturation of the cerebral cortex is undisturbed in the absence of thalamic input as it was demonstrated in the conditional Celsr3 mutant mice (Zhou et al. 2008, 2009, 2010). Although thalamocortical projections do not play an instructive role in the initial basic steps of cortical arealization, they do contribute to many aspects of cortical development. Much of thalamocortical axons' (TCAs') influence is driven by thalamocortical activity. Conditional codeletion of vesicular glutamate transporters *Vglut1* and *Vglut2* from somatosensory thalamus using *Sert-Cre* leads to a complete loss of thalamocortical neurotransmission (Fig. 5A; Li et al. 2013). This causes an absence of TCA clustering into periphery related patterns in barrel cortex, a lack of cytoarchitectonic barrel formation, and an aberrant shift in L4 neurons from stellate to pyramidal morphology. Interestingly, codeletion of synaptic vesicle fusion protein genes *Rim1* and *Rim2* using *Sert-Cre* leads to a partial loss of thalamocortical neurotransmission (Fig. 5B; Narboux-Nême et al. 2012). With this partial loss, L4 neurons invade barrel centers, recapitulating the barrelless phenotype. However, the topography of TCAs is maintained; they cluster correctly into periphery related patterns in barrel cortex.

The reciprocal connectivity between cortex and thalamus enables synchronized electrophysiological patterns across these structures from very early developmental stages. The thalamus is already active prenatally; it is characterized by spontaneous calcium waves that mediate internuclear thalamic communication (Moreno-Juan et al. 2017). Prenatal thalamic activity and spontaneous waves can be blocked by overexpression of the inwardly rectifying potassium channel

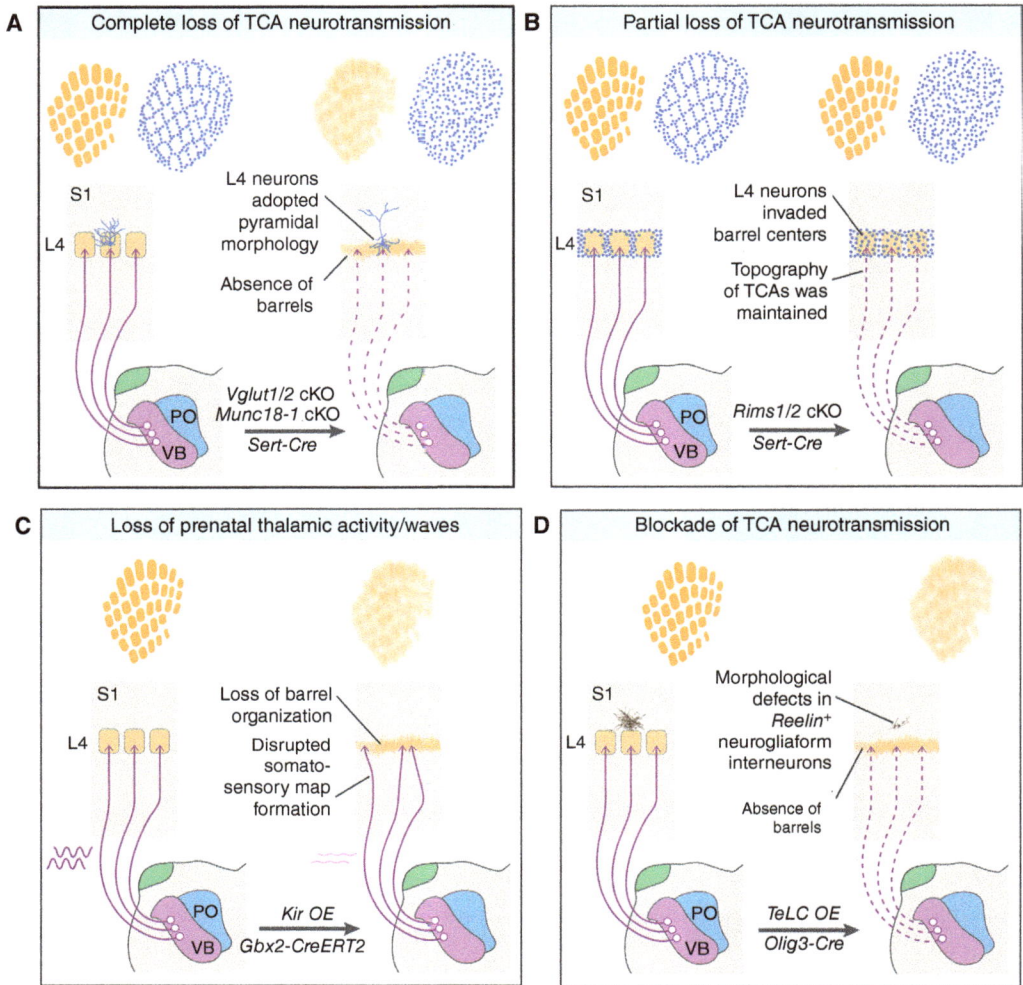

Figure 5. Thalamic regulation of cortical development. (*A–D*) Thalamocortical axons (TCAs) can influence cortical development through its neuronal activity. Changes to thalamocortical activity can disrupt TCAs clustering into barrels, cytoarchitectonic barrel formation, sensory map development, and L4 neuron and interneuron morphologies. (*E–G*) Loss of innervation or sensory activity from primary thalamic nuclei can induce cross hierarchical plasticity in cortex, leading primary cortical areas to adopt the gene expression and connectivity patterns of higher-order cortical areas. (*H*) Early loss of thalamocortical axons can alter cortical neurogenesis. (dLGN) Dorsal lateral geniculate nucleus, (VB) ventrobasal, (PO) posterior nucleus. (*Continued*)

gene *Kir2.1* using *Gbx2-CreERT2* (Fig. 5C; Moreno-Juan et al. 2017). Systemic application of tamoxifen in this model during early development leads to loss of barrel organization and disrupts somatosensory map formation (Fig. 5C; Anton-Bolanos et al. 2018). Interestingly, targeted application of tamoxifen to selectively block spontaneous waves in auditory thalamus leads to altered wave patterns in the ventrobasal (VB) complex and subsequently causes an enlargement of the barrel fields in S1 (Moreno-Juan et al. 2017). Furthermore, optical recording of voltage-sensitive dyes and current source density analysis have demonstrated that thalamocortical projections can deliver sustained depolarization patterns to cortex after thalamic stimulation in

Cite this article as *Cold Spring Harb Perspect Biol* doi: 10.1101/cshperspect.a041503

E Selective ablation of geniculate TCAs

V1 V2
L4
L4 neurons in V1 adopted V2-like gene expression
← V2
dLGN
dLGN
Coup-tf1 cKO
Ror-Cre

F Neonatal unilateral enucleation

V1 V2
L5
L5 neurons in V1 projected to LGN, adopting connectivity of a higher-order area
← V2
dLGN
dLGN
Neonatal unilateral enucleation

G Selective ablation of VB

S1 S2
L4
L4 neurons in S1 adopted S2 gene expression
L4 neurons in S1 showed multimodal activation
← S2
PO
VB
PO
VB
DTA OE
Sert-Cre

H Loss of TCAs reaching cortex

L1
L2/3
L4
L5/6
Reduced upper layer neurogenesis
PO
VB
PO
VB
Gbx2 cKO
Vgf cKO
Olig3-Cre

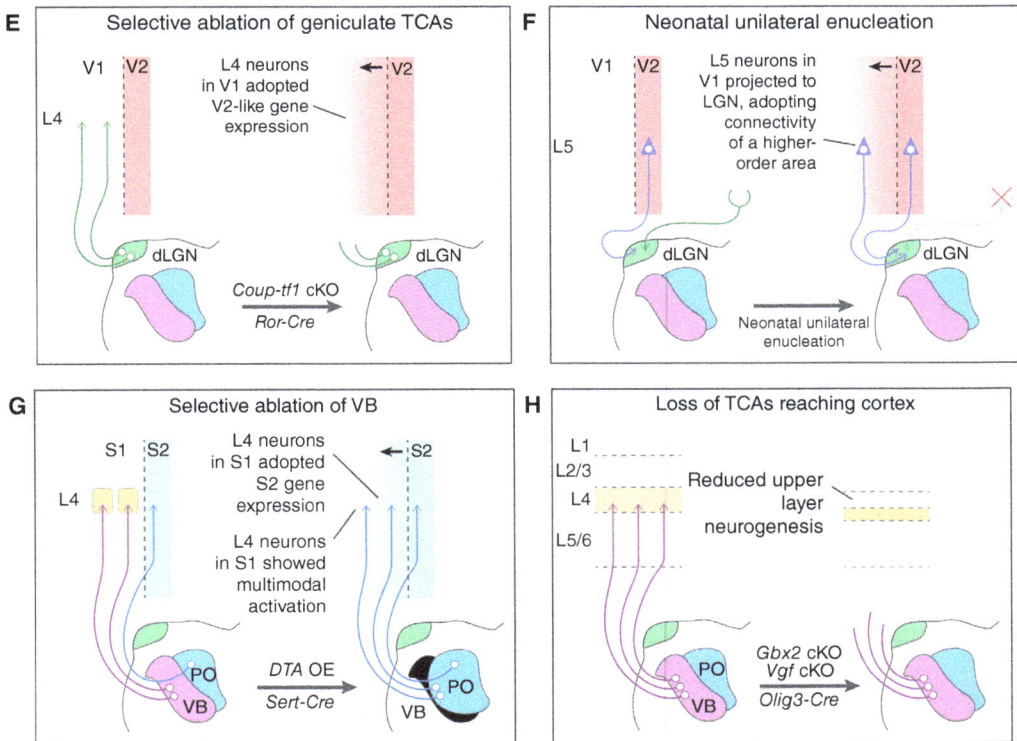

Figure 5. (*Continued*)

embryonic brain slices (Higashi et al. 2002, 2005; Molnár et al. 2003a,b). Together, these experiments show that spontaneous wave-like activity in prenatal thalamus organizes the architecture and territory of the somatosensory map in developing cortex.

Interestingly, thalamocortical activity is further required for the development of inhibitory interneurons. Blockade of TCA neurotransmission by overexpression of the tetanus toxin light chain gene *TeLC* using *Olig3-Cre*, which disrupts development of periphery related patterning of TCAs in barrel cortex and cytoarchitectonic barrels, leads to remarkable morphological defects in Reelin[+] neurogliaform interneurons (Fig. 5D; DeMarco García et al. 2015). Together, these studies demonstrate the importance of thalamocortical neurotransmission in the formation of periphery-related thalamocortical patterning in primary sensory cortex, the cytoarchitectonic formation of layer 4 clusters that form cortical barrels, and the morphological acquisition of excitatory and inhibitor cortical neurons.

Thalamocortical connectivity is dependent on the hierarchy of the given nucleus. First-order thalamic nuclei (e.g., VB) receive direct input from the sensory periphery and relay sensory information to primary cortical areas of the appropriate modality. Higher-order thalamic nuclei (e.g., posterior nucleus [PO]) receive input from higher-order cortical areas and relay back to cortex, thus forming transthalamic (cortico-thalamo-cortical) pathways (Casas-Torremocha et al. 2022). Lineage analysis has revealed that the neurons of first- and higher-order thalamic nuclei are generated from distinct pools of progenitors (Shi et al. 2017). Although thalamic projections largely synapse on L4 neurons, those arising from distinct thalamic nuclei show differences in target layer preference (Sherman and Guillery 2005).

Corticothalamic connectivity is also hierarchical and layer dependent. In mouse, neurons from three cortical layers (L5, L6a, and remnants of subplate neurons in L6b) contribute to corticothalamic projections (Grant et al. 2012; Hoerder-Suabedissen et al. 2018; Molnár 2019). Layer-specific transgenic reporter lines have enabled mapping of layer-dependent corticothalamic targets. L5 and some populations of L6b neurons selectively target higher-order thalamic nuclei, whereas L6a and some populations of L6b cells target both first and higher-order thalamic nuclei (Grant et al. 2012; Hoerder-Suabedissen et al. 2018). The choreography of cortical innervation of thalamus follows a specific algorithm and is sensitive to alterations in sensory input; the allocations of L6 and L5 projections can change after early sensory deprivations (Grant et al. 2016). Corticofugal projections reach thalamus at specific developmental stages; those from subplate neurons arrive first followed by L6a and L5 (Grant et al. 2012). However, the first corticofugal projections to enter dorsal thalamus originate from L5 neurons (Clascá et al. 1995). Corticofugal axons from subplate and L6a neuron reach the proximity of the thalamus early but pause near the reticular nucleus before entering dorsal thalamus (Molnár and Cordery 1999a,b; Grant et al. 2012). Interestingly, corticothalamic axons extend in intermediate zone and internal capsule in an organized fashion, but substantially rearrange near the thalamic reticular nucleus (Mitrofanis and Guillery 1993; Adams et al. 1997; Molnár 1998; Grant et al. 2012). Like the accumulation of thalamocortical projections within the subplate zone, corticofugal projections from L6a and subplate accumulate during an equivalent of the "waiting" period. Indeed, some of these L6a and subplate axons do not enter their targeted thalamic nuclei until after the first postnatal week in mouse (e.g., L6 TCAs enter dorsal lateral geniculate nucleus (dLGN) around P10) (Grant et al. 2012, 2016).

First-order thalamic nuclei innervate primary cortical areas, and this innervation contributes to the refinement of areal identity. Cortical arealization can be disrupted if sensory input is altered or if thalamic input is rewired. Conditional deletion of the transcription factor gene *Coup-tf1* using *Rora-Cre* leads to selective loss of thalamocortical axons from the LGN (Fig. 5E; Chou et al. 2013). In the absence of primary visual input from thalamus, L4 neurons in V1 adopt higher-order visual cortex gene expression. Interestingly, sensory activity not only influences the hierarchy of cortical area, but also the corticofugal connections that are hierarchy dependent. This has been demonstrated for the dLGN in the mouse. The dLGN normally receives corticothalamic projections from L6 neurons in V1, and L5 neurons of V1 only form connections with higher-order thalamic nuclei, such as pulvinar. After neonatal enucleation, however, L5 projections from V1 aberrantly innervate dLGN and form synapses that are maintained into adulthood (Fig. 5F; Giasafaki et al. 2022). Reducing early activity in the retina has similar effects (Grant et al. 2016). Early sensory deprivation can therefore substantially change the relationship between cortical area hierarchy and layer-dependent corticothalamic innervation (Frangeul et al. 2016; Grant et al. 2016; Giasafaki et al. 2022). This form of plasticity where first-order thalamic nuclei, in the absence of peripheral sensory input, acquire connectivity reminiscent of higher-order thalamic nuclei has been termed cross-hierarchical corticothalamic plasticity (Fig. 5F; Grant et al. 2016). This cross-hierarchical plasticity is also associated with gene expression changes that make deafferented first-order thalamic nuclei more like higher-order thalamic nuclei (Frangeul et al. 2016; Giasafaki et al. 2022).

Like V1, S1 and its hierarchy are similarly dependent on VB innervation. Selective ablation of VB thalamocortical neurons using *Rosa-DTA* driven by *Sert-Cre* causes L4 neurons in S1 to adopt S2 gene expression (Fig. 5G; Pouchelon et al. 2014). The prospective S1 in this model is innervated by thalamocortical terminals from the higher-order thalamic nucleus PO and shows multimodal sensory activation normally only present in higher-order cortical areas. Together, these studies demonstrate that first-order sensory thalamic nuclei play an important role in specifying primary sensory cortical areas. In the absence of thalamocortical innervation from sensory thalamus, primary cortical areas undergo a cross-hierarchical remodeling to adopt higher-order areal fate. Both cross-modal and cross-hierarchical plasticities have an impact on the information processing in thalamo-cortico-thalamic networks. However, the exact impact of the cross-hierarchical plasticity is still not understood.

Although the basic pattern of neurogenesis, neuronal migration, and maturation of the cerebral cortex is undisturbed in the absence of thalamic input (Zhou et al. 2008, 2009, 2010), thalamic projections can exert area-specific influences on the developing cortex. In addition to activity, thalamocortical axons can provide other factors that influence cortical development. TCAs reach cortex prior to the completion of cortical neurogenesis and neuronal migration and are therefore poised to influence these processes. At around E14.5, TCAs release a bFGF-like diffusible factor that increases proliferative divisions during the genesis of upper layer cortical neurons (Dehay et al. 2001); TCAs can therefore influence cortical lamination. Gerstmann et al. (2015) demonstrated that thalamic afferents influence cortical progenitors via ephrin A5-EphA4 interactions. More recently, conditional deletion of *Gbx2* from the thalamus using *Olig3-Cre* leads to loss of TCAs reaching the cortex (Fig. 5H; Monko et al. 2022). This thalamus-specific manipulation causes a significant reduction in upper layer neurogenesis. Importantly, conditional deletion of *Vgf* using the same *Olig3-Cre* approach results in similar phenotypic changes, implicating *Vgf* as the thalamus-derived diffusible signal carried by TCAs that exerts influence on upper layer neurogenesis (Monko et al. 2022; Sato et al. 2022). Together, these experiments demonstrate that extrinsic cues from TCAs play an important role in region-specific cortical neurogenesis and laminar development.

CORTICAL REGULATION OF TCA TARGETING

The precise topography by which thalamocortical axons innervate cortical areas is established progressively during development. As discussed above, the earliest thalamocortical interactions

and entry into cortical plate are orchestrated by the subplate, where thalamocortical axons form transient synapses (Allendoerfer and Shatz 1994; Kanold and Luhmann 2010). The transient circuits between subplate neurons, thalamic afferents, and L4 neurons are now widely recognized as constituting a key mechanism for early circuit formation (Kanold and Luhmann 2010; Molnár et al. 2020). Subplate neurons integrate into cortical circuits in an age- and area-dependent fashion (Piñon et al. 2009; Hoerder-Suabedissen and Molnár 2012; Tolner et al. 2012; Viswanathan et al. 2012) and their transient circuits match spontaneous and sensory-driven thalamocortical activity during development (Molnár et al. 2020). After the "waiting" period, thalamocortical axons invade cortical plate and target L4. The recognition of target L4 neurons and maturation of these connections rely on multiple mechanisms, including a layer-specific stop signal independent of regulated transmitter release from thalamocortical projections (Molnár and Blakemore 1991; Blakey et al. 2012; Molnár et al. 2012; Yamamoto and López-Bendito 2012).

Modality-specific thalamocortical axons from VB, LGN, and MGN, respectively, target primary somatosensory (S1), visual (V1), and auditory (A1) cortex with exquisite precision. Genetic manipulations that alter or shift cortical areal identity have provided important insights into how this specificity is achieved. Ectopic caudal expression of the morphogen fibroblast growth factor 8 (FGF8) by in utero electroporation leads to a duplication of both S1 and V1 in the cortex (Fig. 6A; Shimogori and Grove 2005, Assimacopoulos et al. 2012). Remarkably, the resulting duplicated V1s are both innervated by LGN TCAs, and both respond to visual stimuli. Similarly, both duplicated S1s are innervated by VB TCAs and show duplicated somatotopic input, possibly from bifurcated TCAs. Thus, thalamocortical axons innervate cortical regions with a spatial specificity that is not based on mere physical coordinates; rather, this specificity is dependent on genetic specification of the cortex and can be rerouted following shifts in cortical patterning. The capacity for thalamocortical axons to accommodate changes in cortical arealization is remarkable. When the transcription factor

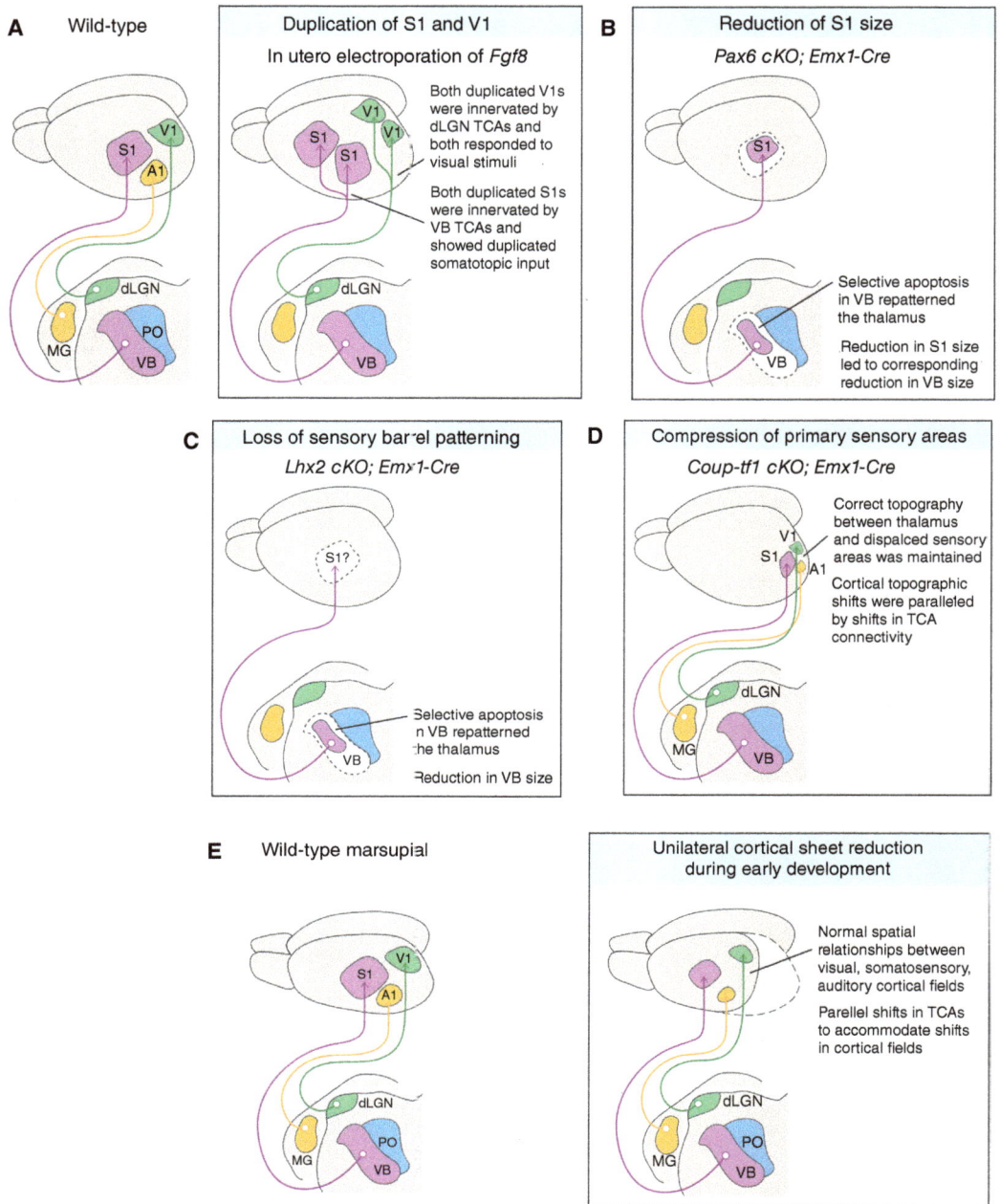

Figure 6. Cortical regulation of thalamocortical axon targeting. (*A–D*) Manipulations in mice that change the size of primary cortical areas or shift their location are paralleled by shifts in thalamocortical axon targeting and changes in thalamic nucleus size; the correct topography between thalamus and cortex is maintained. (*E*) Experiments in marsupials showed a similar developmental plasticity.

gene *Coup-tf1* is conditionally deleted from cortical progenitors using *Emx1-Cre*, a massive expansion of frontal areas compresses primary sensory areas to caudal cortex (Fig. 6C; Armentano et al. 2007). Interestingly, the correct topography between thalamus and the displaced sensory areas is maintained; the dramatic cortical topographic shifts are paralleled by equivalent shifts in TCA connectivity.

In contrast to *Fgf8* manipulations, *Emx1-Cre* mediated deletion of the transcription factor gene *Pax6* from cortical progenitors fails to elicit substantial changes in thalamocortical topography (Piñon et al. 2008). However, cortex-specific deletion of *Pax6* does lead to a miniaturization of S1 (Fig. 6B; Zembrzycki et al. 2013). Following this reduction in S1 size, selective apoptosis in the VB repatterns the thalamus and leads to a corresponding reduction in VB size. Consistent with this finding, recent work shows that conditional deletion of transcription factor gene *Lhx2* using *Emx1-Cre* leads to a loss of sensory barrels, which is accompanied by a reduction in the size of the VB nucleus (Wang et al. 2017). Thus, the patterning of thalamic nuclei is influenced by the cortical areas with which they form reciprocal connections.

In addition to rodents, cortical influence of thalamocortical connections have been studied in marsupials, in which the comparatively immature stage at birth (corresponding to E11 in mouse) enables very early manipulations (Molnár et al. 1998c). Unilateral cortical sheet reduction in the marsupial *Monodelphis domestica* during early development, prior to the arrival of TCAs, leads to a reduction in cortical volume but maintains the normal spatial relationships between V1, S1, and A1 on the remaining cortical sheet (Fig. 6E; Huffman et al. 1999). Subsequently, parallel shifts in TCAs take place to accommodate these shifts in cortical fields. Together, these experiments highlight that cortical areas are first patterned by intrinsic genetic factors. This patterning in turn influences the pathfinding of afferent TCAs. These experiments also reveal the targeting of TCAs is not rigid and not dependent on physical coordinates; they can be flexibly rewired to parallel alterations or shifts in cortical arealization.

CONCLUSIONS

Thalamus and cortex and their interlinking circuits evolved together; thalamo-cortico–thalamo-cortical circuits form functional units that are closely integrated. The development and evolution of these structures are dependent on each other. Thalamic projections reach the cortex by crossing through regions of the forebrain that had substantial modifications during evolution, and the developmental changes at the PSPB reflects the diverging evolutionary paths between mammals and sauropsids. Thalamic projections cross the PSBP by interacting with transient early corticofugal projections from subplate neurons. The earliest cortical neurogenetic and migration programs are largely independent of subcortical input. However, thalamo-cortical projections start to interact with the cortex prior to the completion of cortical neurogenesis and neuronal migration; they mediate some of the first signals that contribute to area-specific cortical circuit formation. The timing and nature of these early interactions can differ in diverse taxa, but they are present in all mammals. Corticofugal projections also interact with the developing thalamus and have an impact on the formation of cortico–thalamo-cortical circuits and the hierarchy of thalamic nuclei. The choreography of cortico-thalamic innervation follows a specific algorithm and is sensitive to alterations in sensory input. The development and plasticity of thalamo–corticothalamic circuits provide flexibility and developmental adaptability tailored to the circumstances of the individual and the evolutionary requirements of the species.

ACKNOWLEDGMENTS

Some of the ideas and concepts reviewed in this work are the updated views articulated in our previous publications: Molnár (1998); Molnár et al. (2012, 2020); Garcia-Moreno and Molnár (2020); and Bandiera and Molnár (2022). We are grateful to all previous and current laboratory members for their comments and discussions. The work in the laboratory of Z.M. has been funded by MRC, BBSRC, Royal Society, Wellcome Trust, Oxford Martin School, Anatomical

Society, St John's College. Z.M. is an Einstein Visiting Fellow at Charité–Universitätsmedizin Berlin, Cluster of Excellence NeuroCure and Institute of Biochemistry (2020–2024). The work in the laboratory of K.Y.K. has been funded by the National Institutes of Health (NIH) grants R01 NS097525, R01 MH126287, and R01 NS125313 and Simons Foundation Autism Research Initiative 402213, 324586.

REFERENCES

Adams NC, Lozsádi DA, Guillery RW. 1997. Complexities in the thalamocortical and corticothalamic pathways. *Eur J Neurosci* **9:** 204–209. doi:10.1111/j.1460-9568 .1997.tb01391.x

Allendoerfer KL, Shatz CJ. 1994. The subplate, a transient neocortical structure: its role in the development of connections between thalamus and cortex. *Annu Rev Neurosci* **17:** 185–218. doi:10.1146/annurev.ne.17.030194 .001153

Antón-Bolaños N, Espinosa A, López-Bendito G. 2018. Developmental interactions between thalamus and cortex: a true love reciprocal story. *Curr Opin Neurobiol* **52:** 33–41. doi:10.1016/j.conb.2018.04.018

Armentano M, Chou SJ, Srubek Tomassy G, Leingärtner A, O'Leary DDM, Studer M. 2007. COUP-TFI regulates the balance of cortical patterning between frontal/motor and sensory areas. *Nat Neurosci* **10:** 1277–1286. doi:10.1038/ nn1958

Assimacopoulos S, Kao T, Issa NP, Grove EA. 2012. Fibroblast growth factor 8 organizes the neocortical area map and regulates sensory map topography. *J Neurosci* **32:** 7191–7201. doi:10.1523/JNEUROSCI.0071-12.2012

Bagnard D, Chounlamountri N, Püschel AW, Bolz J. 2001. Axonal surface molecules act in combination with semaphorin 3a during the establishment of corticothalamic projections. *Cereb Cortex* **11:** 278–285. doi:10.1093/cer cor/11.3.278

Bandiera S, Molnár Z. 2022. Development of the thalamocortical systems. In *The thalamus* (ed. Halassa M). Cambridge University Press, Cambridge.

Blakey D, Wilson MC, Molnár Z. 2012. Termination and initial branch formation of SNAP-25-deficient thalamocortical fibres in heterochronic organotypic co-cultures. *Eur J Neurosci* **35:** 1586–1594. doi:10.1111/j.1460-9568 .2012.08120.x

Braisted JE, Catalano SM, Stimac R, Kennedy TE, Tessier-Lavigne M, Shatz CJ, O'Leary DDM. 2000. Netrin-1 promotes thalamic axon growth and is required for proper development of the thalamocortical projection. *J Neurosci* **20:** 5792–5801. doi:10.1523/JNEUROSCI.20-15-05792 .2000

Bruguier H, Suarez R, Manger P, Hoerder-Suabedissen A, Shelton AM, Oliver DK, Packer AM, Ferran JL, García-Moreno F, Puelles L, et al. 2020. In search of common developmental and evolutionary origin of the claustrum and subplate. *J Comp Neurol* **528:** 2956–2977. doi:10 .1002/cne.24922

Carney RS, Alfonso TB, Cohen D, Dai H, Nery S, Stoica B, Slotkin J, Bregman BS, Fishell G, Corbin JG. 2006. Cell migration along the lateral cortical stream to the developing basal telencephalic limbic system. *J Neurosci* **26:** 11562–11574. doi:10.1523/JNEUROSCI.3092-06.2006

Carney RS, Cocas LA, Hirata T, Mansfield K, Corbin JG. 2009. Differential regulation of telencephalic pallial–subpallial boundary patterning by Pax6 and Gsh2. *Cereb Cortex* **19:** 745–759. doi:10.1093/cercor/bhn123

Carrera I, Ferreiro-Galve S, Sueiro C, Anadón R, Rodríguez-Moldes I. 2008. Tangentially migrating GABAergic cells of subpallial origin invade massively the pallium in developing sharks. *Brain Res Bull* **75:** 405–409. doi:10.1016/j .brainresbull.2007.10.013

Casas-Torremocha D, Rubio-Teves M, Hoerder-Suabedissen A, Hayashi S, Prensa L, Molnár Z, Porrero C, Clascá F. 2022. A combinatorial input landscape in the "higher-order relay" posterior thalamic nucleus. *J Neurosci* **42:** 7757–7781. doi:10.1523/JNEUROSCI.0698-22.2022

Catalano SM, Robertson RT, Killackey HP. 1991. Early ingrowth of thalamocortical afferents to the neocortex of the prenatal rat. *Proc Natl Acad Sci* **88:** 2999–3003. doi:10.1073/pnas.88.8.2999

Chapouton P, Schuurmans C, Guillemot F, Götz M. 2001. The transcription factor neurogenin 2 restricts cell migration from the cortex to the striatum. *Development* **128:** 5149–5159. doi:10.1242/dev.128.24.5149

Chen Y, Magnani D, Theil T, Pratt T, Price DJ. 2012. Evidence that descending cortical axons are essential for thalamocortical axons to cross the pallial–subpallial boundary in the embryonic forebrain. *PLoS ONE* **7:** e33105. doi:10.1371/journal.pone.0033105

Chou SJ, Babot Z, Leingärtner A, Studer M, Nakagawa Y, O'Leary DDM. 2013. Geniculocortical input drives genetic distinctions between primary and higher-order visual areas. *Science* **340:** 1239–1242. doi:10.1126/science .1232806

Clascá F, Angelucci A, Sur M. 1995. Layer-specific programs of development in neocortical projection neurons. *Proc Natl Acad Sci* **92:** 11145–11149. doi:10.1073/pnas.92.24 .11145

Cobos I, Puelles L, Martínez S. 2001. The avian telencephalic subpallium originates inhibitory neurons that invade tangentially the pallium (dorsal ventricular ridge and cortical areas). *Dev Biol* **239:** 30–45. doi:10.1006/dbio.2001.0422

Cordery P, Molnár Z. 1999. Embryonic development of connections in turtle pallium. *J Comp Neurol* **413:** 26–54. doi:10.1002/(SICI)1096-9861(19991011)413:1

De Carlos JA, O'Leary DD. 1992. Growth and targeting of subplate axons and establishment of major cortical pathways. *J Neurosci* **12:** 1194–1211. doi:10.1523/JNEURO SCI.12-04-01194.1992

Dehay C, Savatier P, Cortay V, Kennedy H. 2001. Cell-cycle kinetics of neocortical precursors are influenced by embryonic thalamic axons. *J Neurosci* **21:** 201–214. doi:10 .1523/JNEUROSCI.21-01-00201.2001

De Marco García NV, Priya R, Tuncdemir SN, Fishell G, Karayannis T. 2015. Sensory inputs control the integration of neurogliaform interneurons into cortical circuits. *Nat Neurosci* **18:** 393–401. doi:10.1038/nn.3946

Doyle DZ, Lam MM, Qalieh A, Qalieh Y, Sorel A, Funk OH, Kwan KY. 2021. Chromatin remodeler *Arid1a* regulates subplate neuron identity and wiring of cortical connectivity. *Proc Natl Acad Sci* **118:** e2100686118. doi:10.1073/pnas.2100686118

Dwyer ND, Manning DK, Moran JL, Mudbhary R, Fleming MS, Favero CB, Vock VM, O'Leary DD, Walsh CA, Beier DR. 2011. A forward genetic screen with a thalamocortical axon reporter mouse yields novel neurodevelopment mutants and a distinct emx2 mutant phenotype. *Neural Dev* **6:** 3. doi:10.1186/1749-8104-6-3

Fenk LA, Riquelme JL, Laurent G. 2023. Interhemispheric competition during sleep. *Nature* **616:** 312–318. doi:10.1038/s41586-023-05827-w

Fernandez AS, Pieau C, Repérant J, Boncinelli E, Wassef M. 1998. Expression of the *Emx-1* and *Dlx-1* homeobox genes define three molecularly distinct domains in the telencephalon of mouse, chick, turtle and frog embryos: implications for the evolution of telencephalic subdivisions in amniotes. *Development* **125:** 2099–2111. doi:10.1242/dev.125.11.2099

Frangeul L, Pouchelon G, Telley L, Lefort S, Luscher C, Jabaudon D. 2016. A cross-modal genetic framework for the development and plasticity of sensory pathways. *Nature* **538:** 96–98. doi:10.1038/nature19770

García-Moreno F, Molnár Z. 2020. Variations of telencephalic development that paved the way for neocortical evolution. *Prog Neurol* **194:** 101865. doi:10.1016/j.pneurobio.2020.101865

García-Moreno F, Anderton E, Jankowska M, Begbie J, Encinas JM, Irimia M, Molnár Z. 2018. Absence of tangentially migrating glutamatergic neurons in the developing avian brain. *Cell Rep* **22:** 96–109. doi:10.1016/j.celrep.2017.12.032

Garel S, Yun K, Grosschedl R, Rubenstein JLR. 2002. The early topography of thalamocortical projections is shifted in *Ebf1* and *Dlx1/2* mutant mice. *Development* **129:** 5621–5634. doi:10.1242/dev.00166

Garel S, Huffman KJ, Rubenstein JL. 2003. Molecular regionalization of the neocortex is disrupted in *Fgf8* hypomorphic mutants. *Development* **130:** 1903–1914. doi:10.1242/dev.00416

Gerstmann K, Pensold D, Symmank J, Khundadze M, Hübner CA, Bolz J, Zimmer G. 2015. Thalamic afferents influence cortical progenitors via ephrin A5-EphA4 interactions. *Development* **142:** 140–150. doi:10.1242/dev.104927

Ghosh A, Shatz CJ. 1993. A role for subplate neurons in the patterning of connections from thalamus to neocortex. *Development* **117:** 1031–1047. doi:10.1242/dev.117.3.1031

Giasafaki C, Grant E, Hoerder-Suabedissen A, Hayashi S, Lee S, Molnár Z. 2022. Cross-hierarchical plasticity of corticofugal projections to dLGN after neonatal monocular enucleation. *J. Comp Neurol* **530:** 978–997. doi:10.1002/cne.25304

González-Arnay E, González-Gómez M, Meyer G. 2017. A radial glia fascicle leads principal neurons from the pallial–subpallial boundary into the developing human insula. *Front Neuroanat* **11:** 111. doi:10.3389/fnana.2017.00111

Grant E, Hoerder-Suabedissen A, Molnár Z. 2012. Development of the corticothalamic projections. *Front Neurosci* **6:** 53. doi:10.3389/fnins.2012.00053

Grant E, Hoerder-Suabedissen A, Molnár Z. 2016. The regulation of corticofugal fibre targeting by retinal inputs. *Cereb Cortex* **26:** 1336–1348. doi:10.1093/cercor/bhv315

Halassa MM. 2022. *The thalamus.* Cambridge University Press, Cambridge.

Hanashima C, Molnár Z, Fishell G. 2006. Building bridges to the cortex. *Cell* **125:** 24–27. doi:10.1016/j.cell.2006.03.021

Hevner RF, Miyashita-Lin E, Rubenstein JL. 2002. Cortical and thalamic axon pathfinding defects in Tbr1, Gbx2, and Pax6 mutant mice: evidence that cortical and thalamic axons interact and guide each other. *J Comp Neurol* **447:** 8–17. doi:10.1002/cne.10219

Higashi S, Molnár Z, Kurotani T, Inokawa H, Toyama K. 2002. Functional thalamocortical connections develop during embryonic period in the rat: an optical recording study. *Neuroscience* **115:** 1231–1246. doi:10.1016/S0306-4522(02)00418-9

Higashi S, Hioki K, Kurotani T, Molnár Z. 2005. Functional thalamocortical synapse reorganization from subplate to layer IV during postnatal development in the *Reeler*-like mutant rat (*Shaking Rat Kawasaki*): an optical recording study. *J Neuroscience* **25:** 1395–1406. doi:10.1523/JNEUROSCI.4023-04.2005

Hoerder-Suabedissen A, Molnár Z. 2012. Morphology of mouse subplate cells with identified projection targets changes with age. *J Comp Neurol* **520:** 174–185. doi:10.1002/cne.22725

Hoerder-Suabedissen A, Molnár Z. 2015. Development, evolution and pathology of neocortical subplate neurons. *Nat Rev Neurosci* **16:** 133–146. doi:10.1038/nrn3915

Hoerder-Suabedissen A, Oeschger FM, Krishnan ML, Belgard TG, Wang WZ, Lee S, Webber C, Petretto E, Edwards AD, Molnár Z. 2013. Expression profiling of mouse subplate reveals a dynamic gene network and disease association with autism and schizophrenia. *Proc Natl Acad Sci* **110:** 3555–3560. doi:10.1073/pnas.1218510110

Hoerder-Suabedissen A, Hayashi S, Upton L, Nolan Z, Casas-Torremocha D, Grant E, Viswanathan S, Kanold PO, Clasca F, Kim Y, et al. 2018. Subset of cortical layer 6b neurons selectively innervates higher order thalamic nuclei in mice. *Cereb Cortex* **28:** 1882–1897. doi:10.1093/cercor/bhy036

Horvath T, Hirsch J, Molnár Z. 2022. *Body, brain, behavior, three views and a conversation.* Elsevier, Amsterdam.

Huffman K, Molnár Z, van Dellen A, Khan D, Blakemore C, Krubitzer L. 1999. Compression of sensory fields on a reduced cortical sheet. *J Neurosci* **19:** 9939–9952. doi:10.1523/JNEUROSCI.19-22-09939.1999

Jones EG. 2007. *The thalamus.* Cambridge University Press, Cambridge.

Jones L, López-Bendito G, Gruss P, Stoykova A, Molnár Z. 2002. *Pax6* is required for the normal development of the forebrain axonal connections. *Development* **129:** 5041–5052. doi:10.1242/dev.129.21.5041

Kanold PO, Luhmann HJ. 2010. The subplate and early cortical circuits. *Annu Rev Neurosci* **33:** 23–48. doi:10.1146/annurev-neuro-060909-153244

Kostovic I, Rakic P. 1984. Development of prestriate visual projections in the monkey and human fetal cerebrum revealed by transient cholinesterase staining. *J Neurosci* **4:** 25–42. doi:10.1523/JNEUROSCI.04-01-00025.1984

Kostovic I, Rakic P. 1990. Developmental history of the transient subplate zone in the visual and somatosensory cortex of the macaque monkey and human brain. *J Comp Neurol* **297**: 441–470. doi:10.1002/cne.902970309

Kwan KY, Lam MMS, Krsnik Z, Kawasawa YI, Lefebvre V, Šestan N. 2008. SOX5 postmitotically regulates migration, postmigratory differentiation, and projections of subplate and deep-layer neocortical neurons. *Proc Natl Acad Sci* **105**: 16021–16026. doi:10.1073/pnas.0806791105

Kwan KY, Šestan N, Anton ES. 2012. Transcriptional coregulation of neuronal migration and laminar identity in the neocortex. *Development* **139**: 1535–1546. doi:10.1242/dev.069963

Li H, Fertuzinhos S, Mohns E, Hnasko TS, Verhage M, Edwards R, Sestan N, Crair MC. 2013. Laminar and columnar development of barrel cortex relies on thalamocortical neurotransmission. *Neuron* **79**: 970–986 doi:10.1016/j.neuron.2013.06.043

López-Bendito G, Molnár Z. 2003. Thalamocortical development: how are we going to get there? *Nat Rev Neurosci* **4**: 276–289. doi:10.1038/nrn1075

López-Bendito G, Chan CH, Mallamaci A, Parnavelas J, Molnár Z. 2002. The role of *Emx2* in the development of the reciprocal connectivity between cortex and thalamus. *J Comp Neurol* **451**: 153–169. doi:10.1002/cne.10345

López-Bendito G, Cautinat A, Sánchez JA, Bielle F, Flames N, Garratt AN, Talmage DA, Role LW, Charnay P, Marín O, et al. 2006. Tangential neuronal migration controls axon guidance: a role for neuregulin-1 in thalamocortical axon navigation. *Cell* **125**: 127–142. doi:10.1016/j.cell.2006.01.042

Lund RD, Mustari MJ. 1977. Development of the geniculocortical pathway in rats. *J Comp Neurol* **173**: 289–305. doi:10.1002/cne.901730206

Marín O, Baker J, Puelles L, Rubenstein JLR. 2002. Patterning of the basal telencephalon and hypothalamus is essential for guidance of cortical projections. *Development* **129**: 761–773. doi:10.1242/dev.129.3.761

Métin C, Godement P. 1996. The ganglionic eminence may be an intermediate target for corticofugal and thalamocortical axons. *J Neurosci* **16**: 3219–3235. doi:10.1523/JNEUROSCI.16-10-03219.1996

Métin C, Alvarez C, Moudoux D, Vitalis T, Pieau C, Molnár Z. 2007. Conserved pattern of tangential neuronal migration during forebrain development. *Development* **134**: 2815–2827. doi:10.1242/dev.02869

Mitrofanis J, Guillery RW. 1993. New views of the thalamic reticular nucleus in the adult and the developing brain. *Trends Neurosci* **16**: 240–245. doi:10.1016/0166-2236(93)90163-G

Miyashita-Lin EM, Hevner R, Wassarman KM, Martinez S, Rubenstein JL. 1999. Early neocortical regionalization in the absence of thalamic innervation. *Science* **285**: 906–909. doi:10.1126/science.285.5429.906

Molnár Z. 1998. *Development of thalamocortical connections*, p. 264. Springer, Heidelberg, Germany.

Molnár Z. 2011. Evolution of cerebral cortical development. *Brain Behav Evol* **78**: 94–107. doi:10.1159/000327325

Molnár Z. 2019. Cortical layer with no known function. *Eur J Neurosci* **49**: 957–963. doi:10.1111/ejn.13978

Molnár Z, Blakemore C. 1991. Lack of regional specificity for connections formed between thalamus and cortex in coculture. *Nature* **351**: 475–477. doi:10.1038/351475a0

Molnár Z, Blakemore C. 1995. How do thalamic axons find their way to the cortex? *Trends Neurosci* **18**: 389–397. doi:10.1016/0166-2236(95)93935-Q

Molnár Z, Blakemore C. 1999. Development of signals influencing the growth and termination of thalamocortical axons in organotypic culture. *Exp Neurol* **156**: 363–393. doi:10.1006/exnr.1999.7032

Molnár Z, Butler AB. 2002. The corticostriatal junction: a crucial region for forebrain development and evolution. *Bioessays* **24**: 530–541. doi:10.1002/bies.10100

Molnár Z, Cordery P. 1999a. Connections between cells of the internal capsule, thalamus, and cerebral cortex in embryonic rat. *J Comp Neurol* **413**: 1–25. doi:10.1002/(SICI)1096-9861(19991011)413:1

Molnár Z, Cordery P. 1999b. Connections between cells of the internal capsule, thalamus, and cerebral cortex in embryonic rat. *J Comp Neurol* **413**: 1–25.

Molnár Z, Hannan A. 2000. Development of thalamocortical projections in normal and mutant mice. *Results and problems in cell differentiation, mouse brain development* (ed. Goffinet A, Rakic P), pp. 293–332. Springer, Berlin.

Molnár Z, Adams R, Blakemore C. 1998a. Mechanisms underlying the early establishment of thalamocortical connections in the rat. *J Neurosci* **18**: 5723–5745. doi:10.1523/JNEUROSCI.18-15-05723.1998

Molnár Z, Adams R, Goffinet AM, Blakemore C. 1998b. The role of the first postmitotic cortical cells in the development of thalamocortical innervation in the *Reeler* mouse. *J Neurosci* **18**: 5746–5765. doi:10.1523/JNEUROSCI.18-15-05746.1998

Molnár Z, Knott GW, Blakemore C, Saunders NR. 1998c. Development of thalamocortical projections in the south American gray short-tailed opossum (*Monodelphis domestica*). *J Comp Neuro* **398**: 491–514. doi:10.1002/(SICI)1096-9861(19980907)398:4

Molnár Z, López-Bendito G, Small J, Partridge LD, Blakemore C, Wilson MC. 2002. Normal development of embryonic thalamocortical connectivity in the absence of evoked synaptic activity. *J Neurosci* **22**: 10313–10323. doi:10.1523/JNEUROSCI.22-23-10313.2002

Molnár Z, Higashi S, López-Bendito G. 2003a. Choreography of early thalamocortical development. *Cereb Cortex* **13**: 661–669. doi:10.1093/cercor/13.6.661

Molnár Z, Kurotani T, Higashi S, Yamamoto N, Toyama K. 2003b. Development of functional thalamocortical synapses studied with current source density analysis in whole forebrain slices. *Brain Res Bull* **60**: 355–371. doi:10.1016/S0361-9230(03)00061-3

Molnár Z, Garel S, López-Bendito G, Maness P, Price DJ. 2012. Mechanisms controlling the guidance of thalamocortical axons through the embryonic forebrain. *Eur J Neurosci* **35**: 1573–1585. doi:10.1111/j.1460-9568.2012.08119.x

Molnár Z, Luhmann HJ, Kanold PO. 2020. Transient local and global circuits match spontaneous and sensory driven activity patterns during cortical development. *Science* **370**: eabb2153. doi:10.1126/science.abb2153

Monko T, Rebertus J, Stolley J, Salton SR, Nakagawa Y. 2022. Thalamocortical axons regulate neurogenesis and laminar fates in the early sensory cortex. *Proc Natl Acad Sci* **119:** e2201355119. doi:10.1073/pnas.2201355119

Montiel JF, Vasistha NV, Garcia-Moreno F, Molnár Z. 2016. From sauropsids to mammals and back: new approaches to comparative cortical development. *J Comp Neuro* **52:** 630–645. doi:10.1002/cne.23871

Moreno NA, González S, Rétaux S. 2008. Evidences for tangential migrations in *Xenopus* telencephalon: developmental patterns and cell tracking experiments. *Dev Neurobiol* **68:** 504–520. doi:10.1002/dneu.20603

Moreno N, Morona R, López JM, González A. 2010. Subdivisions of the turtle *Pseudemys scripta* subpallium based on the expression of regulatory genes and neuronal markers. *J Comp Neurol* **518:** 4877–4902. doi:10.1002/cne.22493

Moreno-Juan V, Filipchuk A, Antón-Bolaños N, Mezzera C, Gezelius H, Andrés B, Rodríguez-Malmierca L, Susín R, Schaad O, Iwasato T, et al. 2017. Prenatal thalamic waves regulate cortical area size prior to sensory processing. *Nat Commun* **8:** 14172. doi:10.1038/ncomms14172

Narboux-Nême N, Evrard A, Ferezou I, Erzurumlu RS, Kaeser PS, Lainé J, Rossier J, Ropert N, Südhof TC, Gaspar P. 2012. Neurotransmitter release at the thalamocortical synapse instructs barrel formation but not axon patterning in the somatosensory cortex. *J Neurosci* **32:** 6183–6196. doi:10.1523/JNEUROSCI.0343-12.2012

Norimoto H, Fenk LA, Li HH, Tosches MA, Gallego-Flores T, Hain D, Reiter S, Kobayashi R, Macias A, Arends A, et al. 2020. A claustrum in reptiles and its role in slow-wave sleep. *Nature* **578:** 413–418. doi:10.1038/s41586-020-1993-6

Northcutt G, Kaas JH. 1995. The emergence and evolution of mammalian neocortex. *Trends Neurosci* **18:** 373–379. doi:10.1016/0166-2236(95)93932-N

O'Leary DD, Sahara S. 2008. Genetic regulation of arealization of the neocortex. *Curr Opin Neurobiol* **18:** 90–100. doi:10.1016/j.conb.2008.05.011

Pal S, Dwivedi D, Pramanik T, Godbole G, Iwasato T, Jabaudon D, Bhalla US, Tole S. 2021. An early cortical progenitor-specific mechanism regulates thalamocortical innervation. *J Neurosci* **41:** 6822–6835. doi:10.1523/JNEUROSCI.0226-21.2021

Pedraza M, Hoerder-Suabedissen A, Albert-Maestro MA, Molnár Z, De Carlos JA. 2014. Extracortical origin of some murine subplate cell populations. *Proc Natl Acad Sci* **111:** 8613–8618. doi:10.1073/pnas.1323816111

Piñon MC, Tuoc TC, Ashery-Padan R, Molnár Z, Stoykova A. 2008. Altered molecular regionalization and normal thalamocortical connections in cortex-specific *Pax6* knock-out mice. *J Neurosci* **28:** 8724–8734. doi:10.1523/JNEUROSCI.2565-08.2008

Piñon MC, Jethwa A, Jacobs E, Campagnoni A, Molnár Z. 2009. Dynamic integration of subplate neurons into the cortical barrel field circuitry during postnatal development in the Golli-tau-eGFP (GTE) mouse. *J Physiol* **587:** 1903–1915. doi:10.1113/jphysiol.2008.167767

Pouchelon G, Gambino F, Bellone C, Telley L, Vitali I, Lüscher C, Holtmaat A, Jabaudon D. 2014. Modality-specific thalamocortical inputs instruct the identity of postsynaptic L4 neurons. *Nature* **511:** 471–474. doi:10.1038/nature13390

Puelles L. 2014. Development and evolution of the claustrum. In *The claustrum: structural, functional, and clinical neuroscience*, pp 85–118. Academic Press, San Diego.

Puelles L. 2017. Comments on the updated tetrapartite pallium model in the mouse and chick, featuring a homologous claustro-insular complex. *Brain Behav Evol* **90:** 171–189. doi:10.1159/000479782

Puelles L, Kuwana E, Puelles E, Bulfone A, Shimamura K, Keleher J, Smiga S, Rubenstein JLR. 2000. Pallial and subpallial derivatives in the embryonic chick and mouse telencephalon, traced by the expression of the genes Dlx-2, Emx-1, Nkx-2.1, Pax-6, and Tbr-1. *J Comp Neurol* **424:** 409–438. doi:10.1002/1096-9861(20000828)424:3<409::AID-CNE3>3.0.CO;2-7

Puelles L, Medina L, Borello U, Legaz I, Teissier A, Pierani A, Rubenstein JLR. 2016a. Radial derivatives of the mouse ventral pallium traced with Dbx1-LacZ reporters. *J Chem Neuroanat* **75:** 2–19. doi:10.1016/j.jchemneu.2015.10.011

Puelles L, Medina L, Borello U, Legaz I, Teissier A, Pierani A, Rubenstein JLR. 2016b. Radial derivatives of the mouse ventral pallium traced with Dbx1-LacZ reporters. *J Chem Neuroanat* **75:** 2–19. doi:10.1016/j.jchemneu.2015.10.011

Puelles L, Ayad A, Alonso A, Sandoval JE, Martínez-de-la-Torre M, Medina L, Ferran JL. 2016c. Selective early expression of the orphan nuclear receptor *Nr4a2* identifies the claustrum homolog in the avian mesopallium: impact on sauropsidian/mammalian pallium comparisons. *J Comp Neurol* **524:** 665–703. doi:10.1002/cne.23902

Rakic P. 1974. Neurons in rhesus monkey visual cortex: systematic relation between time of origin and eventual disposition. *Science* **183:** 425–427. doi:10.1126/science.183.4123.425

Rakic P. 1988. Specification of cerebral cortical areas. *Science* **241:** 170–176. doi:10.1126/science.3291116

Rakić S, Davis C, Molnár Z, Nikolić M, Parnavelas JG. 2006. Role of p35/Cdk5 in preplate splitting in the developing cerebral cortex. *Cereb Cortex* **16:** i35–i45. doi:10.1093/cercor/bhj172

Rash BG, Grove EA. 2006. Area and layer patterning in the developing cerebral cortex. *Curr Opin Neurobiol* **16:** 25–34. doi:10.1016/j.conb.2006.01.004

Rueda-Alaña E, Martínez-Garay I, Encinas J, Molnár Z, García-Moreno F. 2018. Dbx1-derived pyramidal neurons are generated locally in the developing murine neocortex. *Front Neurosci* **12:** 792. doi:10.3389/fnins.2018.00792

Sato H, Hatakeyama J, Iwasato T, Araki K, Yamamoto N, Shimamura K. 2022. Thalamocortical axons control the cytoarchitecture of neocortical layers by area-specific supply of VGF. *eLife* **11:** e67549. doi:10.7554/eLife.67549

Shatz CJ, Luskin MB. 1986. Relationship between the geniculocortical afferents and their cortical target cells during development of the cat's primary visual cortex. *J Neurosci* **6:** 3655–3668. doi:10.1523/JNEUROSCI.06-12-03655.1986

Sherman SM, Guillery RW. 2005. *Exploring the thalamus and its role in cortical function*. The MIT Press, Cambridge, MA.

Shi W, Xianyu A, Han Z, Tang X, Li Z, Zhong H, Mao T, Huang K, Shi SH. 2017. Ontogenetic establishment of order-specific nuclear organization in the mammalian thalamus. *Nat Neurosci* **20:** 516–528. doi:10.1038/nn.4519

Shimogori T, Grove EA. 2005. Fibroblast growth factor 8 regulates neocortical guidance of area-specific thalamic innervation. *J Neurosci* **25:** 6550–6560. doi:10.1523/JNEUROSCI.0453-05.2005

Smith-Fernandez A, Pieau C, Repérant J, Boncinelli E, Wassef M. 1998. Expression of the *Emx-1* and *Dlx-1* homeobox genes define three molecularly distinct domains in the telencephalon of mouse, chick, turtle and frog embryos: implications for the evolution of telencephalic subdivisions in amniotes. *Development* **125:** 2099–2111. doi:10.1242/dev.125.11.2099

Spadory T, Duque A, Selemon LD. 2022. Spatial-temporal topography in neurogenesis of the macaque thalamus. *Brain Struct Funct* **227:** 1673–1682. doi:10.1007/s00429-022-02463-4

Suárez R, Paolino A, Fenlon LR, Morcom LR, Kozulin P, Kurniawan ND, Richards LJ. 2018. A pan-mammalian map of interhemispheric brain connections predates the evolution of the corpus callosum. *Proc Natl Acad Sci* **115:** 9622–9627. doi:10.1073/pnas.1808262115

Tolner EA, Sheikh A, Yukin AY, Kail K, Kanold PC. 2012. Subplate neurons promote spindle bursts and thalamocortical patterning in the neonatal rat somatosensory cortex. *J Neurosci* **32:** 692–702. doi:10.1523/JNEUROSCI.1538-11.2012

Tosches MA, Yamawaki TM, Naumann RK, Jacobi AA, Tushev G, Laurent G. 2018. Evolution of pallium, hippocampus, and cortical cell types revealed by single-cell transcriptomics in reptiles. *Science* **360:** 881–888. doi:10.1126/science.aar4237

Tuttle R, Nakagawa Y, Johnson JE, O'Leary DD. 1999. Defects in thalamocortical axon pathfinding correlate with altered cell domains in *mash-1*-deficient mice. *Development* **126:** 1903–1916. doi:10.1242/dev.126.9.1903

Vasistha NA, García-Moreno F, Arora S, Cheung AF, Arnold SJ, Robertson EJ, Molnár Z. 2015. Cortical and clonal contribution of Tbr2 expressing progenitors in the developing mouse brain. *Cereb Cortex* **25:** 3290–3302. doi:10.1093/cercor/bhu125

Viswanathan S, Bandyopadhyay S, Kao JP, Kanold PO. 2012. Changing microcircuits in the subplate of the developing cortex. *J Neurosci* **32:** 1589–1601. doi:10.1523/JNEUROSCI.4748-11.2012

Wang CF, Hsing HW, Zhuang ZH, Wen M-H, Chang W-J, Briz CG, Nieto M, Shyu BC, Chou SJ. 2017. Lhx2 expression in postmitotic cortical neurons initiates assembly of the thalamocortical somatosensory circuit. *Cell Rep* **18:** 849–856. doi:10.1016/j.celrep.2017.01.001

Yamamoto N, López-Bendito G. 2012. Shaping brain connections through spontaneous neural activity. *Eur J Neurosci* **35:** 1595–1604. doi:10.1111/j.1460-9568.2012.08101.x

Zembrzycki A, Chou SJ, Ashery-Padan R, Stoykova A, O'Leary DDM. 2013. Sensory cortex limits cortical maps and drives top-down plasticity in thalamocortical circuits. *Nat Neurosci* **16:** 1060–1067. doi:10.1038/nn.3454

Zhou L, Bar I, Achouri Y, Campbell K, De Backer O, Hebert JM, Jones K, Kessaris N, de Rouvroit CL, O'Leary DD, et al. 2008. Early forebrain wiring: genetic dissection using conditional *Celsr3* mutant mice. *Science* **320:** 946–949. doi:10.1126/science.1155244

Zhou L, Qu Y, Tissir F, Goffinet AM. 2009. Role of the atypical cadherin Celsr3 during development of the internal capsule. *Cerebl Cortex* **19:** i114–i119. doi:10.1093/cercor/bhp032

Zhou L, Gall D, Qu Y, Prigogine C, Cheron G, Tissir F, Schiffmann SN, Goffinet AM. 2010. Maturation of "neocortex isole" in vivo in mice. *J Neurosci* **30:** 7928–7939. doi:10.1523/JNEUROSCI.6005-09.2010

Variability in Neural Circuit Formation

Kevin J. Mitchell

Smurfit Institute of Genetics and Institute of Neuroscience, Trinity College Dublin,
Dublin D02 PN40, Ireland

Correspondence: kevin.mitchell@tcd.ie

The study of neural development is usually concerned with the question of how nervous systems get put together. Variation in these processes is usually of interest as a means of revealing these normative mechanisms. However, variation itself can be an object of study and is of interest from multiple angles. First, the nature of variation in both the processes and the outcomes of neural development is relevant to our understanding of how these processes and outcomes are encoded in the genome. Second, variation in the wiring of the brain in humans may underlie variation in all kinds of psychological and behavioral traits, as well as neurodevelopmental disorders. And third, genetic variation that affects circuit development provides the raw material for evolutionary change. Here, I examine these different aspects of variation in circuit development and consider what they may tell us about these larger questions.

THE CODING PROBLEM

If we want to understand how variation can affect the processes of development, we need to understand how those processes are encoded in the genome. A good deal of research in developmental neurobiology is about this question, or at least is relevant to it. The main approach in experimental developmental neurobiology has been to isolate specific processes or developmental events, such as axons crossing the midline or innervating a muscle or topographically projecting across a target region, and then to try and identify specific molecules involved in these processes and investigate their functions.

This approach has been successful in identifying many molecules involved in directing axon guidance and synaptic target selection (Kolodkin and Tessier-Lavigne 2011; de Wit and Ghosh 2016; Mitchell 2018a; Sanes and Zipursky 2020). However, this reductive, controlled approach risks giving the impression that individual guidance or targeting events are controlled by individual molecules, acting in isolation. The reality in vivo is much messier and more complex —all guidance and connectivity decisions are influenced by multiple factors acting at once (e.g., Winberg et al. 1998; Yu et al. 2000; Xu et al. 2020).

A growth cone will encounter multiple secreted or cell-surface molecular cues at any moment, as well as adhesive and anti-adhesive proteins, the general environment of proteins in the extracellular matrix, and the physical forces presented by the landscape over which the axon is extending (Franze 2020; Breau and Trembleau 2023). The influence of these factors will be determined by the repertoire of receptor and adhesion proteins that the growing neuron itself is expressing, but not in a linear, one-at-a-time fashion (Klumpe et al. 2023). Many guidance and connectivity receptors operate in signaling complexes, with context-dependent interactions in *cis* as well as in *trans* (e.g., Marquardt et al. 2005; Perez-Branguli et al. 2016; for reviews, see Morales and Kania 2017; Stoeckli 2018; Südhof 2018; Zang et al. 2021).

This array of molecular and physical factors collectively generates what can be thought of as an energy landscape, constraining the growth of the axon along specific routes. Conrad Waddington proposed the "epigenetic landscape" as a visual metaphor of cell fate decisions through development (Waddington 1957). His idea was that this landscape was shaped by the functions and interactions of large numbers of genes (what we would now call a gene regulatory network), which collectively constrain the expression profiles of differentiating cells along defined trajectories toward specific attractor states (mature cell types). This idea can be adapted to describe the processes of axon guidance and synaptic target selection, with the added element that the "landscape" is not in abstract "gene expression space," but describes the real physical (and molecular) terrain of the developing nervous system (Fig. 1). This terrain presents a different set of possible growth channels to axons depending on their repertoire of receptors.

Rather than a one-to-one functional mapping between specific cue-receptor pairs and specific guidance decisions, this gives a view of a collective set of constraints that tend to channel growing neurons to their appropriate target regions and tend to direct synaptogenesis with the appropriate target cells or even the appropriate subcellular regions (Hiesinger 2021a,b). In turn, this landscape of cues and the respective repertoires of receptors are generated by a developmental program orchestrated by gene regulatory networks and patterning mechanisms involving dozens or hundreds of signaling proteins and transcription factors (Zarin et al. 2014; Herrera and Escalante 2022).

The distributed, indirect nature of this genetic encoding of wiring patterns has implications for how the resultant (really emergent) patterns are affected by different types of genetic variation. It also highlights the inherently statistical nature of these processes, which are probabilistic at the level of single cells, due to noise and randomness at molecular levels.

The genome does not contain enough information, quantitatively speaking, to specify the projection of every axon and the connections of every neuron (Hassan and Hiesinger 2015; Koulakov et al. 2021). It does not encode endpoint information directly at all, in fact, nor does it directly encode algorithmic information (Hiesinger 2021a), at least not in the sense of a set of instructions that isomorphically and separably specify the processes of development in a decomposable manner. Rather, it encodes a generative model—a set of latent variables (protein and regulatory element sequences) that collectively constrain the self-organizing processes of development to reliably produce a new individual of some species type (KJ Mitchell and N Cheney, unpubl.). Changes to the sequence of the

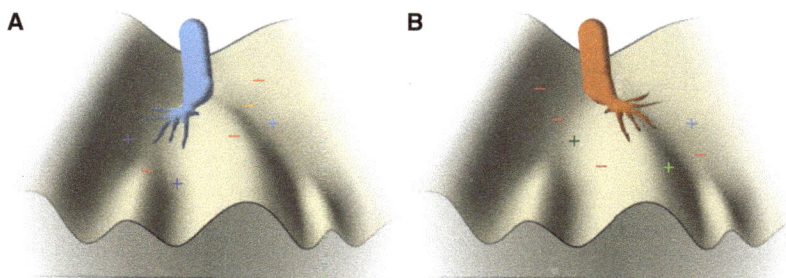

Figure 1. The neurodevelopmental landscape. A reimagined version of Waddington's landscape, showing axonal growth cones being channeled down alternate paths (*A,B*) by virtue of their differential sensitivity to molecular guidance cues expressed across the physical terrain of the developing nervous system.

genome can alter these latent variables and affect the outcome. But the developmental processes themselves also necessarily introduce variability in the detailed outcome (Vogt 2015; Mitchell 2018b).

SOURCES OF VARIATION IN OUTCOME

The outcome of neural wiring will thus vary across individuals of a species for two reasons: inevitable variation in the genetic program and equally inevitable variation in how any individual "run" of any particular version of that program plays out. In addition, exposure to certain environmental stressors, toxins, or other insults can affect neural development in diverse ways. These can be clinically important (e.g., fetal alcohol spectrum disorders; Popova et al. 2023), but will not be discussed further here. Another major source of variation in neural wiring across individuals within a species is sex. Males and females of many species show systematic variation in connectivity of specific circuits, particularly those underlying sexual and reproductive behaviors (Knoedler and Shah 2018; Meeh et al. 2021). Here, we will concern ourselves with sources and consequences of individual differences more generally.

GENETIC VARIATION

On the genetic front, phenotypic variation can arise due to single mutations of large effect and/or due to the combined, polygenic effects of many genetic variants. Experimental work in model organisms typically involves the former, although polygenic background can be an important modulating factor. In humans, both types of genetic influence are at play.

Single Mutations

As anyone who has spent months or years looking for a phenotype in a mutant animal knows, many mutations of single genes—even homozygous complete null mutations—seem to be well tolerated by the developing organism. Even when biochemical and gene expression evidence suggest the likely involvement of some specific protein

in the guidance of some axons or the specification of their synaptic connections, removal of that protein may produce no phenotypic effect, or may cause effects in some anatomical contexts where the gene is expressed, but not in others.

This suggests that either the protein in question is not in fact involved in the processes specifying the circuitry in question or that this function is dispensable, due to redundancy, robustness, or regulation (some reactive compensatory processes). In model organisms, such cryptic functions can often be revealed through analysis of epistatic interactions in genetically sensitized backgrounds. These kinds of screens and analyses have provided the means to identify guidance pathways operating in parallel (e.g., Winberg et al. 1998; Yu et al. 2000; Cate et al. 2016) or to elucidate biochemical pathways connecting cues and receptors to cellular responses (for review, see Zang et al. 2021).

That being said, it is also true that many single mutations do have large effects on neural circuitry development. Forward genetic screens have been highly successful in identifying important molecules that specify neural projections and connectivity. These have been based, for example, on direct anatomical visualization of axonal patterns (e.g., Seeger et al. 1993; van Vactor et al. 1993; Leighton et al. 2001) or on behavioral outcomes (e.g., Hedgecock et al. 1990). This approach is highly powerful in that it lets the system tell you what is important in an unbiased way. But such screens will, as a consequence, identify mutations that affect the phenotypes of interest in possibly quite indirect ways. Indeed, as we will see below, most of the genetic variation affecting any given process probably does so indirectly (i.e., by affecting proteins not directly involved in the cellular process itself) (Boyle et al. 2017).

The opposite approach—reverse genetics—is also a powerful method to test in vivo the importance of some putative guidance or connectivity factor identified through biochemical or molecular means for example (e.g., Mitchell et al. 1996; Serafini et al. 1996; Feldheim et al. 2000). This is a more directed (or biased) approach from the outset, but it is usually not possible to predict in advance how the system will react to disruption of any given gene.

One interesting trend is that, while loss-of-function or removal of a single gene is often reasonably well buffered by the system, gain-of-function mutations or manipulations can often produce far stronger effects. For example, ectopic expression of a cue can often induce much more dramatic effects on the system than its removal (e.g., Nose et al. 1994; Mitchell et al. 1996; Winberg et al. 1998). Similarly, dominant-active or dominant-negative mutations can have wider consequences on biochemical pathways and developmental processes than simple removal of a protein. These kinds of effects are particularly relevant in human genetics.

The Genetics of Neurodevelopment in Humans

Neurodevelopmental disorders in humans are often divided into rare "Mendelian" conditions (many with identified genetic causes, such as Fragile X syndrome, Rett syndrome, Timothy syndrome, etc.), or common disorders with broad diagnostic labels such as autism, schizophrenia, epilepsy, intellectual disability, attention-deficit hyperactivity disorder, and many others. The latter are highly heritable but are taken to be genetically "complex," and most cases until recently have remained idiopathic (no clear cause has been identified).

In fact, this dichotomy is largely artificial for two reasons (Mitchell 2015). First, more rare mutations are being discovered that confer high risk for these common diagnostic categories, revealing underlying genetic heterogeneity of these umbrella terms. And second, because even the supposedly "Mendelian" conditions show important modifying effects of other mutations or of the polygenic background more generally.

Mutations in axon guidance and synaptic connectivity genes have been implicated in both specific neurological conditions and in these broader diagnostic categories. There is a small number of specific neurological conditions in humans that result from mutations in canonical "axon guidance" genes. (The scare quotes denote the fact that these proteins often play roles in other processes and even in other tissues). In some cases, the neuroanatomical pathology and neurological symptoms likely reflect quite direct roles of the genes in question in patterning specific circuits. These include, for example, defects in oculomotor circuit formation in patients with *ROBO3* mutations, resulting in horizontal gaze palsy (Jen et al. 2004), or defects in midline axonal structures in patients with mutations in the *DCC* gene, resulting in "mirror movements" (Srour et al. 2010), as observed in mice (Fazeli et al. 1997). A similar condition is observed in patients with dominant mutations in *ARHGEF7*, which encodes a component of the DCC signal transduction pathway (Schlienger et al. 2023). (Other examples of human disorders arising from mutations in guidance or connectivity genes [including *CNTNAP2*, *L1CAM*, *NTNG1*, and *PCDH19*, for example] are reviewed in Engle 2010; Blockus and Chedotal 2015; Betancur and Mitchell 2015; and Yuasa-Kawada et al. 2023.)

Most of these conditions are recessive, meaning both copies of the gene must be mutated to cause the phenotype (as is commonly observed for axon guidance genes in model organisms). Such conditions are therefore very rare and have usually been identified in consanguineous populations (e.g., for *ELFN1* gene; Dursun et al. 2021). However, while this is true of complete loss-of-function mutations, mutations that result in a truncated protein (especially for transmembrane receptors) can often have dominant effects. Because many guidance proteins interact in complexes with other factors, a truncated protein can have a greater (dominant-negative) impact on developmental processes than simple absence of a protein. For example, single-copy de novo truncating mutations in *SEMA6B* have been implicated in cases of epilepsy (Hamanaka et al. 2020) and intellectual disability (Cordovado et al. 2022), while similar mutations in *PLXNA1* have been implicated in autism cases (Fu et al. 2022). This class of mutation may make an important contribution to the etiology of neurodevelopmental disorders more generally (Torene et al. 2023).

Mutations in some other genes, including some involved in synaptic connectivity such as *NRXN1*, can present with much more variable clinical manifestations and generally increase risk across many common diagnostic categories

 Cite this article as *Cold Spring Harb Perspect Biol* doi: 10.1101/cshperspect.a041504

(Castronovo et al. 2020). Among rare de novo or inherited high-risk mutations in cases with intellectual disability, developmental delay, epilepsy, autism, or schizophrenia, there are examples of genes involved in axon guidance or neuropil patterning (e.g., *DSCAM*, *PLXNA1*, *PLXNB1*) or synaptogenesis (e.g., *NRXN1*, *NLGN4X*) (Satterstrom et al. 2020; Fu et al. 2022; Trost et al. 2022).

Risk genes are generally enriched for genes expressed in fetal brain and involved more broadly in neurodevelopmental processes. These include processes "upstream" of circuit specification, like regulation of gene expression and chromatin function, as well as "downstream" processes like synaptic plasticity or function, which may impact on activity-dependent refinement of neural circuits (e.g., NMDA-receptor genes like *GRIN2A* or synaptic protein genes like *SHANK3* or *SYN-GAP1*; Betancur and Mitchell 2015). However, there are also many genes implicated in these illnesses where neither a direct nor indirect link to the processes of circuit formation or refinement is apparent.

Even for conditions defined clinically by direct neuroanatomical dysconnectivity, such as agenesis of the corpus callosum, most of the implicated genes are not ones that encode specific guidance cues and receptors (Pânzaru et al. 2022). This reflects the fact that the processes of neural circuit formation do not just rely on the genes specifying the "instructions" of which axons project where, but also on those encoding all the proteins required for the more general machinery of regulation of gene expression, signal transduction, cellular motility, and so on. There are, genetically speaking (as observed for all complex traits [Boyle et al. 2017] and in forward genetic screens in model organisms [Mitchell 2018a]), simply many more ways to impair the processes of neural circuit formation indirectly and nonspecifically than the converse.

Some important principles of the genetic architecture of neurodevelopmental disorders are revealed by the study of a particular class of mutations: copy number variants (CNVs) (Mollon et al. 2023). These are deletions or duplications of segments of chromosomes, or, as in Down syndrome, even whole chromosomes. These mutational events can recur at certain sites in the genome, with a very low frequency, but enough to mean that many, many thousands of people carry the effectively identical genetic lesion. CNVs can cause identifiable syndromes, such as 22q11.2 deletion syndrome, 3q29 deletion syndrome, Williams syndrome, and many others (some classically diagnosable based on things like typical facial morphology, for example). But these genetic lesions are also found at much higher frequency in cases with more general neuropsychiatric conditions than in the control population.

A key finding is that such mutations seem to increase risk across these diagnostic categories, reinforcing the view of an overlapping etiology. The same is true for rare, high-risk, single-gene mutations. Thus, many mutations seem to lead to a general developmental brain dysfunction (Moreno-De-Luca et al. 2013), which can manifest in diverse ways. These manifestations include the qualitatively distinct end states that we recognize as "autism," "schizophrenia," "bipolar disorder," and so on, which may be best thought of as representing maladaptive attractor states that the developing brain may end up in. These phenotypes emerge from the way that the developing brain reacts to a very wide variety of possible insults, rather than any direct molecular function of the disrupted genes. This highlights another key principle at play: the ability of the developing system to buffer such insults is also itself affected by genetic variation.

First, it is observed, not surprisingly, that harboring multiple rare mutations (CNVs or single-gene mutations) increases clinical risk substantially (Girirajan et al. 2010; Guo et al. 2019). Such mutations may be inherited (often from clinically unaffected parents) or may arise de novo. Second, risk is also modified by a polygenic background of common genetic variants of the sort that can be identified by genome-wide association studies (GWAS) (Niemi et al. 2018; Bergen et al. 2019; Antaki et al. 2022; Cirnigliaro et al. 2023).

Polygenic Variation

GWAS for conditions like schizophrenia, autism, bipolar disorder, depression, epilepsy, or other neuropsychiatric conditions have identified hundreds of common single-nucleotide polymorphisms (SNPs), which confer increased risk. SNPs are sites in the genome where some percentage of the population has one base (say an "A"), while others have a different base (say a "C"). If one of the versions is significantly more frequent in cases with a particular condition than in controls, then this association implies that those variants increase risk (given that the opposite direction of causation is implausible). Importantly, the increased risk conferred by any single SNP is tiny—almost, but not quite negligible—but the collective risk conferred by the overall polygenic burden of such common risk variants can be substantial.

This kind of polygenic background has been shown to be a contributor to clinical risk in combination with rare mutations in autism and schizophrenia (Bergen et al. 2019; Antaki et al. 2022; Cirnigliaro et al. 2023) and a modifier even of "Mendelian" neurodevelopmental conditions (Niemi et al. 2018). As with the rare variants, most of this polygenic risk seems to be shared across conditions.

For our purposes, what is interesting is that the genes implicated by these common risk variants, as observed for rare, high-risk mutations, are consistently enriched for ones expressed in the embryonic or fetal brain and involved in various neurodevelopmental processes, including synapse organization and synapse assembly, as well as ion channel biology and synaptic transmission (Mallard et al. 2022; Trubetskoy et al. 2022; Als et al. 2023).

In addition to neurodevelopmental disorders, GWAS also implicate neurodevelopmental genes and processes in the genetic variance contributing to general differences in brain morphology and connectivity as well as psychological and behavioral traits in humans and other species. For example, GWAS of structural brain connectivity measures derived from diffusion-weighted neuroimaging found enrichment for genes involved in "neuronal differentiation," "neural migration," "neural projection guidance," and "axon development" (Sha et al. 2023). This suggests, not surprisingly, that variation in such genes in humans can manifest at the macroscopic level of brain-wide structural connectivity.

The same trend emerges for GWAS of cognitive traits (e.g., Davies et al. 2018) and personality traits (e.g., Karlsson Linnér et al. 2019; Belonogova et al. 2021). These consistently show statistical enrichment for genes expressed in fetal brain and involved in a variety of neurodevelopmental processes, such as "neurogenesis," "regulation of nervous system development," "regulation of neuron projection development," and "synapse assembly," among others. (Note, again, that such enrichments are not exclusive—many other kinds of genes can also contribute to such phenotypic differences, presumably less directly.)

In dogs, a very large study identified genetic drivers of diversification of behavioral traits between breeds. These highlighted variants in noncoding regions of genes enriched for fetal brain expression and for rather broad functional categories of "development" and "neurogenesis." Post hoc analyses of a sheepdog cluster identified variants in a cluster of genes involved in axon guidance, including members of the Ephrin, Netrin, slit, and semaphorin pathways (Dutrow et al. 2022).

Collectively, these genetic studies show that variation in genes directly involved in processes of neural circuit assembly contributes to the etiology of specific, rare neurological disorders, as well as common neuropsychiatric conditions. In addition, variation in neurodevelopmental genes contributes to heritable differences in brain structural connectivity and in psychological and behavioral traits across the general population. It is important not to overplay these enrichments, however—the majority of genetic variants affecting neural circuit development will likely do so indirectly. Regardless, the upshot is that genetic differences affecting brain wiring (directly or indirectly) can manifest in important ways.

However, genetic differences are not the only source of variance in neural circuit development and brain wiring. Another important source—often overlooked—is stochasticity in the processes of development themselves.

DEVELOPMENTAL VARIATION

As mentioned above, the genome does not contain enough information to encode precisely the numbers and positions of all the different cell types in the nervous system and the myriad connections they each make. Rather, it encodes a generative model that constrains the self-organizing processes of cell differentiation and migration and the subsequent processes of axon guidance and synaptic target selection. These processes have evolved to robustly and reliably direct development toward an outcome that falls within a species-typical range. However, this is achieved statistically through collective cellular interactions rather than deterministically on a cell-by-cell basis.

The processes involved rely on wet, jiggly, jittery components and are consequently noisy on molecular and cellular levels (Symmons and Raj 2016). There is substantial stochasticity at play in gene expression (Raj and van Oudenaarden 2008), protein–protein interactions, diffusion of molecules, and all other cellular processes (Tsimring 2014). The system has, necessarily, evolved to minimize the impact of molecular and cellular noise on the resultant processes and outcome of development (Masel and Siegal 2009; Wagner 2013). However, this robustness has limits—considerable variation in the microscopic details of neural anatomy and connectivity still arises due to this inherent randomness. The processes of development do not play out in exactly the same manner in any individual "run" of the program, even when starting from the identical genome (Vogt 2015; Mitchell 2018b).

This kind of variability is apparent even in wild-type animals but manifests much more obviously in mutants. The robustness of the system is an evolved property that depends on the integrity of the genomic "program" as a whole. In the presence of genetic variation—either large-effect single mutations or a polygenic background of common variants—this property of distributed robustness is degraded. The consequence, well known since the early days of genetics, is that mutations tend to not only shift the mean of some quantitative phenotype or produce a tightly defined qualitatively novel phenotype, but also increase the phenotypic variance (Waddington 1957). This observation is commonplace in research on model organisms, where mutations

can result in probabilistic outcomes of neurodevelopmental phenotypes.

To take two arbitrary examples (from my own experience), mutations in the *Semaphorin-6A* gene in mice result in defasciculation and misprojection of the fibers of the fornix—the output projections from the hippocampus to basal forebrain regions (Rünker et al. 2011). However, this phenotype is only partially penetrant and manifests in an apparently probabilistic way on either side (or both, or neither) across genetically identical animals (Fig. 2). Similar diversity is observed in the projections of the corticospinal tract across individual *Sema6A* mutant mice (Okada et al. 2019).

At an even finer level, probabilistic effects can be observed in the projections of individual motor axons from segment to segment in the *Drosophila* embryo. For example, in embryos mutant for the *Netrin-B* gene, the RP3 motor neuron sometimes fails to project to and thus innervate the 6/7 body wall muscles (Mitchell et al. 1996). But it does so with an apparently random probability from segment to segment of around 30% (Fig. 2). (This probability can be changed by simultaneously raising or lowering the levels of other guidance cues [Winberg et al. 1998].)

These examples of intra-individual variation strongly argue against the idea that developmental variability observed between genetically identical animals can be ascribed to unknown environmental factors. Rather, they demonstrate both the inherent variability of developmental processes and the fact that this variability is itself a genetic trait.

In some cases, this kind of noise can manifest in a dichotomous manner at the level of macroscopic connectivity. This is due to the highly contingent nature of brain development and the resultant fact that small changes due to noise in some early processes can have cascading effects over later development, resulting in highly divergent trajectories. One well-studied example is the formation of the corpus callosum, the set of axons connecting the two cerebral hemispheres in mammals (Suárez et al. 2014).

The corpus callosum is pioneered by a small set of early-projecting axons (Fig. 3). Their ability to cross the midline depends on the earlier

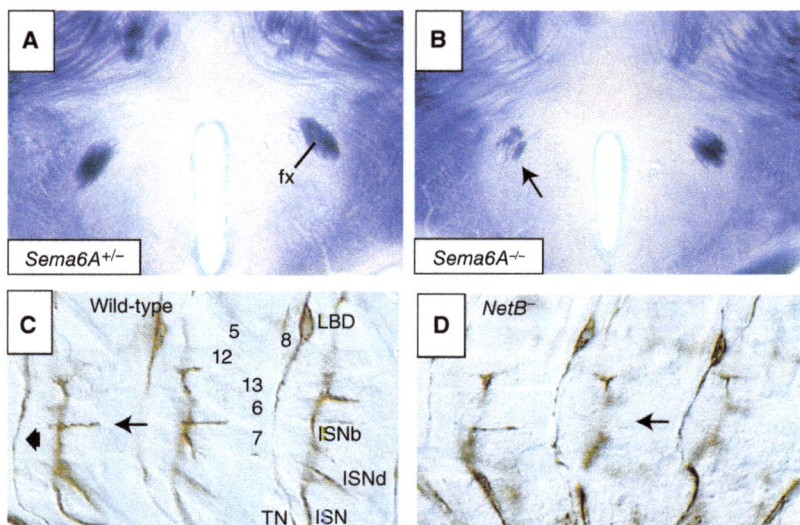

Figure 2. Probabilistic phenotypes. (*A,B*) PLAP-stained adult mouse brain sections from the *Sema6A* gene trap line (Leighton et al. 2001) showing, in coronal cross-section of the tract, a reduced and defasciculated fornix (fx) in (*B*) a homozygous mutant *Sema6A*$^{-/-}$ brain (on one side only; arrow), compared to (*A*) a heterozygous *Sema6A*$^{+/-}$ (phenotypically wild-type) animal. (*C,D*) Photomicrographs of stage 17 *Drosophila* embryos stained with mAb 1D4 (anti-Fas II) to show motor projections across a number of hemi-segments. Anterior is *left*, and dorsal is *up*. (*C*) Wild-type shows normal projection of the RP3 motor axon, which runs in the ISNb nerve, to innervate the cleft between muscles 6 and 7 (arrow). (*D*) In embryos deficient for Netrin-B (normally expressed specifically on muscles 6/7), this innervation is absent in about 30% of segments (arrow). (*A* and *B* reprinted from Rünker et al. 2011 under the terms of the Creative Commons Attribution License; *C* and *D* reprinted, with permission, from Winberg et al. 1998.)

formation of a small astroglial bridge between the two hemispheres (Gobius et al. 2016). If formation of this structure is disrupted, then the pioneer axons cannot cross the midline, and the hundreds of thousands or millions of follower axons that normally comprise the corpus callosum cannot either. Under normal circumstances, this whole process happens highly robustly. But mutations in diverse genes, well studied in both mice and humans, can result in failure of these processes and agenesis of the corpus callosum.

The contingent nature of these processes—which make them highly sensitive to noise at an early stage—results in a (more or less) bimodal distribution of phenotypes across genetically identical individuals (e.g., Ruge and Newland 1996). This has been well studied in mice, where different inbred (and thus isogenic) lines have different frequencies of callosal agenesis—this structure either forms fairly normally or is

completely absent (Wahlsten et al. 2006). This indicates that what is inherited is a certain risk or probability of this phenotype arising, but that the particular outcome depends on the way stochastic developmental processes play out (Fig. 3). There are clear parallels to the inheritance of risk for neuropsychiatric conditions, such as schizophrenia or epilepsy, where even monozygotic twins may differ in clinical manifestation.

These examples show the ubiquity of developmental variation and its importance as a source of phenotypic variance, directly visible at the level of neuroanatomical phenotypes. It is not surprising that such differences in neural circuits can be correlated with behavioral differences, as demonstrated in flies (Linneweber et al. 2020) and humans (Ruge and Newland 1996).

In humans, it seems highly likely that stochastic developmental variation is the source of much of what is called the "non-shared environment" (Mitchell 2018a; Tikhodeyev and

Cite this article as *Cold Spring Harb Perspect Biol* doi: 10.1101/cshperspect.a041504

A

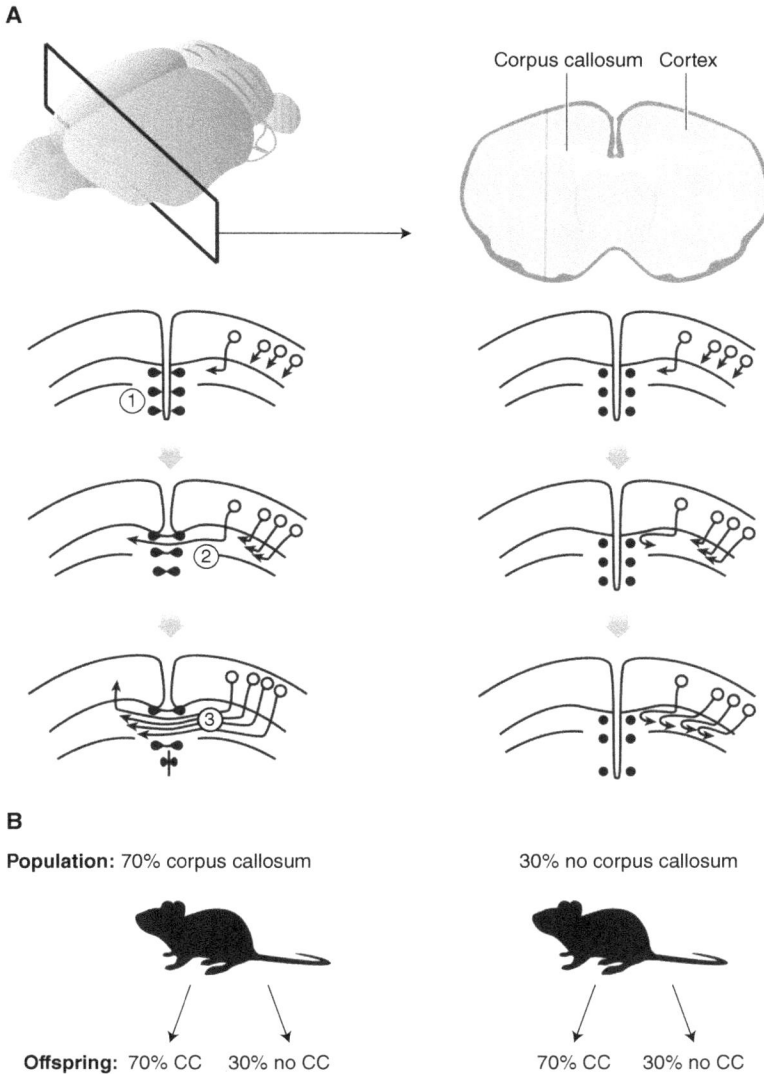

B

Population: 70% corpus callosum   30% no corpus callosum

Offspring: 70% CC   30% no CC   70% CC   30% no CC

Figure 3. Corpus callosum (CC) development. (*A*) (*Top*) A section through an adult mouse brain shows the cerebral cortex covering the two hemispheres and the CC connecting them. (*Bottom left*) The stages of normal CC development. (*1*) Midline cells fuse and form a bridge between the two hemispheres. (*2*) Pioneer axons cross. (*3*) Follower axons cross. (*Bottom right*) When the midline cells fail to fuse, pioneer and follower axons fail to cross, resulting in absence of the CC. (*B*) In some mouse strains, a proportion of animals end up with no CC, despite every animal being genetically identical. This probabilistic effect is inherited regardless of the phenotype of the parent. (Reprinted, with permission, from Mitchell 2018b.)

Shcherbakova 2019). This is the component of phenotypic variance identified from twin and family studies that is attributable to neither genetics nor the family environment. (It is most readily identified as the source of residual differences between identical twins reared together,

but, of course, is a source of variance that affects everyone.) Researchers in behavioral genetics have often ascribed such variance to nonsystematic environmental exposures or idiosyncratic experiences (Plomin 2011) (without much luck in identifying what those might be). An alternative explanation is that this variance is not "environmental" at all, but inherent to the processes of development themselves.

This view is consistent with the interpretation mentioned above of the findings from GWAS of psychiatric conditions. The very general enrichment for neurodevelopmental functions among the genes with "hits" from such studies, rather than any convergence on specific functions, is consistent with the idea that an increasing polygenic burden of such variants decreases developmental robustness (Fig. 4). Individuals with a high polygenic burden are thus less

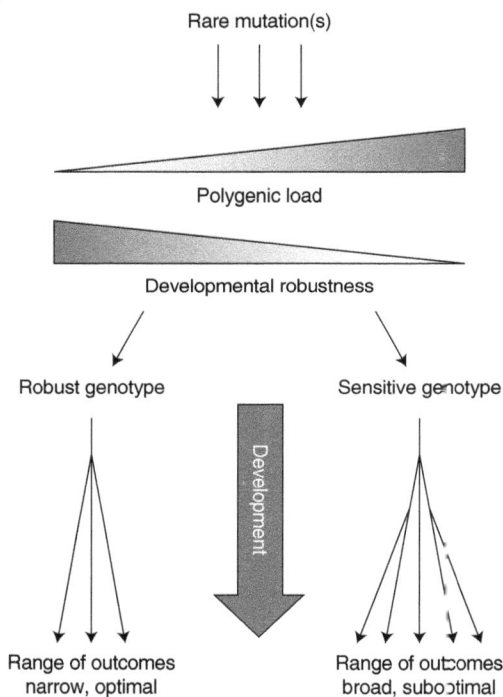

Figure 4. Genetic effects on neural development. Possible effects of rare mutations can be buffered by the distributed robustness of developmental systems. However, this robustness is itself a genetic trait and can be degraded by increasing polygenic load (of other rare mutations or many common genetic variants). Robust genotypes may buffer the effects of rare mutations and produce more optimal outcomes with low variance (across genetically identical individuals), while developmentally sensitive genotypes may produce more variable outcomes, including some with clinical consequences.

able to buffer the effects of new mutations (Mitchell 2015, 2018b), as observed in numerous studies (Niemi et al. 2018; Bergen et al. 2019; Antaki et al. 2022; Cirnigliaro et al. 2023). Thus, not only is developmental robustness a genetic trait, it is a clinically important one.

EVOLUTIONARY DYNAMICS

Understanding the relationship between genotypes and neural circuit phenotypes is essential to understanding how the latter can evolve. Thinking in reductive terms of specific proteins determining specific guidance or connectivity decisions in an isolatable way will naturally lead to the expectation that evolutionary changes in circuit organization should be traceable to specific mutations. There are, in fact, a handful of examples where known mutations affecting protein sequence (e.g., *Robo3*; Zelina et al. 2014) or expression (e.g., *Zic2*; Vigouroux et al. 2021 or *SATB2*; Paolino et al. 2020) are associated with differences in neural connectivity across species (related to midline crossing of various axonal populations).

Whether such mutations were the evolutionary origin of such differences is another matter and much more difficult to test. It is worth examining our expectations about such differences. In general, any new mutation that causes a drastic difference in neural connectivity in some individual is liable to be strongly selected against. It thus does not seem highly likely that single mutations will be solely responsible for large species differences in neural connectivity.

By contrast, the picture outlined above, where neural circuitry and concomitant behavioral phenotypes are affected by thousands of genetic variants across a population, is more conducive to a view of gradual evolutionary change, occurring due to selection or genetic drift as species diverge. Paradoxically, it is the robustness of the system, which clamps down phenotypic deviations in individuals, that is the very thing that allows such change to occur over longer timeframes (Wagner 2013; Hiesinger and Hassan 2018). This distributed robustness allows genetic variations to accumulate in the population while still generating phenotypes within a viable range.

This generates a substrate of genetic variance on which evolution can act, gradually pushing traits along separate dimensions, without making any individuals radically different from their parents or their conspecifics.

Importantly, independent selection on distinct phenotypes can occur in an emergently modular fashion, even when all traits are highly polygenic and individual genetic variants are pleiotropic (Clune et al. 2013; Kouvaris et al. 2017; KJ Mitchell and N Cheney, in press). The contingent nature of development, outlined above, may even mean that qualitatively distinct outcomes can arise as a consequence of such polygenic selection. That is, novel phenotypes may be reachable even by gradual genetic change. In addition, in populations undergoing this kind of change, perhaps in a novel environment, new mutations may be much more likely to be beneficial if they reinforce the direction of phenotypic travel.

SUMMARY

The relationship between genotypes and phenotypes reflects the distributed, collective nature of the genomic encoding of neural connectivity and the probabilistic, statistical nature of the developmental processes that realize it. At a genomic level, rare mutations, which can have large individual effects, and the collective effects of thousands of common genetic variants, can influence the phenotype. The magnitude of any such effects is often buffered by the distributed robustness of the whole developmental system, but this robustness is itself a genetic trait and is degraded by increasing mutational load. Developmental variability is thus inevitable and can lead to dichotomous outcomes.

In turn, variation in neural architecture contributes to variation in psychological and behavioral traits across the population and to risk of neuropsychiatric disorders, both rare and common. It is thus of enormous clinical importance to understand both the sources and the consequences of such variation. Over much longer timeframes, the underlying genetic variation also provides the substrate for evolutionary change, with natural selection acting as the critic

of any behavioral manifestations. Understanding the nature of the encoding of neural connectivity and the relationships between genotypes and phenotypes is thus essential to understanding how circuits evolve and how this underlies variation in species-typical behaviors.

REFERENCES

Als TD, Kurki MI, Grove J, Voloudakis G, Therrien K, Tasanko E, Nielsen TT, Naamanka J, Veerapen K, Levey DF, et al. 2023. Depression pathophysiology, risk prediction of recurrence and comorbid psychiatric disorders using genome-wide analyses. *Nat Med* **29:** 1832–1844. doi:10.1038/s41591-023-02352-1

Antaki D, Guevara J, Maihofer AX, Klein M, Gujral M, Grove J, Carey CE, Hong O, Arranz MJ, Hervas A, et al. 2022. A phenotypic spectrum of autism is attributable to the combined effects of rare variants, polygenic risk and sex. *Nat Genet* **54:** 1284–1292. doi:10.1038/s41588-022-01064-5

Belonogova NM, Zorkoltseva IV, Tsepilov YA, Axenovich TI. 2021. Gene-based association analysis identifies 190 genes affecting neuroticism. *Sci Rep* **11:** 2484. doi:10.1038/s41598-021-82123-5

Bergen SE, Ploner A, Howrigan D; CNV Analysis Group and the Schizophrenia Working Group of the Psychiatric Genomics Consortium; O'Donovan MC, Smoller JW, Sullivan PF, Sebat J, Neale B, Kendler KS. 2019. Joint contributions of rare copy number variants and common SNPs to risk for schizophrenia. *Am J Psychiatry* **176:** 29–35. doi:10.1176/appi.ajp.2018.17040467

Betancur C, Mitchell KJ. 2015. Synaptic disorders. In *The genetics of neurodevelopmental disorders* (ed. Mitchell KJ), pp. 195–238. Wiley, Hoboken, NJ.

Blockus H, Chedotal A. 2015. Disorders of axon guidance. In *The genetics of neurodevelopmental disorders* (ed. Mitchell KJ), pp. 155–194. Wiley, Hoboken, NJ.

Boyle EA, Li YI, Pritchard JK. 2017. An expanded view of complex traits: from polygenic to omnigenic. *Cell* **169:** 1177–1186. doi:10.1016/j.cell.2017.05.038

Breau MA, Trembleau A. 2023. Chemical and mechanical control of axon fasciculation and defasciculation. *Semin Cell Dev Biol* **140:** 72–81. doi:10.1016/j.semcdb.2022.06.014

Castronovo P, Baccarin M, Ricciardello A, Picinelli C, Tomaiuolo P, Cucinotta F, Frittoli M, Lintas C, Sacco R, Persico AM. 2020. Phenotypic spectrum of *NRXN1* mono- and bi-allelic deficiency: a systematic review. *Clin Genet* **97:** 125–137. doi:10.1111/cge.13537

Cate MS, Gajendra S, Alsbury S, Raabe T, Tear G, Mitchell KJ. 2016. Mushroom body defect is required in parallel to Netrin for midline axon guidance in *Drosophila*. *Development* **143:** 972–977. doi:10.1242/dev.129684

Cirnigliaro M, Chang TS, Arteaga SA, Pérez-Cano L, Ruzzo EK, Gordon A, Bicks LK, Jung JY, Lowe JK, Wall DP, et al. 2023. The contributions of rare inherited and polygenic risk to ASD in multiplex families. *Proc Natl Acad Sci* **120:** e2215632120. doi:10.1073/pnas.2215632120

Clune J, Mouret JB, Lipson H. 2013. The evolutionary origins of modularity. *Proc Biol Sci* **280**: 20122863. doi:10.1098/rspb.2012.2863

Cordovado A, Schaettin M, Jeanne M, Panasenkava V, Denommé-Pichon AS, Keren B, Mignot C, Doco-Fenzy M, Rodan L, Ramsey K, et al. 2022. *SEMA6B* variants cause intellectual disability and alter dendritic spine density and axon guidance. *Hum Mol Genet* **31**: 3325–3340 doi:10.1093/hmg/ddac114

Davies G, Lam M, Harris SE, Trampush JW, Luciano M, Hill WD, Hagenaars SP, Ritchie SJ, Marioni RE, Fawns-Ritchie C, et al. 2018. Study of 300,486 individuals identifies 148 independent genetic loci influencing general cognitive function. *Nat Commun* **9**: 2098. doi:10.1038/s41467-018-04362-x

de Wit J, Ghosh A. 2016. Specification of synaptic connectivity by cell surface interactions. *Nat Rev Neurosci* **17**: 22–35. doi:10.1038/nrn.2015.3

Dursun A, Yalnizoglu D, Yilmaz DY, Oguz KK, Gülbakan B, Koşukcu C, Akar HT, Kahraman AB, Acar NV, Günbey C, et al. 2021. Biallelic mutations in ELFN1 gene associated with developmental and epileptic encephalopathy and joint laxity. *Eur J Med Genet* **64**: 104340. doi:10.1016/j.ejmg.2021.104340

Dutrow EV, Serpell JA, Ostrander EA. 2022. Domestic dog lineages reveal genetic drivers of behavioral diversification. *Cell* **185**: 4737–4755.e18. doi:10.1016/j.cell.2022.11.003

Engle EC. 2010. Human genetic disorders of axon guidance. *Cold Spring Harb Perspect Biol* **2**: a001784. doi:10.1101/cshperspect.a001784

Fazeli A, Dickinson SL, Hermiston ML, Tighe RV, Steen RG, Small CG, Stoeckli ET, Keino-Masu K, Masu M, Rayburn H, et al. 1997. Phenotype of mice lacking functional *Deleted* in colorectal cancer (Dcc) gene. *Nature* **386**: 796–804. doi:10.1038/386796a0

Feldheim DA, Kim YI, Bergemann AD, Frisén J, Barbacid M, Flanagan JG. 2000. Genetic analysis of ephrin-A2 and ephrin-A5 shows their requirement in multiple aspects of retinocollicular mapping. *Neuron* **25**: 563–574. doi:10.1016/s0896-6273(00)81060-0

Franze K. 2020. Integrating chemistry and mechanics: the forces driving axon growth. *Annu Rev Cell Dev Biol* **36**: 61–83. doi:10.1146/annurev-cellbio-100818-125157

Fu JM, Satterstrom FK, Peng M, Brand H, Collins RL Dong S, Wamsley B, Klei L, Wang L, Hao SP, et al. 2022. Rare coding variation provides insight into the genetic architecture and phenotypic context of autism. *Nat Genet* **54**: 1320–1331. doi:10.1038/s41588-022-01104-0

Girirajan S, Rosenfeld JA, Cooper GM, Antonacci F, Siswara P, Itsara A, Vives L, Walsh T, McCarthy SE, Baker C, et al. 2010. A recurrent 16p12.1 microdeletion supports a two-hit model for severe developmental delay. *Nat Genet* **42**: 203–209. doi:10.1038/ng.534

Gobius I, Morcom L, Suárez R, Bunt J, Bukshpun P, Reardon W, Dobyns WB, Rubenstein JL, Barkovich AJ, Sherr EH, et al. 2016. Astroglial-mediated remodeling of the interhemispheric midline is required for the formation of the corpus callosum. *Cell Rep* **17**: 735–747. doi:10.1016/j.celrep.2016.09.033

Guo H, Duyzend MH, Coe BP, Baker C, Hoekzema K, Gerdts J, Turner TN, Zody MC, Beighley JS, Murali SC, et al 2019. Genome sequencing identifies multiple deleterious variants in autism patients with more severe phenotypes. *Genet Med* **21**: 1611–1620. doi:10.1038/s41436-018-0380-2

Hamanaka K, Imagawa E, Koshimizu E, Miyatake S, Tohyama J, Yamagata T, Miyauchi A, Ekhilevitch N, Nakamura F, Kawashima T, et al. 2020. De novo truncating variants in the last exon of SEMA6B cause progressive myoclonic epilepsy. *Am J Hum Genet* **106**: 549–558. doi:10.1016/j.ajhg.2020.02.011

Hassan BA, Hiesinger PR. 2015. Beyond molecular codes: simple rules to wire complex brains. *Cell* **163**: 285–291. doi:10.1016/j.cell.2015.09.031

Hedgecock EM, Culotti JG, Hall DH. 1990. The *unc*-5, *unc*-6, and *unc*-40 genes guide circumferential migrations of pioneer axons and mesodermal cells on the epidermis in *C. elegans*. *Neuron* **4**: 61–85. doi:10.1016/0896-6273(90)90444-k

Herrera E, Escalante A. 2022. Transcriptional control of axon guidance at midline structures. *Front Cell Dev Biol* **10**: 840005. doi:10.3389/fcell.2022.840005

Hiesinger PR. 2021a. *The self-assembling brain. How neural networks grow smarter*. Princeton University Press, Princeton, NJ.

Hiesinger PR. 2021b. Brain wiring with composite instructions. *Bioessays* **43**: e2000166. doi:10.1002/bies.202000166

Hiesinger PR, Hassan BA. 2018. The evolution of variability and robustness in neural development. *Trends Neurosci* **41**: 577–586. doi:10.1016/j.tins.2018.05.007

Jen JC, Chan WM, Bosley TM, Wan J, Carr JR, Rüb U, Shattuck D, Salamon G, Kudo LC, Ou J, et al. 2004. Mutations in a human *ROBO* gene disrupt hindbrain axon pathway crossing and morphogenesis. *Science* **304**: 1509–1513. doi:10.1126/science.1096437

Karlsson Linnér R, Biroli P, Kong E, Meddens SFW, Wedow R, Fontana MA, Lebreton M, Tino SP, Abdellaoui A, Hammerschlag AR, et al. 2019. Genome-wide association analyses of risk tolerance and risky behaviors in over 1 million individuals identify hundreds of loci and shared genetic influences. *Nat Genet* **51**: 245–257. doi:10.1038/s41588-018-0309-3

Klumpe HE, Garcia-Ojalvo J, Elowitz MB, Antebi YE. 2023. The computational capabilities of many-to-many protein interaction networks. *Cell Syst* **14**: 430–446. doi:10.1016/j.cels.2023.05.001

Knoedler JR, Shah NM. 2018. Molecular mechanisms underlying sexual differentiation of the nervous system. *Curr Opin Neurobiol* **53**: 192–197. doi:10.1016/j.conb.2018.09.005

Kolodkin AL, Tessier-Lavigne M. 2011. Mechanisms and molecules of neuronal wiring: a primer. *Cold Spring Harb Perspect Biol* **3**: a001727. doi:10.1101/cshperspect.a001727

Koulakov A, Shuvaev S, Lachi D, Zador A. 2021. Encoding innate ability through a genomic bottleneck. bioRxiv doi:10.1101/2021.03.16.435261

Kouvaris K, Clune J, Kounios L, Brede M, Watson RA. 2017. How evolution learns to generalise: using the principles of learning theory to understand the evolution of developmental organisation. *PLoS Comput Biol* **13**: e1005358. doi:10.1371/journal.pcbi.1005358

Leighton PA, Mitchell KJ, Goodrich LV, Lu X, Pinson K, Scherz P, Skarnes WC, Tessier-Lavigne M. 2001. Defining brain wiring patterns and mechanisms through gene trapping in mice. *Nature* **410:** 174–179. doi:10.1038/35065539

Linneweber GA, Andriatsilavo M, Dutta SB, Bengochea M, Hellbruegge L, Liu G, Ejsmont RK, Straw AD, Wernet M, Hiesinger PR, et al. 2020. A neurodevelopmental origin of behavioral individuality in the *Drosophila* visual system. *Science* **367:** 1112–1119. doi:10.1126/science.aaw7182

Mallard TT, Linnér RK, Grotzinger AD, Sanchez-Roige S, Seidlitz J, Okbay A, de Vlaming R, Meddens SFW, Bipolar Disorder Working Group of the Psychiatric Genomics Consortium, Palmer AA, et al. 2022. Multivariate GWAS of psychiatric disorders and their cardinal symptoms reveal two dimensions of cross-cutting genetic liabilities. *Cell Genom* **2:** 100140. doi:10.1016/j.xgen.2022.100140

Marquardt T, Shirasaki R, Ghosh S, Andrews SE, Carter N, Hunter T, Pfaff SL. 2005. Co-expressed EphA receptors and ephrin-A ligands mediate opposing actions on growth cone navigation from distinct membrane domains. *Cell* **121:** 127–139. doi:10.1016/j.cell.2005.01.020

Masel J, Siegal ML. 2009. Robustness: mechanisms and consequences. *Trends Genet* **25:** 395–403. doi:10.1016/j.tig.2009.07.005

Meeh KL, Rickel CT, Sansano AJ, Shirangi TR. 2021. The development of sex differences in the nervous system and behavior of flies, worms, and rodents. *Dev Biol* **472:** 75–84. doi:10.1016/j.ydbio.2021.01.010

Mitchell KJ. 2015. The genetic architecture of neurodevelopmental disorders. In *The genetics of neurodevelopmental disorders* (ed. Mitchell KJ), pp. 1–28. Wiley, Hoboken, NJ.

Mitchell KJ. 2018a. Revealing the genetic instructions for nervous system wiring. *Trends Neurosci* **41:** 407–409. doi:10.1016/j.tins.2018.04.008

Mitchell KJ. 2018b. *Innate: how the wiring of our brains shapes who we are*. Princeton University Press, Princeton, NJ.

Mitchell KJ, Doyle JL, Serafini T, Kennedy TE, Tessier-Lavigne M, Goodman CS, Dickson BJ. 1996. Genetic analysis of *Netrin* genes in *Drosophila*: Netrins guide CNS commissural axons and peripheral motor axons. *Neuron* **17:** 203–215. doi:10.1016/s0896-6273(00)80153-1

Mollon J, Almasy L, Jacquemont S, Glahn DC. 2023. The contribution of copy number variants to psychiatric symptoms and cognitive ability. *Mol Psychiatry* **28:** 1480–1493. doi:10.1038/s41380-023-01978-4

Morales D, Kania A. 2017. Cooperation and crosstalk in axon guidance cue integration: additivity, synergy, and fine-tuning in combinatorial signaling. *Dev Neurobiol* **77:** 891–904. doi:10.1002/dneu.22463

Moreno-De-Luca A, Myers SM, Challman TD, Moreno-De-Luca D, Evans DW, Ledbetter DH. 2013. Developmental brain dysfunction: revival and expansion of old concepts based on new genetic evidence. *Lancet Neurol* **12:** 406–414. doi:10.1016/S1474-4422(13)70011-5

Niemi MEK, Martin HC, Rice DL, Gallone G, Gordon S, Kelemen M, McAloney K, McRae J, Radford EJ, Yu S, et al. 2018. Common genetic variants contribute to risk of rare severe neurodevelopmental disorders. *Nature* **562:** 268–271. doi:10.1038/s41586-018-0566-4

Nose A, Takeichi M, Goodman CS. 1994. Ectopic expression of connectin reveals a repulsive function during growth cone guidance and synapse formation. *Neuron* **13:** 525–539. doi:10.1016/0896-6273(94)90023-x

Okada T, Keino-Masu K, Suto F, Mitchell KJ, Masu M. 2019. Remarkable complexity and variability of corticospinal tract defects in adult Semaphorin 6A knockout mice. *Brain Res* **1710:** 209–219. doi:10.1016/j.brainres.2018.12.041

Pânzaru MC, Popa S, Lupu A, Gavrilovici C, Lupu VV, Gorduza EV. 2022. Genetic heterogeneity in corpus callosum agenesis. *Front Genet* **13:** 958570. doi:10.3389/fgene.2022.958570

Paolino A, Fenlon LR, Kozulin P, Haines E, Lim JWC, Richards LJ, Suárez R. 2020. Differential timing of a conserved transcriptional network underlies divergent cortical projection routes across mammalian brain evolution. *Proc Natl Acad Sci* **117:** 10554–10564. doi:10.1073/pnas.1922412117

Perez-Branguli F, Zagar Y, Shanley DK, Graef IA, Chédotal A, Mitchell KJ. 2016. Reverse signaling by semaphorin-6A regulates cellular aggregation and neuronal morphology. *PLoS ONE* **11:** e0158686. doi:10.1371/journal.pone.0158686

Plomin R. 2011. Commentary: why are children in the same family so different? Non-shared environment three decades later. *Int J Epidemiol* **40:** 582–592. doi:10.1093/ije/dyq144

Popova S, Charness ME, Burd L, Crawford A, Hoyme HE, Mukherjee RAS, Riley EP, Elliott EJ. 2023. Fetal alcohol spectrum disorders. *Nat Rev Dis Primers* **9:** 11. doi:10.1038/s41572-023-00420-x

Raj A, van Oudenaarden A. 2008. Nature, nurture, or chance: stochastic gene expression and its consequences. *Cell* **135:** 216–226. doi:10.1016/j.cell.2008.09.050

Ruge JR, Newland TS. 1996. Agenesis of the corpus callosum: female monozygotic triplets. case report. *J Neurosurg* **85:** 152–156. doi:10.3171/jns.1996.85.1.0152

Rünker AE, O'Tuathaigh C, Dunleavy M, Morris DW, Little GE, Corvin AP, Gill M, Henshall DC, Waddington JL, Mitchell KJ. 2011. Mutation of semaphorin-6A disrupts limbic and cortical connectivity and models neurodevelopmental psychopathology. *PLoS ONE* **6:** e26488. doi:10.1371/journal.pone.0026488

Sanes JR, Zipursky SL. 2020. Synaptic specificity, recognition molecules, and assembly of neural circuits. *Cell* **181:** 536–556. doi:10.1016/j.cell.2020.04.008

Satterstrom FK, Kosmicki JA, Wang J, Breen MS, De Rubeis S, An JY, Peng M, Collins R, Grove J, Klei L, et al. 2020. Large-scale exome sequencing study implicates both developmental and functional changes in the neurobiology of autism. *Cell* **180:** 568–584.e23. doi:10.1016/j.cell.2019.12.036

Schlienger S, Yam PT, Balekoglu N, Ducuing H, Michaud JF, Makihara S, Kramer DK, Chen B, Fasano A, Berardelli A, et al. 2023. Genetics of mirror movements identifies a multifunctional complex required for Netrin-1 guidance and lateralization of motor control. *Sci Adv* **9:** eadd5501. doi:10.1126/sciadv.add5501

Seeger M, Tear G, Ferres-Marco D, Goodman CS. 1993. Mutations affecting growth cone guidance in *Drosophila*: genes necessary for guidance toward or away from the midline. *Neuron* **10:** 409–426. doi:10.1016/0896-6273(93)90330-t

Serafini T, Colamarino SA, Leonardo ED, Wang H, Bedding-ton R, Skarnes WC, Tessier-Lavigne M. 1996. Netrin-1 is required for commissural axon guidance in the developing vertebrate nervous system. *Cell* **87**: 1001–1014. doi:10.1016/s0092-8674(00)81795-x.

Sha Z, Schijven D, Fisher SE, Francks C. 2023. Genetic architecture of the white matter connectome of the human brain. *Sci Adv* **9**: eadd2870. doi:10.1126/sciadv.add2870

Singh T, Poterba T, Curtis D, Akil H, Al Eissa M, Barchas JD, Bass N, Bigdeli TB, Breen G, Bromet EJ, et al. 2022. Rare coding variants in ten genes confer substantial risk for schizophrenia. *Nature* **604**: 509–516. doi:10.1038/s41586-022-04556-w

Srour M, Rivière JB, Pham JM, Dubé MP, Girard S, Morin S, Dion PA, Asselin G, Rochefort D, Hince P, et al. 2010. Mutations in *DCC* cause congenital mirror movements. *Science* **328**: 592. doi:10.1126/science.1186463

Stoeckli ET. 2018. Understanding axon guidance: are we nearly there yet? *Development* **145**: dev151415. doi:10.1242/dev.151415

Suárez R, Gobius I, Richards LJ. 2014. Evolution and development of interhemispheric connections in the vertebrate forebrain. *Front Hum Neurosci* **8**: 497. doi:10.3389/fnhum.2014.00497

Südhof TC. 2018. Towards an understanding of synapse formation. *Neuron* **100**: 276–293. doi:10.1016/j.neuron.2018.09.040

Symmons O, Raj A. 2016. What's luck got to do with it: single cells, multiple fates, and biological nondeterminism. *Mol Cell* **62**: 788–802. doi:10.1016/j.molcel.2016.05.023

Tikhodeyev ON, Shcherbakova OV. 2019. The problem of non-shared environment in behavioral genetics. *Behav Genet* **49**: 259–269. doi:10.1007/s10519-019-09950-1

Torene RI, Guillen Sacoto MJ, Millan F, Zhang Z, McGee S, Oetjens M, Heise E, Chong K, Sidlow R, et al. 2023. Systematic analysis of variants escaping nonsense-mediated decay uncovers candidate Mendelian diseases. *Am J Hum Genet* doi:10.1016/j.ajhg.2023.11.007

Trost B, Thiruvahindrapuram B, Chan AJS, Engchuan W, Higginbotham EJ, Howe JL, Loureiro LO, Reuter MS, Roshandel D, Whitney J, et al. 2022. Genomic architecture of autism from comprehensive whole-genome sequence annotation. *Cell* **185**: 4409–4427. doi:10.1016/j.cell.2022.10.009

Trubetskoy V, Pardiñas AF, Qi T, Panagiotaropoulou G, Awasthi S, Bigdeli TB, Bryois J, Chen CY, Dennison CA, Hall LS, et al. 2022. Mapping genomic loci implicates genes and synaptic biology in schizophrenia. *Nature* **604**: 502–508. doi:10.1038/s41586-022-04434-5

Tsimring LS. 2014. Noise in biology. *Rep Prog Phys* **77**: 026601. doi:10.1088/0034-4885/77/2/026601

van Vactor D, Sink H, Fambrough D, Tsoo R, Goodman CS. 1993. Genes that control neuromuscular specificity in *Drosophila*. *Cell* **73**: 1137–1153. doi:10.1016/0092-8674(93)90643-5

Vigouroux RJ, Duroure K, Vougny J, Albadri S, Kozulin P, Herrera E, Nguyen-Ba-Charvet K, Braasch I, Suárez R, Del Bene F, et al. 2021. Bilateral visual projections exist in non-teleost bony fish and predate the emergence of tetrapods. *Science* **372**: 150–156. doi:10.1126/science.abe7790

Vogt G. 2015. Stochastic developmental variation, an epigenetic source of phenotypic diversity with far-reaching biological consequences. *J Biosci* **40**: 159–204. doi:10.1007/s12038-015-9506-8

Waddington C. 1957. *The strategy of the genes*. Macmillan, New York.

Wagner A. 2013. *Robustness and evolvability in living systems*. Princeton University Press, Princeton, NJ.

Wahlsten D, Bishop KM, Ozaki HS. 2006. Recombinant inbreeding in mice reveals thresholds in embryonic corpus callosum development. *Genes Brain Behav* **5**: 170–188. doi:10.1111/j.1601-183X.2005.00153.x

Winberg ML, Mitchell KJ, Goodman CS. 1998. Genetic analysis of the mechanisms controlling target selection: complementary and combinatorial functions of netrins, semaphorins, and IgCAMs. *Cell* **93**: 581–591. doi:10.1016/s0092-8674(00)81187-3

Xu C, Theisen E, Maloney R, Peng J, Santiago I, Yapp C, Werkhoven Z, Rumbaut E, Shum B, Tarnogorska D, et al. 2020. Control of synaptic specificity by establishing a relative preference for synaptic partners. *Neuron* **106**: 355. doi:10.1016/j.neuron.2020.04.007

Yu HH, Huang AS, Kolodkin AL. 2000. Semaphorin-1a acts in concert with the cell adhesion molecules fasciclin II and connectin to regulate axon fasciculation in *Drosophila*. *Genetics* **156**: 723–731. doi:10.1093/genetics/156.2.723

Yuasa-Kawada J, Kinoshita-Kawada M, Tsuboi Y, Wu JY. 2023. Neuronal guidance genes in health and diseases. *Protein Cell* **14**: 238–261. doi:10.1093/procel/pwac030

Zang Y, Chaudhari K, Bashaw GJ. 2021. New insights into the molecular mechanisms of axon guidance receptor regulation and signaling. *Curr Top Dev Biol* **142**: 147–196. doi:10.1016/bs.ctdb.2020.11.008

Zarin AA, Asadzadeh J, Labrador JP. 2014. Transcriptional regulation of guidance at the midline and in motor circuits. *Cell Mol Life Sci* **71**: 419–432. doi:10.1007/s00018-013-1434-x

Zelina P, Blockus H, Zagar Y, Péres A, Friocourt F, Wu Z, Rama N, Fouquet C, Hohenester E, Tessier-Lavigne M, et al. 2014. Signaling switch of the axon guidance receptor Robo3 during vertebrate evolution. *Neuron* **84**: 1258–1272. doi:10.1016/j.neuron.2014.11.004

Cite this article as *Cold Spring Harb Perspect Biol* doi: 10.1101/cshperspect.a041504

Bringing Chandelier Cells Out of the Shadows: Exploring the Development of a Unique Neuron Type in the Brain

Clara Lenherr,[1,2] Guilherme Neves,[1,2] Marcio Guiomar de Oliveira,[1,2] and Juan Burrone[1,2]

[1]Centre for Developmental Neurobiology; [2]MRC Centre for Neurodevelopmental Disorders, Kings College London, New Hunts House, Guys Hospital Campus, London SE1 1UL, United Kingdom

Correspondence: juan.burrone@kcl.ac.uk

Chandelier cells (ChCs) represent a unique GABAergic interneuron in the cortex, yet our knowledge of this sparsely populated cell type has remained equally sparse for many years. New tools, however, have brought ChCs out of the shadows, shedding light on their development and function in the rodent brain and, gradually, gaining insights into their properties in primates. This review will focus on the developmental mechanisms that define ChCs as a unique cell type and, where possible, draw parallels to studies in primates, particularly to work in human tissue. What emerges is a picture of a highly plastic neuron with a unique developmental trajectory that appears to be genetically and functionally conserved in the primate brain.

The adult mammalian cortex is formed of essentially two main types of neurons: excitatory pyramidal neurons that release glutamate and inhibitory interneurons that release GABA and modulate pyramidal cell activity. GABAergic interneurons belong to a heterogeneous population of cells with varying morphologies, physiological characteristics, and gene expression profiles (Lim et al. 2018), collectively responsible for dictating the overall levels of activity and synchrony of neuronal circuits. Two important GABAergic cell types that directly innervate pyramidal cells are the somatostatin (SST) and parvalbumin (PV) expressing interneurons. PV cells, typically considered to be fast-spiking cells, can be further divided into two subclasses of interneurons: basket cells (BCs) and chandelier cells (ChCs). Whereas SST cells (SCs) preferentially target dendrites, locally influencing the integration of excitatory inputs, BCs mainly target the soma, and ChCs the axon initial segments (AISs) of pyramidal neurons, directly controlling their output. Arguably, ChCs are one of the least understood interneurons in the brain. Although the presence of synapses along the AIS had already been described in early electron microscopy (EM) work in the 1960s (Fig. 1D), it was not until the mid-1970s that the neurons responsible for this innervation were identified (Szentágothai and Arbib 1974; Jones 1975; Szentágothai 1975; Somogyi 1977). Even then, our knowledge of this

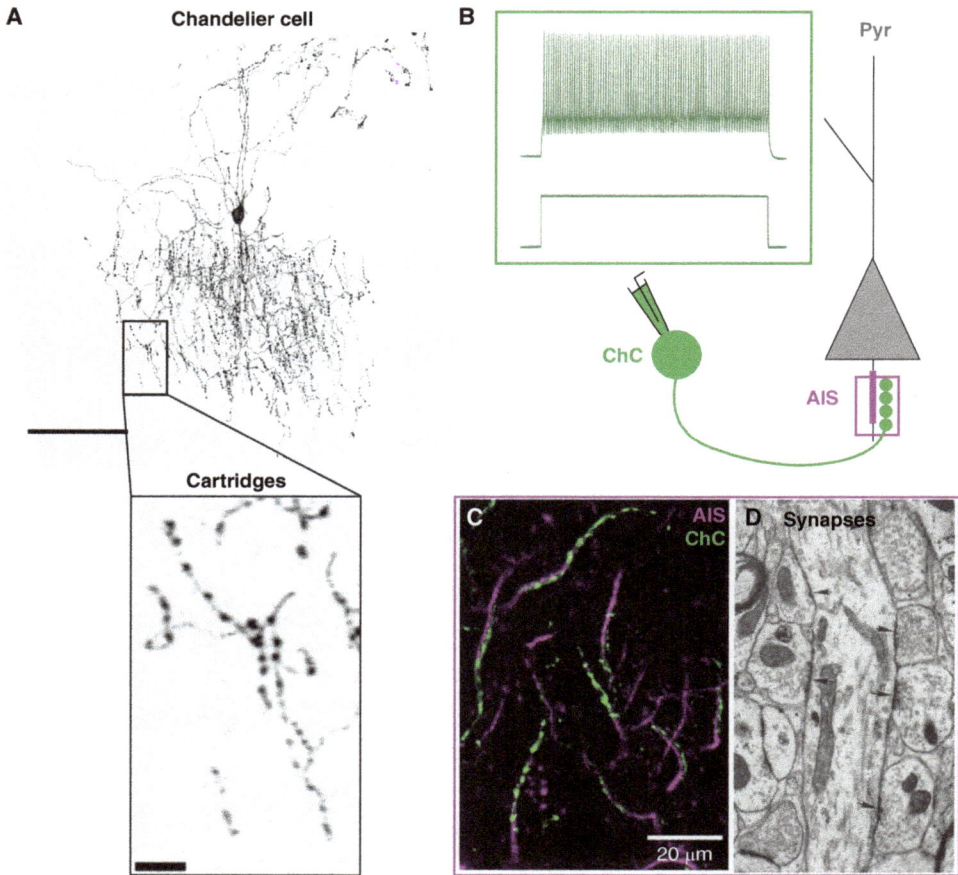

Figure 1. Chandelier cells (ChCs) are highly specialized interneurons that exclusively form synapses along the axon initial segment (AIS) of pyramidal cells. (*A*) Confocal image of a green fluorescent protein (GFP) expressing ChC from the L2/3 somatosensory cortex of an adult mouse. Cartridges of ChC axo-axonic boutons are enlarged from the area highlighted with a rectangle. (*B*) Example whole-cell current clamp recording from a ChC showing the typical fast-spiking phenotype. Obtained from the Allen Brain Atlas Cell Types Database (celltypes.brain-map.org/experiment/electrophysiology/614777438). (*C*) Confocal image of ChC boutons (green) forming synapses along the axon initial segment of pyramidal neurons (magenta). (*D*) Transmission electron micrograph from the sensory cortex of the cat showing numerous synaptic boutons forming synapses (labeled with arrows) on the axon initial segment. Scale bars, 100 µm (*A*, *top*); 10 µm (*A*, *bottom*). (*D*, reprinted, with permission, from Jones and Powell 1969.)

interneuron remained sparse until the more recent development of transgenic mice that allowed the labeling of ChCs (Taniguchi et al. 2013) and began to shed light on their properties. ChCs are fast-spiking interneurons with a distinct morphology consisting of a highly arborized and densely packed axon, forming rows of synapses, termed cartridges, perfectly aligned along the AIS of hundreds of neighboring pyra-

midal neurons, resembling candlesticks on a chandelier (Fig. 1; Fairén and Valverde 1980; Peters et al. 1982; Somogyi et al. 1982; Freund et al. 1983; DeFelipe et al. 1985). The high connection probability of ChC axons, together with the selective targeting of the AIS, a subcellular compartment responsible for the initiation of action potentials and therefore crucial for integrating inputs to generate an output (Grubb and

Burrone 2010a; Kole and Brette 2018), suggest that ChCs likely play an important role in network activity (Compans and Burrone 2023). This review charts our current understanding of ChCs based on work in rodents and, where possible, primates, and highlights the features that make them an attractive cell type to study neuronal development and function.

DEFINING CHANDELIER CELLS

Defining a cell type is not an easy task (Dance 2024), but ChCs appear to represent a particularly good example of a neuron that is reliably identified as a unique type of cell across different species. In fact, classifying different interneurons into distinct groups has been an important area of research that has generated intense debate for many years (Petilla Interneuron Nomenclature Group et al. 2008; Zeng and Sanes 2017; Huang and Paul 2019; Yuste et al. 2020) and remains a work in progress (Gouwens et al. 2020; Miller et al. 2020; Yuste et al. 2020). Based mostly on their morphology and to some extent their physiology, ChCs can be readily distinguished from other interneuron types. In addition to these defining features, ChCs also show specific gene expression profiles. Recent unbiased clustering approaches that make use of high-throughput single-cell transcriptomics (scRNAseq) have revolutionized the classification of neuron types and identified new molecular markers for specific cells in the brain. Modeled on the HUGO Gene Nomenclature Committee (HGNC) system, each transcriptomic cell cluster is assigned a unique and meaningful accession ID, analogous to a gene HGNC symbol (Miller et al. 2020), where all information collected on each cell type across multiple laboratories can be aggregated. Using this molecular classification, ChCs appear to fall into two distinct groups: those located in upper layers, classified as 051—Pvalb Chandelier Gaba, currently the best-characterized group and the main focus of this review, and those in deeper layers, classified as 050 Lamp5 Lhx6 Gaba (Fig. 2A; Yao et al. 2023). In recent studies using Patch-Seq methodology (Fuzik et al. 2016), the properties of Pvalb Chandelier Gaba cells, and to a lesser extent Lamp5 Lhx6 Gaba cells, could be directly matched with an independent classification that combined morphological, electrophysiological, and molecular measurements (Gouwens et al. 2020). Compared with other transcriptionally defined cell subtypes, Pvalb Chandelier Gaba cells had highly homogeneous morphological and electrophysiological properties (Scala et al. 2021). Whether both sets of ChCs are part of the same neuron type is an ongoing topic of debate (Huang and Paul 2019). In this review, we will concentrate on upper layer Pvalb Chandelier cells since they have been the main focus of past work. For simplicity, we will refer to them as ChCs from here on.

One exciting line of investigation that these approaches have enabled is the study of the evolution of cellular diversity. Of particular relevance is the identification of homologous cell classes between mouse and primate cortex. Recent work used scRNAseq analysis to identify putative ChCs in the gorilla, macaque, and chimpanzee cortex (Suresh et al. 2023). In line with this, an early study found that ChCs were one of seven closely matched transcriptional cell types shared between human and mouse cortices (Hodge et al. 2019), confirming their status as a unique prototypical neuron class, while also suggesting their function is conserved between mouse and human. More recently, epigenomic data such as chromatin accessibility and DNA methylation have been integrated into the transcriptomic-based cell classification, allowing the analysis of cell type evolution between mouse, marmoset, and human primary motor cortex (Bakken et al. 2021). Again, ChCs could be identified with very high accuracy across species. Confirming previous studies, only 25 differentially expressed genes between chandelier and BCs were common across all three species, despite similar overall gene expression signatures. Among these DEGs, transcription factors (e.g., RORA and NFIB) and extracellular receptors (e.g., UNC5B) were good candidates as master regulators of ChC-specific properties across species. In summary, ChCs represent one of the most homogeneous and conserved of all studied interneuron subtypes to date, a feature that can be exploited to study cellular properties while minimizing the confounds inherent to cell heterogeneity.

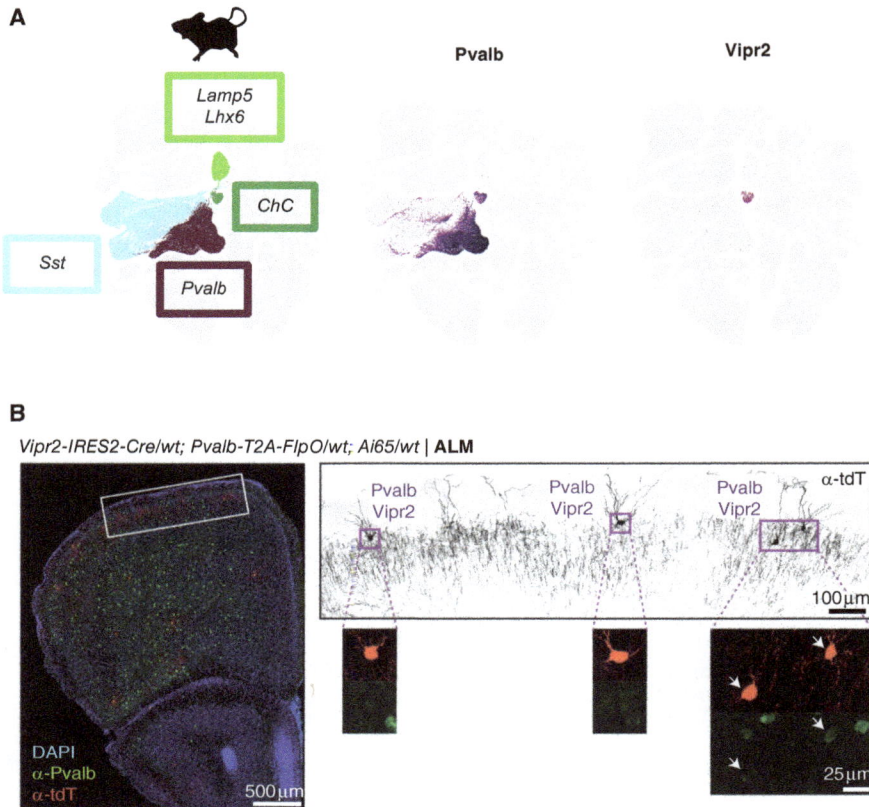

Figure 2. Cortical chandelier cells (ChCs) represent two molecularly distinct GABAergic cell type. (*A*) Uniform Manifold Approximation and Projection (UMAP) dimensional reduction representation of data obtained from RNA sequencing data of 611,000 single mouse telencephalic GABAergic neurons obtained from the Allen Brain Atlas Knowledge Platform ABC Atlas (knowledge.brain-map.org/abcatlas). The four classes that have been shown to be derived from the Medial Ganglionic Eminences (MGE) are highlighted in different colors as in Yao et al. (2023). Layer 2/3 ChCs (ChC) are labeled in dark green and Layer 5/6 ChCs (Lamp5 Lhx6) in light green. *Right-hand* panels show the expression of selected marker genes in individual MGE cells (darker shades of purple represent higher expression levels) superimposed in the UMAP representation (light gray). Selected genes are *Pvalb* (Parvalbumin), *Vipr2* (Vasoactive Intestinal Peptide Receptor 2). (*B*) A new tool, Vipr2-IRES2-Cre; Pvalb-T2A-FlpO can be used to label Layer 2/3 ChCs in the motor cortex. (*Left*) Coronal sections (100 μm) of Vipr2-IRES2-cre/wt; Pvalb-T2A-FlpO/wt; Ai65/wt brains immunolabeled using antibodies against Pvalb (green) and tdTomato (red or black in the *top right* panel). (ALM) Anterior lateral motor cortex. (*B*, reprinted, with permission, from Tasic et al. 2018.)

EMBRYONIC ORIGINS OF CHANDELIER CELLS

Cortical interneurons (cINs) are generated in the subpallium, away from the developing cortex, and undertake a long tangential migration to reach the cortex. Neurogenesis of cINs is concentrated in transient subpallial germinal zones known as the ganglionic eminences: the medial, lateral, and caudal ganglionic eminences (MGE, LGE, and CGE, respectively), and, to a lesser extent, in the preoptic area (Hu et al. 2017; Lim et al. 2018). ChCs are mostly generated in the ventral MGE at relatively late stages of embryonic development (Inan et al. 2012; Taniguchi et al. 2013). The MGE is molecularly defined by the expression of NK2 homeobox 1 (Nkx2-1), a key determinant of MGE identity, in the ventricular zone (Lim et al. 2018). In a groundbreaking study that introduced the first method to label ChCs during development, temporally controlled genetic fate mapping of the Nkx2-1 lineage revealed that many ChCs are generated at embryonic day (E)16–17 in the ventral germinal zone, which is thought to be a remnant of the MGE that retains Nkx2-1 expression (Taniguchi et al. 2013). ChC progenitors were found to migrate tangentially along the lateral ventricle wall to reach the cortical subventricular zone (SVZ) where they disperse before crossing the developing cortical plate to reach layer 1, where they continue to disperse before migrating radially to settle at the border of layers 1 and 2 (Taniguchi et al. 2013). ChCs that populate layers 1/2 are generated earlier than those that populate layers 5/6 (Taniguchi et al. 2013; Sultan et al. 2018; Kelly et al. 2019). These studies suggest that ChCs populate the cortex in an outside-in fashion, contrary to what is known both for glutamatergic cells and other cINs (Greig et al. 2013). As discussed above, despite their shared progenitors, the differences between upper and deeper layer ChCs are an ongoing topic of debate, as their transcriptional, morphological, and electrophysiological profiles are somewhat distinct (Greig et al. 2013; Tasic et al. 2018; Huang and Paul 2019; Gouwens et al. 2020). Although the generation of ChCs is dependent on the MGE genetic program, spearheaded by Nkx2-1, the genetic mechanisms that drive ChC differentiation are still unclear. Subtype specification of cINs starts as soon as these cells become postmitotic and distinct early transcriptional identities can be detected for some cIN subtypes that express PV, SST, VIP, and NDNF, but not specifically for ChCs (Mi et al. 2018). The relatively late embryonic stage at which ChCs are generated and their protracted development might explain why an early transcriptional identity remains elusive (Mayer et al. 2018).

Primate cINs seem to have a similar origin and develop in a comparable way to those of mice (Petanjek et al. 2009b) but the specific origins of ChCs in humans and nonhuman primates have not been directly shown. However, the molecular similarity of the subpallial progenitor regions across species (Petanjek et al. 2009a; Hansen et al. 2013; Ma et al. 2013; Casalia et al. 2021), combined with the transcriptomic similarities of ChCs across species (see previous section and Shi et al. 2021), make it likely that primate ChCs are also generated in the MGE in a similar fashion. More tools to label ChCs in species other than mice are needed to properly address these open questions.

LABELING AND MANIPULATING CHANDELIER CELLS IN THE MOUSE BRAIN

One important consequence of generating high-throughput scRNAseq data sets is the ability to identify differentially expressed genes among cell clusters. This information is crucial for the generation of tools for visualizing, measuring, and perturbing ChCs, particularly in mice (Tasic et al. 2018; Raudales et al. 2024). A perfect mouse line would allow targeting in a specific, as well as comprehensive fashion. The most widely used transgenic mouse line to label ChCs is the NKx2.1 Cre–ERT2 line (Nkx2-1tm1.1(cre/ERT2)Zjh/J; Taniguchi et al. 2013). In this line, tamoxifen-inducible Cre recombinase expression is regulated by the endogenous promoter/enhancer elements of the Nkx2-1 gene. By timing the tamoxifen administration to late embryogenic or early postnatal stages at the peak of ChC generation, sparse labeling of ChCs can be achieved. Although this is the preferred line for tracking ChCs during development, not all ChCs are labeled, and not all labeled cells are ChCs (Kelly et al. 2019). Despite these limitations, the spectacular advances in our understanding of ChC physiology enabled by this line highlight the importance of developing improved approaches to target and label ChCs (Fig. 2B; Tasic et al. 2018; Raudales et al. 2024).

Based on scRNAseq data, new lines have been created that rely on the differential expression of multiple genes using intersectional strategies. One example uses the combination of Pvalb and Vipr2 expression (Vipr2-IRES2-cre; Pvalb-T2A-FlpO.2B) (Tasic et al. 2018), and labels upper layer ChCs in the motor cortex, but not in the visual cortex. More recently, the intersection of Nkx2.1 expression during embryogenesis with either of the mature ChC markers Unc5b or PthIh has been shown to label a large population of neurons targeting AISs of pyramidal cells across several brain structures (Raudales et al. 2024). Further characterization of these newly generated mouse lines will help shine a light on the function of this elusive cell type. A recent initiative launched at the Allen Brain Institute aims to combine different epigenomic data sets to select putative mouse and human short enhancer sequences to drive gene expression selectively in defined cell types, including ChCs (Ben-Simon et al. 2024). If successful, it will pave the way for targeting ChCs with genetically encoded labels, activity reporters, or actuators using AAV technologies in both mice and primates, enabling in vivo visualization and perturbation of these cells and illuminating discoveries regarding their role in the intact brain.

BUILDING THE CHANDELIER

The characteristic highly branched axonal morphology of ChCs suggests that despite populating the cortex sparsely, they may actually play an important role at the circuit-wide level. In adults, a single ChC will contact 30%–50% of neighboring pyramidal cells within a ~200 µm radius (Inan et al. 2013). Considering the spatial overlap of the axonal fields of ChCs, 80% of pyramidal cells in the somatosensory cortex will receive at least one axo-axonic input (Schneider-Mizell et al. 2021). Although ChCs appear to densely innervate cortical circuits, they also show a remarkable heterogeneity—a given pyramidal cell in the mouse cortex will receive anywhere from 0 to 30 axo-axonic synapses and be contacted by varying numbers of ChCs, ranging from 0 to as many as 9 (DeFelipe et al. 1985; Somogyi et al. 1985; Inan et al. 2013; Veres et al. 2014; Schneider-Mizell et al. 2021). Such variance is intriguing and may be evidence of ongoing plasticity by ChC outputs (discussed below) and/or of selectivity in the wiring principles of ChCs, whereby the number of synapses

is biased toward specific populations of cells. Evidence for the latter has emerged from the prelimbic cortex where it was shown that ChCs do indeed preferentially target pyramidal neurons that project to the basolateral amygdala (BLA), rather than those projecting contralaterally (Lu et al. 2017). Similarly, ChCs preferentially target pyramidal neurons with intracortical projections in the visual and auditory cortex (Sloper and Powell 1979; Fairén and Valverde 1980; De Carlos et al. 1985; DeFelipe et al. 1985; Farinas and DeFelipe 1991) and, in the cat visual cortex, ChCs were biased toward contacting callosal or ipsilateral corticocortical projecting neurons rather than corticothalamic projecting neurons (Farinas and DeFelipe 1991).

The overall morphology and subcellular innervation of ChCs are features that are maintained in all primates that have been studied so far, although ChCs are typically larger and appear to have more complex axonal arbors in humans (DeFelipe et al. 1989; Anderson et al. 1995; Inda et al. 2007; Fish et al. 2013). A high-resolution serial EM reconstruction of pyramidal cell inhibitory inputs across mouse, macaque, and human brains showed that while axo-somatic inputs were lower in macaque and human pyramidal neurons compared to mice, axo-axonic synapses increased in number by ~40% (Fig. 3A; Loomba et al. 2022). A larger number of axo-axonic synapses measured with immunocytochemistry was also observed in human cortical neurons when compared to mouse neurons, which may reflect longer AISs in humans (Ostos et al. 2023). Overall, it appears that despite being few in number, ChCs remain an integral part of inhibitory circuits in rodents and primates alike.

Although ChC axons show some targeting specificity at the cellular level, one of the most striking aspects of ChC innervation is the exquisite subcellular specificity with which it contacts pyramidal neurons. Their highly branched axons directly target the AIS of pyramidal neurons, a proximal region of the axon that is only 20–60 µm long (Grubb and Burrone 2010b; Huang and Rasband 2018; Leterrier 2018), representing not more than 0.1% of the entire surface area of a typical mouse layer 2/3 pyramidal

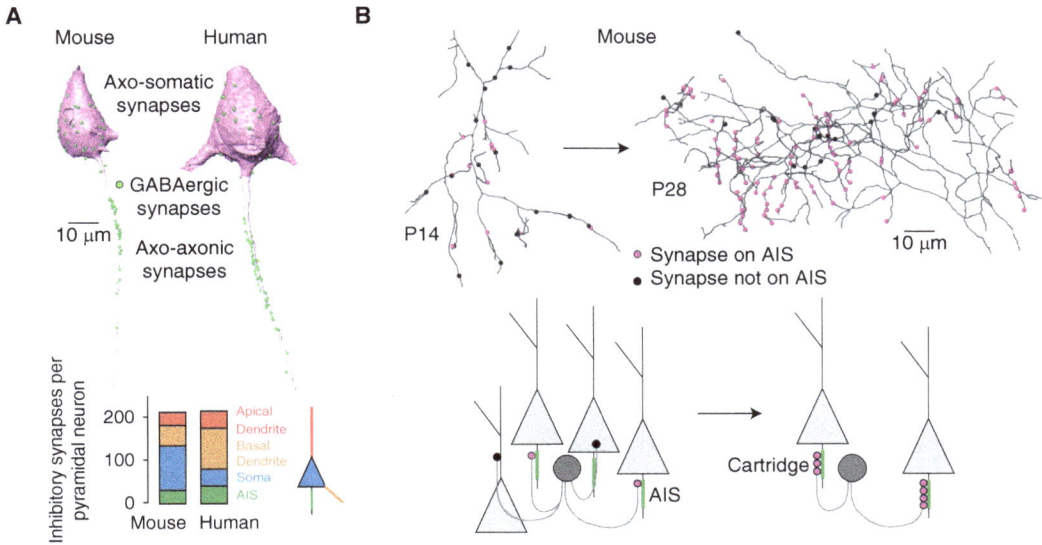

Figure 3. Formation of axo-axonic synapses along the axon initial segment (AIS). (*A*) A comparison of GABAergic synapses across distinct subcellular compartments for pyramidal neurons from the mouse temporal cortex and human superior temporal gyrus. (*Top*) Examples of a 3D reconstruction of a mouse and a human pyramidal neuron (purple) showing axo-somatic and axo-axonic GABAergic synapses (green). (*Bottom*) The overall number of inhibitory synapses is similar in human and mouse neurons but their subcellular distribution varies. Axo-axonic synapse numbers (green) are slightly increased in human neurons. (Data from Loomba et al. 2022.) (*B*) Formation of axo-axonic synapses in mouse primary somatosensory cortex (S1). (*Top*) Example reconstructions of mouse ChC axons across development (P14 and P28) showing boutons contacting either the AIS (magenta) or other subcellular compartments (black). (*Bottom*) Schematic summarizing main findings and describing potential mechanism for target specificity. At P14, boutons that contact the AIS (magenta) are kept, while others (black) are lost. Cartridges are proposed to emerge from the pioneer boutons that target the AIS early in development. (Figure reprinted and modified, with permission, from Gour et al. 2021, The American Association for the Advancement of Science.)

neuron. How this is achieved remains unclear, but recent work charting the formation and refinement of ChC axons and their axo-axonic synapses during development in mice has begun to uncover some basic principles that may provide key mechanistic insights. A distinguishable feature of ChC synapse formation is its timing—axo-axonic synapses only begin to form after most other GABAergic synapses have been established. In vivo imaging of the dynamics of this process (Taniguchi et al. 2013; Pan-Vazquez et al. 2020) showed that whereas the dendrites of ChCs are formed early in development, axonal arborization and the formation of synapses at the AIS takes place during a tight temporal window, from ~P12 to P18, in agreement with other reports (Fazzari et al. 2010; Steinecke et al.

2017). This work also provided the first hints that ChC axons may refine their connectivity through some form of pruning, initially extending branches that are later retracted (Taniguchi et al. 2013). Since these pruned branches did not typically contain synapses, they were proposed to provide a mechanism for shaping axonal morphology rather than being involved in subcellular targeting specificity. More recent work using serial EM reconstructions of cortical neurons during development began to bridge the gaps between the appearance of axo-axonic boutons, cartridges, and the axonal arbor (Gour et al. 2021). At P14, ChCs lacked complete subcellular specificity, with only 60% of synapses forming onto the AIS and the rest onto the dendritic shaft or soma of pyramidal neurons. During this period, no cartridges were observed—instead single synapses were typically formed across all compartments. By P28, when all axo-axonic synapses were formed, 90% of all ChC contacts targeted an AIS and were mostly arranged in cartridges. One possible model to explain these findings is that ChCs initially extend axons to form isolated contacts with neighboring pyramidal neurons—those that form onto the AIS are kept and act as a tether to form cartridges, while those onto other subcellular compartments are lost (Fig. 3B). However, this elegant model does not fully explain the clear initial bias for ChC axons toward the AIS, even at early stages of synaptogenesis, and leaves open the intriguing question of why ChC synapses at the AIS are selectively kept and others lost. These questions, which have been at the heart of studies trying to identify the molecules responsible for the formation of axo-axonic synapses (Tai et al. 2014, 2019; Favuzzi et al. 2019; Hayano et al. 2021), are an active area of research that has been reviewed elsewhere (Gallo et al. 2020; Compans and Burrone 2023) and will, therefore, not be discussed here.

REFINING THE CHANDELIER—ACTIVITY-DEPENDENT PLASTICITY DURING DEVELOPMENT

ChC development, particularly the formation of axo-axonic synapses, is also guided by activity-dependent mechanisms. As discussed above, a cursory glance at the number of axo-axonic synapses received by a given pyramidal neuron in the cortex immediately shows high levels of heterogeneity. A closer look shows that this heterogeneity appears to depend on other features, such as the laminar depth of pyramidal neuron somas, the level of perisomatic inhibition from non-ChC sources, and soma size (Fig. 4A; De-Felipe et al. 1985; Schneider-Mizell et al. 2021; Jung et al. 2022), which may reflect the ability of ChCs to regulate their innervation based on the characteristics of their postsynaptic targets. In the monkey prefrontal cortex, PV-immunoreactive terminals contacting the AIS in layer 2/3 follow a postnatal developmental trajectory that mirrors that of pyramidal neuron spine

density, rapidly increasing in the first 3 months of life, peaking ∼1 year, and then progressively declining to adult levels by 4 years (Anderson et al. 1995). Although this may be attributed to changes in the expression levels of PV that is susceptible to extrinsic factors (Vogt Weisenhorn et al. 1998), the temporal coincidence in the changes observed for axo-axonic synapses and dendritic spines, suggests that ChCs may adjust their innervation based on the activity levels of pyramidal cells.

Recent work has shown that axo-axonic synapses can undergo bidirectional changes in their number and density following chemogenetically induced increases in cortical network activity (Pan-Vazquez et al. 2020). Specifically, hyperactivating layer 2/3 of the somatosensory cortex during the period of synapse formation (P12–18) decreased the number of axo-axonic boutons at the AIS, whereas the same manipulation in adult mice increased their number (Fig. 4B). It is, therefore, possible that ChCs follow plasticity rules that track the polarity of GABAergic neurotransmission, which is thought to switch from depolarizing early in development to hyperpolarizing later in life (Ben-Ari 2002; Ben-Ari et al. 2007), as discussed in the following section.

More recently, studies have investigated the plasticity of axo-axonic synapses in response to naturalistic activity modulation during visual or motor learning tasks in adult mice. Specifically, repeated exposure to visual stimuli in a virtual tunnel increased the fraction of AISs contacted by ChCs, representing changes in the selectivity of ChC outputs following learning (Seignette et al. 2024). Another study found that mice trained in a motor reward learning task exhibited a larger heterogeneity of axo-axonic synapse number across pyramidal neurons in the prelimbic cortex compared to mice that explored the environment without goal-driven reward learning, suggesting that ChCs may differentially adjust the degree of innervation to individual postsynaptic targets during motor learning (Fig. 4C; Jung et al. 2023). In both cases, learning contributed to an increased sensory (orientation or direction) selectivity of the neuronal population. Indeed, direction selectivity and goal-di-

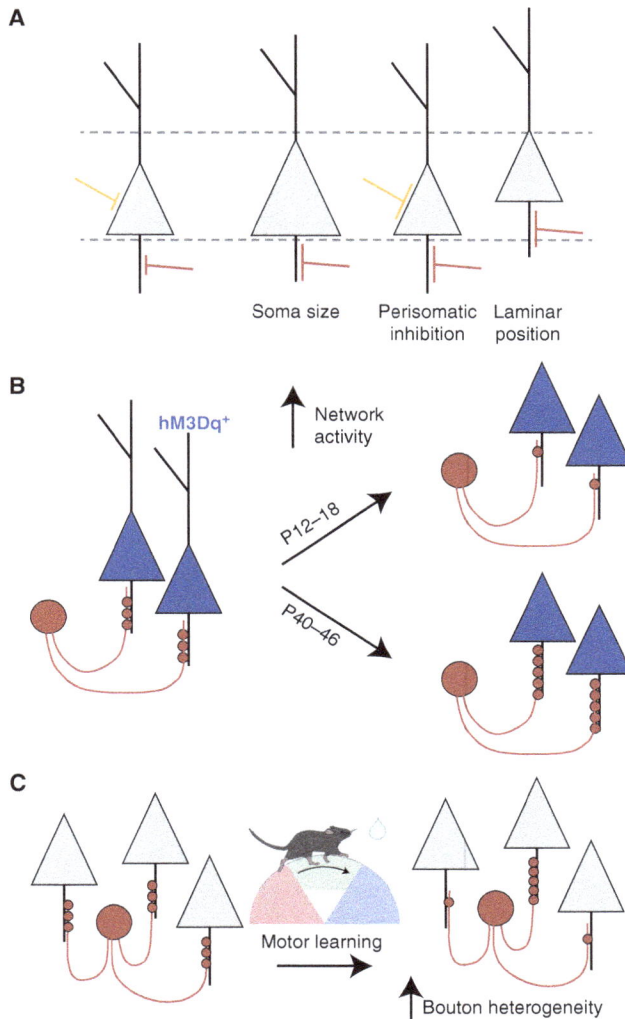

Figure 4. Activity-dependent plasticity of chandelier cells (ChCs). (*A*) The level of ChC innervation correlates with soma size, the amount of non-ChC perisomatic inhibition, and soma laminar position of postsynaptic targets; (*B*) activity-dependent axo-axonic plasticity following chemogenetic enhancement of pyramidal cell activity in layer 2/3 of the somatosensory cortex decreases the number of axo-axonic synapses when activity is manipulated during the period of ChC synapse formation (P12–18), but increases axo-axonic synapses in adult (P40–46) mice; and (*C*) the number of ChC boutons per axon initial segment (AIS) becomes more heterogenous across pyramidal neurons in the premotor cortex following a reward-based motor learning task, indicative of fine-tuned control of ChCs over individual pyramidal neurons. (*B* is adapted from Pan-Vazquez et al. 2020 and *C* is adapted from Jung et al. 2023.)

rected navigation performance were disrupted when ChCs were chemogenetically silenced in the latter study, raising the possibility that axo-axonic synapse plasticity in adulthood functions to modulate the sensory selectivity of cortical neurons. Given that the strength of axo-axonic inhibition has been shown to correlate with the number of boutons contacting the AIS (Veres et al. 2014), the plasticity of axo-axonic synapses may represent a mechanism to cater axo-axonic innervation to the ongoing activity levels of postsynaptic neurons. This, in turn, may contribute to the heterogeneity in axo-axonic synapse number found across cells in the cortex (Schneider-Mizell et al. 2021). Besides the ability for axo-axonic synapses to change in response to the activity levels of their postsynaptic targets, a recent study explored how axo-axonic inputs themselves regulate the properties of the AIS on pyramidal neurons. They showed that ablating ChCs or inactivating axo-axonic neurotransmission for prolonged periods of time in adult mice led to a decrease in AIS length and neuronal excitability, while chronic chemogenetic activation of ChCs led to opposing increases in AIS length and neuronal excitability, suggesting that ChC activity may regulate postsynaptic firing properties in a bidirectional manner (Zhao et al. 2024). Although clearly plastic, whether axo-axonic synapses are sensitive only to network-wide changes in activity, as in the chemogenetic activity manipulations (Pan-Vazquez et al. 2020), or respond to ongoing changes in the activity of single pyramidal neurons remains to be established. Experiments assessing the spatial sensitivity of ChCs would provide much-needed information on the rules that fine-tune ChC connectivity in the brain.

In addition to these ongoing synaptic plasticity events, activity has also been shown to modulate the overall number of interneurons in the cortex during development. In fact, up to 50% of all generated cINs in the mouse cortex are eliminated via apoptosis during early postnatal development (P5–10) (Southwell et al. 2012). More importantly, increasing levels of network activity during this period promoted interneuron survival via cell-intrinsic mechanisms, suggesting that the proportions of cell types in the cortex can be fine-tuned (Denaxa et al. 2018; Priya et al. 2018; Wong et al. 2018). ChCs also follow this program of developmental apoptosis during a similar critical window. However, just a few days later in the binocular visual cortex, coinciding with the arrival of callosal projections but before eye opening, ChCs undergo further activity-dependent apoptosis driven by spontaneous retinal waves and contralateral innervation (Wang et al. 2021). Preventing the removal of ChCs during the second postnatal week led to severe deficits in binocular vision, underscoring the important role of circuit refinement through cell death on brain function.

CANDLESTICKS OF A CHANDELIER— EXPLORING AXO-AXONIC SYNAPSE FUNCTION

ChCs modulate pyramidal cell function through the release of the neurotransmitter GABA. A fascinating feature of GABAergic neurotransmission in general is that it switches polarity throughout development, from depolarizing early on to hyperpolarizing later in circuit wiring. This switch, which depends on the postsynaptic chloride concentration, is thought to drive network activity when GABA is depolarizing (Represa and Ben-Ari 2005; Sipilä et al. 2005; Cellot and Cherubini 2013), and provide inhibition later in life, when GABA is hyperpolarizing. While a developmental decrease in the chloride concentration of neocortical pyramidal neurons has been shown (Tyzio et al. 2003, 2007) and attributed to changes in the expression of chloride cotransporters (Fig. 5A; Yamada et al. 2004; Leonzino et al. 2016), the polarity and developmental trajectory of GABA at the AIS remains somewhat controversial.

A surprising finding in recent years has been the idea that the switch in GABA polarity at different subcellular domains may follow different temporal rules. In particular, this switch appears to be delayed at the AIS. Hints of this delay were first described in gramicidin perforated patch recordings from pyramidal cells, where the chloride reversal potential for GABA (E_{GABA}) was shown to remain depolarizing in experiments carried out relatively late in development. Specifically, whereas GABAergic inputs onto the somatodendritic compartment are depolarizing in early development, mostly during the first postnatal week, those onto the AIS continue to be depolarizing until much later, beyond the third postnatal week in rodents (Szabadics et al. 2006; Khirug et al. 2008; Woodruff et al. 2009). In fact, a direct comparison between the developmental trajectory of E_{GABA}

Cite this article as *Cold Spring Harb Perspect Biol* doi: 10.1101/cshperspect.a041506

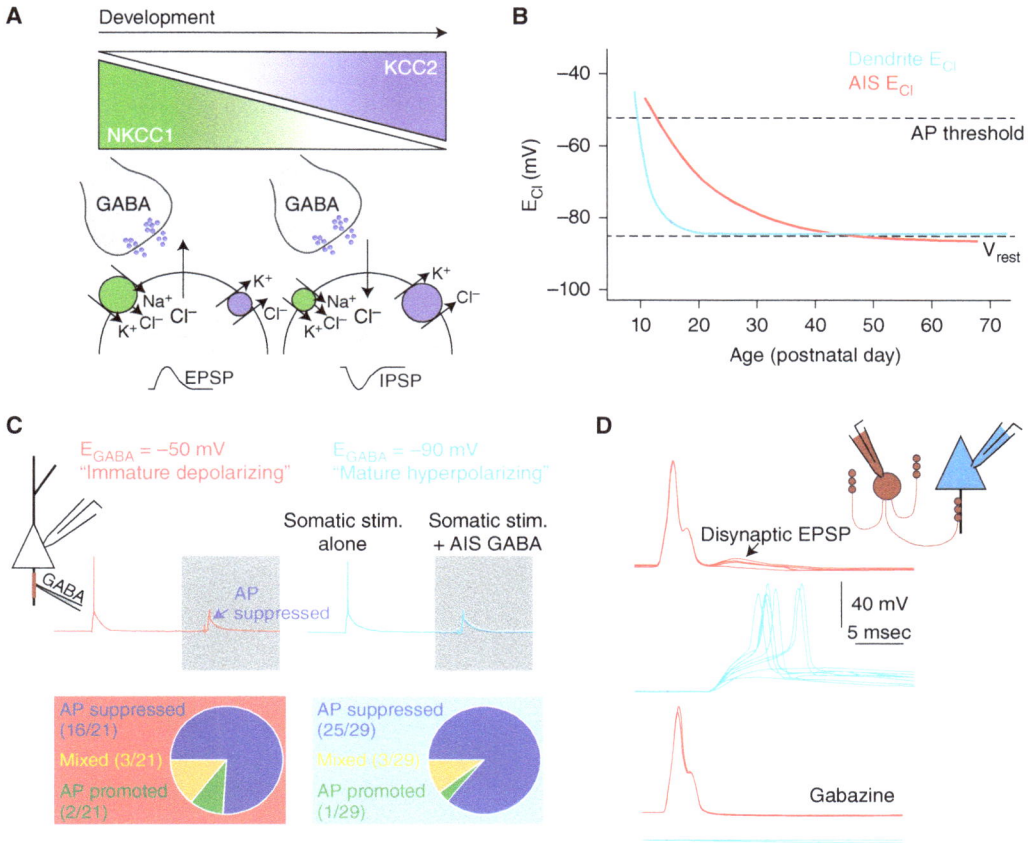

Figure 5. Functions of GABAergic axo-axonic synapses. (*A*) The developmental shift in GABA polarity is attributed to changes in the expression of chloride cotransporters. Levels of the chloride importer NKCC1 decrease throughout development, whereas levels of the chloride exporter KCC2 increase throughout development. This contributes to high intracellular chloride early on, making GABA depolarizing, whereas low intracellular chloride in adulthood leads to hyperpolarizing GABA. (*B*) The chloride reversal potential (E_{Cl}) gradually becomes more negative throughout development, contributing to a polarity shift in GABA from depolarizing to hyperpolarizing as the E_{Cl} approaches the resting membrane potential (V_{rest}), but is temporally delayed at the axon initial segment (AIS) compared to in dendrites. (*C*) GABA iontophoresis onto the AIS still suppresses action potentials when GABA is depolarizing, when cells have an immature (high) intracellular chloride concentration. (*D*) Whole-cell recordings from a human axo-axonic cell (red) showed that inducing an action potential elicited disynaptic excitatory postsynaptic potentials (EPSPs) back to the same cell and GABA-mediated third-order spikes in a neighboring cell (blue) that were blocked by gabazine. (*B*, modified from Rinetti-Vargas et al. 2017, reprinted under the Creative Commons CC-BY License; *C*, data in Lipkin and Bender 2023, © 2023 the authors; *D*, modified from Szabadics et al. 2006, The American Association for the Advancement of Science.)

in dendrites and the AIS showed a clear developmental delay in the E_{GABA} switch at the AIS (Fig. 5B; Rinetti-Vargas et al. 2017). Thus, the polarity shift for axo-axonic synapses at the AIS appears to be delayed in relation to other dendrite- or soma-targeting GABAergic interneuron types, a feature that intriguingly mirrors the temporal delay in synapse formation (Taniguchi et al. 2013; Gour et al. 2021; Wang et al. 2021).

The exact timing of the GABA polarity shift at the AIS, however, remains controversial. GABA likely remains depolarizing at the AIS up to the third postnatal week, as seen in voltage imaging studies where targeted GABA ionto-phoresis onto the AIS revealed a depolarizing effect of GABA in the mouse cortex between P14 and P18 (Pan-Vazquez et al. 2020). Studies using gramicidin perforated patch recordings have demonstrated both depolarizing (Szabadics et al. 2006; Khirug et al. 2008; Woodruff et al. 2011) as well as hyperpolarizing (Veres et al. 2014) effects of GABA in rodents aged between three to five postnatal weeks. Conversely, extracellular recordings carried out in 4- to 6-week-old rat hippocampal slices, a later period in development, showed that ChC stimulation resulted in direct hyperpolarization of pyramidal cells resulting in clear inhibition (Glickfeld et al. 2009). In mice above 6 weeks of age, optogenetically induced stimulation of axo-axonic synapses at the AIS resulted in hyperpolarization and a decrease in postsynaptic spiking probability (Wang et al. 2014). Overall, the emerging picture is one where axo-axonic synapses show a delayed switch from depolarizing to hyperpolarizing during development, although the exact timing of this switch is unclear. It is also important to keep in mind that discrepancies across studies could reflect differences in brain region or species tested, as well as the experimental paradigms used to assess polarity (Table 1). A more pragmatic possibility is that the ChC function depends on the membrane potential state of postsynaptic neurons and its relation to E_{GABA}. In this case, ChCs could act bidirectionally, hyperpolarizing cells in a depolarized state (above E_{GABA}) but depolarizing cells in a hyperpolarized state (below E_{GABA}). In support of this, the influence of ChC activity on spike probability was shown to differ under quiescent compared to a dynamic in vivo–like membrane potential state (Woodruff et al. 2011). However, direct evidence for this, especially in vivo, is still lacking.

Perhaps the most controversial question in the field is whether depolarizing GABA is actually excitatory. Although some evidence exists in support of an excitatory role for GABA released by ChCs in rat and human brain slices (Szabadics et al. 2006), recent work in acute mouse slices has shown that depolarizing GABA at

the AIS still dampened action potential firing through shunting inhibition (Fig. 5C; Lipkin and Bender 2023). These findings suggest that despite its depolarized E_{GABA}, axo-axonic synapses may be mostly inhibitory throughout development. In the BLA of P18–24 mice, inducing action potentials in ChCs during a sinusoidal current pulse in postsynaptic cells significantly reduced the firing probability of pyramidal cells, suggesting that ChCs directly inhibit action potential generation (Veres et al. 2014). Overall, whether ChCs are primarily inhibitory or excitatory during development remains to be properly understood. Future work tracking ChC modulation of network activity in vivo during development may begin to resolve these open questions.

LIGHTING THE CHANDELIER—THE ROLE OF CHANDELIER CELLS IN CIRCUIT FUNCTION

Despite the increasing number of experiments examining ChC activity in the brain, we still know little about their function in vivo. Recent in vivo imaging studies have exploited the use of transgenic mouse lines that enable the selective labeling of ChCs (Taniguchi et al. 2011; Raudales et al. 2024). While ChCs are poorly tuned to visual or somatosensory stimuli, several studies have consistently reported arousal-related activity (Bienvenu et al. 2012; Massi et al. 2012; Dudok et al. 2021; Schneider-Mizell et al. 2021; Bugeon et al. 2022; Jung et al. 2023). Indeed, although superficial ChCs in sensory cortices receive some sensory inputs, the majority of their inputs are from local cortical cells, mainly from layer 5, and from long-range sources including the basal forebrain and thalamus (Lu et al. 2017; Seignette et al. 2024). Cortical ChCs also exhibit highly correlated activity in vivo (Dudok et al. 2021; Schneider-Mizell et al. 2021; Jung et al. 2023; Seignette et al. 2024). Interestingly, however, ChC activity was shown to decorrelate following both motor and visual learning (Jung et al. 2023; Seignette et al. 2024) and was accompanied by improvements in direction and orientation selectivity, respectively, suggesting that decorrelating ChC activity may contribute to learned feature selectivity in

Table 1. Summary of studies on GABA polarity at the axon initial segment (AIS)

References	Species	Age	Brain region	Method	AIS GABA polarity
Szabadics et al. (2006)	Rats	P20–35	Somatosensory cortex LAYER 2/3	Gramicidin perforated patch	Depolarizing
Khirug et al. (2008)	Mice and rats	P16–20	Hippocampus dentate gyrus	Gramicidin perforated patch	Depolarizing
Woodruff et al. (2009)	Mice	P15–23	Somatosensory cortex LAYER 2/3	Gramicidin perforated patch and cell-attached recordings	Depolarizing
Woodruff et al. (2011)	Mice	P18–25	Motor/somatosensory cortex LAYER 2/3	Gramicidin perforated patch	Depolarizing
Pan-Vazquez et al. (2020)	Mice	P14–18	Somatosensory cortex LAYER 2/3	Voltage imaging	Depolarizing
Rinetti-Vargas et al. (2017)	Mice	P11–75	Prefrontal cortex LAYER 2/3	Gramicidin perforated patch	Depolarizing from P11–21; hyperpolarizing from P21
Veres et al. (2014)	Mice	P18–24	Basolateral amygdala	Gramicidin perforated patch	Hyperpolarizing
Glickfeld et al. (2009)	Rats	P28–42	Hippocampus CA1	Extracellular recordings	Hyperpolarizing
Wang et al. (2014)	Mice	P42–56	Piriform cortex	Gramicidin perforated patch	Hyperpolarizing

pyramidal neurons. In the hippocampus, ChC activity has been correlated with spontaneous oscillatory activity, implicating a functional role in pacing pyramidal activity. Notably, hippocampal ChCs preferentially fire at the peak or descending phase of movement-related theta oscillations, before other interneurons and when pyramidal cell activity decreases to low levels (Klausberger et al. 2004, 2005; Tukker et al. 2007; Viney et al. 2013; Varga et al. 2014). Overall, it appears that ChCs are preferentially recruited during states of arousal and may function to coordinate or refine neuronal activity at a network-wide level. In line with this, acutely enhancing network activity by applying a GABA$_A$ antagonist in vivo during whole-cell recordings significantly increased ChC firing rates more than other cortical fast-spiking or pyramidal neurons (Zhu et al. 2004).

As well as catering their activity to the state of the surrounding network, ChCs have also been shown to exert circuit-wide effects on neuronal activity through multisynaptic events. In the cortex, for example, ChCs were found to elicit disynaptic glutamatergic events both in nearby ChCs and back onto themselves (Taniguchi

et al. 2013). Similarly, in the mouse BLA, direct stimulation of ChCs evoked responses in pyramidal neurons consisting of GABAergic and glutamatergic components, as well as feedback excitation onto themselves (Woodruff et al. 2006; Spampanato et al. 2016). A more recent study showed that polysynaptic activity in several surrounding neurons, evoked by a single pyramidal input to a ChC, contributed to the orchestration of slow-wave ripples in the BLA (Perumal et al. 2021). While there is very limited evidence for ChC output in human tissue, a notable study that carried out patch-clamp recordings in adult human cortical tissue did demonstrate that direct stimulation of ChCs elicited disynaptic excitatory postsynaptic potentials (EPSPs) in ChCs and third-order depolarizing events in surrounding neurons (Fig. 5D; Szabadics et al. 2006). Thus, in both mouse and human brain tissue, ChCs are capable of inducing complex multisynaptic events that propagate through the local circuit. The contribution of ChC activity to network-wide events in mice may also be aided by both synaptic (Jiang et al. 2015) and electrical coupling between ChCs. In particular, there is strong evidence to support electrical coupling via gap junctions. Dendro-dendritic gap junctions have been structurally identified between ChCs in the mouse cortex and rat hippocampus (Baude et al. 2007; Schneider-Mizell et al. 2021), and paired whole-cell recordings revealed bidirectional electrical coupling in >80% of ChC pairs tested (Woodruff et al. 2011). Although the mechanism by which this is achieved is not yet properly understood, the importance of ChC function in modulating network activity is underscored by the sensitivity of axo-axonic synapses to seizures in rodents and monkeys (Ribak 1985; Dinocourt et al. 2003; Sayin et al 2003) as well as in epileptic foci of human patients (DeFelipe et al. 1993; Marco et al. 1996, 1997; DeFelipe 1999). In summary, although the impact of ChC function within cortical networks remains controversial and appears to be complex, their activity has been consistently shown to be state-dependent and to correlate well with arousal.

Overall, ChCs represent a well-defined cellular subclass with distinct transcriptomic and morphological features, as well as highly specific subcellular connectivity. By exploiting the relative homogeneity of ChCs and the advent of new tools to label these cells across different species, we hope to gain new insights into their development and function in the brain.

COMPETING INTEREST STATEMENT

The authors declare no competing interests.

ACKNOWLEDGMENTS

This research was funded in whole, or in part, by the Wellcome Trust (215508/Z/19/Z to J.B.). This work was also supported by a BBSRC project grant (BB/S000526/1 to J.B.) and by the MRC Centre for Neurodevelopmental Disorders, King's College London (MR/N026063/1). C.L. was funded by an MRC-ITND PhD studentship (MR/P502108/1).

REFERENCES

Anderson SA, Classey JD, Condé F, Lund JS, Lewis DA. 1995. Synchronous development of pyramidal neuron dendritic spines and parvalbumin-immunoreactive chandelier neuron axon terminals in layer III of monkey prefrontal cortex. *Neuroscience* 67: 7–22. doi:10.1016/0306-4522(95)00051-J

Bakken TE, Jorstad NL, Hu Q, Lake BB, Tian W, Kalmbach BE, Crow M, Hodge RD, Krienen FM, Sorensen SA, et al. 2021. Comparative cellular analysis of motor cortex in human, marmoset and mouse. *Nature* 598: 111–119. doi:10.1038/s41586-021-03465-8

Baude A, Bleasdale C, Dalezios Y, Somogyi P, Klausberger T. 2007. Immunoreactivity for the GABA$_A$ receptor α1 subunit, somatostatin and Connexin36 distinguishes axo-axonic, basket, and bistratified interneurons of the rat hippocampus. *Cereb Cortex* 17: 2094–2107. doi:10.1093/cercor/bhl117

Ben-Ari Y. 2002. Excitatory actions of gaba during development: the nature of the nurture. *Nat Rev Neurosci* 3: 728–739. doi:10.1038/nrn920

Ben-Ari Y, Gaiarsa JL, Tyzio R, Khazipov R. 2007. GABA: a pioneer transmitter that excites immature neurons and generates primitive oscillations. *Physiol Rev* 87: 1215–1284. doi:10.1152/physrev.00017.2006

Ben-Simon Y, Hooper M, Narayan S, Daigle T, Dwivedi D, Way SW, Oster A, Stafford DA, Mich JK, Taormina MJ, et al. 2024. A suite of enhancer AAVs and transgenic mouse lines for genetic access to cortical cell types. bioRxiv doi:10.1101/2024.06.10.597244

Bienvenu TC, Busti D, Magill PJ, Ferraguti F, Capogna M. 2012. Cell-type-specific recruitment of amygdala interneurons to hippocampal theta rhythm and noxious stimuli in vivo. *Neuron* 74: 1059–1074. doi:10.1016/j.neuron.2012.04.022

Bugeon S, Duffield J, Dipoppa M, Ritoux A, Prankerd I, Nicoloutsopoulos D, Orme D, Shinn M, Peng H, Forrest H, et al. 2022. Publisher correction: a transcriptomic axis predicts state modulation of cortical interneurons. *Nature* 609: E10. doi:10.1038/s41586-022-05209-8

Casalia ML, Li T, Ramsay H, Ross PJ, Paredes MF, Baraban SC. 2021. Interneuron origins in the embryonic porcine medial ganglionic eminence. *J Neurosci* 41: 3105–3119. doi:10.1523/JNEUROSCI.2738-20.2021

Cellot G, Cherubini E. 2013. Functional role of ambient GABA in refining neuronal circuits early in postnatal development. *Front Neural Circuits* 7: 136. doi:10.3389/fncir.2013.00136

Compans B, Burrone J. 2023. Chandelier cells shine a light on the formation of GABAergic synapses. *Curr Opin Neurobiol* 80: 102697. doi:10.1016/j.conb.2023.102697

Dance A. 2024. What is a cell type, really? The quest to categorize life's myriad forms. *Nature* 633: 754–756. doi:10.1038/d41586-024-03073-2

De Carlos JA, Lopez-Mascaraque L, Valverde F. 1985. Development, morphology and topography of chandelier cells in the auditory cortex of the cat. *Brain Res* 354: 293–300. doi:10.1016/0165-3806(85)90182-8

DeFelipe J. 1999. Chandelier cells and epilepsy. *Brain* 122: 1807–1822. doi:10.1093/brain/122.10.1807

DeFelipe J, Hendry SH, Jones EG, Schmechel D. 1985. Variability in the terminations of GABAergic chandelier cell axons on initial segments of pyramidal cell axons in the monkey sensory-motor cortex. *J Comp Neurol* 231: 364–384. doi:10.1002/cne.902310307

DeFelipe J, Hendry SH, Jones EG. 1989. Visualization of chandelier cell axons by parvalbumin immunoreactivity in monkey cerebral cortex. *Proc Natl Acad Sci* **86:** 2093–2097. doi:10.1073/pnas.86.6.2093

DeFelipe J, Garcia Sola R, Marco P, del Río MR, Pulido P, Ramón y Cajal S. 1993. Selective changes in the micro-organization of the human epileptogenic neocortex revealed by parvalbumin immunoreactivity. *Cereb Cortex* **3:** 39–48. doi:10.1093/cercor/3.1.39

Denaxa M, Neves G, Rabinowitz A, Kemlo S, Liodis P, Burrone J, Pachnis V. 2018. Modulation of apoptosis controls inhibitory interneuron number in the cortex. *Cell Rep* **22:** 1710–1721. doi:10.1016/j.celrep.2018.01.064

Dinocourt C, Petanjek Z, Freund TF, Ben-Ari Y, Esclapez M. 2003. Loss of interneurons innervating pyramidal cell dendrites and axon initial segments in the CA1 region of the hippocampus following pilocarpine-induced seizures. *J Comp Neurol* **459:** 407–425. doi:10.1002/cne.10622

Dudok B, Szoboszlay M, Paul A, Klein PM, Liao Z, Hwaun E, Szabo GG, Geiller T, Vancura B, Wang B-S, et al. 2021. Recruitment and inhibitory action of hippocampal axo-axonic cells during behavior. *Neuron* **109:** 3838–3850.e8. doi:10.1016/j.neuron.2021.09.033

Fairén A, Valverde F. 1980. A specialized type of neuron in the visual cortex of cat: a Golgi and electron microscope study of chandelier cells. *J Comp Neurol* **194:** 761–779. doi:10.1002/cne.901940405

Farinas I, DeFelipe J. 1991. Patterns of synaptic input on corticocortical and corticothalamic cells in the cat visual cortex. II: The axon initial segment. *J Comp Neurol* **304:** 70–77. doi:10.1002/cne.903040106

Favuzzi E, Deogracias R, Marques-Smith A, Maeso P, Jezequel J, Exposito-Alonso D, Balia M, Kroon T, Hinojosa AJ, Maraver E F, et al. 2019. Distinct molecular programs regulate synapse specificity in cortical inhibitory circuits. *Science* **363:** 413–417. doi:10.1126/science.aau8977

Fazzari P, Paternain AV, Valiente M, Pla R, Luján R, Lloyd K, Lerma J, Marín O, Rico B. 2010. Control of cortical GABA circuitry development by Nrg1 and ErbB4 signalling. *Nature* **464:** 1376–1380. doi:10.1038/nature08928

Fish KN, Hoftman GD, Sheikh W, Kitchens M, Lewis DA. 2013. Parvalbumin-containing chandelier and basket cell boutons have distinctive modes of maturation in monkey prefrontal cortex. *J Neurosci* **33:** 8352–8358. doi:10.1523/JNEUROSCI.0306-13.2013

Freund TF, Martin KA, Smith AD, Somogyi P. 1983. Glutamate decarboxylase-immunoreactive terminals of Golgi-impregnated axoaxonic cells and of presumed basket cells in synaptic contact with pyramidal neurons of the cat's visual cortex. *J Comp Neurol* **221:** 263–278. doi:10.1002/cne.902210303

Fuzik J, Zeisel A, Máté Z, Calvigioni D, Yanagawa Y, Szabó G, Linnarsson S, Harkany T. 2016. Integration of electro-physiological recordings with single-cell RNA-seq data identifies neuronal subtypes. *Nat Biotechnol* **34:** 175–183. doi:10.1038/nbt.3443

Gallo NB, Paul A, Van Aelst L. 2020. Shedding light on chandelier cell development, connectivity, and contribution to neural disorders. *Trends Neurosci* **43:** 565–580. doi:10.1016/j.tins.2020.05.003

Glickfeld LL, Roberts JD, Somogyi P, Scanziani M. 2009. Interneurons hyperpolarize pyramidal cells along their entire somatodendritic axis. *Nat Neurosci* **12:** 21–23. doi:10.1038/nn.2230

Gour A, Boergens KM, Heike N, Hua Y, Laserstein P, Song K, Helmstaedter M. 2021. Postnatal connectomic development of inhibition in mouse barrel cortex. *Science* **371:** eabb4534. doi:10.1126/science.abb4534

Gouwens NW, Sorensen SA, Baftizadeh F, Budzillo A, Lee BR, Jarsky T, Alfiler L, Baker K, Barkan E, Berry K, et al. 2020. Integrated morphoelectric and transcriptomic classification of cortical GABAergic cells. *Cell* **183:** 935–953.e19. doi:10.1016/j.cell.2020.09.057

Greig LC, Woodworth MB, Galazo MJ, Padmanabhan H, Macklis JD. 2013. Molecular logic of neocortical projection neuron specification, development and diversity. *Nat Rev Neurosci* **14:** 755–769. doi:10.1038/nrn3586

Grubb MS, Burrone J. 2010a. Activity-dependent relocation of the axon initial segment fine-tunes neuronal excitability. *Nature* **465:** 1070–1074. doi:10.1038/nature09160

Grubb MS, Burrone J. 2010b. Building and maintaining the axon initial segment. *Curr Opin Neurobiol* **20:** 481–488. doi:10.1016/j.conb.2010.04.012

Hansen DV, Lui JH, Flandin P, Yoshikawa K, Rubenstein JL, Alvarez-Buylla A, Kriegstein AR. 2013. Non-epithelial stem cells and cortical interneuron production in the human ganglionic eminences. *Nat Neurosci* **16:** 1576–1587. doi:10.1038/nn.3541

Hayano Y, Ishino Y, Hyun JH, Orozco CG, Steinecke A, Potts E, Oisi Y, Thomas CI, Guerrero-Given D, Kim E, et al. 2021. IgSF11 homophilic adhesion proteins promote layer-specific synaptic assembly of the cortical interneuron subtype. *Sci Adv* **7:** eabf1600. doi:10.1126/sciadv.abf1600

Hodge RD, Bakken TE, Miller JA, Smith KA, Barkan ER, Graybuck LT, Close JL, Long B, Johansen N, Penn O, et al. 2019. Conserved cell types with divergent features in human versus mouse cortex. *Nature* **573:** 61–68. doi:10.1038/s41586-019-1506-7

Hu JS, Vogt D, Sandberg M, Rubenstein JL. 2017. Cortical interneuron development: a tale of time and space. *Development* **144:** 3867–3878. doi:10.1242/dev.132852

Huang ZJ, Paul A. 2019. The diversity of GABAergic neurons and neural communication elements. *Nat Rev Neurosci* **20:** 563–572. doi:10.1038/s41583-019-0195-4

Huang CY, Rasband MN. 2018. Axon initial segments: structure, function, and disease. *Ann NY Acad Sci* **1420:** 46–61. doi:10.1111/nyas.13718

Inan M, Welagen J, Anderson SA. 2012. Spatial and temporal bias in the mitotic origins of somatostatin- and parvalbumin-expressing interneuron subgroups and the chandelier subtype in the medial ganglionic eminence. *Cereb Cortex* **22:** 820–827. doi:10.1093/cercor/bhr148

Inan M, Blázquez-Llorca L, Merchán-Pérez A, Anderson SA, DeFelipe J, Yuste R. 2013. Dense and overlapping innervation of pyramidal neurons by chandelier cells. *J Neurosci* **33:** 1907–1914. doi:10.1523/JNEUROSCI.4049-12.2013

Inda MC, Defelipe J, Muñoz A. 2007. The distribution of chandelier cell axon terminals that express the GABA plasma membrane transporter GAT-1 in the human neocortex. *Cereb Cortex* **17:** 2060–2071. doi:10.1093/cercor/bhl114

Jiang X, Shen S, Cadwell CR, Berens P, Sinz F, Ecker AS, Patel S, Tolias AS. 2015. Principles of connectivity among morphologically defined cell types in adult neocortex. *Science* **350:** aac9462. doi:10.1126/science.aac9462

Jones EG. 1975. Varieties and distribution of non-pyramidal cells in the somatic sensory cortex of the squirrel monkey. *J Comp Neurol* **160:** 205–267. doi:10.1002/cne.901600204

Jones EG, Powell TP. 1969. Synapses on the axon hillocks and initial segments of pyramidal cell axons in the cerebral cortex. *J Cell Sci* **5:** 495–507. doi:10.1242/jcs.5.2.495

Jung K, Choi Y, Kwon H-B. 2022. Cortical control of chandelier cells in neural codes. *Front Cell Neurosci* **16:** 992409.

Jung K, Chang M, Steinecke A, Burke B, Choi Y, Oisi Y, Fitzpatrick D, Taniguchi H, Kwon HB. 2023. An adaptive behavioral control motif mediated by cortical axo-axonic inhibition. *Nat Neurosci* **26:** 1379–1393. doi:10.1038/s41593-023-01380-x

Kelly SM, Raudales R, Moissidis M, Kim G, Huang J. 2019. Multipotent radial glia progenitors and fate-restricted intermediate progenitors sequentially generate diverse cortical interneuron types. bioRxiv doi:10.1101/735019

Khirug S, Yamada J, Afzalov R, Voipio J, Khiroug L, Kaila K. 2008. GABAergic depolarization of the axon initial segment in cortical principal neurons is caused by the Na-K-2Cl cotransporter NKCC1. *J Neurosci* **28:** 4635–4639. doi:10.1523/JNEUROSCI.0908-08.2008

Klausberger T, Márton LF, Baude A, Roberts JD, Magill PJ, Somogyi P. 2004. Spike timing of dendrite-targeting bistratified cells during hippocampal network oscillations in vivo. *Nat Neurosci* **7:** 41–47. doi:10.1038/nn1159

Klausberger T, Marton LF, O'Neill J, Huck JH, Dalezios Y, Fuentealba P, Suen WY, Papp E, Kaneko T, Watanabe M, et al. 2005. Complementary roles of cholecystokinin- and parvalbumin-expressing GABAergic neurons in hippocampal network oscillations. *J Neurosci* **25:** 9782–9793. doi:10.1523/JNEUROSCI.3269-05.2005

Kole MH, Brette R. 2018. The electrical significance of axon location diversity. *Curr Opin Neurobiol* **51:** 52–59 doi:10.1016/j.conb.2018.02.016

Leonzino M, Busnelli M, Antonucci F, Verderio C, Mazzanti M, Chini B. 2016. The timing of the excitatory-to-inhibitory GABA switch is regulated by the oxytocin receptor via KCC2. *Cell Rep* **15:** 96–103. doi:10.1016/j.celrep.2016.03.013

Leterrier C. 2018. The axon initial segment: an updated viewpoint. *J Neurosci* **38:** 2135–2145. doi:10.1523/JNEUROSCI.1922-17.2018

Lim L, Mi D, Llorca A, Marín O. 2018. Development and functional diversification of cortical interneurons. *Neuron* **100:** 294–313. doi:10.1016/j.neuron.2018.10.009

Lipkin AM, Bender KJ. 2023. Axon initial segment GABA inhibits action potential generation throughout periadolescent development. *J Neurosci* **43:** 6357–6368. doi:10.1523/JNEUROSCI.0605-23.2023

Loomba S, Straehle J, Gangadharan V, Heike N, Khalifa A, Motta A, Ju N, Sievers M, Gempt J, Meyer HS, et al. 2022. Connectomic comparison of mouse and human cortex. *Science* **377:** eabo0924. doi:10.1126/science.abo0924

Lu J, Tucciarone J, Padilla-Coreano N, He M, Gordon JA, Huang ZJ. 2017. Selective inhibitory control of pyramidal neuron ensembles and cortical subnetworks by chandelier cells. *Nat Neurosci* **20:** 1377–1383. doi:10.1038/nn.4624

Ma T, Wang C, Wang L, Zhou X, Tian M, Zhang Q, Zhang Y, Li J, Liu Z, Cai Y, et al. 2013. Subcortical origins of human and monkey neocortical interneurons. *Nat Neurosci* **16:** 1588–1597. doi:10.1038/nn.3536

Marco P, Sola RG, Pulido P, Alijarde MT, Sánchez A, Ramón y Cajal S, DeFelipe J. 1996. Inhibitory neurons in the human epileptogenic temporal neocortex. An immunocytochemical study. *Brain* **119:** 1327–1347. doi:10.1093/brain/119.4.1327

Marco P, Sola RG, Ramón y Cajal S, DeFelipe J. 1997. Loss of inhibitory synapses on the soma and axon initial segment of pyramidal cells in human epileptic peritumoural neocortex: implications for epilepsy. *Brain Res Bull* **44:** 47–66. doi:10.1016/S0361-9230(97)00090-7

Massi L, Lagler M, Hartwich K, Borhegyi Z, Somogyi P, Klausberger T. 2012. Temporal dynamics of parvalbumin-expressing axo-axonic and basket cells in the rat medial prefrontal cortex in vivo. *J Neurosci* **32:** 16496–16502. doi:10.1523/JNEUROSCI.3475-12.2012

Mayer C, Hafemeister C, Bandler RC, Machold R, Batista Brito R, Jaglin X, Allaway K, Butler A, Fishell G, Satija R. 2018. Developmental diversification of cortical inhibitory interneurons. *Nature* **555:** 457–462.

Mi D, Li Z, Lim L, Li M, Moissidis M, Yang Y, Gao T, Hu TX, Pratt T, Price DJ, et al. 2018. Early emergence of cortical interneuron diversity in the mouse embryo. *Science* **360:** 81–85.

Miller JA, Gouwens NW, Tasic B, Collman F, van Velthoven CT, Bakken TE, Hawrylycz MJ, Zeng H, Lein ES, Bernard A. 2020. Common cell type nomenclature for the mammalian brain. *eLife* **9:** e59928. doi:10.7554/eLife.59928

Ostos S, Aparicio G, Fernaud-Espinosa I, DeFelipe J, Muñoz A. 2023. Quantitative analysis of the GABAergic innervation of the soma and axon initial segment of pyramidal cells in the human and mouse neocortex. *Cereb Cortex* **33:** 3882–3909. doi:10.1093/cercor/bhac314

Pan-Vazquez A, Wefelmeyer W, Gonzalez Sabater V, Neves G, Burrone J. 2020. Activity-dependent plasticity of axo-axonic synapses at the axon initial segment. *Neuron* **106:** 265–276.e6. doi:10.1016/j.neuron.2020.01.037

Perumal MB, Latimer B, Xu L, Stratton P, Nair S, Sah P. 2021. Microcircuit mechanisms for the generation of sharp-wave ripples in the basolateral amygdala: a role for chandelier interneurons. *Cell Rep* **35:** 109106. doi:10.1016/j.celrep.2021.109106

Petanjek Z, Berger B, Esclapez M. 2009a. Origins of cortical GABAergic neurons in the cynomolgus monkey. *Cereb Cortex* **19:** 249–262. doi:10.1093/cercor/bhn078

Petanjek Z, Kostović I, Esclapez M. 2009b. Primate-specific origins and migration of cortical GABAergic neurons. *Front Neuroanat* **3:** 26. doi:10.3389/neuro.05.026.2009

Peters A, Proskauer CC, Ribak CE. 1982. Chandelier cells in rat visual cortex. *J Comp Neurol* **206:** 397–416. doi:10.1002/cne.902060408

Petilla Interneuron Nomenclature Group; Ascoli GA, Alonso-Nanclares L, Anderson SA, Barrionuevo G, Benavides-Piccione R, Burkhalter A, Buzsáki G, Cauli B, Defelipe J, et al. 2008. Petilla terminology: nomenclature of features of GABAergic interneurons of the cerebral cortex. *Nat Rev Neurosci* **9:** 557–568. doi:10.1038/nrn 2402

Priya R, Paredes MF, Karayannis T, Yusuf N, Liu X, Jaglin X, Graef I, Alvarez-Buylla A, Fishell G. 2018. Activity regulates cell death within cortical interneurons through a calcineurin-dependent mechanism. *Cell Rep* **22:** 1695–1709. doi:10.1016/j.celrep.2018.01.007

Raudales R, Kim G, Kelly SM, Hatfield J, Guan W, Zhao S, Paul A, Qian Y, Li B, Huang ZJ. 2024. Specific and comprehensive genetic targeting reveals brain-wide distribution and synaptic input patterns of GABAergic axo-axonic interneurons. *eLife* **13:** RP93481. doi:10.7554/eLife .93481

Represa A, Ben-Ari Y. 2005. Trophic actions of GABA on neuronal development. *Trends Neurosci* **28:** 278–283. doi:10.1016/j.tins.2005.03.010

Ribak CE. 1985. Axon terminals of GABAergic chandelier cells are lost at epileptic foci. *Brain Res* **326:** 251–260. doi:10.1016/0006-8993(85)90034-4

Rinetti-Vargas G, Phamluong K, Ron D, Bender KJ. 2017. Periadolescent maturation of GABAergic hyperpolarization at the axon initial segment. *Cell Rep* **20:** 21–29. doi:10 .1016/j.celrep.2017.06.030

Sayin U, Osting S, Hagen J, Rutecki P, Sutula T. 2003. Spontaneous seizures and loss of axo-axonic and axo-somatic inhibition induced by repeated brief seizures in kindled rats. *J Neurosci* **23:** 2759–2768. doi:10.1523/JNEUROSCI .23-07-02759.2003

Scala F, Kobak D, Bernabucci M, Bernaerts Y, Cadwell CR, Castro JR, Hartmanis L, Jiang X, Laturnus S, Miranda E, et al. 2021. Phenotypic variation of transcriptomic cell types in mouse motor cortex. *Nature* **598:** 144–150. doi:10 .1038/s41586-020-2907-3

Schneider-Mizell CM, Bodor AL, Collman F, Brittain D, Bleckert A, Dorkenwald S, Turner NL, Macrina T, Lee K, Lu R, et al. 2021. Structure and function of axo-axonic inhibition. *eLife* **10:** e73783. doi:10.7554/eLife.73783

Seignette K, Jamann N, Papale P, Terra H, Porneso RO, de Kraker L, van der Togt C, van der Aa M, Neering P, Ruimschotel E, et al. 2024. Experience shapes chandelier cell function and structure in the visual cortex. *eLife* **12:** RP91153. doi:10.7554/eLife.91153

Shi Y, Wang M, Mi D, Lu T, Wang B, Dong H, Zhong S, Chen Y, Sun L, Zhou X, et al. 2021. Mouse and human share conserved transcriptional programs for interneuron development. *Science* **374:** eabj6641. doi:10.1126/science.ab j6641

Sipilä ST, Huttu K, Soltesz I, Voipio J, Kaila K. 2005. Depolarizing GABA acts on intrinsically bursting pyramidal neurons to drive giant depolarizing potentials in the immature hippocampus. *J Neurosci* **25:** 5280–5289. doi:10 .1523/JNEUROSCI.0378-05.2005

Sloper JJ, Powell TP. 1979. A study of the axon initial segment and proximal axon of neurons in the primate motor and somatic sensory cortices. *Philos Trans R Soc Lond B Biol Sci* **285:** 173–197. doi:10.1098/rstb.1979.0004

Somogyi P. 1977. A specific "axo-axonal" interneuron in the visual cortex of the rat. *Brain Res* **136:** 345–350. doi:10 .1016/0006-8993(77)90808-3

Somogyi P, Freund TF, Cowey A. 1982. The axo-axonic interneuron in the cerebral cortex of the rat, cat and mon-

key. *Neuroscience* **7:** 2577–2607. doi:10.1016/0306-4522 (82)90086-0

Somogyi P, Freund TF, Hodgson AJ, Somogyi J, Beroukas D, Chubb IW. 1985. Identified axo-axonic cells are immunoreactive for GABA in the hippocampus and visual cortex of the cat. *Brain Res* **332:** 143–149. doi:10.1016/0006-8993(85)90397-X

Southwell DG, Paredes MF, Galvao RP, Jones DL, Froemke RC, Sebe JY, Alfaro-Cervello C, Tang Y, Garcia-Verdugo JM, Rubenstein JL, et al. 2012. Intrinsically determined cell death of developing cortical interneurons. *Nature* **491:** 109–113. doi:10.1038/nature11523

Spampanato J, Sullivan RK, Perumal MB, Sah P. 2016. Development and physiology of GABAergic feedback excitation in parvalbumin expressing interneurons of the mouse basolateral amygdala. *Physiol Rep* **4:** e12664. doi:10.14814/phy2.12664

Steinecke A, Hozhabri E, Tapanes S, Ishino Y, Zeng H, Kamasawa N, Taniguchi H. 2017. Neocortical chandelier cells developmentally shape axonal arbors through reorganization but establish subcellular synapse specificity without refinement. *eNeuro* **4:** ENEURO.0057-17.2017. doi:10.1523/ENEURO.0057-17.2017

Sultan KT, Liu WA, Li ZL, Shen Z, Li Z, Zhang XJ, Dean O, Ma J, Shi SH. 2018. Progressive divisions of multipotent neural progenitors generate late-born chandelier cells in the neocortex. *Nat Commun* **9:** 4595. doi:10.1038/s41467-018-07055-7

Suresh H, Crow M, Jorstad N, Hodge R, Lein E, Dobin A, Bakken T, Gillis J. 2023. Comparative single-cell transcriptomic analysis of primate brains highlights human-specific regulatory evolution. *Nat Ecol Evol* **7:** 1930–1943. doi:10.1038/s41559-023-02126-7

Szabadics J, Varga C, Molnár G, Oláh S, Barzó P, Tamás G. 2006. Excitatory effect of GABAergic axo-axonic cells in cortical microcircuits. *Science* **311:** 233–235. doi:10.1126/ science.1121325

Szentágothai J. 1975. The "module-concept" in cerebral cortex architecture. *Brain Res* **95:** 475–496. doi:10.1016/ 0006-8993(75)90122-5

Szentágothai J, Arbib MA. 1974. Conceptual models of neural organization. *Neurosci Res Program Bull* **12:** 305–510.

Tai Y, Janas JA, Wang CL, Van Aelst L. 2014. Regulation of chandelier cell cartridge and bouton development via DOCK7-mediated ErbB4 activation. *Cell Rep* **6:** 254–263. doi:10.1016/j.celrep.2013.12.034

Tai Y, Gallo NB, Wang M, Yu JR, Van Aelst L. 2019. Axo-axonic innervation of neocortical pyramidal neurons by GABAergic chandelier cells requires ankyrinG-associated L1CAM. *Neuron* **102:** 358–372.e9. doi:10.1016/j.neuron .2019.02.009

Taniguchi H, He M, Wu P, Kim S, Paik R, Sugino K, Kvitsiani D, Fu Y, Lu J, Lin Y, et al. 2011. A resource of Cre driver lines for genetic targeting of GABAergic neurons in cerebral cortex. *Neuron* **71:** 995–1013. doi:10.1016/j .neuron.2011.07.026

Taniguchi H, Lu J, Huang ZJ. 2013. The spatial and temporal origin of chandelier cells in mouse neocortex. *Science* **339:** 70–74. doi:10.1126/science.1227622

Tasic B, Yao Z, Graybuck LT, Smith KA, Nguyen TN, Bertagnolli D, Goldy J, Garren E, Economo MN, Viswanathan S, et al. 2018. Shared and distinct transcriptomic cell

types across neocortical areas. *Nature* **563:** 72–78. doi:10
.1038/s41586-018-0654-5

Tukker JJ, Fuentealba P, Hartwich K, Somogyi P, Klausberger T. 2007. Cell type-specific tuning of hippocampal interneuron firing during γ oscillations in vivo. *J Neurosci* **27:** 8184–8189. doi:10.1523/JNEUROSCI.1685-07.2007

Tyzio R, Ivanov A, Bernard C, Holmes GL, Ben-Ari Y, Khazipov R. 2003. Membrane potential of CA3 hippocampal pyramidal cells during postnatal development. *J Neurophysiol* **90:** 2964–2972. doi:10.1152/jn.00172.2003

Tyzio R, Holmes GL, Ben-Ari Y, Khazipov R. 2007. Timing of the developmental switch in GABA(A) mediated signaling from excitation to inhibition in CA3 rat hippocampus using gramicidin perforated patch and extracellular recordings. *Epilepsia* **48:** 96–105. doi:10.1111/j.1528-1167.2007.01295.x

Varga C, Oijala M, Lish J, Szabo GG, Bezaire M, Marchionni I, Golshani P, Soltesz I. 2014. Functional fission of parvalbumin interneuron classes during fast network events. *eLife* **3:** e04006. doi:10.7554/eLife.04006

Veres JM, Nagy GA, Vereczki VK, Andrási T, Hájos N. 2014. Strategically positioned inhibitory synapses of axo-axonic cells potently control principal neuron spiking in the basolateral amygdala. *J Neurosci* **34:** 16194–16206. doi:10.1523/JNEUROSCI.2232-14.2014

Viney TJ, Lasztoczi B, Katona L, Crump MG, Tukker JJ, Klausberger T, Somogyi P. 2013. Network state-dependent inhibition of identified hippocampal CA3 axo-axonic cells in vivo. *Nat Neurosci* **16:** 1802–1811. doi:10.1038/nn.3550

Vogt Weisenhorn DM, Celio MR, Rickmann M. 1998. The onset of parvalbumin-expression in interneurons of the rat parietal cortex depends upon extrinsic factor(s). *Eur J Neurosci* **10:** 1027–1036. doi:10.1046/j.1460-9568.1998.00120.x

Wang X, Hooks BM, Sun QQ. 2014. Thorough GABAergic innervation of the entire axon initial segment revealed by an optogenetic "laserspritzer." *J Physiol* **592:** 4257–4276. doi:10.1113/jphysiol.2014.275719

Wang BS, Bernardez Sarria MS, An X, He M, Alam NM, Prusky GT, Crair MC, Huang ZJ. 2021. Retinal and callosal activity-dependent chandelier cell elimination shapes binocularity in primary visual cortex. *Neuron* **109:** 502–515.e7. doi:10.1016/j.neuron.2020.11.004

Wong FK, Bercsenyi K, Sreenivasan V, Portalés A, Fernández-Otero M, Marín O. 2018. Pyramidal cell regulation of interneuron survival sculpts cortical networks. *Nature* **557:** 668–673. doi:10.1038/s41586-018-0139-6

Woodruff AR, Monyer H, Sah P. 2006. GABAergic excitation in the basolateral amygdala. *J Neurosci* **26:** 11881–11887. doi:10.1523/JNEUROSCI.3389-06.2006

Woodruff A, Xu Q, Anderson SA, Yuste R. 2009. Depolarizing effect of neocortical chandelier neurons. *Front Neural Circuits* **3:** 15. doi:10.3389/neuro.04.015.2009

Woodruff AR, McGarry LM, Vogels TP, Inan M, Anderson SA, Yuste R. 2011. State-dependent function of neocortical chandelier cells. *J Neurosci* **31:** 17872–17886. doi:10.1523/JNEUROSCI.3894-11.2011

Yamada J, Okabe A, Toyoda H, Kilb W, Luhmann HJ, Fukuda A. 2004. Cl⁻ uptake promoting depolarizing GABA actions in immature rat neocortical neurones is mediated by NKCC1. *J Physiol* **557:** 829–841. doi:10.1113/jphysiol.2004.062471

Yao Z, van Velthoven CTJ, Kunst M, Zhang M, McMillen D, Lee C, Jung W, Goldy J, Abdelhak A, Aitken M, et al. 2023. A high-resolution transcriptomic and spatial atlas of cell types in the whole mouse brain. *Nature* **624:** 317–332. doi:10.1038/s41586-023-06812-z

Yuste R, Hawrylycz M, Aalling N, Aguilar-Valles A, Arendt D, Armañanzas R, Ascoli GA, Bielza C, Bokharaie V, Bergmann TB, et al. 2020. A community-based transcriptomics classification and nomenclature of neocortical cell types. *Nat Neurosci* **23:** 1456–1468. doi:10.1038/s41593-020-0685-8

Zeng H, Sanes JR. 2017. Neuronal cell-type classification: challenges, opportunities and the path forward. *Nat Rev Neurosci* **18:** 530–546. doi:10.1038/nrn.2017.85

Zhao R, Ren B, Xiao Y, Tian J, Zou Y, Wei J, Qi Y, Hu A, Xie X, Huang ZJ, et al. 2024. Axo-axonic synaptic input drives homeostatic plasticity by tuning the axon initial segment structurally and functionally. *Sci Adv* **10:** eadk4331. doi:10.1126/sciadv.adk4331

Zhu Y, Stornetta RL, Zhu JJ. 2004. Chandelier cells control excessive cortical excitation: characteristics of whisker-evoked synaptic responses of layer 2/3 nonpyramidal and pyramidal neurons. *J Neurosci* **24:** 5101–5108. doi:10.1523/JNEUROSCI.0544-04.2004

 Cite this article as *Cold Spring Harb Perspect Biol* doi: 10.1101/cshperspect.a041506

Characterizing Large-Scale Human Circuit Development with In Vivo Neuroimaging

Tomoki Arichi[1,2,3]

[1]Centre for the Developing Brain, School of Biomedical Engineering and Imaging Sciences, King's College London, St Thomas' Hospital, London SE1 7EH, United Kingdom

[2]MRC Centre for Neurodevelopmental Disorders, King's College London, New Hunt's House, Guy's Campus, London SE1 1UL, United Kingdom

[3]Children's Neurosciences, Evelina London Children's Hospital, Guy's and St Thomas' NHS Foundation Trust, London SE1 7EH, United Kingdom

Correspondence: tomoki.arichi@kcl.ac.uk

Large-scale coordinated patterns of neural activity are crucial for the integration of information in the human brain and to enable complex and flexible human behavior across the life span. Through recent advances in noninvasive functional magnetic resonance imaging (fMRI) methods, it is now possible to study this activity and how it emerges in the living fetal brain across the second half of human gestation. This work has demonstrated that functional activity in the fetal brain has several features in keeping with highly organized networks of activity, which are undergoing a highly programmed and rapid sequence of development before birth, in which long-range connections emerge and core features of the mature functional connectome (such as hub regions and a gradient organization) are established. In this review, the findings of these studies are summarized, their relationship to the known changes in developmental neurobiology is considered, and considerations for future work in the context of limitations to the fMRI approach are presented.

Over the last century, pioneering postmortem studies have provided detailed information about the dramatic cellular and anatomical changes that occur in the human brain during the time leading to birth. While such studies have provided fundamental insights into how the structure of the fetal brain evolves on the micro- and mesoscale across this period, it is only recently that the critically important role that activity plays in early brain development has begun to be understood. Furthering this knowledge is vital, as it is increasingly recognized that altered patterns of activity before birth (either through genetic mechanisms or acquired lesions) frequently lead to permanent changes in brain structure, circuit organization, and function (Miguel et al. 2019). This has key implications for behavior and neurological function, with converging evidence now suggesting that early disruptions in brain connectivity (how distinct brain regions are structurally and functionally connected to one another) are a key pathological feature underlying neurodevelopmental conditions that are lifelong but manifest in childhood such as autism (Testa-Silva et al. 2012; Deneault et al. 2018; Ciarrusta et al. 2020) and mental health disorders (e.g., schizophrenia) (Gilmore et al. 2010; Sigurdsson 2016).

In its earliest stages, neural activity occurs spontaneously in local circuits within clusters of developing neurons, and in doing so helps to enhance and/or modulate further neurogenesis, progenitor cell differentiation, and synaptogenesis (Luhmann et al. 2016). These processes have been studied in cutting edge in vitro and animal work, which are detailed in other articles in this special collection. Here, we instead focus on the large-scale patterns of coordinated neural activity that emerge in the second half of human gestation as the cortex and its macroscopic framework of axonal connections are established (Kostović and Jovanov-Milošević 2006). Importantly, much of the previous knowledge about these processes has been inferred from preterm-born infants studied in the equivalent period to the third trimester of gestation (28 to 40 postmenstrual weeks). However, information derived from infants that have been prematurely exposed to the ex utero environment is unlikely to be truly representative of "normal" fetal brain development. This can now be overcome through recent methodological advances in noninvasive neuroimaging, which for the first time enable in vivo study of the emergence of larger-scale neural "circuits" in human fetuses. Here the methods themselves are described, benefits and limitations are considered, findings are described in the context of known developmental neurobiology, and future directions for study are presented.

CHARACTERIZING BRAIN FUNCTION AND CONNECTIVITY IN THE BRAIN WITH NEUROIMAGING

Although the importance of integration and cooperative patterns of neuronal activity for brain function has long been established in neuroscience (Hebb 1949), understanding of the importance of large-scale neural circuitry and the concept of "functional connectivity" (long-range temporal correlations in activity between spatially distinct brain areas with neuroimaging) is relatively recent (Friston et al. 1993). The latter is generally considered to represent the ability of the brain to share information between different regions, each of which with their own specific processing role. The resulting correlated patterns of large-scale brain activity are thought to enable complex human behavior by providing the framework needed for integration and exchange of information during both extrinsically and intrinsically generated functions (van den Heuvel and Hulshoff Pol 2010). In addition to mapping the brain's functional connections, the "structural connectivity" of the white matter axonal pathways between brain regions can also be noninvasively mapped using diffusion magnetic resonance imaging (MRI) methods (Le Bihan et al. 2001). The resulting structural connectivity measures have been shown to significantly predict patterns of functional connectivity, suggesting that they represent the anatomical framework on which large-scale patterns of activity propagate (Honey et al. 2010).

Correlated patterns of neural activity can also be studied with a variety of neurophysiological (i.e., electroencephalography [EEG], magnetoencephalography [MEG]) and neuroimaging methods (i.e., positron emission tomography (PET), near-infrared spectroscopy (NIRS)). However, the majority of these methods are not easily applied to studying the in utero fetal brain as they measure signals either through the scalp or require injection of a radioactive tracer. In contrast, functional magnetic resonance imaging (fMRI) can provide an entirely noninvasive and safe measure of fetal brain activity, with relatively good whole-brain spatial sensitivity (usually a few millimeters cubed) and temporal resolution (usually a few seconds). The sampled blood oxygen level–dependent (BOLD) fMRI signal is an indirect measure of neural activity, as the contrast mechanism is generated by sampling temporal signal fluctuations arising from localized changes in the relative proportion of paramagnetic deoxygenated hemoglobin and diamagnetic oxygenated hemoglobin (Ogawa et al. 1990). These change due to the local alterations in cerebral blood flow that are associated with neural activity, through the carefully controlled neurovascular coupling cascade (Attwell and Iadecola 2002). In the seminal work of

Cite this article as *Cold Spring Harb Perspect Biol* doi: 10.1101/cshperspect.a041496

Logothetis et al. (2001), fluctuations in the fMRI BOLD signal were found to most closely relate temporally to cortical local field potentials (LFPs) and thus represent the sum of synaptic inputs within a population of neurons, rather than their spiking output.

In the mature brain, fMRI has been widely applied to spatially map areas of activity in the brain associated with particular tasks, both within putative primary processing regions and more widely across the engaged network of associative regions (Bandettini et al. 1992; Kwong et al. 1992; Ogawa et al. 1992). However, a major advance in understanding the functional organization of the brain occurred when it was discovered that patterns of correlated low frequency (0.01 to 0.1 Hz) activity could also be reproducibly identified even at rest (i.e., in the absence of a particular task or stimulus), particularly between functional homolog regions in each hemisphere (Biswal et al. 1995). The energy cost of this resting activity is significant, with early PET studies demonstrating that transient periods of task induced neural activity are associated with only very small rises in oxygen metabolism from that which is already needed during the baseline resting condition (Fox et al. 1988). The spatial organization of resting patterns of activity across specific brain areas have been termed "resting state networks" (Snyder and Raichle 2012). In addition to recapitulating the spatial patterns of correlated activity induced by a particular stimulus or task (Cole et al. 2014), resting state networks in adults have been found to be highly reproducible both within a given subject and across large populations (Shen et al. 2018), suggesting that they are an intrinsic brain property that is preserved across behavioral states and people. While the repertoire of these networks continues to grow, the classical complement includes those covering the primary motor and sensory cortical regions in both hemispheres (the motor, somatosensory, primary visual, lateral visual, auditory networks), in addition to those often considered to be "higher order" networks incorporating the medial and lateral frontal regions, insular cortices, and anterior cingulate gyri (Beckmann et al. 2005; Damoiseaux et al. 2006). Of particular interest has been the so-called "default mode network," which encompasses the medial prefrontal cortex, precuneus, and the bilateral posteroinferior parietal lobes, and has been proposed to have a key role in facets of complex human brain processing such as self-referential thought (Raichle et al. 2001; Greicius et al. 2003).

The application of the above fMRI methods into the study of newborn infants demonstrated that even shortly after birth, resting state networks resembling those seen in the adult brain could also be reliably identified (Fransson et al. 2007). Of particular interest, extension of study populations into preterm-born infants imaged before the time of normal birth found a clear pattern of maturation, with the topology of resting state networks seen to progress from simple unilateral clusters of local connectivity in a single hemisphere in the youngest infants (<28 wk postmenstrual age) to distributed bilateral networks with long-range interhemispheric or anteroposterior patterns of connectivity by term equivalent age (Doria et al. 2010; Smyser et al. 2010). These results highlight a clear pattern of emerging functional connectivity, suggesting that the foundations of the brain's lifelong network architecture are established during the equivalent period to the third trimester of gestation. The key importance of this period is further emphasized by studies that have shown that altered functional connectivity in preterm-born infants is associated with later adverse neurodevelopmental outcome (Linke et al. 2018; Eyre et al. 2021; Cyr et al. 2022) and is sustained into later childhood and adulthood (Papini et al. 2016; Wehrle et al. 2018; Hadaya and Nosarti 2020).

NEUROBIOLOGICAL DEVELOPMENT UNDERLYING THE ESTABLISHMENT OF LARGE-SCALE CIRCUITRY IN THE FETAL BRAIN

Across the 40 wk of gestation, the human cortex undergoes an extremely rapid but highly programmed sequence of microstructural and macrostructural maturation that is presumed to lay the anatomical framework needed for the aforementioned patterns of long-range connectivity (for review, see Pöpplau and Hanganu-Opatz

2024; as well as Kostović et al. 2019; Molnár et al. 2019). The formation of the cortex in this time is a protracted process that is characterized by proliferation and tangential/radial migration of neural progenitor cells from the ventricular and outer subventricular zones (Marin and Rubenstein 2003). These cells first reach their final location on the outer cortical surface from 12 wk of gestation, with this process largely complete by approximately 30 wk, although it continues even up to 2 yr of age in specific cortical regions (Cadwell et al. 2019). Migration occurs earlier in the dorsal brain (peaking in the occipital lobe at 20 wk of gestation) compared with parietal (peaking at 23 wk) and frontal regions (peaking at 26 wk) (Trivedi et al. 2009; Paredes et al. 2016). During this time, genetic mechanisms and signaling pathways direct neuronal differentiation, leading to the characteristic anatomical features seen in the mature cortex such as lamination (Cadwell et al. 2019), which is seen first in the primary sensory and motor cortices at 25 wk of gestation, with the full adult complement of distinct lamina seen by 32 wk (Bystron et al. 2008; Kostović et al. 2019). Cortical folding rapidly proceeds across the third trimester of gestation such that the majority of the mature brain's sulcal landmarks can be identified by full term (van der Knaap et al. 1996; Yun et al. 2020). During this time, endogenously generated synchronous neural activity occurs in early circuits and is critical during the aforementioned processes by guiding fundamental processes including synaptogenesis, neuronal maturation, and dendritic arborization (Khazipov and Luhmann 2006). Ex vivo studies suggest this evolves from spontaneous events that spread locally in the form of oscillatory calcium waves and giant depolarizing potentials, before further neurochemical maturation (specifically that of the GABA and glutamate neurotransmitter systems) enables large amplitude bursting events in the latter half of gestation (Khazipov and Luhmann 2006).

The thalami are of particular interest as the emerging thalamocortical axonal pathways are known to provide key inputs into the developing cortex, additionally helping to guide cortical areal differentiation and establish the circuitry underlying sensory integration across the life span (Sur and Rubenstein 2005; Price et al. 2006; Kostović and Judaš 2010; Krsnik et al. 2017). The sequence of thalamocortical maturation begins with fiber outgrowth at 8–9.5 gestational weeks, pathfinding at 9–14 wk, "waiting" in the cortical subplate region between 14 and 22 wk, and finally ingrowth into the cortical plate at 23–24 wk (Krsnik et al. 2017). Thalamic afferents and early-generated transient subplate neurons synapse during the waiting period (Wess et al. 2017). These synapses play a key role in forming a functional template for the development of thalamocortical networks and overall cortical architecture (Ohtaka-Maruyama et al. 2018; Molnár et al. 2020). A fundamental feature of these developing neural circuits is spontaneous activity, which begins in the subplate neurons even before the establishment of cortical layers (Luhmann et al. 2022). The critical importance of this activity has been demonstrated by selective surgical ablation of the subplate in rodents, which abolishes spontaneous cortical activity and disrupts permanent cortical organization (Tolner et al. 2012).

The above developmental changes in tissue microstructure and structural connectivity can be characterized in the postmortem and living fetal brain with diffusion MRI (Takahashi et al. 2012; Huang and Vasung 2014; Vasung et al. 2019; Wilson et al. 2023; Zheng et al. 2023). This includes visualizing the dissolution of the developing cortical plate's radial organization in the second trimester (which persists longer within the gyral crests in comparison to the sulcal depths) (Takahashi et al. 2012). In preterm infants, systematic changes in diffusion MRI-derived microstructural metrics are suggestive of a predominant increase in dendritic arborization and neurite growth in cortical gray matter between 25 and 38 wk of gestation (Batalle et al. 2019). Related methods have also been used with in utero MRI data to characterize maturational changes in microstructure seen within transient tissue layers (intermediate zone, subplate, and cortical plate) as they grow and dissolve in fetuses, with specific developmental trajectories associated with distinct thalamocortical white matter pathways (Wilson et al. 2023). These pathways are presumed to represent bun-

Cite this article as *Cold Spring Harb Perspect Biol* doi: 10.1101/cshperspect.a041496

dles of emerging premyelinated axonal fibers that (in keeping with histological studies) mature in a tract-specific manner: with the commissural and projection fibers (corpus callosal and corticospinal tracts) visible by 22 wk gestation and the optic radiations becoming more evident later (Wilson et al. 2021). Although already visible, the microstructure of these tracts is still significantly changing and becoming more organized, consistent with the hypothesis that the anatomical framework underlying the brain's long-range connectivity is still being established in the fetal period (Jakab et al. 2017; Jaimes et al. 2020; Machado-Rivas et al. 2021; Wilson et al. 2021).

EXPLORING CHANGES IN LARGE-SCALE FUNCTIONAL CONNECTIVITY IN THE FETAL BRAIN

Despite its clear appeal as a noninvasive and safe means of studying in utero whole brain neural activity, the application of fMRI to study fetal brain activity is relatively recent and thus there are relatively few published studies. Such work is essential as the existing knowledge derived from preterm infants is unlikely to be truly representative of "normal" in utero brain development. The potential of studying in utero brain activity in fetuses using fMRI was first described in pioneering studies that showed simple and spatially indistinct areas of activity in response to auditory stimulation (Moore et al. 2001), visual stimulation (Fulford et al. 2003), and at rest (Schöpf et al. 2012). With advances in both MR image acquisition and processing strategies, the field has expanded significantly in the last 10 yr, particularly in studies characterizing the functional organization of the fetal brain in the resting state over the third trimester of human gestation.

Broadly speaking, the results of these studies (summarized in Fig. 1) have shown that:

1. Resting state networks can be readily identified in the fetal brain from at least 19–20 wk gestation.

2. As seen in preterm infants, resting-state networks progress from single areas of activity in

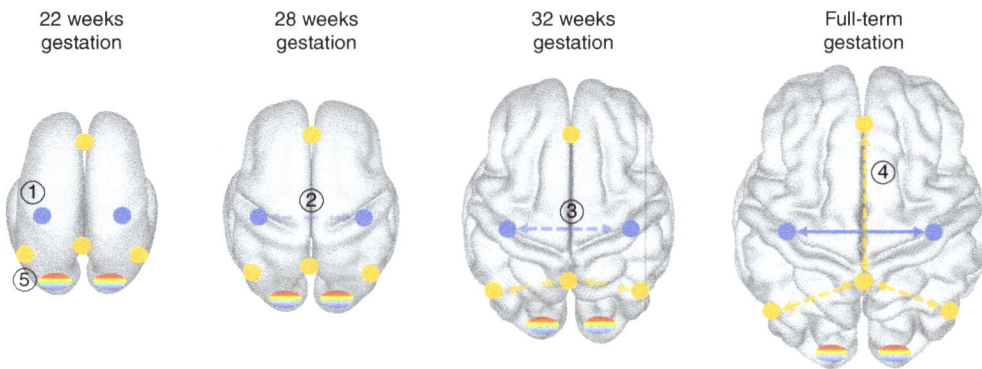

① Resting state networks can be seen at the start of the third trimester as areas of local connectivity only.

② Long-range patterns of connectivity including interhemispheric connections emerge across the third trimester.

③ Long-range patterns of connectivity are established first in primary sensory and motor networks.

④ Connectivity is established later in higher-order and associative networks.

⑤ Complex network features such as gradient organization are already established in fetal networks.

Figure 1. Summary of changes in large-scale connectivity seen in the fetal brain across the second half of gestation with functional magnetic resonance imaging (MRI). Functional MRI (fMRI) can be used to identify reproducible spatial patterns of correlated brain activity, which are distributed across specific brain regions (indicated by colored circles in the figure) into "resting state networks." These can be readily identified in the fetal brain from the start of the second half of gestation and show specific features suggesting that they are highly organized and undergo a systematic pattern of maturation through to full-term gestation.

a single hemisphere to more distinct adult-like topographies encompassing both hemispheres as the brain's long-range patterns of connectivity emerge.

3. Maturation of resting state network connectivity occurs earlier in the primary sensory and motor systems compared to those associated with higher-order associative and cognitive systems.

4. Specific features of mature complex network structure can already be identified in the fetal period, suggesting that their establishment is fundamental to the brain's functional organization across the life span.

Patterns of significant functional connectivity have been studied in human fetuses from as young as 19–20 wk gestation (Turk et al. 2019; De Asis-Cruz et al. 2021a). In general, these early resting state networks are seen as localized clusters of activity that isolate to single brain regions rather than the long-range patterns of long-range connectivity characteristically seen in the mature brain, suggesting that the underlying spontaneous activity occurs initially across local processing units as the associated short-range connectivity emerges (Schöpf et al. 2012; Thomason et al. 2013, 2015; Ferrazzi et al. 2014). These may relate to evolving forms of spontaneous neuronal activity events that are endogenously generated and then propagate across the surrounding cortex, and are considered to be a hallmark of the developing mammalian brain in the perinatal period (Luhmann et al. 2016). An important alternative explanation is that the correlated fluctuations in the fMRI BOLD signal may be non-neural in origin, and may instead only represent isolated temporal variations in cerebral blood flow as the mechanisms underlying neurovascular coupling may be too immature to support neural activity in the same way as it does in the mature brain (Kozberg and Hillman 2016b; Kozberg et al. 2016). However, this hypothesis is not in keeping with the reproducible finding of several distinct resting state networks in fetuses each with their own unique low frequency time courses and topographies, which appear independent from the time course of cardiovascular

pulsations and the anatomical location of early blood vessels alone (Thomason et al. 2013, 2015; Ferrazzi et al. 2014; Kim et al. 2023b) and spatially resemble those seen in preterm infants (Doria et al. 2010). Furthermore, large-scale patterns of resting electrical neural activity can also be readily seen in fetuses with MEG (Eswaran et al. 2007; Sheridan et al. 2010).

A further key observation across studies is the evolution and establishment of long-range patterns of connectivity across human gestation, with initially immature local patterns of activity maturing toward the adult-like resting-state network topology seen in full-term neonates (Eyre et al. 2021). This is most striking when looking at interhemispheric functional connectivity between homolog regions (e.g., the primary motor cortices), which increases linearly across the third trimester (Thomason et al. 2013, 2015). As with preterm infants and later childhood (Doria et al. 2010; Gilmore et al. 2018), this is seen to occur first in the primary motor and sensory cortices, before within the frontal and associative regions. These themes are also apparent when using an alternative approach in which the networks themselves are defined by age-related changes in their constituent functional connectivity, as opposed to the traditional method of characterizing "average" networks across the entire study population (Karolis et al. 2023). In keeping with emerging patterns of long-range connectivity across gestation, the identified fetal "matnets" have symmetrical spatial distributions encompassing functional homolog regions. Together, these changes result in a maturational decrease in global synchrony and increasing lateralization, which is significantly predictive of gestational age (Kim et al. 2023a; Taymourtash et al. 2023). Importantly, these developmental changes are also accompanied during the same period by increases in structural connectivity measures within key white matter pathways including the corpus collosum, inferior longitudinal fasculi, and thalamocortical tracts (Jaimes et al. 2020; Machado-Rivas et al. 2021; Wilson et al. 2021, 2023).

Functional connectivity in the mature brain has been found to have specific characteristics that optimize efficient information exchange and can support the dynamic, flexible processing required for adaptive environmental interaction and complex sociocognitive functioning (Sporns 2022). This is supported by a "gradient" organization, whereby the underlying patterns of functional and structural connectivity topographically vary in a continuous manner across the entire cortical surface (Margulies et al. 2016; Bernhardt et al. 2022). Functional connectivity within resting state networks is similarly nonuniform between constituent regions and varies in a graduated manner (Haak et al. 2018). Surprisingly, this seemingly complex but fundamental property can also be robustly identified within fetal resting state networks from as early as 25 wk gestation (Willers Moore et al. 2023). A further key feature of the mature functional "connectome" is the presence of hub regions that have a critically important role in information integration and efficient processing (van den Heuvel and Sporns 2013). These densely connected hub regions are also present in the fetal brain within specific parts of the primary and associative cortices (van den Heuvel et al. 2018; Turk et al. 2019). As these largely recapitulate those seen in preterm infants and later across the life span, this suggests that their establishment is a fundamental developmental process that is intrinsically generated and directed across early life (van den Heuvel et al. 2015). Further detailed studies have found that this development is also reflected in age-related changes of specific graph theory metrics including small-world index, normalized clustering and path length, global and local efficiency, and modularity (De Asis-Cruz et al. 2021b) and changes in thalamocortical connectivity (Taymourtash et al. 2023). The developmental trajectories of these measures appear to undergo a transition at 30–31 wk gestation, perhaps reflecting the developmental switch from endogenously generated to sensory-driven activity (Luhmann et al. 2016).

CONSIDERATIONS FOR STUDYING THE FETAL BRAIN WITH fMRI AND FUTURE DIRECTIONS

There are several significant challenges for the acquisition of in utero fMRI data, not only due to those associated with ensuring safety for the mother and fetus, but also due to the problems inherent to acquiring MR images from an uncooperative subject inside a unique uterine environment (Manganaro et al. 2023). Image-acquisition sequences must work within the constraints required to ensure appropriate levels of energy deposition and noise, to account for the effects of the maternal tissue and organs on magnetic field inhomogeneity, and are generally associated with reduced signal-to-noise ratio due to the physical distance between the fetus and the receive coil (Christiaens et al. 2019). Perhaps the most significant challenge is overcoming the considerable effects of motion artifact that arise due to unavoidable maternal (breathing and body movements) and spontaneous fetal movements during image acquisition (You et al. 2016; Sobotka et al. 2022). This is particularly important as it has long been known that head motion leads to nonneural changes in the fMRI signal, which can significantly affect the identification of activity and lead to spurious patterns of functional connectivity (Hajnal et al. 1994; Power et al. 2014; Satterthwaite et al. 2019).

One relatively common approach for addressing this potential problem has been to identify and completely exclude data time points corrupted by motion artifact (Thomason et al. 2013, 2015, 2017, 2018; van den Heuvel et al. 2018; Turk et al. 2019). While this approach has had some success in exploring the early emergence of functional connectivity in fetuses, such "motion scrubbing" of data is relatively inefficient (as sometimes a large amount of the collected data is discarded), does not address associated geometric image distortions, and importantly limits studies to specific behavioral states when the fetus is inactive. As a result, several recent studies have now described comprehensive frameworks that encompass tailored image acquisition, processing, and analysis strategies, which together have been designed specifically to address the aforementioned limitations (Seshamani et al. 2013; Ferrazzi et al. 2014; You et al. 2016; Scheinost et al. 2018; Sobotka et al. 2022; Taymourtash et al. 2022; Karolis et al. 2023). While a detailed review of these methodologies is beyond the scope of this article, they have been reviewed (e.g., van den Heuvel and Thomason 2016;

Rajagopalan et al. 2021) and some of the processing strategies empirically evaluated elsewhere (Ji et al. 2022).

A further key consideration is the indirect nature of the fMRI BOLD signal, as the relationship between neural activity and dynamic changes in cerebral blood flow and the oxygen-binding state of hemoglobin is unlikely to be stable across early human brain development (Harris et al. 2011; Kozberg and Hillman 2016a). This complicates interpretation, as one cannot assume that changes in BOLD signal amplitude or localization have the same meaning in fetuses as is accepted in adult fMRI studies. To overcome this, detailed studies are needed to combine data from other MR contrasts (anatomical, tissue composition and microstructure, vascular density, and blood flow) together with information derived from animal models. While there are spatial similarities and developmental trends between the patterns of activity seen in fetuses and preterm infants with fMRI, studies are also needed for systematic comparison. Such work is likely to be nontrivial due to inherent differences in data acquisition and artifacts in studies of the two populations, resulting in discrete processing and analysis strategies. Last, further work is needed to understand how changes in the fetal functional connectome are influenced by maternal and environmental factors such as toxins and stress (Thomason et al. 2019, 2021; van den Heuvel et al. 2021; Hendrix et al. 2022) relate to behavior (Thomason et al. 2018; Ji et al. 2023), disease, and potentially can predict later neurodevelopmental outcome.

CONCLUSIONS

In the second half of gestation, neural activity in the human brain undergoes a marked transition as its lifelong framework of long-range circuitry is established. This is reflected in the emergence of topographically organized and reproducible patterns of functional connectivity, which can be noninvasively studied in the womb, using recent advances in methods like fMRI. These studies hold great promise not only for characterizing the fundamental developmental processes that occur in this juncture, but also for providing

much needed new insight into how these processes are altered by environmental factors and disease, and ultimately may lead to difficulties later in life.

ACKNOWLEDGMENTS

The author thanks Professor Maria Fitzgerald for invaluable discussion prior to the preparation of this article and proofreading; and Jucha Willers Moore and Dr. Slava Karolis for review and advice. T.A. receives support from the Medical Research Council UK via a Transition Support Award [MR/V036874/1] and the Centre for Neurodevelopmental Disorders, King's College London [MR/N026063/1].

REFERENCES

*Reference is also in this subject collection.

Attwell D, Iadecola C. 2002. The neural basis of functional brain imaging signals. *Trends Neurosci* **25**: 621–625. doi:10.1016/S0166-2236(02)02264-6

Bandettini PA, Wong EC, Hinks RS, Tikofsky RS, Hyde JS. 1992. Time course EPI of human brain function during task activation. *Magn Reson Med* **25**: 390–397. doi:10.1002/mrm.1910250220

Batalle D, O'Muircheartaigh J, Makropoulos A, Kelly CJ, Dimitrova R, Hughes EJ, Hajnal JV, Zhang H, Alexander DC, Edwards AD, et al. 2019. Different patterns of cortical maturation before and after 38 weeks gestational age demonstrated by diffusion MRI in vivo. *Neuroimage* **185**: 764–775. doi:10.1016/j.neuroimage.2018.05.046

Beckmann CF, DeLuca M, Devlin JT, Smith SM. 2005. Investigations into resting-state connectivity using independent component analysis. *Philos Trans R Soc Lond B Biol Sci* **360**: 1001–1013. doi:10.1098/rstb.2005.1634

Bernhardt BC, Smallwood J, Keilholz S, Margulies DS. 2022. Gradients in brain organization. *Neuroimage* **251**: 118987. doi:10.1016/j.neuroimage.2022.118987

Biswal B, Yetkin FZ, Haughton VM, Hyde JS. 1995. Functional connectivity in the motor cortex of resting human brain using echo-planar MRI. *Magn Reson Med* **34**: 537–541. doi:10.1002/mrm.1910340409

Bystron I, Blakemore C, Rakic P. 2008. Development of the human cerebral cortex: Boulder Committee revisited. *Nat Rev Neurosci* **9**: 110–122. doi:10.1038/nrn2252

Cadwell CR, Bhaduri A, Mostajo-Radji MA, Keefe MG, Nowakowski TJ. 2019. Development and arealization of the cerebral cortex. *Neuron* **103**: 980–1004. doi:10.1016/j.neuron.2019.07.009

Christiaens D, Slator PJ, Cordero-Grande L, Price AN, Deprez M, Alexander DC, Rutherford M, Hajnal JV, Hutter J. 2019. In utero diffusion MRI: challenges, advances, and applications. *Top Magn Reson Imaging* **28**: 255–264. doi:10.1097/RMR.0000000000000211

Cite this article as *Cold Spring Harb Perspect Biol* doi: 10.1101/cshperspect.a041496

Ciarrusta J, Dimitrova R, Batalle D, O'Muircheartaigh J, Cordero-Grande L, Price A, Hughes E, Kangas J, Perry E, Javed A, et al. 2020. Emerging functional connectivity differences in newborn infants vulnerable to autism spectrum disorders. *Transl Psychiatry* **10:** 131. doi:10.1038/s41398-020-0805-y

Cole MW, Bassett DS, Power JD, Braver TS, Petersen SE. 2014. Intrinsic and task-evoked network architectures of the human brain. *Neuron* **83:** 238–251. doi:10.1016/j.neuron.2014.05.014

Cyr PEP, Lean RE, Kenley JK, Kaplan S, Meyer DE, Neil JJ, Alexopoulos D, Brady RG, Shimony JS, Rodebaugh TL, et al. 2022. Neonatal motor functional connectivity and motor outcomes at age two years in very preterm children with and without high-grade brain injury. *Neuroimage Clin* **36:** 103260. doi:10.1016/j.nicl.2022.103260

Damoiseaux JS, Rombouts SA, Barkhof F, Scheltens P, Stam CJ, Smith SM, Beckmann CF. 2006. Consistent resting-state networks across healthy subjects. *Proc Natl Acad Sci* **103:** 13848–13853. doi:10.1073/pnas.0601417103

De Asis-Cruz J, Andersen N, Kapse K, Khrisnamurthy D, Quistorff J, Lopez C, Vezina G, Limperopoulos C. 2021a. Global network organization of the fetal functional connectome. *Cereb Cortex* **31:** 3034–3046. doi:10.1093/cercor/bhaa410

De Asis-Cruz J, Barnett SD, Kim JH, Limperopoulos C. 2021b. Functional connectivity-derived optimal gestational-age cut points for fetal brain network maturity. *Brain Sci* **11:** 921. doi:10.3390/brainsci11070921

Deneault E, White SH, Rodrigues DC, Ross PJ, Faheem M, Zaslavsky K, Wang Z, Alexandrova R, Pellecchia G, Wei W, et al. 2018. Complete disruption of autism-susceptibility genes by gene editing predominantly reduces functional connectivity of isogenic human neurons. *Stem Cell Reports* **11:** 1211–1225. doi:10.1016/j.stemcr.2018.10.003

Doria V, Beckmann CF, Arichi T, Merchant N, Groppo M, Turkheimer FE, Counsell SJ, Murgasova M, Aljabar P, Nunes RG, et al. 2010. Emergence of resting state networks in the preterm human brain. *Proc Natl Acad Sci* **107:** 20015–20020. doi:10.1073/pnas.1007921107

Eswaran H, Haddad NI, Shihabuddin BS, Preissl H, Siegel ER, Murphy P, Lowery CL. 2007. Non-invasive detection and identification of brain activity patterns in the developing fetus. *Clin Neurophysiol* **118:** 1940–1946. doi:10.1016/j.clinph.2007.05.072

Eyre M, Fitzgibbon SP, Ciarrusta J, Cordero-Grande L, Price AN, Poppe T, Schuh A, Hughes E, O'Keeffe C, Brandon J, et al. 2021. The Developing Human Connectome Project: typical and disrupted perinatal functional connectivity. *Brain* **144:** 2199–2213. doi:10.1093/brain/awab118

Ferrazzi G, Kuklisova Murgasova M, Arichi T, Malamateniou C, Fox MJ, Makropoulos A, Allsop J, Rutherford M, Malik S, Aljabar P, et al. 2014. Resting State fMRI in the moving fetus: a robust framework for motion, bias field and spin history correction. *Neuroimage* **101:** 555–568. doi:10.1016/j.neuroimage.2014.06.074

Fox PT, Raichle ME, Mintun MA, Dence C. 1988. Nonoxidative glucose consumption during focal physiologic neural activity. *Science* **241:** 462–464. doi:10.1126/science.3260686

Fransson P, Skiöld B, Horsch S, Nordell A, Blennow M, Lagercrantz H, Åden U. 2007. Resting-state networks in the infant brain. *Proc Natl Acad Sci* **104:** 15531–15536. doi:10.1073/pnas.0704380104

Friston KJ, Frith CD, Liddle PF, Frackowiak RS. 1993. Functional connectivity: the principal-component analysis of large (PET) data sets. *J Cereb Blood Flow Metab* **13:** 5–14. doi:10.1038/jcbfm.1993.4

Fulford J, Vadeyar SH, Dodampahala SH, Moore RJ, Young P, Baker PN, James DK, Gowland PA. 2003. Fetal brain activity in response to a visual stimulus. *Hum Brain Mapp* **20:** 239–245. doi:10.1002/hbm.10139

Gilmore JH, Kang C, Evans DD, Wolfe HM, Smith JK, Lieberman JA, Lin W, Hamer RM, Styner M, Gerig G. 2010. Prenatal and neonatal brain structure and white matter maturation in children at high risk for schizophrenia. *Am J Psychiatry* **167:** 1083–1091. doi:10.1176/appi.ajp.2010.09101492

Gilmore JH, Knickmeyer RC, Gao W. 2018. Imaging structural and functional brain development in early childhood. *Nat Rev Neurosci* **19:** 123–137. doi:10.1038/nrn.2018.1

Greicius MD, Krasnow B, Reiss AL, Menon V. 2003. Functional connectivity in the resting brain: a network analysis of the default mode hypothesis. *Proc Natl Acad Sci* **100:** 253–258. doi:10.1073/pnas.0135058100

Haak KV, Marquand AF, Beckmann CF. 2018. Connectopic mapping with resting-state fMRI. *Neuroimage* **170:** 83–94. doi:10.1016/j.neuroimage.2017.06.075

Hadaya L, Nosarti C. 2020. The neurobiological correlates of cognitive outcomes in adolescence and adulthood following very preterm birth. *Semin Fetal Neonatal Med* **25:** 101117. doi:10.1016/j.siny.2020.101117

Hajnal JV, Myers R, Oatridge A, Schwieso JE, Young IR, Bydder GM. 1994. Artifacts due to stimulus correlated motion in functional imaging of the brain. *Magn Reson Med* **31:** 283–291. doi:10.1002/mrm.1910310307

Harris JJ, Reynell C, Attwell D. 2011. The physiology of developmental changes in BOLD functional imaging signals. *Dev Cogn Neurosci* **1:** 199–216. doi:10.1016/j.dcn.2011.04.001

Hebb D. 1949. *The organization of behavior*. McGill University, Wiley, New York.

Hendrix CL, Srinivasan H, Feliciano I, Carré JM, Thomason ME. 2022. Fetal hippocampal connectivity shows dissociable associations with maternal cortisol and self-reported distress during pregnancy. *Life (Basel)* **12:** 943. doi:10.3390/life12070943

Honey CJ, Thivierge JP, Sporns O. 2010. Can structure predict function in the human brain? *Neuroimage* **52:** 766–776. doi:10.1016/j.neuroimage.2010.01.071

Huang H, Vasung L. 2014. Gaining insight of fetal brain development with diffusion MRI and histology. *Int J Dev Neurosci* **32:** 11–22. doi:10.1016/j.ijdevneu.2013.06.005

Jaimes C, Machado-Rivas F, Afacan O, Khan S, Marami B, Ortinau CM, Rollins CK, Velasco-Annis C, Warfield SK, Gholipour A. 2020. In vivo characterization of emerging white matter microstructure in the fetal brain in the third trimester. *Hum Brain Mapp* **41:** 3177–3185. doi:10.1002/hbm.25006

Jakab A, Tuura R, Kellenberger C, Scheer I. 2017. In utero diffusion tensor imaging of the fetal brain: a reproducibility study. *Neuroimage Clin* **15**: 601–612. doi:10.1016/j.nicl.2017.06.013

Ji L, Hendrix CL, Thomason ME. 2022. Empirical evaluation of human fetal fMRI preprocessing steps. *Netw Neurosci* **6**: 702–721. doi:10.1162/netn_a_00254

Ji L, Majbri A, Hendrix CL, Thomason ME. 2023. Fetal behavior during MRI changes with age and relates to network dynamics. *Hum Brain Mapp* **44**: 1683–1694. doi:10.1002/hbm.26167

Karolis VR, Fitzgibbon SP, Cordero-Grande L, Farahibozorg SR, Price AN, Hughes EJ, Fetit AE, Kyriakopoulou V, Pietsch M, Rutherford MA, et al. 2023. Maturational networks of human fetal brain activity reveal emerging connectivity patterns prior to ex-utero exposure. *Commun Biol* **6**: 661. doi:10.1038/s42003-023-04969-x

Khazipov R, Luhmann HJ. 2006. Early patterns of electrical activity in the developing cerebral cortex of humans and rodents. *Trends Neurosci* **29**: 414–418. doi:10.1016/j.tins.2006.05.007

Kim JH, De Asis-Cruz J, Cook KM, Limperopoulos C. 2023a. Gestational age-related changes in the fetal functional connectome: in utero evidence for the global signal. *Cereb Cortex* **33**: 2302–2314. doi:10.1093/cercor/bhac209

Kim JH, De Asis-Cruz J, Krishnamurthy D, Limperopoulos C. 2023b. Toward a more informative representation of the fetal-neonatal brain connectome using variational autoencoder. *eLife* **12**: e80878. doi:10.7554/eLife.80878.sa2

Kostović I, Jovanov-Milošević N. 2006. The development of cerebral connections during the first 20-45 weeks' gestation. *Semin Fetal Neonatal Med* **11**: 415–422. doi:10.1016/j.siny.2006.07.001

Kostović I, Judaš M. 2010. The development of the subplate and thalamocortical connections in the human foetal brain. *Acta Paediatr* **99**: 1119–1127. doi:10.1111/j.1651-2227.2010.01811.x

Kostović I, Sedmak G, Judaš M. 2019. Neural histology and neurogenesis of the human fetal and infant brain. *Neuroimage* **188**: 743–773. doi:10.1016/j.neuroimage.2018.12.043

Kozberg M, Hillman E. 2016a. Neurovascular coupling and energy metabolism in the developing brain. *Prog Brain Res* **225**: 213–242. doi:10.1016/bs.pbr.2016.02.002

Kozberg MG, Hillman EM. 2016b. Neurovascular coupling develops alongside neural circuits in the postnatal brain. *Neurogenesis (Austin)* **3**: e1244439. doi:10.1080/23262133.2016.1244439

Kozberg MG, Ma Y, Shaik MA, Kim SH, Hillman EM. 2016. Rapid postnatal expansion of neural networks occurs in an environment of altered neurovascular and neurometabolic coupling. *J Neurosci* **36**: 6704–6717. doi:10.1523/JNEUROSCI.2363-15.2016

Krsnik Z, Majić V, Vasung L, Huang H, Kostović I. 2017. Growth of thalamocortical fibers to the somatosensory cortex in the human fetal brain. *Front Neurosci* **11**: 233. doi:10.3389/fnins.2017.00233

Kwong KK, Belliveau JW, Chesler DA, Goldberg IE, Weisskoff RM, Poncelet BP, Kennedy DN, Hoppel BE, Cohen MS, Turner R, et al. 1992. Dynamic magnetic resonance imaging of human brain activity during primary sensory stimulation. *Proc Natl Acad Sci* **89**: 5675–5679. doi:10.1073/pnas.89.12.5675

Le Bihan D, Mangin JF, Poupon C, Clark CA, Pappata S, Molko N, Chabriat H. 2001. Diffusion tensor imaging: concepts and applications. *J Magn Reson Imaging* **13**: 534–546. doi:10.1002/jmri.1076

Linke AC, Wild C, Zubiaurre-Elorza L, Herzmann C, Duffy H, Han VK, Lee DSC, Cusack R. 2018. Disruption to functional networks in neonates with perinatal brain injury predicts motor skills at 8 months. *Neuroimage Clin* **18**: 399–406. doi:10.1016/j.nicl.2018.02.002

Logothetis NK, Pauls J, Augath M, Trinath T, Oeltermann A. 2001. Neurophysiological investigation of the basis of the fMRI signal. *Nature* **412**: 150–157. doi:10.1038/35084005

Luhmann HJ, Sinning A, Yang JW, Reyes-Puerta V, Stuttgen MC, Kirischuk S, Kilb W. 2016. Spontaneous neuronal activity in developing neocortical networks: from single cells to large-scale interactions. *Front Neural Circuits* **10**: 40. doi:10.3389/fncir.2016.00040

Luhmann HJ, Kanold PO, Molnár Z, Vanhatalo S. 2022. Early brain activity: translations between bedside and laboratory. *Prog Neurobiol* **213**: 102268. doi:10.1016/j.pneurobio.2022.102268

Machado-Rivas F, Afacan O, Khan S, Marami B, Velasco-Annis C, Lidov H, Warfield SK, Gholipour A, Jaimes C. 2021. Spatiotemporal changes in diffusivity and anisotropy in fetal brain tractography. *Hum Brain Mapp* **42**: 5771–5784. doi:10.1002/hbm.25653

Manganaro L, Capuani S, Gennarini M, Miceli V, Ninkova R, Balba I, Galea N, Cupertino A, Maiuro A, Ercolani G, et al. 2023. Fetal MRI: what's new? A short review. *Eur Radiol Exp* **7**: 41. doi:10.1186/s41747-023-00358-5

Margulies DS, Ghosh SS, Goulas A, Falkiewicz M, Huntenburg JM, Langs G, Bezgin G, Eickhoff SB, Castellanos FX, Petrides M, et al. 2016. Situating the default-mode network along a principal gradient of macroscale cortical organization. *Proc Natl Acad Sci* **113**: 12574–12579. doi:10.1073/pnas.1608282113

Marin O, Rubenstein JL. 2003. Cell migration in the forebrain. *Annu Rev Neurosci* **26**: 441–483. doi:10.1146/annurev.neuro.26.041002.131058

Miguel PM, Pereira LO, Silveira PP, Meaney MJ. 2019. Early environmental influences on the development of children's brain structure and function. *Dev Med Child Neurol* **61**: 1127–1133. doi:10.1111/dmcn.14182

Molnár Z, Clowry GJ, Šestan N, Alzu'bi A, Bakken T, Hevner RF, Hüppi PS, Kostović I, Rakic P, Anton ES, et al. 2019. New insights into the development of the human cerebral cortex. *J Anat* **235**: 432–451. doi:10.1111/joa.13055

Molnár Z, Luhmann HJ, Kanold PO. 2020. Transient cortical circuits match spontaneous and sensory-driven activity during development. *Science* **370**: eabb2153. doi:10.1126/science.abb2153

Moore RJ, Vadeyar S, Fulford J, Tyler DJ, Gribben C, Baker PN, James D, Gowland PA. 2001. Antenatal determination of fetal brain activity in response to an acoustic stimulus using functional magnetic resonance imaging. *Hum Brain Mapp* **12**: 94–99. doi:10.1002/1097-0193(200102)12:2<94::AID-HBM1006>3.0.CO;2-E

Ogawa S, Lee TM, Kay AR, Tank DW. 1990. Brain magnetic resonance imaging with contrast dependent on blood oxygenation. *Proc Natl Acad Sci* **87**: 9868–9872. doi:10.1073/pnas.87.24.9868

Cite this article as *Cold Spring Harb Perspect Biol* doi: 10.1101/cshperspect.a041496

Ogawa S, Tank DW, Menon R, Ellermann JM, Kim SG, Merkle H, Ugurbil K. 1992. Intrinsic signal changes accompanying sensory stimulation: functional brain mapping with magnetic resonance imaging. *Proc Natl Acad Sci* **89:** 5951–5955. doi:10.1073/pnas.89.13.5951

Ohtaka-Maruyama C, Okamoto M, Endo K, Oshima M, Kaneko N, Yura K, Okado H, Miyata T, Maeda N. 2018. Synaptic transmission from subplate neurons controls radial migration of neocortical neurons. *Science* **360:** 313–317. doi:10.1126/science.aar2866

Papini C, White TP, Montagna A, Brittain PJ, Froudist-Walsh S, Kroll J, Karolis V, Simonelli A, Williams SC, Murray RM, et al. 2016. Altered resting-state functional connectivity in emotion-processing brain regions in adults who were born very preterm. *Psychol Med* **46:** 3025–3039. doi:10.1017/S0033291716001604

Paredes MF, James D, Gil-Perotin S, Kim H, Cotter JA, Ng C, Sandoval K, Rowitch DH, Xu D, McQuillen PS, et al. 2016. Extensive migration of young neurons into the infant human frontal lobe. *Science* **354:** aaf7073. doi:10.1126/science.aaf7073

* Pöpplau JA, Hanganu-Opatz IL. 2023. Development of prefrontal circuits and cognitive abilities. *Cold Spring Harb Perspect Biol* doi:10.1101/cshperspect.a041502

Power JD, Mitra A, Laumann TO, Snyder AZ, Schlaggar BL, Petersen SE. 2014. Methods to detect, characterize, and remove motion artifact in resting state fMRI. *Neuroimage* **84:** 320–341. doi:10.1016/j.neuroimage.2013.08.048

Price DJ, Kennedy H, Dehay C, Zhou L, Mercier M, Jossin Y, Goffinet AM, Tissir F, Blakey D, Molnár Z. 2006. The development of cortical connections. *Eur J Neurosci* **23:** 910–920. doi:10.1111/j.1460-9568.2006.04620.x

Raichle ME, MacLeod AM, Snyder AZ, Powers WJ, Gusnard DA, Shulman GL. 2001. A default mode of brain function. *Proc Natl Acad Sci* **98:** 676–682. doi:10.1073/pnas.98.2.676

Rajagopalan V, Deoni S, Panigrahy A, Thomason ME. 2021. Is fetal MRI ready for neuroimaging prime time? An examination of progress and remaining areas for development. *Dev Cogn Neurosci* **51:** 100999. doi:10.1016/j.dcn.2021.100999

Satterthwaite TD, Ciric R, Roalf DR, Davatzikos C, Bassett DS, Wolf DH. 2019. Motion artifact in studies of functional connectivity: characteristics and mitigation strategies. *Hum Brain Mapp* **40:** 2033–2051. doi:10.1002/hbm.23665

Scheinost D, Onofrey JA, Kwon SH, Cross SN, Sze G, Ment LR, Papademetris X. 2018. A fetal fMRI specific motion correction algorithm using 2nd order edge features. In *2018 IEEE 15th international symposium on biomedical imaging (ISBI 2018)*, pp. 1288–1292.

Schöpf V, Kasprian G, Brugger PC, Prayer D. 2012. Watching the fetal brain at "rest." *Int J Dev Neurosci* **30:** 11–17. doi:10.1016/j.ijdevneu.2011.10.006

Seshamani S, Fogtmann M, Cheng X, Thomason M, Gatenby C, Studholme C. 2013. Cascaded slice to volume registration for moving fetal FMRI. In *2013 IEEE 10th International Symposium on Biomedical Imaging*, pp. 796–799.

Shen X, Cox SR, Adams MJ, Howard DM, Lawrie SM, Ritchie SJ, Bastin ME, Deary IJ, McIntosh AM, Whalley HC. 2018. Resting-state connectivity and its association with cognitive performance, educational attainment, and household income in the UK Biobank. *Biol Psychiatry Cogn Neurosci*

Neuroimaging **3:** 878–886. doi:10.1016/j.bpsc.2018.06.007

Sheridan CJ, Matuz T, Draganova R, Eswaran H, Preissl H. 2010. Fetal magnetoencephalography—achievements and challenges in the study of prenatal and early postnatal brain responses: a review. *Infant Child Dev* **19:** 80–93. doi:10.1002/icd.657

Sigurdsson T. 2016. Neural circuit dysfunction in schizophrenia: insights from animal models. *Neuroscience* **321:** 42–65. doi:10.1016/j.neuroscience.2015.06.059

Smyser CD, Inder TE, Shimony JS, Hill JE, Degnan AJ, Snyder AZ, Neil JJ. 2010. Longitudinal analysis of neural network development in preterm infants. *Cereb Cortex* **20:** 2852–2862. doi:10.1093/cercor/bhq035

Snyder AZ, Raichle ME. 2012. A brief history of the resting state: the Washington University perspective. *Neuroimage* **62:** 902–910. doi:10.1016/j.neuroimage.2012.01.044

Sobotka D, Ebner M, Schwartz E, Nenning KH, Taymourtash A, Vercauteren T, Ourselin S, Kasprian G, Prayer D, Langs G, et al. 2022. Motion correction and volumetric reconstruction for fetal functional magnetic resonance imaging data. *Neuroimage* **255:** 119213. doi:10.1016/j.neuroimage.2022.119213

Sporns O. 2022. The complex brain: connectivity, dynamics, information. *Trends Cogn Sci* **26:** 1066–1067. doi:10.1016/j.tics.2022.08.002

Sur M, Rubenstein JL. 2005. Patterning and plasticity of the cerebral cortex. *Science* **310:** 805–810. doi:10.1126/science.1112070

Takahashi E, Folkerth RD, Galaburda AM, Grant PE. 2012. Emerging cerebral connectivity in the human fetal brain: an MR tractography study. *Cereb Cortex* **22:** 455–464. doi:10.1093/cercor/bhr126

Taymourtash A, Kebiri H, Schwartz E, Nenning K-H, Tourbier S, Kasprian G, Prayer D, Bach Cuadra M, Langs G. 2022. Spatio-temporal motion correction and iterative reconstruction of in-utero fetal fMRI. In *Medical image computing and computer assisted intervention—MICCAI 2022* (ed. Wang L, Dou Q, Fletcher PT, Speidel S, Li S), pp. 603–612. Springer Nature, Cham, Switzerland.

Taymourtash A, Schwartz E, Nenning KH, Sobotka D, Licandro R, Glatter S, Diogo MC, Golland P, Grant E, Prayer D, et al. 2023. Fetal development of functional thalamocortical and cortico-cortical connectivity. *Cereb Cortex* **33:** 5613–5624. doi:10.1093/cercor/bhac446

Testa-Silva G, Loebel A, Giugliano M, de Kock CP, Mansvelder HD, Meredith RM. 2012. Hyperconnectivity and slow synapses during early development of medial prefrontal cortex in a mouse model for mental retardation and autism. *Cereb Cortex* **22:** 1333–1342. doi:10.1093/cercor/bhr224

Thomason ME, Dassanayake MT, Shen S, Katkuri Y, Alexis M, Anderson AL, Yeo L, Mody S, Hernandez-Andrade E, Hassan SS, et al. 2013. Cross-hemispheric functional connectivity in the human fetal brain. *Sci Transl Med* **5:** 173ra124. doi:10.1126/scitranslmed.3004978

Thomason ME, Grove LE, Lozon TA, Vila AM, Ye Y, Nye MJ, Manning JH, Pappas A, Hernandez-Andrade E, Yeo L, et al. 2015. Age-related increases in long-range connectivity in fetal functional neural connectivity networks in utero. *Dev Cogn Neurosci* **11:** 96–104. doi:10.1016/j.dcn.2014.09.001

Thomason ME, Scheinost D, Manning JH, Grove LE, Hect J, Marshall N, Hernandez-Andrade E, Berman S, Pappas A, Yeo L, et al. 2017. Weak functional connectivity in the human fetal brain prior to preterm birth. *Sci Rep* **7**: 39286. doi:10.1038/srep39286

Thomason ME, Hect J, Waller R, Manning JH, Stacks AM, Beeghly M, Boeve JL, Wong K, van den Heuvel MI, Hernandez-Andrade E, et al. 2018. Prenatal neural origins of infant motor development: associations between fetal brain and infant motor development. *Dev Psychopathol* **30**: 763–772. doi:10.1017/S095457941800072X

Thomason ME, Hect JL, Rauh VA, Trentacosta C, Wheelock MD, Eggebrecht AT, Espinoza-Heredia C, Burt SA. 2019. Prenatal lead exposure impacts cross-hemispheric and long-range connectivity in the human fetal brain. *Neuroimage* **191**: 186–192. doi:10.1016/j.neuroimage.2019.02.017

Thomason ME, Hect JL, Waller R, Curtin P. 2021. Interactive relations between maternal prenatal stress, fetal brain connectivity, and gestational age at delivery. *Neuropsychopharmacology* **46**: 1839–1847. doi:10.1038/s41386-021-01066-7

Tolner EA, Sheikh A, Yukin AY, Kaila K, Kanold PO. 2012. Subplate neurons promote spindle bursts and thalamocortical patterning in the neonatal rat somatosensory cortex. *J Neurosci* **32**: 692–702. doi:10.1523/JNEUROSCI.1538-11.2012

Trivedi R, Gupta RK, Husain N, Rathore RK, Saksena S, Srivastava S, Malik GK, Das V, Pradhan M, Sarma MK, et al. 2009. Region-specific maturation of cerebral cortex in human fetal brain: diffusion tensor imaging and histology. *Neuroradiology* **51**: 567–576. doi:10.1007/s00234-009-0533-8

Turk E, van den Heuvel MI, Benders MJ, de Heus R, Franx A, Manning JH, Hect JL, Hernandez-Andrade E, Hassan SS, Romero R, et al. 2019. Functional connectome of the fetal brain. *J Neurosci* **39**: 9716–9724. doi:10.1523/JNEUROSCI.2891-18.2019

van den Heuvel MP, Hulshoff Pol HE. 2010. Exploring the brain network: a review on resting-state fMRI functional connectivity. *Eur Neuropsychopharmacol* **20**: 519–534. doi:10.1016/j.euroneuro.2010.03.008

van den Heuvel MP, Sporns O. 2013. Network hubs in the human brain. *Trends Cogn Sci* **17**: 683–696. doi:10.1016/j.tics.2013.09.012

van den Heuvel MI, Thomason ME. 2016. Functional connectivity of the human brain in utero. *Trends Cogn Sci* **20**: 931–939. doi:10.1016/j.tics.2016.10.001

van den Heuvel MP, Kersbergen KJ, de Reus MA, Keunen K, Kahn RS, Groenendaal F, de Vries LS, Benders MJ. 2015. The neonatal connectome during preterm brain development. *Cereb Cortex* **25**: 3000–3013. doi:10.1093/cercor/bhu095

van den Heuvel MI, Turk E, Manning JH, Hect J, Hernandez-Andrade E, Hassan SS, Romero R, van den Heuvel MP, Thomason ME. 2018. Hubs in the human fetal brain network. *Dev Cogn Neurosci* **30**: 108–115. doi:10.1016/j.dcn.2018.02.001

van den Heuvel MI, Hect JL, Smarr BL, Qawasmeh T, Kriegsfeld LJ, Barcelona J, Hijazi KE, Thomason ME. 2021. Maternal stress during pregnancy alters fetal cortico-cerebellar connectivity in utero and increases child sleep problems after birth. *Sci Rep* **11**: 2228. doi:10.1038/s41598-021-81681-y

van der Knaap MS, van Wezel-Meijler G, Barth PG, Barkhof F, Ader HJ, Valk J. 1996. Normal gyration and sulcation in preterm and term neonates: appearance on MR images. *Radiology* **200**: 389–396. doi:10.1148/radiology.200.2.8685331

Vasung L, Charvet CJ, Shiohama T, Gagoski B, Levman J, Takahashi E. 2019. Ex vivo fetal brain MRI: recent advances, challenges, and future directions. *Neuroimage* **195**: 23–37. doi:10.1016/j.neuroimage.2019.03.034

Wehrle FM, Michels L, Guggenberger R, Huber R, Latal B, O'Gorman RL, Hagmann CF. 2018. Altered resting-state functional connectivity in children and adolescents born very preterm short title. *Neuroimage Clin* **20**: 1148–1156. doi:10.1016/j.nicl.2018.10.002

Wess JM, Isaiah A, Watkins PV, Kanold PO. 2017. Subplate neurons are the first cortical neurons to respond to sensory stimuli. *Proc Natl Acad Sci* **114**: 12602–12607. doi:10.1073/pnas.1710793114

Willers Moore J, Wilson S, Oldenhinkel M, Cordero-Grande L, Uus A, Kyriakopoulou V, Duff EP, O'Muircheartaigh J, Rutherford MA, Andreae LC, et al. 2023. Gradient organisation of functional connectivity within resting state networks is present from 25 weeks gestation in the human fetal brain. *eLife* **12**: RP90536. doi:10.7554/eLife.90536.1

Wilson S, Pietsch M, Cordero-Grande L, Price AN, Hutter J, Xiao J, McCabe L, Rutherford MA, Hughes EJ, Counsell SJ, et al. 2021. Development of human white matter pathways in utero over the second and third trimester. *Proc Natl Acad Sci* **118**: e2023598118. doi:10.1073/pnas.2023598118

Wilson S, Pietsch M, Cordero-Grande L, Christiaens D, Uus A, Karolis VR, Kyriakopoulou V, Colford K, Price AN, Hutter J, et al. 2023. Spatiotemporal tissue maturation of thalamocortical pathways in the human fetal brain. *eLife* **12**: e83727. doi:10.7554/eLife.83727

Yun HJ, Vasung L, Tarui T, Rollins CK, Ortinau CM, Grant PE, Im K. 2020. Temporal patterns of emergence and spatial distribution of sulcal pits during fetal life. *Cereb Cortex* **30**: 4257–4268. doi:10.1093/cercor/bhaa053

You W, Evangelou IE, Zun Z, Andescavage N, Limperopoulos C. 2016. Robust preprocessing for stimulus-based functional MRI of the moving fetus. *J Med Imaging (Bellingham)* **3**: 026001. doi:10.1117/1.JMI.3.2.026001

Zheng W, Zhao L, Zhao Z, Liu T, Hu B, Wu D. 2023. Spatiotemporal developmental gradient of thalamic morphology, microstructure, and connectivity from the third trimester to early infancy. *J Neurosci* **43**: 559–570. doi:10.1523/JNEUROSCI.0874-22.2022

Cite this article as *Cold Spring Harb Perspect Biol* doi: 10.1101/cshperspect.a041496

Modeling Normal and Abnormal Circuit Development with Recurrent Neural Networks

Daniel Zavitz,[1] ShiNung Ching,[2] and Geoffrey Goodhill[1]

[1]Departments of Developmental Biology and Neuroscience, Washington University in St. Louis School of Medicine, St. Louis, Missouri 63110, USA

[2]Departments of Electrical and Systems Engineering and Biomedical Engineering, Washington University in St. Louis, St. Louis, Missouri 63130, USA

Correspondence: g.goodhill@wustl.edu

Neural development must construct neural circuits that can perform the computations necessary for survival. However, many theoretical models of development do not explicitly address the computational goals of the resulting networks, or computations that evolve in time. Recurrent neural networks (RNNs) have recently come to prominence as both models of neural circuit computation and building blocks of powerful artificial intelligence systems. Here, we review progress in using RNNs for understanding how developmental processes lead to effective computations, and how abnormal development disrupts these computations.

The development of a functioning nervous system proceeds through multiple complex stages, including neurulation, regional specification, neurogenesis, cell fate determination, cell migration, axon guidance and dendritic development, and synapse formation and pruning, followed by ongoing plasticity and refinement. Across different species, the timescales involved in these processes range from hours to years. Interesting theoretical problems abound at all these stages (van Ooyen 2011; Goodhill 2018). For instance, understanding how molecular and/or mechanical cues promote tissue folding, regional identity, and axon guidance, especially given the presence of unavoidable noise in the measurement of concentration (e.g., Gregor et al. 2007; Bicknell et al. 2015; Tkačik et al. 2015). On the other hand, research in the neural network community has focused mostly on the formation of connections between neurons, and in particular on how connection strengths are coded/learned within a fixed architecture so that the network performs specific computations.

Mathematical instantiations of Hebb's rule (Hebb 1949) have proven effective at reproducing receptive field structures found in early sensory areas such as the visual system in response to appropriate activity patterns (von der Malsburg 1973; Linsker 1986; Goodhill 1993), including the effects of altered sensory input such as monocular deprivation (Miller et al. 1989). These approaches are unsupervised, and attempt to address developmental processes fairly explicitly. In contrast in supervised approaches, learning takes place by giving the network a target output for every input pattern and adjusting the weights in the network to achieve this outcome, usually via backpropagation (Rumelhart et al. 1986). An early example of the application of backpropagation to understand neural circuits was the reproduction of neural response properties in area 7a of the posterior parietal cortex of monkeys (Zipser and Andersen 1988). A more recent example is Yamins et al. (2014), which matched response properties through many layers of a deep network with those in the primate visual system. Supervised approaches show how learning a particular computational task can lead to structure matching that found biologically, but do not claim to reproduce actual developmental processes.

Until recently supervised learning approaches to understanding neural computation were primarily focused on layered, feedforward networks, where inputs and outputs do not have a time dimension (e.g., the examples above). However, in the past few years, recurrent neural networks (RNNs) have gained prominence through their ability to accept time-varying inputs and produce time-varying outputs. Such

networks have proved very useful for understanding computations involving the accrual of information over time, such as when animals are trained to respond in specific ways to cues occurring at specific times. This opens up a much more complex repertoire of task possibilities. While the focus in this regard has most commonly been on adult learning occurring on relatively rapid timescales, the question naturally arises of whether such networks can also be applied to understand the emergence of network computations on developmental timescales. We argue that this provides a promising direction for understanding both normal and abnormal development, but that several important issues remain to addressed.

RECURRENT NEURAL NETWORKS

Basic Neurobiological Components of RNNs

RNNs can be interpreted as firing rate models of neural circuits (Fig. 1; Sussillo 2014). This approach allows RNNs to approximate the differ-

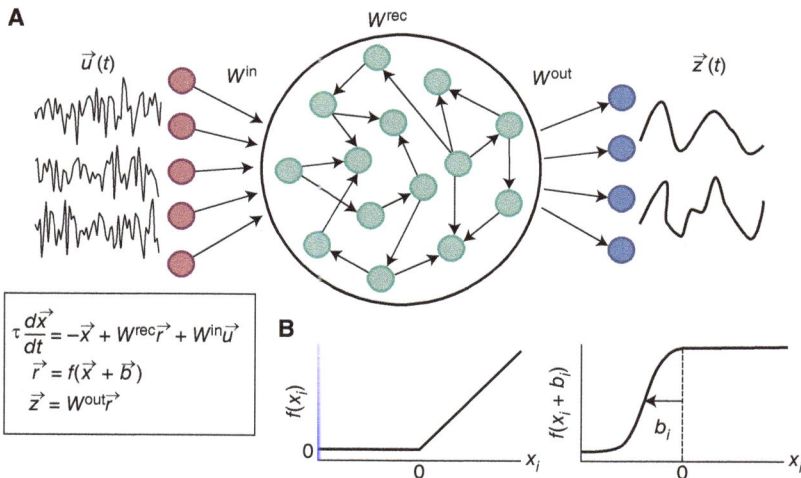

Figure 1. Basic principles of recurrent neural networks (RNNs). (*A*) An RNN transforms time-varying inputs $\vec{u}(t)$ into time-varying outputs $\vec{z}(t)$ using firing rate–type dynamics. The neurons' synaptic current $\vec{x}(t)$ and firing rate $\vec{r}(t)$ represent the RNNs' internal state that captures long-term relationships between inputs and outputs and τ sets the timescale of individual neurons' activity. Trainable parameters include the input W^{in}, recurrent W^{rec}, and output W^{out} weight matrices and the neurons' biases $\vec{b}$. (*B*) Currents are transformed into firing rates by an activation function (or *F–I* curve). Popular choices include nonsaturating functions like a rectified linear unit (ReLu, *left*) or saturating functions like tanh (·) (*right*). The bias b_i determines the amount of current required to activate the neuron.

ential equations of a typical firing rate model (Dayan and Abbott 2005):

$$\tau \frac{d\vec{x}}{dt} = -\vec{x} + W^{\text{rec}}\vec{r} + W^{\text{in}}\vec{u},$$
$$\vec{r} = f(\vec{x} + \vec{b}), \tag{1}$$
$$\vec{y} = W^{\text{out}}\vec{r},$$

where each entry i of the vector $\vec{x}$ is the net, low-pass filtered synaptic current of neuron i, $\vec{r}$ are corresponding instantaneous firing rates, and τ describes the timescale of activity of individual neurons. The feedforward firing rate input and readout of the system are $\vec{u}$ and $\vec{y}$, respectively. The activation function $f(\cdot)$ can be interpreted as an F–I curve (Izhikevich 2006) that mediates the relationship between injected current (I) and neuronal firing rate (F). Trainable parameters include W^{rec}, W^{in}, and W^{out}, the weight matrices that determine the efficacies of recurrent, input, and readout connections, respectively, and $\vec{b}$, the neurons' biases that set their activation threshold (analogous to rheobase). The resulting model is typically used to process sequential data such that the state of each neuron at time t is driven by the input data at t and previous neural states. Thus, RNNs have "memories" in which sustained activity represents information from prior input data by connection weights. This feature is key to learning long-term dependencies in sequence data.

Following the McCulloch–Pitts model (McCulloch and Pitts 1943), $f(\cdot)$ is a threshold-like activation function. Common choices for $f(\cdot)$ can be roughly divided into two categories: sigmoid functions that saturate as current increases (e.g., $\tanh(\cdot)$; Fig. 1B, right) and those that do not (e.g., rectified linear units—ReLu; Fig. 1B, left). F–I curves of both types can be found in vivo (McCormick et al. 1985; Guan et al. 2014) and can change during development (Oswald and Reyes 2008). The functional significance of the specific activation function used remains poorly understood, hence a common approach is to analyze many RNNs with different activation functions (e.g., Yang et al. 2019; Driscoll et al. 2022). Note that this F–I curve classification differs from the taxonomy of Hodgkin–Huxley models (Hodgkin and Huxley 1952) in which neural dynamics are classified by the bifurcation at the activation

threshold (Rinzel and Ermentrout 1989; Izhikevich 2006).

Although synaptic connection weights are primarily determined by training (discussed later), biological features can be imposed by constraining their values. For instance, an RNN can be divided into distinct excitatory and inhibitory subpopulations (Song et al. 2016). While a neuron can be made excitatory or inhibitory by simply giving all its outgoing connections the appropriate sign a priori, when weights are determined only by a training algorithm, neurons have outgoing connections of both signs. Recent work by Song et al. (2016) bypasses this problem by decomposing weight matrices such that $W = W^+ M$, where the connection strength is determined by training the nonnegative matrix W^+ and excitatory/inhibitory identity is fixed by a prior in the diagonal matrix M. This allows standard training algorithms to optimize the network weights while respecting the subpopulation identities.

Additional constraints can also be added to RNN circuit models (Molano-Mazón et al. 2023). Advances in large-scale neural recordings (Urai et al. 2022) have revealed that neurons distributed across many brain regions act in concert to produce behaviors and process sensory information (e.g., Musall et al. 2019; Steinmetz et al. 2019; Stringer et al. 2019). Recent RNN models applied to this spatial scale have incorporated brain region organization by imposing a block structure (Holland et al. 1983) on recurrent connections in which intra-area connections are dense and inter-area connections are sparse (Pinto et al. 2019; Kleinman et al. 2020; Perich and Rajan 2020). Furthermore, the input and output connection weights (W^{in} and W^{out}) can be structured to mirror brain region organization (Song et al. 2016). This allows distinct "sensory" and "motor" areas where external "sensory" input is routed to a subpopulation of neurons and the outgoing "motor" response is read exclusively from a different subpopulation.

Once the architecture of an RNN circuit model is determined, it can be trained to perform a specific task. This is often formulated as a supervised learning problem in which the data set consists of ordered pairs $D = \{(\vec{u}_i, \vec{y}_i^{\text{label}})\}_{i=1}^N$ and the RNN learns to produce a set of desired outputs $\vec{y}_i^{\text{label}}$ when driven by a corresponding input $\vec{u}_i$. Similar to feedforward networks, RNNs learn by adjusting their weights and biases. This can be formulated as an optimization problem:

$$\theta^* = \arg \min_{\theta} L(D; \theta), \qquad (2)$$

where θ and θ^* represent all trainable parameters before and after training, respectively, typically W^{rec}, W^{in}, W^{out}, and $\vec{b}$, and L is a loss function that measures the differences between the RNN's actual and desired outputs. Loss is calculated for each item in the data set and parameters are adjusted to minimize the total loss. Despite their topological differences, the algorithms used to train both feedforward neural networks and RNNs are similar. Most notably, backpropagation can be implemented in RNNs as backpropagation-through-time by transforming the RNN into a deep feedforward network where each layer represents a timestep (Werbos 1990). Many additional training approaches are available including reinforcement learning (Song et al. 2017) and recursive least squares (Sussillo and Abbott 2009), and carry different assumptions that are amenable to specific types of tasks and can heavily influence the resulting neural dynamics (Mikhaeil et al. 2022).

Computations Mediated by RNNs

RNNs maintain an internal state that is ideal for performing computations involving sequences, and so can be trained to perform tasks with a temporal component (although see Wang et al. 2021). In a neuroscience context, RNNs have been used to study a wide variety of tasks, but cognition and motor control have been a particular focus (Vyas et al. 2020; Yang and Wang 2020). Within the cognitive task framework, RNNs have been trained to perform tasks that have been investigated experimentally (e.g., using a random dot motion approach) such as perceptual (Song et al. 2016) or context-dependent decision making (Mante et al. 2013) and working memory (Masse et al. 2019; Orhan and Ma 2019; Ghazizadeh and Ching 2021). The flexibility demonstrated by the cognitive task paradigm

demonstrates the primary advantage of RNNs as a neural circuit model: A single training protocol can typically be reused to train an RNN on many tasks that would require a human to start over afresh when manually constructing circuit models.

RNNs' applicability to motor control tasks stems from a second, related advantage: RNNs are capable of approximating any dynamical system (Schäfer and Zimmermann 2006). For instance, hypotheses about the relationship between neural activity and motor control (Shenoy et al. 2013) have been explored by training RNNs to produce output signals that match experimental recordings of muscle actions and then comparing with corresponding neural recordings from the motor cortex (Hennequin et al. 2014; Sussillo et al. 2015; Saxena et al. 2022). Related brain–computer interface studies have demonstrated that RNNs can effectively and rapidly decode motor commands from neural activity (Sussillo et al. 2012; Saxena et al. 2022).

MODELING NEURAL CIRCUIT DEVELOPMENT WITH RNNs

Structurally Dynamic RNNs

In addition to the synaptic efficacy fine-tuning typical of adult learning, neurodevelopmental processes also include much broader-scale structural changes driven by neurogenesis, synaptogenesis, and pruning. RNN circuit models typically do not make this structure/efficacy distinction and implicitly assume that constituent neurons have all-to-all structural connections whose efficacy is tuned to perform specific computations. However, RNNs with structurally dynamic learning rules have granted insight into the functional significance of specific neurodevelopmental mechanisms, as we describe below (Fig. 2A–C).

Neurogenesis

The recurrent cascade correlation network (Fahlman 1990) was the first RNN to learn via neurogenesis. The network learns mappings between input and output sequences by iterating

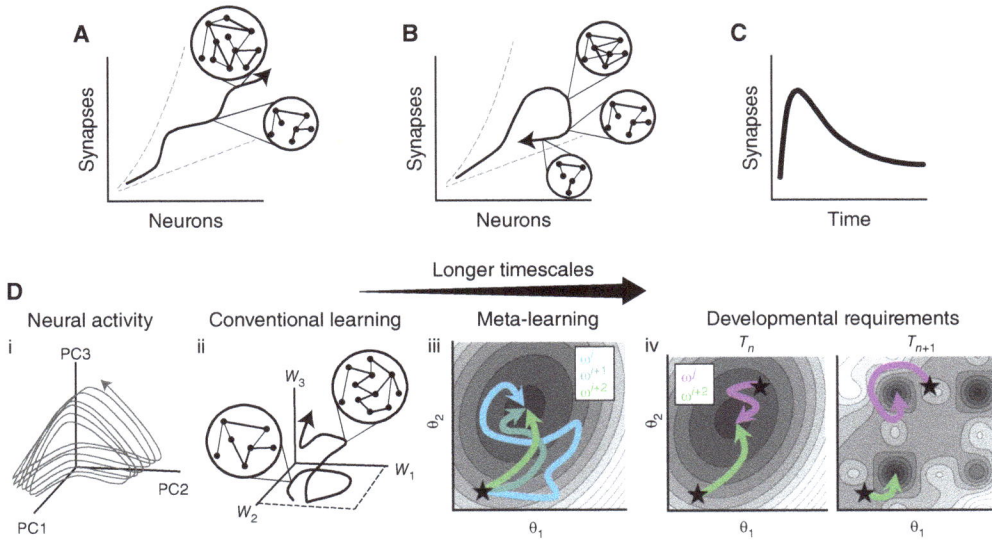

Figure 2. Recurrent neural network (RNN) models of neural development. (*A*) Some RNN models have incorporated structural dynamics into learning. Neurogenesis and synaptogenesis allow learning but restrict exploration of the possible search space (dashed lines indicate bounds on the number of synapses; Fahlman 1990). (*B*) In contrast grow-when-required learning allows more flexible model configurations by enabling synaptic pruning and apoptosis (Parisi et al. 2018). (*C*) Hebbian associative learning mediated by synaptogenesis and pruning can exhibit a phase of rapid growth followed by a phase of net synaptic pruning, which is characteristic of development (Millán et al. 2018). (*D*) Multiple timescales are relevant to neural computations and learning. (*i*) Computations are executed by neural activity at fast timescales. (*ii*) Conventional learning operates at an intermediate timescale. Learning can result both from traversing existing parameter space (e.g., the plane formed by synaptic weights w_1 and w_2) by adjusting synaptic efficacies and expanding parameter space through neuro and synaptogenesis (adding a new synapse with weight w_3). (*iii*) Metalearning occurs at a longer but still intermediate timescale where learning strategies, or metaknowledge ω^*, can be learned iteratively (blue to green arrows finding the minima of a loss function) resulting in more efficient conventional learning of model parameters θ_1, θ_2 (e.g., Goudar et al. 2023). (*iv*) At the longest timescales, the set of tasks needed to be performed is dynamic (T_n to T_{n+1}). Different structural priors (stars), possibly resulting from different learning strategies (ω^{i+2}, ω^j) developed over shorter timescales, may result in different solutions (Molano-Mazón et al. 2023).

over training data, each time reducing a loss function by adding an additional neuron with a recurrent self-connection and feedforward connections from existing neurons. This approach yields an RNN with the minimum number of neurons required to perform a computation when constructed using the specified rule. However, the resulting architecture is restricted to a small region of the possible search space and so has a limited capacity to learn.

A subsequent approach termed GNARL (generalized acquisition of recurrent links) used both neurogenesis and pruning to construct RNNs nonmonotonically with minimal restrictions to both RNN size and architecture (Angeline et al. 1994). This allowed the RNN to explore a much larger region of the parameter space by updating both its topology and its edge weights during learning. However, GNARL is an evolutionary algorithm in which populations of RNNs are initialized, ranked according to their ability to perform a computation (fitness), and then used to construct the next RNN generation according to this ranking, limiting its relevance to neural development.

A more recent approach used neurogenesis and pruning to implement lifelong learning (i.e., learning from a continuous stream of information) (Parisi et al. 2018). In neural networks, the primary obstacle to lifelong learning is catastrophic forgetting. The authors created a two-RNN circuit model based on complementary learning systems theory (McClelland et al. 1995) in which the hippocampus mediates episodic memories that are consolidated into semantic memories in the neocortex. Both RNNs learn via a grow-when-required algorithm (Marsland et al. 2002; Parisi et al. 2017) such that new neurons are created by a Hebbian learning rule only when necessary, and removed if they are no longer used. This allows the network to avoid catastrophic forgetting while adapting to nonstationary inputs. Together these RNN approaches illuminate the computational potential of apoptosis and synaptic pruning and their synergy with neural growth.

Synaptogenesis and Pruning

Neural development in many organisms including mammals can be characterized by initial rapid synaptic growth followed by a period of synaptic pruning that eventual plateaus at a more stable value (Huttenlocher 1979; Markus and Petit 1987; Bourgeois and Rakic 1993; White et al. 1997). Both activity-independent processes reliant on genetic, mechanical, and molecular mechanisms, and activity-dependent processes reliant on learning, sensory experiences, and spontaneous activity, are thought to play a role (Faust et al. 2021). Graph–theoretic analyses of neural development have indicated that overproduction-then-pruning algorithms can simultaneously enhance circuit robustness and efficiency, key measures of global network structure (Navlakha et al. 2018). However, the precise contributions of activity-dependent and independent processes to synaptic architecture development and their computational significance remain unclear.

Recent studies using RNNs have uncovered potential relationships between activity-dependent synaptogenesis and pruning and memory formation (Johnson et al. 2010; Millán et al. 2018, 2019, 2021). This approach modifies Hopfield networks, a classic RNN model capable of associative learning (Hopfield 1982), to learn via preferential attachment and detachment. More specifically, the RNN implements a stochastic, activity-dependent learning rule in which neu-

rons gain and lose edges according to empirically derived probability distributions defined by the current each neuron receives and the number of synaptic connections in the RNN. The resulting RNNs exhibit synaptic overproduction-then-pruning that matches human and mouse data. Analysis revealed that strong coupling between activity and structure is necessary for memory formation, synaptic pruning can optimize neural circuits using local plasticity rules, and that an initial period of dense synaptic connectivity can enhance memory stability. Together, this approach synthesizes methods from machine learning, network science, and statistical physics to reveal how feedback between the dynamics of circuit structure and activity may interact throughout development to enhance computational performance.

Development of Circuit Function

Computational and behavioral requirements change over development. As a result, understanding the development of computational function is critical to understanding neural development. Complementary to the mechanism-first approaches described above, neural development can be investigated in a computation-first manner, where RNNs are trained to match the development of circuit function and then analyzed to reveal possible neurodevelopmental mechanisms. This approach leverages advances in machine learning to explore the functional principles of neural circuit development and produce hypotheses about the underlying mechanisms.

Development of Cognitive Function

Cognitive abilities improve during neural development. For example, monkeys exhibit an improvement in working memory tasks from adolescence to adulthood, that is accompanied by an increase in activity in prefrontal cortex neurons during the task's delay period (Zhou et al. 2016a). However, specific, causal relationships between neurobiological changes and cognitive improvement remain difficult to uncover. RNNs can be used to investigate this relationship by first rep-

 Cite this article as *Cold Spring Harb Perspect Biol* doi: 10.1101/cshperspect.a041507

licating specific computations across a developmental trajectory and then dissecting the underlying neural dynamics.

Recent work has taken this approach to link experimentally observed changes in neural activity to improvement in specific cognitive functions throughout development (Liu et al. 2021). The authors trained RNNs to match the performance of adolescent and adult monkeys (35%–65% and >65% correct trials, respectively) on working memory and response inhibition task variants (Zhou et al. 2013, 2016b). At each stage of development artificial and prefrontal cortex neural activity were compared to explore the extent to which computational optimization, rather than specific (possibly unrelated) biological processes, can explain changes in neural dynamics resulting from development. Intriguingly, RNN activity dynamics in both working memory and response inhibition tasks mirrored specific changes observed in prefrontal cortex neural activity and hypothesized to drive task improvement. Although not a full, causal explanation, this approach implicates the computational function of specific neurodevelopmental changes and demonstrates the utility of top-down RNN approaches to modeling neural circuit development. RNNs' flexibility allows this approach to serve as a promising template for exploring the computational relevance of other neurodevelopmental changes across model systems.

Modeling Considerations

Three key factors require special consideration when using RNN circuit models to investigate neural development. First, changes observed in circuit structure and activity during learning are not necessarily related to the task (Hennig et al. 2021). Even in a nondevelopmental context, neural representations in adult animals performing familiar tasks are well known to "drift" over time (Driscoll et al. 2017) due to many separate biological processes. The variety and scale of biological and cognitive changes present in neural development may obscure the relationship between RNN and neural circuit learning. Second, catastrophic forgetting (Ratcliff 1990; Kudithipudi et al. 2022) is a particular problem for modeling

neural development since this involves learning a large amount of information over much longer time periods relative to learning problems considered in most RNN applications.

Third, neural circuits develop to concurrently perform a wide variety of tasks. Recent advances in multitask machine learning (Zhang and Yang 2018) enable individual RNN circuit models to learn many tasks (Yang et al. 2019). Understanding the distinctions between how tasks are represented within RNNs in these settings is an active area of research (Sucholutsky et al. 2023). Studies exploring the resulting neural circuits (in a nondevelopmental context) have observed compositionality in task variable representations: RNNs tend to develop functionally specialized subcircuits to execute computations that are reused when performing many tasks (Yang et al. 2019; Driscoll et al. 2022). However, not all task sets require this functional specialization, and may instead produce computations that are executed by the collective dynamics of all neurons (Dubreuil et al. 2022). Together, these studies indicate that "more is different" (Anderson 1972) and interactions between tasks within a task set may have a substantial impact on the computational mechanisms learned by RNNs. As a result, improving a task set's approximation of behavioral and computational requirements during development could be critical to constructing rich and informative theories about the underlying mechanisms of circuit development.

Learning at Multiple Timescales

Metalearning

Neural development involves learning at disparate timescales. For example, a specific task T (e.g., a single stimulus-reward association) can be learned on a shorter timescale, but recognizing commonalities in a sequence of tasks $\{T_1, T_2, T_3, \ldots\}$ may require learning over a much longer timescale. However, learning at different timescales is not an independent process. Rather, learning at longer timescales often benefits from ongoing increases in efficiency, indicating the presence of learning-to-learn (Harlow 1949).

Metalearning (Hospedales et al. 2021), offers a flexible framework to couple learning across different timescales by learning-to-learn (Fig. 2D). Conventional learning algorithms make specific assumptions about how to learn, including hyperparameter values of the learning algorithm or initial values of the model's parameters. The metalearning paradigm seeks to improve these assumptions. Formally, metalearning can be viewed as a nested optimization problem:

$$\theta^* = \arg \min_{\theta} L(D; \theta, \omega), \qquad (3)$$

$$\omega^* = \arg \min_{\omega} L^{\mathrm{meta}}(D; \theta^*(\omega), \omega). \qquad (4)$$

The inner optimization problem (Equation 3) constitutes conventional supervised learning and ω represents an assumption of the learning algorithm (e.g., a specific hyperparameter). The outer, metalearning problem (Equation 4) seeks to optimize ω given θ^*, the parameters resulting from Equation 3, using its own learning algorithm and loss function L^{meta} (Hospedales et al. 2021; Wang 2021). Within this paradigm, learning at the fast (inner) timescale shapes learning at the slow (outer), developmentally relevant timescale, and vice versa.

Metalearning in RNNs

RNN metalearning studies can be categorized by their metarepresentations. Mirroring advances in machine learning (Finn et al. 2017), metalearning the RNN weight initialization has yielded insight into the computational process underlying learning over the disparate timescales relevant to neural development (Goudar et al. 2023; Molano-Mazón et al. 2023). Within this framework, RNNs are pretrained on a set of tasks $\{T_1, T_2, \ldots, T_n\}$ and then the RNN's ability to generalize to novel tasks $\{T_{n+1}, T_{n+2}, T_{n+3}, \ldots\}$ is assessed. The specific parameter configuration induced by pretraining corresponds to "structural priors," embodying preexisting knowledge learned over the course of development or possibly longer, evolutionary timescales (Zador 2019; Koulakov et al. 2022; Barabási et al. 2023). Recent analysis has demonstrated that

RNNs with structural priors induced by pretraining on naturalistic tasks match suboptimal behavior exhibited by rats in a two alternative forced choice task (Molano-Mazón et al. 2023). In contrast, RNNs trained only on the two alternative force choice tasks exceeded the rats' performance. This indicates the importance of incorporating learning across different timescales in producing biologically relevant RNN models, but does not differentiate between development and evolution. From a neural manifold perspective, metalearning weight initializations have been shown to enable rapid generalization to new problems through the construction schema (Goudar et al. 2023), neural representations that abstract commonalities across previous experience and play a key role in developmental psychology (Piaget 2005). This metalearned solution minimizes the weight changes necessary to learn additional tasks, thus linking learning acceleration to possible wiring constraints resulting from the biophysical mechanisms of neural development.

From both a neurobiological (Doya 2002) and machine learning perspective (Li et al. 2017), the learning algorithm's hyperparameters (e.g., the learning rate) are a natural metarepresentation choice. Neuromodulators have been shown to selectively regulate learning across a wide variety of circumstances (Marder 2011). Work on the neural basis of reward-driven learning has linked neuromodulators to specific parameters and hyperparameters in reinforcement and metareinforcement learning algorithms (Schultz et al. 1997; Doya 2002). More recently, neuromodulation has been explicitly incorporated into RNN circuit models as an additional mechanism of experience-dependent plasticity that modulates neural activity throughout tasks (Wang et al. 2018; Jiang and Litwin-Kumar 2021). In this approach, RNN weights are trained via gradient descent to produce computations that track a changing environment using neuromodulation mechanisms. Consequently, the standard roles of synaptic weights as parameters and neuromodulators as hyperparameters are effectively transposed. RNN weights act as hyperparameters that are metalearned on a long, neural development-relevant timescale, while neuro-

Cite this article as *Cold Spring Harb Perspect Biol* doi: 10.1101/cshperspect.a041507

modulation mechanisms implement parameters that learn by integrating reward information on a short, behaviorally relevant timescale.

RNN MODELS OF NEURODEVELOPMENTAL DISORDERS

A great variety of mental health conditions are neurodevelopmental in origin. Examples include schizophrenia, autism spectrum disorder (ASD), and attention-deficit/hyperactivity disorder. These cause deficits spanning a wide range of domains, including social, emotional, communicative, intellectual, and sensory functions (Morris-Rosendahl and Crocq 2020). Numerous studies investigating underlying mechanisms have revealed changes at the genetic, neuronal, and circuit levels (Sahin and Sur 2015). However, precise, causal relationships between observed pathologies and disruptions to circuit function remain poorly understood (Mizusaki and O'Donnell 2021; Hitchcock et al. 2022). By more accurately accounting for development in RNN paradigms, we gain the ability to make inferences about diseases in which specific mechanisms are known (Fig. 3).

Computational models of circuit dysfunction have spanned a range of levels, including networks of Hodgkin–Huxley neurons (O'Donnell et al. 2017; Onasch and Gjorgjieva 2020), leaky integrate-and-fire neurons (Cano-Colino and Compte 2012; Cavanagh et al. 2020; Lam et al. 2022), and population firing rate models (Murray et al. 2017). These generally follow a three-step process (Fig. 3A). First, the circuit model's parameters are tuned to fit a healthy state. Second, specific parameters are altered to match abnormalities found in a neurodevelopmental disorder. Third, the resulting computational deficits are compared to the symptoms of the disorder. In principle, this approach could be applied to any neurodevelopmental disorder (or psychiatric disorders more generally) but studies so far have focused primarily on ASD (Sahin and Sur 2015) and schizophrenia (Owen et al. 2011). Parameter alterations used to induce a disease state can be grouped into three broad categories (Lanillos et al. 2020): disconnection, where the network has an atypical decrease or increase in connections (Stevens 1992; Friston and Frith 1995), E/I imbalance, where the relative amounts of excitation and inhibition are abnormal (Rubenstein and Merzenich 2003), and hypopriors, in which circuits exhibit an abnormal reliance on sensory information relative to top-down predictions based on prior experiences (Pellicano and Burr 2012).

The advantages of RNNs as a model of neural circuit function described earlier also make RNNs suitable to investigate circuit dysfunction. The heterogeneous symptoms of neurodevelopmental disorders require a flexible modeling framework capable of diverse computations, and the ability of RNNs to learn many computations using the same training process, often within the same RNN (Yang et al. 2019), is well aligned to this goal. Similarly, the loss function of RNNs allows the exact quantification of computational errors across tasks. As a result, RNNs are equipped with a natural method to measure computational deficits induced by a parameter perturbation. Additionally, RNNs are universal approximators (Schäfer and Zimmermann 2006), often producing rich neural dynamics that can be analyzed to provide mechanistic explanations of computational deficits.

Schizophrenia

The earliest use of RNNs to investigate neurodevelopmental disorders applied Hopfield networks to explain how symptoms of schizophrenia might arise during memory formation (Hoffman 1987). The memory capacity of Hopfield networks is ~15% of the number of neurons (Hopfield 1982). Rather than applying a parameter alteration corresponding to a proposed circuit mechanism of schizophrenia, Hoffman (1987) induced a disease state in the network by overloading the number of stored memories. Surprisingly, the resulting errors were qualitatively different from those observed below the overloading threshold and were interpretable in the context of schizophrenia. In the subthreshold regime, memories were either recalled correctly or resulted in a "generalization" error in which the final state stabilized near a cluster of similar memories. In contrast, the overloaded

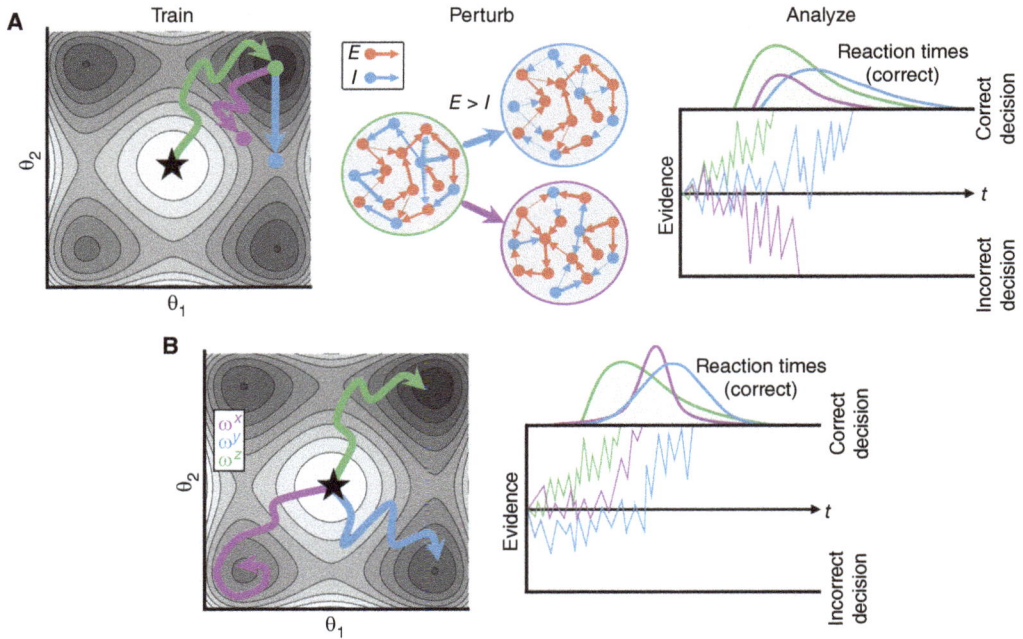

Figure 3. Recurrent neural network (RNN) modeling of neurodevelopmental disorders. (*A*) The train, perturb, analyze approach to RNN models of neurodevelopmental disorders. First, the RNN's parameters (θ_1 and θ_2) are trained (green line) such that a loss function is minimized and the RNN is in a "healthy" state (green dot). Then the parameters are perturbed (purple and blue lines) away from the optimal configuration to "disease" states. These perturbations may be systematic and easily interpretable, such as E/I imbalances in models of autism (e.g., weaking inhibitory connections, blue; Echeveste et al. 2022) or more abstract such as memory overloading in models of schizophrenia (purple; Hoffman 1987). The resulting computational changes can be analyzed to link circuit structure to computational deficits. For example, in a two alternative forced choice task, perturbations will result in lower performance (*lower* areas under the purple and blue distributions) but may also induce specific changes to how computations are performed (e.g., increasing the mean reaction time, i.e., translation of reaction time distributions to the *right*). (*B*) The metalearning perturbation approach to RNN models of neurodevelopmental disorders. Changes (ω^x, purple, and ω^y, blue) to the metalearned healthy state (ω^z, green) may result in distinct solutions to the learning problem (Philippsen and Nagai 2018). These solutions may have similar task performance (areas under the green, blue, and purple distributions are comparable) but perform computations in radically different ways (blue and gold distributions have different means and higher-order moments).

state produced two types of pathological errors: misperceptions, in which the network exactly recalled an incorrect memory, and more severe dysfunction—similar to delusions or hallucinations—in which many inputs recalled the same, nonmemory state. The insights into schizophrenia provided by this study were limited by the direction of its inference: observing a dysfunctional state in an RNN and then mapping it onto a psychiatric disorder without an underlying biological mechanism. Despite this limitation, its findings demonstrated the applicability

of RNNs to psychiatric disorders (Lanillos et al. 2020).

Subsequent studies examined the ability of excessive synaptic pruning to account for symptoms of schizophrenia. Hoffman and Dobscha (1989) trained Hopfield networks to recall a set of memories and then removed connections from weakest to strongest while tracking recall performance using the same error classes described above. Surprisingly, they found that synaptic pruning induced "delusions" and "hallucinations" as described above, although not until ~80% of synapses were removed. This demonstrated the feasibility of a biologically motivated alteration to an RNN as a model for a psychiatric disorder. Furthermore, it is aligned with significant, experimental observed pruning during healthy development (Huttenlocher 1979). However, the model predicts that delusions and hallucinations in schizophrenia are accompanied by memory loss, which is at odds with clinically observed symptoms.

A second series of RNN schizophrenia models addressed this mechanism/symptom inconsistency by analyzing an alternative theory due to Stevens (1992) in which recurrent connectivity strength in the hippocampus overcompensates for degraded external inputs (Horn and Ruppin 1995; Ruppin et al. 1996). The authors used a more biologically motivated Hopfield model variant (Tsodyks 1988; Tsodyks and Feigel'man 1988) with directed connections and a more realistic level of neural activity. Mean field theoretic analysis and simulations revealed the functional consequences of both proposed biological mechanisms. Weakening the model's external input impaired memory function and strengthening recurrent connections restored memory function at the cost of inducing hallucination-like memory retrieval errors. Combined, these models of schizophrenia highlight the ability of RNNs to parse redundant circuit mechanisms in psychiatric disorders (Mizusaki and O'Donnell 2021).

Autism Spectrum Disorder

The proportion of excitation and inhibition neurons receive is critical to sensory processing (Anderson et al. 2000; Wehr and Zador 2003) and is a fundamental description of the activity in neuronal networks (van Vreeswijk and Sompolinsky 1996; Ahmadian and Miller 2021). Similarly, an imbalance in E/I signaling is a prominent hypothesis for the mechanism underlying ASDs (Rubenstein and Merzenich 2003). However, the extent to which E/I imbalances can explain the symptoms of autism is not well understood (O'Donnell et al. 2017; Antoine et al. 2019).

A recent ASD study (Echeveste et al. 2022) used RNNs to provide a mechanistic link between the E/I imbalance theory of circuit dysfunction and the hypopriors theory of ASD, a variation of the Bayesian framework for perception (Pellicano and Burr 2012). The hypopriors theory of ASD proposes that higher uncertainty in prior distributions results in an overreliance on sensory information. To explore the neural basis of hypopriors, Echeveste et al. (2022) built on previous work that trained an excitatory–inhibitory RNN circuit model to implement Bayesian perceptual inference such that optimal estimates of the environmental state were encoded in the activity of the excitatory neurons (Echeveste et al. 2020). E/I imbalances were then induced in the trained model by weakening inhibitory connections, and the effect on perception was measured from the resulting posterior distributions. Critically, the E/I imbalances induced an overreliance on sensory information, as predicted by the hypopriors theory of ASD. This result represents one of the clearest links between circuit dysfunction and computational deficits and highlights the potential of RNNs to reveal and clarify the neural basis of developmental disorders.

Using the language of metalearning, the train-alter-compare approach induces a disease state by shifting the RNN out of an optimal, learned state (i.e., changing θ in Equation 3). Alternatively, differences in cognitive processes observed in neurodevelopmental disorders could be a distinct, locally optimal state resulting from changes throughout development. As a result, neurodevelopmental disorders could be more appropriately modeled by shifting the RNN out of a metalearned state (i.e., changing ω^* in Equation 4; Fig. 3B). Philippsen and Nagai (2018) took the latter approach, training an RNN to perform time series forecasting by estimating the mean and variance of an input signal. The authors separately altered two learning hyperparameters, external contribution and aberrant precision, and measured the resulting differences in task performance and internal task variable representation after a fixed number of training steps. The external contribution determined the amount of sensory information used when building its internal representation of the time series, and the aberrant precision determined the RNNs' estimate of the signal variance. While most parameters allowed the RNN to learn the task, the degree of structure in the internal task variable representations varied independently of task performance. This highlights that differences in cognitive processes may not align with decreases in task performance. As a result, investigation of specific behaviors alone may not be sufficient for understanding the mechanisms and cognitive processes underlying autism and other neurodevelopmental disorders. Similarly, limits inherent to train–alter–compare studies may be addressed by metalearning models of neurodevelopmental disorders.

CONCLUSIONS

Theories of neural development require a framework capable of extensive changes to circuit structure and function spread across multiple timescales. RNNs can fit all these criteria and have already yielded crucial insight into the codevelopment of neural circuit structure, dynamics, and computation. Incorporation of metalearning, often paired with reinforcement learning, as an explicit model of multiple developmental timescales, represents a promising method for gaining insight into both normal (Wang et al. 2018; Jiang and Litwin-Kumar 2021; Goudar et al. 2023) and abnormal development (Philippsen and Nagai 2018). In addition, the implementation of structurally dynamic learning rules has yielded insight into the computational significance of neurodevelopmental mechanisms.

How can RNN models of neural development work synergistically with RNN circuit models more broadly? Major challenges in understanding neural development and neurodevelopmental disorders parallel challenges in RNN analysis of mature circuits. Although RNNs are often used to hypothesize the relationship between circuit structure and function (Sussillo 2014), alternative, mechanistically distinct RNN hypotheses often yield nearly identical solutions to the same problem (Vyas et al. 2020; O'Shea et al. 2022). Similarly, redundancy in neural circuit function (Fig. 3B) represents a major challenge in understanding the neural ba-

sis of neurodevelopmental disorders (Mizusaki and O'Donnell 2021). Because these are two sides of the same coin, techniques that identify the mechanisms of neurodevelopmental disorders could be used to guide analysis that differentiates the RNN hypothesis. Recent advances in multitask learning in RNNs (Yang et al. 2019) offer a potential solution to this problem. By training the same RNN to perform many computations, the functional consequences of circuit differences—regardless of origin—can be compared across tasks.

In summary, the ability of RNNs to flexibly perform complex computations using the basic components of neural circuits makes them an appealing model for understanding not just the brain in general, but also the emergence of network computations on developmental timescales.

REFERENCES

Ahmadian Y, Miller KD. 2021. What is the dynamical regime of cerebral cortex? *Neuron* **109**: 3373–3391. doi:10.1016/j.neuron.2021.07.031

Anderson PW. 1972. More is different. *Science* **177**: 393–396. doi:10.1126/science.177.4047.393

Anderson JS, Carandini M, Ferster D. 2000. Orientation tuning of input conductance, excitation, and inhibition in cat primary visual cortex. *J Neurophysiol* **84**: 909–926. doi:10.1152/jn.2000.84.2.909

Angeline PJ, Saunders GM, Pollack JB. 1994. An evolutionary algorithm that constructs recurrent neural networks. *IEEE Trans Neural Netw* **5**: 54–65. doi:10.1109/72.265960

Antoine MW, Langberg T, Schnepel P, Feldman DE. 2019. Increased excitation-inhibition ratio stabilizes synapse and circuit excitability in four autism mouse models. *Neuron* **101**: 648–661.e4. doi:10.1016/j.neuron.2018.12.026

Barabási DL, Beynon T, Katona Á, Perez-Nieves N. 2023. Complex computation from developmental priors. *Nat Commun* **14**: 2226. doi:10.1038/s41467-023-37980-1

Bicknell BA, Dayan P, Goodhill GJ. 2015. The limits of chemosensation vary across dimensions. *Nat Commun* **6**: 7468. doi:10.1038/ncomms8468

Bourgeois J, Rakic P. 1993. Changes of synaptic density in the primary visual cortex of the macaque monkey from fetal to adult stage. *J Neurosci* **13**: 2801–2820. doi:10.1523/JNEUROSCI.13-07-02801.1993

Cano-Colino M, Compte A. 2012. A computational model for spatial working memory deficits in schizophrenia. *Pharmacopsychiatry* **45**(Suppl 1): S49–S56. doi:10.1055/s-0032-1306314

Cavanagh SE, Lam NH, Murray JD, Hunt LT, Kennerley SW. 2020. A circuit mechanism for decision-making biases and

Cite this article as *Cold Spring Harb Perspect Biol* doi: 10.1101/cshperspect.a041507

NMDA receptor hypofunction. *eLife* **9**: e53664. doi:10 .7554/eLife.53664

Dayan P, Abbott LF. 2005. *Theoretical neuroscience: computational and mathematical modeling of neural systems.* MIT Press, Cambridge, MA.

Doya K. 2002. Metalearning and neuromodulation. *Neural Netw* **15**: 495–506. doi:10.1016/S0893-6080(02)00044-8

Driscoll LN, Pettit NL, Minderer M, Chettih SN, Harvey CD. 2017. Dynamic reorganization of neuronal activity patterns in parietal cortex. *Cell* **170**: 986–999.e16. doi:10 .1016/j.cell.2017.07.021

Driscoll L, Shenoy K, Sussillo D. 2022. Flexible multitask computation in recurrent networks utilizes shared dynamical motifs. bioRxiv doi:10.1101/20220815503870

Dubreuil A, Valente A, Beiran M, Mastrogiuseppe F, Ostojic S. 2022. The role of population structure in computations through neural dynamics. *Nat Neurosci* **25**: 783–794. doi:10.1038/s41593-022-01088-4

Echeveste R, Aitchison L, Hennequin G, Lengyel M. 2020. Cortical-like dynamics in recurrent circuits optimized for sampling-based probabilistic inference. *Nat Neurosci* **23**: 1138–1149. doi:10.1038/s41593-020-0671-1

Echeveste R, Ferrante E, Milone DH, Samengo I. 2022. Bridging physiological and perceptual views of autism by means of sampling-based Bayesian inference. *Netw Neurosci* **6**: 196–212. doi:10.1162/netn_a_00219

Fahlman S. 1990. The recurrent cascade-correlation architecture. *Adv Neural Inf Process Syst* **3**: 190–196.

Faust TE, Gunner G, Schafer DP. 2021. Mechanisms governing activity-dependent synaptic pruning in the developing mammalian CNS. *Nat Rev Neurosci* **22**: 657–673. doi:10 .1038/s41583-021-00507-y

Finn C, Abbeel P, Levine S. 2017. Model-agnostic meta-learning for fast adaptation of deep networks. In *International Conference on Machine Learning, Vol. 70 of Proceedings of Machine Learning Research*, pp. 1126–1135. PMLR, Sydney, Australia.

Friston KJ, Frith CD. 1995. Schizophrenia: a disconnection syndrome. *Clin Neurosci* **3**: 89–97.

Ghazizadeh E, Ching S. 2021. Slow manifolds within network dynamics encode working memory efficiently and robustly. *PLoS Comput Biol* **17**: e1009366. doi:10.1371/journal .pcbi.1009366

Goodhill GJ. 1993. Topography and ocular dominance: a model exploring positive correlations. *Biol Cybern* **69**: 109–118. doi:10.1007/BF00226194

Goodhill GJ. 2018. Theoretical models of neural development. *iScience* **8**: 183–199. doi:10.1016/j.isci.2018.09.017

Goudar V, Peysakhovich B, Freedman DJ, Buffalo EA, Wang X-J. 2023. Schema formation in a neural population subspace underlies learning-to-learn in flexible sensorimotor problem-solving. *Nat Neurosci* **26**: 879–890. doi:10.1038/ s41593-023-01293-9

Gregor T, Tank DW, Wieschaus EF, Bialek W. 2007. Probing the limits to positional information. *Cell* **130**: 153–164. doi:10.1016/j.cell.2007.05.025

Guan D, Armstrong WE, Foehring RC. 2014. Electrophysiological properties of genetically identified subtypes of layer 5 neocortical pyramidal neurons: Ca²⁺ dependence and differential modulation by norepinephrine. *J Neurophysiol* **113**: 2014–2032. doi:10.1152/jn.00524.2014

Harlow HF. 1949. The formation of learning sets. *Psychol Rev* **56**: 51–65. doi:10.1037/h0062474

Hebb DO. 1949. *The organization of behavior: a neuropsychological theory.* Psychology, New York.

Hennequin G, Vogels TP, Gerstner W. 2014. Optimal control of transient dynamics in balanced networks supports generation of complex movements. *Neuron* **82**: 1394–1406. doi:10.1016/j.neuron.2014.04.045

Hennig JA, OBY ER, Losey DM, Batista AP, Yu BM, Chase SM. 2021. How learning unfold in the brain: toward and optimization view. *Neuron* **109**: 3720–3735. doi:10.1016/ j.neuron.2021.09.005

Hitchcock PF, Fried EI, Frank MJ. 2022. Computational psychiatry needs time and context. *Annu Rev Psychol* **73**: 243–270. doi:10.1146/annurev-psych-021621-124910

Hodgkin AL, Huxley AF. 1952. A quantitative description of membrane current and its application to conduction and excitation in nerve. *J Physiol* **117**: 500–544. doi:10.1113/ jphysiol.1952.sp004764

Hoffman RE. 1987. Computer simulations of neural information processing and the schizophrenia-mania dichotomy. *Arch Gen Psychiatry* **44**: 178. doi:10.1001/archpsyc.1987 .01800140090014

Hoffman RE, Dobscha SK. 1989. Cortical pruning and the development of schizophrenia: a computer model. *Schizophr Bull* **15**: 477–490. doi:10.1093/schbul/15.3.477

Holland PW, Laskey KB, Leinhardt S. 1983. Stochastic blockmodels: first steps. *Soc Netw* **5**: 109–137. doi:10.1016/ 0378-8733(83)90021-7

Hopfield JJ. 1982. Neural networks and physical systems with emergent collective computational abilities. *Proc Natl Acad Sci* **79**: 2554–2558. doi:10.1073/pnas.79.8.2554

Horn D, Ruppin E. 1995. Compensatory mechanisms in an attractor neural network model of schizophrenia. *Neural Comput* **7**: 182–205. doi:10.1162/neco.1995.7.1.182

Hospedales T, Antoniou A, Micaelli P, Storkey A. 2022. Meta-learning in neural networks: a survey. *IEEE Trans Pattern Anal Machine Intell* **44**: 5149–5169. doi:10.1109/TPAMI .2021.3079209

Huttenlocher PR. 1979. Synaptic density in human frontal cortex—developmental changes and effects of aging. *Brain Res* **163**: 195–205. doi:10.1016/0006-8993(79)90349-4

Izhikevich EM. 2006. *Dynamical systems in neuroscience.* MIT Press, Cambridge, MA.

Jiang L, Litwin-Kumar A. 2021. Models of heterogeneous dopamine signaling in an insect learning and memory center. *PLoS Comput Biol* **17**: e1009205. doi:10.1371/jour nal.pcbi.1009205

Johnson S, Marro J, Torres JJ. 2010. Evolving networks and the development of neural systems. *J Stat Mech* **2010**: P03003. doi:10.1088/1742-5468/2010/03/P03003

Kleinman M, Chandrasekaran C, Kao JC. 2020. Recurrent neural network models of multi-area computation underlying decision-making. bioRxiv doi:10.1101/798553

Koulakov A, Shuvaev S, Lachi D, Zador A. 2022. Encoding innate ability through a genomic bottleneck. bioRxiv doi:10.1101/20210316435261

Kudithipudi D, Aguilar-Simon M, Babb J, Bazhenov M, Blackiston D, Bongard J, Brna AP, Chakravarthi Raja S, Cheney N, Clune J, et al. 2022. Biological underpinnings for lifelong learning machines. *Nat Mach Intell* **4:** 196–210. doi:10.1038/s42256-022-00452-0

Lam NH, Borduqui T, Hallak J, Roque A, Anticevic A Krystal JH, Wang XJ, Murray JD. 2022. Effects of altered excitation-inhibition balance on decision making in a cortical circuit model. *J Neurosci* **42:** 1035–1053. doi:10.1523/JNEUROSCI.1371-20.2021

Lanillos P, Oliva D, Philippsen A, Yamashita Y, Nagai Y, Cheng G. 2020. A review on neural network models of schizophrenia and autism spectrum disorder. *Neural Netw* **122:** 338–363. doi:10.1016/j.neunet.2019.10.014

Li Z, Zhou F, Chen F, Li H. 2017. Meta-SGD: learning to learn quickly for few-shot learning. arXiv doi:10.48550/ARXIV170709835

Linsker R. 1986. From basic network principles to neural architecture: emergence of spatial-opponent cells. *Proc Natl Acad Sci* **83:** 7508–7512. doi:10.1073/pnas.83.19.7508

Liu YH, Zhu J, Constantinidis C, Zhou X. 2021. Emergence of prefrontal neuron maturation properties by training recurrent neural networks in cognitive tasks. *iScience* **24:** 103178. doi:10.1016/j.isci.2021.103178

Mante V, Sussillo D, Shenoy KV, Newsome WT. 2013. Context-dependent computation by recurrent dynamics in prefrontal cortex. *Nature* **503:** 78–84. doi:10.1038/nature12742

Marder E. 2011. Variability, compensation, and modulation in neurons and circuits. *Proc Natl Acad Sci* **108:** 15542–15548. doi:10.1073/pnas.1010674108

Markus EJ, Petit TL. 1987. Neocortical synaptogenesis, aging, and behavior: lifespan development in the motor-sensory system of the rat. *Exp Neurol* **96:** 262–278. doi:10.1016/0014-4886(87)90045-8

Marsland S, Shapiro J, Nehmzow U. 2002. A self-organising network that grows when required. *Neural Netw* **15:** 1041–1058. doi:10.1016/S0893-6080(02)00078-3

Masse NY, Yang GR, Song HF, Wang XJ, Freedman DJ. 2019. Circuit mechanisms for the maintenance and manipulation of information in working memory. *Nat Neurosci* **22:** 1159–1167. doi:10.1038/s41593-019-0414-3

McClelland JL, McNaughton BL, O'Reilly RC. 1995. Why there are complementary learning systems in the hippocampus and neocortex: insights from the successes and failures of connectionist models of learning and memory. *Psychol Rev* **102:** 419–457. doi:10.1037/0033-295X.102.3.419

McCormick DA, Connors BW, Lighthall JW, Prince DA. 1985. Comparative electrophysiology of pyramidal and sparsely spiny stellate neurons of the neocortex. *J Neurophysiol* **54:** 782–806. doi:10.1152/jn.1985.54.4.782

McCulloch WS, Pitts W. 1943. A logical calculus of the ideas immanent in nervous activity. *Bull Math Biophys* **5:** 115–133. doi:10.1007/BF02478259

Mikhaeil J, Monfared Z, Durstewitz D. 2022. On the difficulty of learning chaotic dynamics with RNNs. In *Advances in neural information processing systems* (ed. Koyejo S, et al.), Vol. 35, pp. 11297–11312. Curran Associates, Red Hook, NY.

Millán AP, Torres JJ, Johnson S, Marro J. 2018. Concurrence of form and function in developing networks and its role in synaptic pruning. *Nat Commun* **9:** 2236. doi:10.1038/s41467-018-04537-6

Millán AP, Torres JJ, Marro J. 2019. How memory conforms to brain development. *Front Comput Neurosci* **13:** 22. doi:10.3389/fncom.2019.00022

Millán AP, Torres JJ, Johnson S, Marro J. 2021. Growth strategy determines the memory and structural properties of brain networks. *Neural Netw* **142:** 44–56. doi:10.1016/j.neunet.2021.04.027

Miller KD, Keller JB, Stryker MP. 1989. Ocular dominance column development: analysis and simulation. *Science* **245:** 605–615. doi:10.1126/science.2762813

Mizusaki BEP, O'Donnell C. 2021. Neural circuit function redundancy in brain disorders. *Curr Opin Neurobiol* **70:** 74–80. doi:10.1016/j.conb.2021.07.008

Molano-Mazón M, Shao Y, Duque D, Yang GR, Ostojic S, de la Rocha J. 2023. Recurrent networks endowed with structural priors explain suboptimal animal behavior. *Curr Biol* **33:** 622–638.e7. doi:10.1016/j.cub.2022.12.044

Morris-Rosendahl DJ, Crocq MA. 2020. Neurodevelopmental disorders—the history and future of a diagnostic concept. *Dialogues Clin Neurosci* **22:** 65–72. doi:10.31887/DCNS.2020.22.1/macrocq

Murray JD, Jaramillo J, Wang XJ. 2017. Working memory and decision-making in a frontoparietal circuit model. *J Neurosci* **37:** 12167–12186. doi:10.1523/JNEUROSCI.0343-17.2017

Musall S, Kaufman MT, Juavinett AL, Gluf S, Churchland AK. 2019. Single-trial neural dynamics are dominated by richly varied movements. *Nat Neurosci* **22:** 1677–1686. doi:10.1038/s41593-019-0502-4

Navlakha S, Bar-Joseph Z, Barth AL. 2018. Network design and the brain. *Trends Cogn Sci* **22:** 64–78. doi:10.1016/j.tics.2017.09.012

O'Donnell C, Gonç Alves T, Portera-Cailliau C, Sejnowski TJ. 2017. Beyond excitation/inhibition imbalance in multidimensional models of neural circuit changes in brain disorders. *eLife* **6:** e26724. doi:10.7554/eLife.26724

Onasch S, Gjorgjieva J. 2020. Circuit stability to perturbations reveals hidden variability in the balance of intrinsic and synaptic conductances. *J Neurosci* **40:** 3186–3202. doi:10.1523/JNEUROSCI.0985-19.2020

Orhan AE, Ma WJ. 2019. A diverse range of factors affect the nature of neural representations underlying short-term memory. *Nat Neurosci* **22:** 275–283. doi:10.1038/s41593-018-0314-y

O'Shea DJ, Duncker L, Goo W, Sun X, Vyas S, Trautmann EM, Diester I, Ramakrishnan C, Deisseroth K, Sahani M. 2022. Direct neural perturbations reveal a dynamical mechanism for robust computation. bioRxiv doi:10.1101/2022.12.16.520768

Oswald AMM, Reyes AD. 2008. Maturation of intrinsic and synaptic properties of layer 2/3 pyramidal neurons in mouse auditory cortex. *J Neurophysiol* **99:** 2998–3008. doi:10.1152/jn.01160.2007

Owen MJ, O'Donovan MC, Thapar A, Craddock N. 2011. Neurodevelopmental hypothesis of schizophrenia. *Br J Psychiatry* **198:** 173–175. doi:10.1192/bjp.bp.110.084384

Parisi GI, Tani J, Weber C, Wermter S. 2017. Lifelong learning of human actions with deep neural network self-organiza-

Cite this article as *Cold Spring Harb Perspect Biol* doi: 10.1101/cshperspect.a041507

tion. *Neural Netw* **96:** 137–149. doi:10.1016/j.neunet.2017 .09.001

Parisi GI, Tani J, Weber C, Wermter S. 2018. Lifelong learning of spatiotemporal representations with dual-memory recurrent self-organization. *Front Neurorobot* **12:** 78. doi:10 .3389/fnbot.2018.00078

Pellicano E, Burr D. 2012. When the world becomes "too real": a Bayesian explanation of autistic perception. *Trends Cogn Sci* **16:** 504–510. doi:10.1016/j.tics.2012.08.009

Perich MG, Rajan K. 2020. Rethinking brain-wide interactions through multi-region "network of networks" models. *Curr Opin Neurobiol* **65:** 146–151. doi:10.1016/j.conb .2020.11.003

Philippsen A, Nagai Y. 2018. Understanding the cognitive mechanisms underlying autistic behavior: a recurrent neural network study. In *2018 Joint IEEE 8th International Conference on Development and Learning and Epigenetic Robotics (ICDL-EpiRob)*, pp. 84–90. IEEE, Piscataway, NJ.

Piaget J. 1997. *Language and thought of the child*. Routledge, London. doi:10.4324/9780203992739

Pinto L, Rajan K, DePasquale B, Thiberge SY, Tank DW, Brody CD. 2019. Task-dependent changes in the large-scale dynamics and necessity of cortical regions. *Neuron* **104:** 810–824.e9. doi:10.1016/j.neuron.2019.08.025

Ponce-Alvarez A, Mochol G, Hermoso-Mendizabal A, de la Rocha J, Deco G. 2020. Cortical state transitions and stimulus response evolve along stiff and sloppy parameter dimensions, respectively. *eLife* **9:** e53268. doi:10.7554/eLife .53268

Ratcliff R. 1990. Connectionist models of recognition memory: constraints imposed by learning and forgetting functions. *Psychol Rev* **97:** 285–308. doi:10.1037/0033-295X.97 .2.285

Rinzel J, Ermentrout GB. 1989. Analysis of neural excitability and oscillations. In *Methods in neuronal modeling: from synapses to networks*, pp. 135–169. MIT Press, Cambridge, MA.

Rubenstein JLR, Merzenich MM. 2003. Model of autism: increased ratio of excitation/inhibition in key neural systems. *Genes Brain Behav* **2:** 255–267. doi:10.1034/j.1601-183X .2003.00037.x

Rumelhart DE, Hinton GE, Williams RJ. 1986. Learning representations by back-propagating errors. *Nature* **323:** 533–536. doi:10.1038/323533a0

Ruppin E, Reggia JA, Horn D. 1996. Pathogenesis of schizophrenic delusions and hallucinations: a neural model. *Schizophr Bull* **22:** 105–121. doi:10.1093/schbul/22.1.105

Sahin M, Sur M. 2015. Genes, circuits, and precision therapies for autism and related neurodevelopmental disorders. *Science* **350:** aab3897. doi:10.1126/science.aab3897

Saxena S, Russo AA, Cunningham J, Churchland MM. 2022. Motor cortex activity across movement speeds is predicted by network-level strategies for generating muscle activity. *eLife* **11:** e67620. doi:10.7554/eLife.67620

Schäfer AM, Zimmermann HG. 2006. Recurrent neural networks are universal approximators. In *Artificial Neural Networks—ICANN 2006*. Lecture Notes in Computer Science, Vol. 4131. Springer, Berlin.

Schultz W, Dayan P, Montague PR. 1997. A neural substrate of prediction and reward. *Science* **275:** 1593–1599. doi:10 .1126/science.275.5306.1593

Shenoy KV, Sahani M, Churchland MM. 2013. Cortical control of arm movements: a dynamical systems perspective. *Annu Rev Neurosci* **36:** 337–359. doi:10.1146/annurev-neuro-062111-150509

Song HF, Yang GR, Wang XJ. 2016. Training excitatory-inhibitory recurrent neural networks for cognitive tasks: a simple and flexible framework. *PLoS Comput Biol* **12:** e1004792. doi:10.1371/journal.pcbi.1004792

Song HF, Yang GR, Wang XJ. 2017. Reward-based training of recurrent neural networks for cognitive and value-based tasks. *eLife* **6:** e21492. doi:10.7554/eLife.21492

Steinmetz NA, Zatka-Haas P, Carandini M, Harris KD. 2019. Distributed coding of choice, action and engagement across the mouse brain. *Nature* **576:** 266–273. doi:10 .1038/s41586-019-1787-x

Stevens JR. 1992. Abnormal reinnervation as a basis for schizophrenia: a hypothesis. *Arch Gen Psychiatry* **49:** 238. doi:10 .1001/archpsyc.1992.01820030070009

Stringer C, Pachitariu M, Steinmetz N, Reddy CB, Carandini M, Harris KD. 2019. Spontaneous behaviors drive multidimensional, brainwide activity. *Science* **364:** 225. doi:10 .1126/science.aav7893

Sucholutsky I, Muttenthaler L, Weller A, Peng A, Bobu A, Kim B, Love BC, Grant E, Groen I, Achterberg J, et al. 2023. Getting aligned on representational alignment. arXiv doi:10.48550/arXiv.2310.13018

Sussillo D. 2014. Neural circuits as computational dynamical systems. *Curr Opin Neurobiol* **25:** 156–163. doi:10.1016/j .conb.2014.01.008

Sussillo D, Abbott LF. 2009. Generating coherent patterns of activity from chaotic neural networks. *Neuron* **63:** 544–557. doi:10.1016/j.neuron.2009.07.018

Sussillo D, Nuyujukian P, Fan JM, Kao JC, Stavisky SD, Ryu S, Shenoy K. 2012. A recurrent neural network for closed-loop intracortical brain–machine interface decoders. *J Neural Eng* **9:** 026027. doi:10.1088/1741-2560/9/2/ 026027

Sussillo D, Churchland MM, Kaufman MT, Shenoy KV. 2015. A neural network that finds a naturalistic solution for the production of muscle activity. *Nat Neurosci* **18:** 1025–1033. doi:10.1038/nn.4042

Tkačik G, Dubuis JO, Petkova MD, Gregor T. 2015. Positional information, positional error, and readout precision in morphogenesis: a mathematical framework. *Genetics* **199:** 39–59. doi:10.1534/genetics.114.171850

Tsodyks MV. 1988. Associative memory in asymmetric diluted network with low level of activity. *Europhys Lett* **7:** 203–208. doi:10.1209/0295-5075/7/3/003

Tsodyks MV, Feigel'man MV. 1988. The enhanced storage capacity in neural networks with low activity level. *Europhys Lett* **6:** 101–105. doi:10.1209/0295-5075/ 6/2/002

Urai AE, Doiron B, Leifer AM, Churchland AK. 2022. Large-scale neural recordings call for new insights to link brain and behavior. *Nat Neurosci* **25:** 11–19. doi:10.1038/ s41593-021-00980-9

van Ooyen A. 2011. Using theoretical models to analyse neural development. *Nat Rev Neurosci* **12:** 311–326. doi:10 .1038/nrn3031

van Vreeswijk C, Sompolinsky H. 1996. Chaos in neuronal networks with balanced excitatory and inhibitory activity.

Science **274:** 1724–1726. doi:10.1126/science.274.5293 .1724

von der Malsburg C. 1973. Self-organization of orientation sensitive cells in the striate cortex. *Kybernetik* **14:** 85–100. doi:10.1007/BF00288907

Vyas S, Golub MD, Sussillo D, Shenoy KV. 2020. Computation through neural population dynamics. *Annu Rev Neurosci* **43:** 249–275. doi:10.1146/annurev-neuro-092619-094115

Wang JX. 2021. Meta-learning in natural and artificial intelligence. *Curr Opin Behav Sci* **38:** 90–95. doi:10.1016/j .cobeha.2021.01.002

Wang JX, Kurth-Nelson Z, Kumaran D, Tirumala D, Soyer H, Leibo JZ, Hassabis D, Botvinick M. 2018. Prefrontal cortex as a meta-reinforcement learning system. *Nat Neurosci* **21:** 860–868. doi:10.1038/s41593-018-0147-8

Wang PY, Sun Y, Axel R, Abbott LF, Yang GR. 2021. Evolving the olfactory system with machine learning. *Neuron* **109:** 3879–3892.e5. doi:10.1016/j.neuron.2021.09.010

Wehr M, Zador AM. 2003. Balanced inhibition underlies tuning and sharpens spike timing in auditory cortex. *Nature* **426:** 442–446. doi:10.1038/nature02116

Werbos PJ. 1990. Backpropagation through time: what it does and how to do it. *Proc IEEE* **78:** 1550–1560. doi:10.1109/5 .58337

White E, Weinfeld L, Lev D. 1997. A survey of morphogenesis during the early postnatal period in PMBSF barrels of mouse SmI cortex with emphasis on barrel D4. *Somatosens Mot Res* **14:** 34–55. doi:10.1080/08990229771204

Yamins DLK, Hong H, Cadieu CF, Solomon EA, Seibert D, DiCarlo JJ. 2014. Performance-optimized hierarchical models predict neural responses in higher visual cortex.

Proc Natl Acad Sci **111:** 8619–8624. doi:10.1073/pnas .1403112111

Yang GR, Wang X-J. 2020. Artificial neural networks for neuroscientists: a primer. *Neuron* **107:** 1048–1070. doi:10 .1016/j.neuron.2020.09.005

Yang GR, Joglekar MR, Song HF, Newsome WT, Wang XJ. 2019. Task representations in neural networks trained to perform many cognitive tasks. *Nat Neurosci* **22:** 297–306. doi:10.1038/s41593-018-0310-2

Zador AM. 2019. A critique of pure learning and what artificial neural networks can learn from animal brains. *Nat Commun* **10:** 3770. doi:10.1038/s41467-019-11786-6

Zhang Y, Yang Q. 2018. An overview of multi-task learning. *Natl Sci Rev* **5:** 30–43. doi:10.1093/nsr/nwx105

Zhou X, Zhu D, Qi XL, Lees CJ, Bennett AJ, Salinas E, Stanford TR, Constantinidis C. 2013. Working memory performance and neural activity in prefrontal cortex of peripubertal monkeys. *J Neurophysiol* **110:** 2648–2660. doi:10 .1152/jn.00370.2013

Zhou X, Qi XL, Constantinidis C. 2016a. Distinct roles of the prefrontal and posterior parietal cortices in response inhibition. *Cell Rep* **14:** 2765–2773. doi:10.1016/j.celrep.2016 .02.072

Zhou X, Zhu D, King SG, Lees CJ, Bennett AJ, Salinas E, Stanford TR, Constantinidis C. 2016b. Behavioral response inhibition and maturation of goal representation in prefrontal cortex after puberty. *Proc Natl Acad Sci* **113:** 3353–3358. doi:10.1073/pnas.1518147113

Zipser D, Andersen RA. 1988. A back-propagation programmed network that simulates response properties of a subset of posterior parietal neurons. *Nature* **331:** 679–684. doi:10.1038/331679a0

Reimagining Cortical Connectivity by Deconstructing Its Molecular Logic into Building Blocks

Xiaoyin Chen

Allen Institute for Brain Science, Seattle, Washington 98109, USA

Correspondence: xiaoyin.chen@alleninstitute.org

Comprehensive maps of neuronal connectivity provide a foundation for understanding the structure of neural circuits. In a circuit, neurons are diverse in morphology, electrophysiology, gene expression, activity, and other neuronal properties. Thus, constructing a comprehensive connectivity map requires associating various properties of neurons, including their connectivity, at cellular resolution. A commonly used approach is to use the gene expression profiles as an anchor to which all other neuronal properties are associated. Recent advances in genomics and anatomical techniques dramatically improved the ability to determine and associate the long-range projections of neurons with their gene expression profiles. These studies revealed unprecedented details of the gene–projection relationship, but also highlighted conceptual challenges in understanding this relationship. In this article, I delve into the findings and the challenges revealed by recent studies using state-of-the-art neuroanatomical and transcriptomic techniques. Building upon these insights, I propose an approach that focuses on understanding the gene–projection relationship through basic features in gene expression profiles and projections, respectively, that associate with underlying cellular processes. I then discuss how the developmental trajectories of projections and gene expression profiles create additional challenges and necessitate interrogating the gene–projection relationship across time. Finally, I explore complementary strategies that, together, can provide a comprehensive view of the gene–projection relationship.

In the nervous system, the connectivity of neurons with diverse properties, such as morphology, gene expression profiles, and electrophysiological properties, determines the structure of a circuit. Unraveling the wiring logic of diverse neurons is not only a central goal for understanding circuit structure, but also essential for constructing realistic functional circuit models. These detailed circuit models make specific predictions regarding neuronal activity in behaving animals and how perturbation affects behavior, and are critical for understanding how circuits function (e.g., see Chalfie et al. 1985; Zeng and Sanes 2017; Turner-Evans et al. 2020).

A key challenge in building comprehensive circuit models is to find an anchor of neuronal identities with which all other neuronal properties, including connectivity, can be associated—a

Rosetta Stone. Gene expression profile is especially suited for anchoring neuronal properties because of several characteristics. First, gene expression profiles, and by extension transcriptomically defined cell types, are strongly associated with morphology, connectivity, and other neuronal properties. Second, some single-cell and spatial transcriptomic techniques are highly scalable and can generate nearly complete atlases of cellular gene expression profiles across the entire brain. Notably, comprehensive transcriptomic atlases have been generated or are currently being generated not only in traditional model organisms such as mice (Chen et al. 2022b; Langlieb et al. 2023; Shi et al. 2023; Yao et al. 2023b; Zhang et al. 2023b) and nonhuman primates but also in amphibians (Lust et al. 2022; Wei et al. 2022; Woych et al. 2022) and reptiles (Tosches et al. 2018; Hain et al. 2022). Third, many emerging techniques make it feasible to associate gene expression profiles of single neurons with other neuronal properties, such as dendritic morphology, electrophysiological properties (Cadwell et al. 2016; Gouwens et al. 2020; Scala et al. 2021), behavior-evoked activity (Bugeon et al. 2022; Condylis et al. 2022), long-range projections (Chen et al. 2019; Sun et al. 2021; Zhao et al. 2022; Zhang et al. 2023a), and synaptic connectivity (Zhang et al. 2023a). Finally, genetic tools that provide access to defined subpopulations of neurons often rely on specific patterns in gene expression, so identifying patterns in gene expression that correspond to other neuronal properties provides a rational path for designing genetic tools to access neuronal populations defined by these properties. Thus, associating gene expression profiles with other neuronal properties, including connectivity, not only lays the foundation for a comprehensive understanding of circuit structure, but also informs the design of genetic tools to access subpopulations of neurons defined by other properties.

Recent advances in systematic mesoscale projection mapping (Hunnicutt et al. 2014; Oh et al. 2014; Zingg et al. 2014; Harris et al. 2019), single-neuron reconstruction (Winnubst et al. 2019; Peng et al. 2021; Gao et al. 2022), and barcoding-based neuroanatomical techniques (Kebschull et al. 2016; Han et al. 2018; Chen et al. 2019; Zhang et al. 2023a) have substantially increased our understanding of the relationship between long-range axonal projections of neurons and their gene expression profiles, especially in the mouse cortex (Chen et al. 2019; Klingler et al. 2021; Sun et al. 2021). These studies revealed a complex relationship: Although major transcriptomic types correspond to clear differences in projections, fine-grained transcriptomic types appear to have a many-to-many relationship to projections. This lack of clear correspondence across fine-grained subpopulations of neurons raises many questions about whether and how gene expression profiles reflect the heterogeneity in projections. What is the underlying biological basis for this many-to-many relationship? What biological factors enable and constrain the association between gene expression and projections? And what data and experimental strategies are needed to unravel this gene–projection relationship? Many developmental studies have revealed biological processes that contribute to the organization of projections (O'Leary 1987; Innocenti and Price 2005; Molyneaux et al. 2007; Custo Greig et al. 2013), but a comprehensive view of how projections are organized in adult neurons remains unclear.

Answering these questions requires not only collecting new data that associate gene expression profiles with projections, but also establishing conceptual frameworks under which these data are analyzed. Here, I propose that gene coexpression and collateralization of projections are associated with underlying biological processes and constraints. Thus, linking these features in single neurons, especially during development, may provide valuable information for understanding the gene–projection relationship. I will start by comparing different definitions of neuronal types based on projections and transcriptomics, respectively, and summarizing recent findings regarding the gene–projection relationship in the mouse cortex. I will then explore how focusing on gene coexpression and projection collateralization can provide insights into this relationship. Finally, I propose a developmental approach to resolve this relationship using complementary tools.

EVOLVING DEFINITION OF CELL TYPES OF CORTICAL PROJECTION NEURONS

Cataloging neurons into distinct cell types simplifies the cellular diversity of the brain, facilitates the reproducibility of studies, and enables the integration of findings across studies. Although defining cell types is not essential for understanding the relationship between gene expression profiles and projections, the simplicity of grouping neurons into types makes it an appealing starting point for discussing the gene–projection relationship. Historically, neuronal types have been defined using a variety of cellular properties, such as morphology, transmitter types, electrophysiological properties, and connectivity (for reviews, see Zeng and Sanes 2017; Yuste et al. 2020; Zeng 2022). In the cortex, there are two commonly used approaches to defining projection neurons: a classic projection-based definition and a more recent classification based on transcriptomics. Although these two systems are largely consistent and even share similar names

for some cell types, detailed classifications of several minor cell types differ between them. In this article, I will refer to the transcriptomically defined neuronal types extensively and use them as a basis for discussion. To avoid confusing readers who are not familiar with the transcriptomic definitions of neuronal types in the cortex and the correspondence to the classic naming conventions, here I provide a correspondence between the two systems and elaborate on why I use the transcriptomically defined types as a basis for further discussion.

Most neurons in the cerebral cortex belong to two large groups, glutamatergic excitatory neurons and GABAergic inhibitory neurons. Except for some rare types of inhibitory neurons (Tomioka and Rockland 2007; Higo et al. 2009; Lee et al. 2014; He et al. 2016; Tasic et al. 2016; Cho et al. 2023), most cortical projection neurons are excitatory. In the classic projection-based system, excitatory neurons are divided into three major classes based on the presence or absence of projections to certain brain regions (Fig. 1;

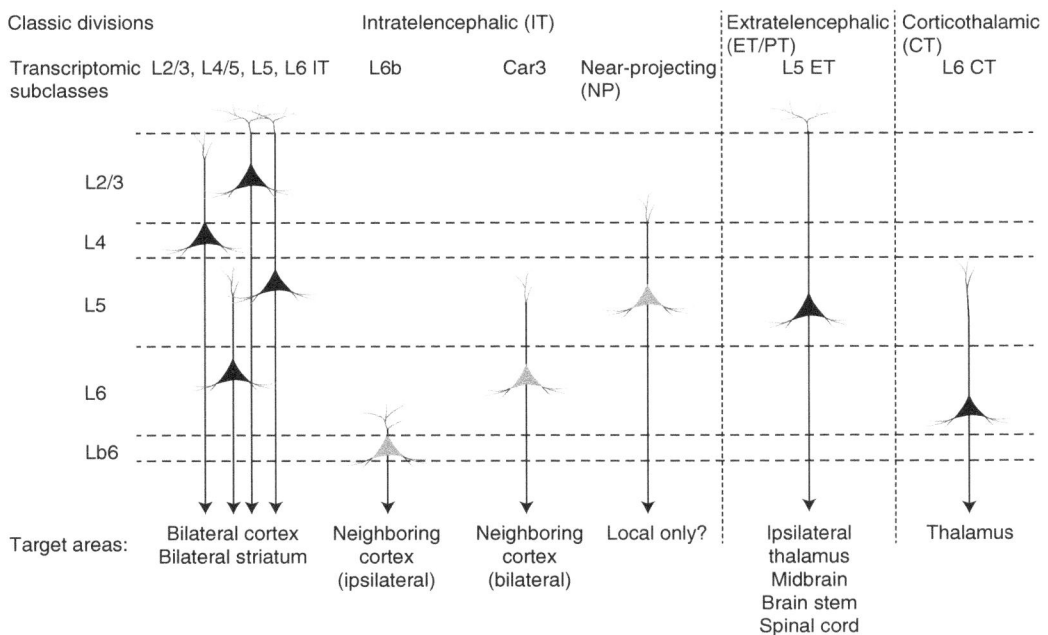

Figure 1. Summary of cortical excitatory neuron types. Major types of cortical excitatory neurons are shown with their classic division names, transcriptomic subclasses, laminar positions of somas, and major projection target areas. Cells in gray are transcriptomic types that are combined with other types in the classic divisions; these cells would likely have been considered as intratelencephalic (IT) neurons based on their projection patterns.

Cite this article as *Cold Spring Harb Perspect Biol* doi: 10.1101/cshperspect.a041509

Harris and Shepherd 2015). Intratelencephalic (IT) neurons project largely within the telencephalon, which includes the cortex, other cortex-like structures, and the striatum. Neurons in layer 4 were thought to project only locally and were sometimes considered distinct from IT neurons. The layer 5 extratelencephalic (L5 ET) neurons, also known as the pyramidal tract neurons, project to subcortical areas, including the thalamus, the midbrain, the brain stem, and/or the spinal cord. The layer 6 corticothalamic (L6 CT) neurons project mostly to the thalamus. These three classes are broad divisions, and neurons within a class remain incredibly diverse in their projections.

The transcriptomics-based system, in contrast, groups neurons not by the presence or absence of individual genes, but by how similar the neurons are in their overall gene expression profiles. In practice, this is done by clustering gene expression profiles measured in single-cell and single-nucleus RNA-seq experiments. Recent studies using these approaches further refine the classic tripartite of IT/L5 ET/L6 CT neurons into additional transcriptomic types (Zeisel et al. 2015; Tasic et al. 2016, 2018; Yao et al. 2021a,b, 2023b). In these transcriptomic studies, the divisions of IT/L5 ET/L6 CT were usually named "subclasses," with IT neurons being further divided into four subclasses based on their gene expression profiles. These four IT subclasses are named by the layers that they are largely found in: L2/3 IT, L4/5 IT, L5 IT, and L6 IT. These studies also revealed that the classically defined L4 neurons, which lack long-range projections, are similar in gene expression profiles to other L4/5 IT neurons (Tasic et al. 2016, 2018; Yao et al. 2021b) that do project to other cortical areas (Harris et al. 2019). In addition, these studies revealed three rarer transcriptomic types that were distinct from IT neurons but would have been classified as IT neurons based on their projections (Fig. 1). Car3 neurons are concentrated in deep layers of lateral cortical areas (Yao et al. 2021a,b; Chen et al. 2022b) and project to neighboring cortical areas and homotopic contralateral areas, but not the striatum (Peng et al. 2021). Car3 neurons are transcriptomically similar to neurons in the claustrum. L6b neurons are relat-

ed to subplate neurons during cortical development (Chun and Shatz 1989; Kanold and Luhmann 2010) and form a continuum with L6 CT neurons transcriptomically (Yao et al. 2021b). Recent large-scale projection mapping experiments revealed a thin layer of neurons in layer 6b that project to neighboring ipsilateral cortical areas (Muñoz-Castañeda et al. 2021), although whether this population corresponds to the transcriptomically defined L6b neurons requires further examination. Near-projecting neurons are mostly found in layer 5 and sometimes in layer 6 (Tasic et al. 2016, 2018; Sun et al. 2021; Zhang et al. 2021a). They are thought to project to nearby cortical areas (Tasic et al. 2016), although their exact projection patterns remain unclear. Beyond these subclass-level divisions, neurons are further divided into "supertypes" and/or "types" that reflect nuanced differences in their gene expression profiles (Yao et al. 2021b).

There is increasing evidence, gathered using multiple approaches, that supports the hypothesis that the transcriptomic subclasses and types reflect differences in various cellular properties across neuronal populations in the cortex. First, transcriptomic divisions fully capture the classic divisions of IT/L5 ET/L6 CT neurons, so all projection differences that are used to define the three classes of neurons are also reflected in their gene expression profiles at this granularity. Second, beyond the classic divisions, spatial transcriptomic studies (Zhang et al. 2021a, 2023b; Chen et al. 2022b) revealed that transcriptomically defined neuronal types, both at the subclass level and type level, are generally enriched at characteristic layers, sublayers, and combinations of areas in the cortex. Third, Patch-seq studies have shown that gene expression differences at the type level largely correspond to morphological and electrophysiological differences (Gouwens et al. 2020; Berg et al. 2021; Kalmbach et al. 2021; Scala et al. 2021). Finally, projection mapping studies using Cre-dependent anterograde tracing (Harris et al. 2019), single-neuron reconstruction (Peng et al. 2021), retro-seq (Zhang et al. 2021a), and barcoding-based approaches (Chen et al. 2019) have all revealed different preferences in projections across IT subclasses, Car3 neurons, and NP neurons (see below for more details).

 Cite this article as *Cold Spring Harb Perspect Biol* doi: 10.1101/cshperspect.a041509

To summarize, the transcriptomic types are largely consistent with the classic projection-based divisions at a coarse granularity, but they can also resolve cellular differences beyond the classic divisions. This higher resolution in distinguishing cell populations makes them an appealing foundation for discussing detailed gene–projection relationships. Basing further discussions on transcriptomic types, however, does not imply whether transcriptomic types should or should not be treated as the "ground truth" for cardinal cell types (e.g., see Paul et al. 2017). Because transcriptomic types reflect differences in various neuronal properties, it is meaningful to ask how transcriptomic types correspond to projections regardless of whether they reflect cardinal cell types. Throughout this article, I use "cell type" or "neuronal type" when referring to the general concept of categorizing cell populations, and use "transcriptomic type" or "transcriptomically defined neuronal type" when specifically referring to populations defined based on gene expression profiles.

CORTICAL PROJECTIONS HAVE A COMPLEX RELATIONSHIP WITH GENE EXPRESSION PROFILES

Beyond these high-level divisions, the projection patterns of cortical neurons exhibit a remarkable degree of diversity and organization. Individual neurons project to a range of possible targets seemingly stochastically, and the probability of projections to each target varies across subpopulations of neurons. Moreover, many neurons project to multiple target areas, and the probability that a neuron coinnervates a set of target areas (i.e., its collateralization pattern) also varies depending on the combination of target areas. Several types of differences in collateralization have so far been observed among cortical neurons. First, collateralization patterns can be over-represented or underrepresented for neurons in a single area (Fig. 2A). For example, in the mouse primary visual cortex, neurons are more likely to project to both AM and PM, or both AL and LM, than expected by chance; in contrast, neurons that project to both PM and AL, or both PM and LM, are depleted (Han et al. 2018). Similar

preferences for collateralization involving specific sets of areas are also observed among other neuronal subpopulations and cortical areas (Chen et al. 2019; Sun et al. 2021), but the enriched or depleted collateralization patterns vary across different layers (Fig. 2B). For instance, in the superficial layers of the primary motor cortex, IT neurons that project to either the ipsilateral secondary motor cortex or the ipsilateral secondary somatosensory cortex are less likely to project to the contralateral primary motor cortex, whereas such a relationship is not observed in the middle layers (Muñoz-Castañeda et al. 2021). Second, some neurons send collaterals more broadly than other neurons (Fig. 2C). For instance, layer 5 IT neurons project to similar sets of target areas as IT neurons in layer 2/3, but a single layer 5 neuron usually projects to more target areas (Harris et al. 2019; Muñoz-Castañeda et al. 2021). Finally, the areas involved in collateralization patterns can themselves be connected in a way that corresponds to the collateralized inputs. For example, the main olfactory bulb and the olfactory cortex are wired in a way that results in parallel circuits across the olfaction system. Individual mitral cells in the main olfactory bulb send collaterals to both the olfactory cortex and subsets of extrapiriform olfactory areas. For each mitral cell, the projections to the olfactory cortex are concentrated at a certain anteroposterior location, and olfactory cortical neurons at those locations usually target the same subsets of extrapiriform olfactory areas that the mitral cell coinnervates (Fig. 2D; Chen et al. 2022c). This match in the collateral projections of the mitral cells and the projections of the olfactory cortex thus forms closed-loop parallel circuits, which were proposed to process distinct olfactory information (Chen et al. 2022c).

This remarkable diversity in projections is matched by a complex relationship with gene expression profiles. Recent studies focusing on understanding this relationship have mostly used three strategies: by anterograde axonal tracing from a genetically defined subpopulation of neurons, by interrogating gene expression profiles in retrogradely labeled cells, and by using barcoding-based, single-cell projection mapping techniques (Table 1).

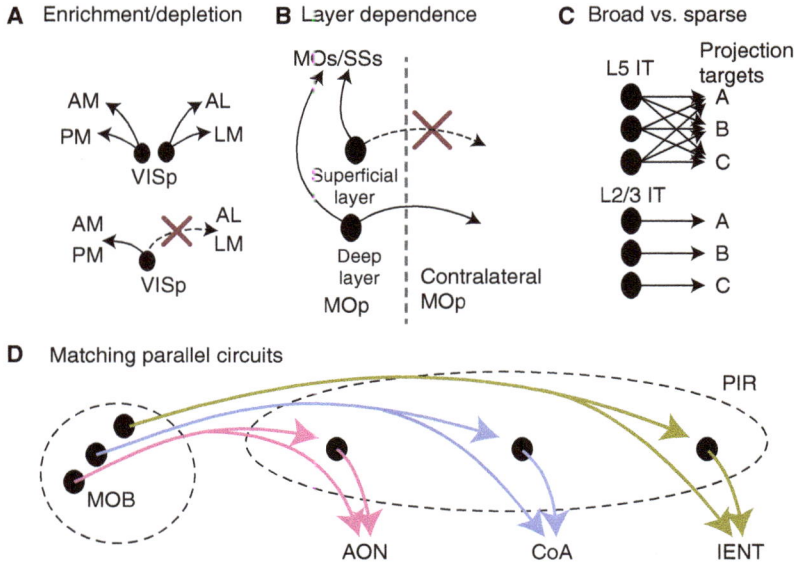

Figure 2. Projection collateralization is diverse across neuronal subpopulations. (*A*) Enriched (*top*) and depleted (*bottom*) projection collateralization patterns for single neurons in VISp, based on data in Han et al. (2018). (*B*) Preferences for collateralization patterns are distinct across neurons in different layers of MOp (Muñoz-Castañeda et al. 2021). (*C*) L5 intratelencephalic (IT) neurons and L2/3 IT neurons project to similar sets of target areas, but each L5 IT neuron projects more broadly (Harris et al. 2019; Muñoz-Castañeda et al. 2021). (*D*) Projections of mitral cells in the main olfactory bulb form parallel circuits with matching projections from the piriform cortex (Chen et al. 2022c). (VISp) Primary visual cortex, (AM, PM, AL, LM) higher visual areas, (MOp) primary motor cortex, (MOs) secondary motor cortex, (SSs) secondary somatosensory cortex, (MOB) main olfactory bulb, (PIR) piriform cortex, (AON) anterior olfactory nucleus, (CoA) cortical amygdala, (lENT) lateral entorhinal cortex.

In the anterograde tracing strategy (Fig. 3A), genetically defined subpopulations of neurons are labeled with a fluorescent protein, and their axonal patterns are determined by imaging and tracing throughout the brain either as a population (Harris et al. 2019; Muñoz-Castañeda et al. 2021) or individually (Winnubst et al. 2019; Peng et al. 2021; Gao et al. 2022). These studies consistently revealed several features of projections across transcriptomic subclasses of neurons, especially IT neurons. First, different subclasses of IT neurons from the same cortical area project to similar sets of potential targets, but the probability of projections to each area is different across subclasses. For example, L5 IT neurons are more likely to project to the striatum than other IT neurons (Shepherd 2013). Second, individual neurons project to only a subset of all targets that are possible for neurons of that transcriptomic type. Third, despite high heterogeneity in

their projection targets, individual IT neurons project largely symmetrically across the two hemispheres. Fourth, individual IT neurons project to different numbers of targets depending on transcriptomic subclasses. Finally, the laminar patterns of their axons are distinct across subclasses. These differences in the laminar patterns of axons are largely consistent with the divisions of feedforward and feedback projections and are used to define hierarchy across cortical areas (Felleman and Van Essen 1991; Harris et al. 2019; Peng et al. 2021). The anterograde tracing approach is usually limited in its ability to distinguish fine-grained neuronal types (Harris et al. 2019; Yao et al. 2023a), because existing genetic tools, such as Cre lines (He et al. 2016; Daigle et al. 2018; Matho et al. 2021), enhancer AAVs (Mich et al. 2021), and ADAR (adenosine deaminases acting on RNA)-based labeling tools (Qian et al. 2022), can often only label transcrip-

Table 1. Summary of strategies for associating gene expression profiles with projections

	Genetic labeling-based anterograde tracing		Retro-seq	Barcoded neuroanatomy
	Bulk	Single cell		
Cellular resolution		Single cell	Single cell	Single cell
Mapping resolution	High (determined by imaging)		Low (limited by injection)	Low (dissection-based) to high (in situ sequencing-based)
Gene-resolving power	Low (limited by genetic tools)		Medium (spatial transcriptomics) to high (single-cell RNA-seq)	None (bulk RNA-seq) to medium (in situ sequencing) to high (single-cell RNA-seq)
Scalability constraint	Highly scalable (limited only by the number of animals)	Sparse labeling and tracing	Imaging and/or sequencing throughput	In situ sequencing/single-cell RNA-seq throughput (excluding MAPseq), and/or number of target areas sampled
Main limitations	Low resolution in gene expression, no cellular resolution	Low resolution in gene expression, limited number of cells, difficult to apply to species with large brains	Can only map projections to a small number of targets, difficult to examine collateralizations	Mapping resolution is limited by dissection, in situ sequencing of whole axonal patterns is time consuming
Advantages	Highly scalable; easy to establish in a laboratory, can label cells by developmental gene expression profiles	Gold-standard axon morphology, high spatial resolution, can label cells by developmental gene expression profiles	Direct association with detailed gene expression profiles, epi-retro-seq can reveal epigenomic properties	Direct association to gene expression profiles in single cells (in situ), scalable to species with large brains, cost-effective, ideal for revealing collateralization patterns
Select references	Hunnicutf et al. 2014; Oh et al. 2014; Zingg et al. 2014; Harris et al. 2019; Muñoz-Matho et al. 2021; Muñoz-Castañeda et al. 2021	Winnubst et al. 2019; Muñoz-Castañeda et al. 2021; Peng et al. 2021; Gao et al. 2022	Tasic et al. 2018; Zhang et al. 2021a,b	Zador et al. 2012; Kebschull et al. 2016; Han et al. 2018; Chen et al. 2019; Klingler et al. 2021; Sun et al. 2021; Yuan et al. 2023

tomic types of cortical excitatory neurons at a relatively coarse granularity.

Performing single-cell RNA-seq (Chevée et al. 2018; Economo et al. 2018; Tasic et al. 2018; Kim et al. 2020; Peng et al. 2021; Zhao et al. 2022) or applying spatial transcriptomic techniques (Zhang et al. 2021a) on retrogradely labeled neurons (i.e., retro-seq) can resolve neuronal subpopulations beyond the limited transcriptomic resolution of genetic tools (Fig. 3B). Because these transcriptomics-based approaches can read out gene expression profiles more comprehensively, studies using these approaches revealed a detailed gene–projection relationship that is distinct across IT, L5 ET, and L6 CT neurons. Economo et al. (2018) found two subpopulations of L5 ET neurons in the anterior lateral motor cortex, one projecting to the thalamus and the other to the medulla. By combining retrograde tracing, single-cell tracing, and single-cell RNA-seq, this study found that these two subpopulations of neurons occupy different sublaminar positions in L5b and correspond to distinct transcriptomic types. Similar subpopulations of L5 ET neurons were also found in the primary motor cortex (Muñoz-Castañeda et al. 2021; Peng et al. 2021; Zhang et al. 2021a). Therefore, L5 ET neurons likely consist of multiple transcriptomic types with distinct projections. In contrast, both L6 CT neurons (Chevée et al.

2018) and IT neurons exhibit a more complex relationship between gene expression profiles and projections. Tasic et al. (2018) and Zhang et al. (2021a) found that the probabilities of projections to different target areas vary across transcriptomic types of IT neurons, but the correspondence between projections and transcriptomic types is not one-to-one. For example, all transcriptomic types of IT neurons in the primary motor cortex can project to the somatosensory cortex, but the fraction of neurons that do so varies widely across types (Zhang et al. 2021a). This lack of clear correspondence is unlikely to be due to suboptimal clustering when defining the transcriptomic types. Kim et al. (2020) found more than 800 differentially expressed genes between neurons from the primary visual cortex that project to either higher visual area AL or PM, but the gene expression profiles of these two populations of neurons, which were defined by projections and not by clustering their gene expression profiles, still overlapped considerably. Interestingly, Zhang et al. (2021b) conducted single-nucleus methylome sequencing on retrogradely labeled IT neurons and found that the DNA methylation profiles can predict the projection targets of neurons more precisely. Because the methylation patterns of DNA reflect both current gene expression (Mo et al. 2015; Luo et al. 2017) and the developmental his-

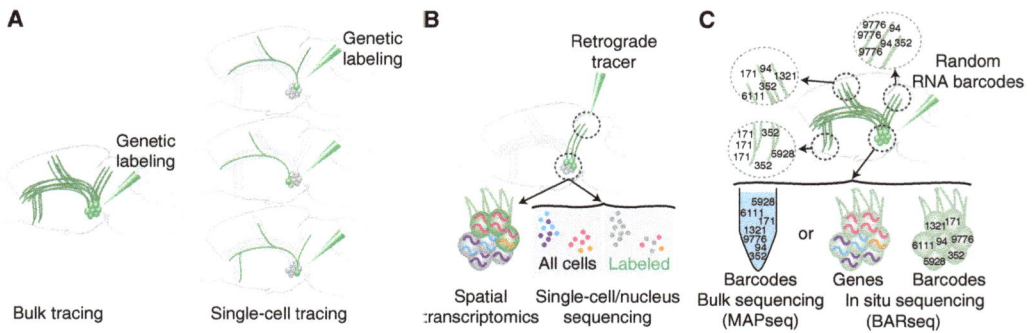

Figure 3. Three strategies for associating gene expression profiles with projections. (*A*) In anterograde tracing, neurons are genetically labeled, and the whole population (*left*) or individual neurons (*right*) are imaged and traced. Tracing of individual neurons requires sparse labeling. (*B*) Neurons are retrogradely labeled by a tracer injected in a projection target area. The gene expression profiles of the labeled neurons are either interrogated in situ using various spatial transcriptomic techniques or by single-cell/single-nucleus RNA-seq. (*C*) In barcoding-based approaches, neurons are labeled with random RNA barcodes. Barcodes and genes in the somas are sequenced in situ in BARseq, whereas only barcodes are sequenced in bulk in MAPseq. Matching barcodes in axons to those in somas reveals the projection patterns of individual neurons.

tory of the cell (Lister et al. 2013), the association between the methylome and projections hints at a strong link between developmental history and projections. Although these retrograde labeling-based studies revealed gene expression differences that are associated with projections to individual targets, these targets are usually part of enriched and/or depleted collateral patterns (e.g., coinnervation of AL and PM, two areas labeled in Kim et al. [2020], are depleted based on Han et al. [2018]). Thus, differences in projections to these individual targets likely reflect broader changes in projection collateralization.

Because retrograde labeling-based approaches only provide a binary and probabilistic readout of the uptake of tracers that are injected into target areas, it can only reveal projections to one or a small number of target areas at a time (but also see Zhao et al. 2022). Thus, it is difficult to use these approaches to interrogate collateralization patterns. Barcoding-based neuroanatomical techniques can overcome this limitation by mapping projections to many areas in large numbers of neurons. In two common implementations of this strategy, MAPseq and BARseq (Fig. 3C), random RNA barcodes are encoded in a Sindbis virus genome and uniquely label each neuron. As the barcode RNAs replicate, some are transported to the axons by an engineered carrier protein. Sequencing and matching barcodes in the axons to those in the somas then reveal the projection patterns of neurons. Because neurons are uniquely identified by RNA barcodes, which are read out by high-throughput sequencing, barcoding-based approaches can be applied to map brain-wide projections of tens of thousands of neurons altogether (Huang et al. 2020). The first-generation technique using this concept, MAPseq, uses tissue dissection followed by bulk RNA sequencing to read out the barcodes. Later variations use in situ sequencing (BARseq) (Chen et al. 2019; Sun et al. 2021) or single-cell RNA-seq (Klingler et al. 2021) to read out barcodes in the somas. Because both in situ sequencing and single-cell RNA-seq can also read out gene expression profiles in the same cells, these approaches can associate projections with gene expression at cellular resolution. For example, Chen et al. (2019) used a combination of laminar positions and a small number of cell-type markers to distinguish IT neurons at the subclass level and found that L6 IT neurons have different preferences of collateralization compared to IT neurons in other layers. Sun et al. (2021) used in situ sequencing to associate collateral patterns with the expression of classical cadherins and found that IT neurons in the auditory cortex project to similar sets of areas as IT neurons in the motor cortex if they express similar sets of cadherins. In situ sequencing cannot only read out gene expression profiles and barcodes in somas, but also barcodes in axons so that projections can be resolved at a higher spatial resolution (Yuan et al. 2023). Klingler et al. (2021) used single-cell RNA-seq to associate gene expression profiles with projections from the primary somatosensory cortex and found that the timing of SOX11 expression determines whether neurons project to the motor cortex or secondary somatosensory cortex. Thus, these approaches can reveal an association between gene expression profiles and both projections to individual target areas and collateralization patterns in adult neurons and/or in development. Barcoding combined with in situ sequencing has also been recently adapted for mapping synaptic connectivity (Zhang et al. 2023a) using barcoded rabies virus (Clark et al. 2021; Saunders et al. 2022). Because in situ sequencing is sufficiently fast and cost-effective for brain-wide interrogation (Chen et al. 2022b), this approach provides a promising path to comprehensively map the complex connectivity patterns of transcriptomic types of neurons on a brain-wide scale.

To summarize, beyond the high-level divisions of IT/L5 ET/L6 CT neurons, projections are associated with gene expression profiles in a complex manner. Differences in projections can correspond to either transcriptomic types (L5 ET neurons) or to the expression of sets of genes that do not separate neurons into distinct types (IT and L6 CT neurons). This association can involve not only projections to individual target areas, but also their collateralization patterns. Because collateralization cannot be revealed by interrogating projections to individual areas, achieving a full understanding of the complex relationship between gene expression profiles and projections requires comprehensive mapping of whole axonal patterns of neurons and their gene expression profiles at single-cell resolution.

INFERRING THE RELATIONSHIP BETWEEN PROJECTIONS AND GENE EXPRESSION PROFILES THROUGH UNDERLYING BIOLOGICAL PROCESSES

Because different transcriptomic types of neurons are associated with projections to varying degrees, transcriptomic types are likely oversimplified to fully capture the complex relationship between gene expression and projections. The link between gene expression and projections is indirect. The transcription of any particular gene is unlikely to directly bias the targeting of axons, but its protein product may be part of cellular processes that have a direct impact on projections. These cellular processes include, for example, signaling pathways that mediate axonal guidance, growth, and/or pruning. Thus, a better understanding of how gene expression relates to projections requires finding features in gene expression profiles and projections that reflect the underlying cellular processes. Here, I argue that many biological processes are reflected in sets of coexpressed genes, but not necessarily transcriptomic types. I will explore a similar approach for projections and discuss that projection collateralization, which is analogous to gene coexpression, reflects biological constraints on projections. Gene coexpression and projection collateralization, rather than type-based classification, can be thought of as the building blocks of neuronal identity.

Ideally, a neuronal type is defined by many cellular properties such as morphology, electrophysiological properties, developmental history, gene expression, and connectivity. Thus, transcriptomic types, which are defined solely by gene expression profiles, are not guaranteed to reflect differences in other neuronal properties. In practice, however, transcriptomic types show remarkable association with many neuronal properties (Cadwell et al. 2016; Paul et al. 2017; Economo et al. 2018; Tasic et al. 2018; Gouwens et al. 2020; Berg et al. 2021; Kalmbach et al. 2021;

Scala et al. 2021; Zeng 2022). The exact nature of this association still requires further exploration, but it may be partially explained by how gene expression specifies and maintains cell identities (Fig. 4A; Hobert 2008). During development, cell identities are specified and maintained through the continuous expression of transcription factors, named "terminal selector genes" (Hobert 2008). Many terminal selector genes function in combinations (Reilly et al. 2020; Özel et al. 2022) to sustain both their own expression (Way and Chalfie 1989; Mitani et al. 1993; Wu et al. 2001; Wenick and Hobert 2004; Etchberger et al. 2007; Topalidou and Chalfie 2011; Doitsidou et al. 2013; Zheng et al. 2015; Galazo et al. 2016; Zheng and Chalfie 2016) and to drive the expression of many "effector genes" that are needed for the functions of specialized cell types (Chalfie et al. 1981; Way and Chalfie 1988; Finney and Ruvkun 1990; Wenick and Hobert 2004; Etchberger et al. 2007). Terminal selector genes also suppress competing cell fates by inhibiting the expression of other transcription factors (Custo Greig et al. 2013; Gordon and Hobert 2015; Tsyporin et al. 2021; Galazo et al. 2022). Thus, the identity of a cell is determined by the binary and self-sustained expression of terminal selector genes. Phenotypically, however, the identity of a cell is reflected in a collection of cellular properties, and these cellular properties are associated with the coexpression of effector genes.

Across cell types and species, the coexpression of genes is remarkably well-conserved (Harris et al. 2021; Crow et al. 2022). One factor that contributes to this conservation is that the protein products of coexpressed genes usually function synergistically. For example, many ion channels (e.g., DEG/ENaC channels) (Árnadóttir and Chalfie 2010) and signaling pathways (e.g., Wnt signaling) (Komiya and Habas 2008) require multiple subunits and/or components to function, so disrupting the expression of any component would render the remaining expressed genes nonfunctional. Thus, the conservation of coexpression is based on the functional relationship of their protein products, not by the terminal selector genes. In fact, the relationship between cell identity and coexpressed effector genes is less constrained both in evolution and across cell types.

Cite this article as *Cold Spring Harb Perspect Biol* doi: 10.1101/cshperspect.a041509

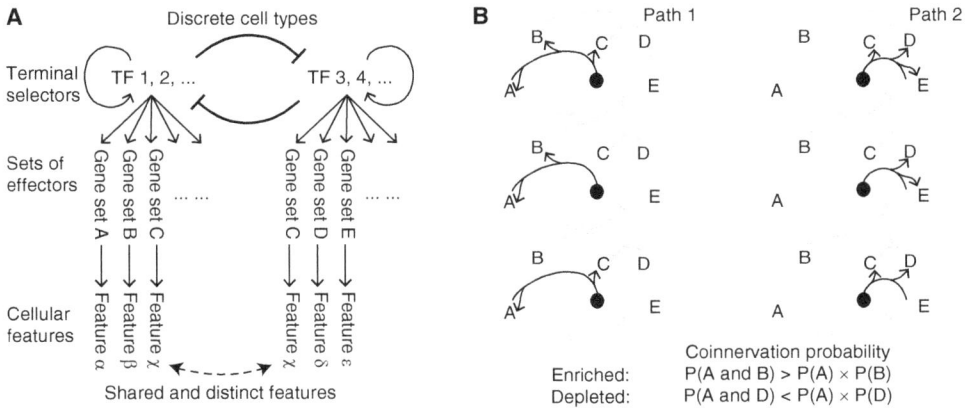

Figure 4. Gene coexpression and projection coinnervation patterns reflect underlying biological processes. (*A*) Terminal selector genes (Hobert 2008) sustain their own expression and suppress alternative sets of terminal selectors to maintain stable cell identity. Terminal selector genes drive the expression of sets of effector genes, and each coexpressed set of genes is needed for different cellular features. Coexpressed genes and corresponding features can be shared across cell types. (TF) Transcription factor. (*B*) Axonal paths determine potential coinnervation patterns. Two sets of neurons send their axons along two paths that pass by different sets of potential target areas. Although the exact targets a neuron innervates vary across neurons, areas along the same path are more likely to be coinnervated by the same neuron. This coinnervation leads to a higher probability of projections to sets of targets along the same path and a lower probability of projections to areas along different paths.

Most genomic regions that show significant evolutionary changes in humans (Pollard et al. 2006; Prabhakar et al. 2006) contain gene regulatory elements (Capra et al. 2013), suggesting that the regulatory logic of gene expression is highly variable across species. In addition, whereas cell types are, by definition, mutually exclusive, the same coexpression patterns of genes can be seen across multiple cell types (Fig. 4A; Harris et al. 2021). Thus, although the expression of effector genes is mechanistically driven by transcription factors such as terminal selector genes, their coexpression is kept stable in evolution and across cell types by the related functions of their protein products. In a simple analogy, gene coexpression can be thought of as the basic building blocks of the transcriptomic identities of neurons. In this analogy, the terminal selector genes dictate a list of building blocks—that is, effector genes—a cell has, and transcriptomic types are common combinations of building blocks. This analogy by no means diminishes the importance of cell types in understanding complex biological systems; rather, it describes an organization at the molecular level beyond cell types.

Understanding the organization of projections requires overcoming some similar challenges to understanding gene expression. Like gene expression profiles, projections are highly diverse and seemingly probabilistic. Thus, identifying features of projections that reflect underlying biological constraints beyond the apparent stochasticity can be extremely useful. The stochasticity in projections likely reflects both technical noise in measurements and biological variability (see next section), but many physical and molecular factors affect which target areas a neuron can innervate. For example, an axon can only project to target areas along its path. Although the axonal path does not guarantee projections to any single target area, it increases the likelihood that areas along the path are coinnervated and thereby determines preferred collateralization patterns (Fig. 4B). Although short-range horizontal projections of cortical neurons to neighboring areas appear to take a wide range of paths, axons that go through major fiber tracts take highly stereotypical paths as seen in single-cell tracing experiments (Winnubst et al. 2019). Because these stereotypical axonal paths are subject to cellular processes that achieve stringent developmental control (Custo Greig et al. 2013; Lodato and Arlotta 2015), the axonal paths are likely associated with coexpressed genes that are involved in the specification of projections during development. The collateralization patterns are biased not only by the axonal paths, but also by other developmental mechanisms that could be considered stochastic in nature. As a toy example, consider projections to two target areas, A and B, which go through activity-dependent competition and elimination, so that mature neurons keep only one of the two projections. In this case, although whether a neuron projects to either A or B would appear to be random, the coinnervation of both A and B would be depleted. Thus, collateralization patterns can be associated with developmental constraints that may not result in the specificity of projections to individual target areas.

Collateralization patterns are also similar to gene coexpression in that a neuron may contain a combination of collateralization patterns. For example, in a simplified view, IT neurons in the somatosensory cortex may be considered as having subsets of projections to the following areas: a callosal projection targeting multiple areas on the contralateral cortex, a striatal projection that targets the striatum bilaterally, a rostral projection to sets of frontal areas, and/or a lateral projection to higher somatosensory areas and lateral association areas. Thus, analogous to how gene coexpression can be thought of as the building blocks of the transcriptomic identities of cells, the collateralization patterns are the building blocks of projections. Because both projection and transcriptomic building blocks can reflect cellular and developmental processes, associating collateralization and coexpressed genes may reveal these underlying relationships that are obscured at the transcriptomic type level (Kim et al. 2020; Sun et al. 2021).

A DEVELOPMENTAL PERSPECTIVE CAN UNRAVEL GENE–PROJECTION RELATIONSHIP

A key challenge in unraveling the relationship between gene expression profiles and projections is that projections are specified during development and are usually not affected by gene expression in the adult. This is different from many other neuronal properties, which require sustained gene expression in adult neurons to maintain. In many species, gene expression profiles of neurons and glial cells go through an hourglass-shaped transformation in development: transcriptomic identities first converge, then diversify again in later developmental stages, and this secondary diversification can occur through differential expression of sets of genes that do not distinguish neuronal populations before convergence (Fig. 5A; Li et al. 2018b; Zeisel et al. 2018; Zhu et al. 2018; Di Bella et al. 2021). Thus, the expression of genes that are associated with distinct projections can be obscured in adult (Fig. 5Ba; Li et al. 2017). Many cell surface molecules, which play critical roles in specifying projections, are reused for other developmental processes and ultimately expressed in a different population of neurons (e.g., cadherins [Hayano et al. 2014; Killen et al. 2017; Sun et al. 2021] and plexins [Li et al. 2018a; Guajardo et al. 2019]). The timing of gene expression itself can also determine projection targets (Fig. 5Bb; Klingler et al. 2021). In addition to these genetically determined differences in projections, cortical projections are also massively reshaped by pruning of exuberant projections (Fig. 5Bc). For example, L5 ET neurons in different areas may or may not project to the spinal cord, and this difference is achieved by pruning the corticospinal axon (Stanfield et al. 1982; O'Leary and Stanfield 1985, 1986; Stanfield and O'Leary 1985a,b; O'Leary 1987). In IT neurons, axons initially target broad regions in the cortex, but the targeting is refined substantially during postnatal development (Innocenti 1981; Price and Blakemore 1985; Webster et al. 1991; Assal and Innocenti 1993; Aggoun-Zouaoui and Innocenti 1994; Houzel et al. 1994; Callaway 1998). Because this pruning depends on neuronal activity and axon–axon competition instead of cell-autonomous genetic programs (Innocenti and Price 2005), the resulting projection patterns are unlikely to be associated with gene expression profiles even during development. Thus, the challenges brought about by the developmental

Cite this article as *Cold Spring Harb Perspect Biol* doi: 10.1101/cshperspect.a041509

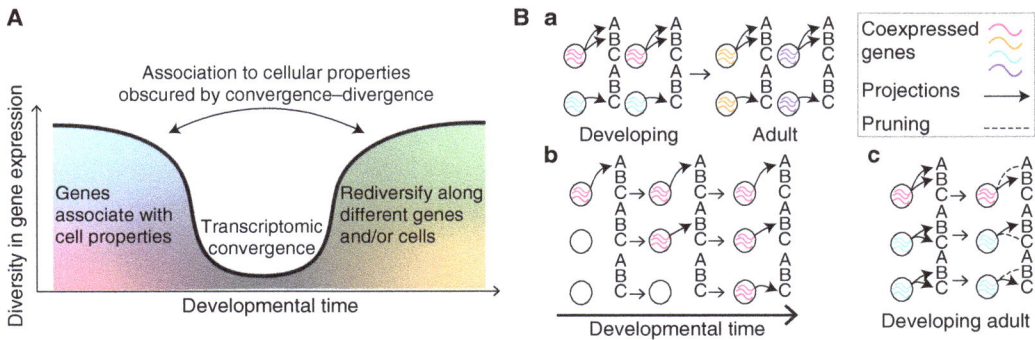

Figure 5. Multiple developmental processes can obscure the association between projections and gene expression profiles. (*A*) Gene expression profiles of neurons go through an hourglass-shaped dynamic in development. In many cell types, circuits, and species, gene expression profiles converge, and then diversify again later in development. For some cell types, this secondary diversification occurs across different cell populations from those that can be distinguished before the convergence. (*B*) Developmental changes in gene expression profiles and/or projections can obscure the gene–projection relationship in adult neurons. (*a*) Gene expression profiles can converge and/or diverge during development, and genes that associate with projections during development are absent in adult neurons. (*b*) The timing of gene expression is associated with different projections. (*c*) Projections can be pruned extensively, and this process obscures the gene–projection relationship when the projections are specified. Colored squiggles indicate coexpressed genes. Arrows indicate projections, and dashed lines indicate pruned projections.

processes of projections involve at least two issues: The relationship between gene expression profiles and projections can be transient, and gene expression only partially determines the target areas a neuron projects to.

Thus, to understand the relationship between gene expression profiles and projections, it is important to capture the transient association between them and to distinguish features of projections that are determined by genetic and nongenetic processes. A coexpression and collateralization-based approach can be easily applied to uncover developmental relationships. Several developmental questions will be of interest to understanding the gene–projection relationship. For example, what features of projections are determined by the expression of specific genes? Identifying enriched and/or depleted collateralization patterns that strongly associate with gene expression, followed by developmental manipulation to establish causality, may provide a comprehensive view of the extent to which genetic programs determine projections. How do gene expression profiles progress after projections are specified? Determining the trajectory of gene expression can provide a link

between collateralization patterns that are formed in development and gene coexpression in adult neurons. And for projection features that are not determined solely by gene expression during development, how do nongenetic components shape these features? A wide range of techniques can be used to generate complementary insights into these questions. Comprehensively interrogating both gene expression profiles and projections in developing neurons can provide a wholistic view of the relationship between the two. Creating such a developmental map of gene expression and projections at densely sampled developmental time points will likely use scalable single-cell/single-nucleus RNA-seq techniques (Rosenberg et al. 2018; Hagemann-Jensen et al. 2022; Yao et al. 2023b; Zhang et al. 2023a), spatial transcriptomic techniques (Sun et al. 2021; Bugeon et al. 2022; Chen et al. 2022b; Borm et al. 2023; Zhang et al. 2023b), barcoded neuroanatomical techniques (Chen et al. 2019; Huang et al. 2020; Sun et al. 2021; Zhang et al. 2023a), and sequencing-based lineage tracing (Frieda et al. 2017; Kalhor et al. 2018; Raj et al. 2018; Bandler et al. 2022; Fang et al. 2022). This systematic but correlational approach can be complemented by perturbation studies using either conventional developmental manipulations or multiplexed approaches (e.g., Perturb-seq) (Jin et al. 2020), which can establish causal relationships between key sets of coexpressed genes and projections. Techniques that directly associate gene expression profiles and projections with neuronal activity can provide additional information on biologically relevant variations in projections that are not determined solely by gene expression. Toward this goal, spatial transcriptomics is emerging as a powerful tool to directly associate gene expression profiles with neuronal activities measured by two-photon calcium imaging (Bugeon et al. 2022; Condylis et al. 2022). A similar approach based on in situ sequencing and barcoding has the potential to associate projections with neuronal activity more comprehensively than previous approaches based on retrograde mapping (Movshon and Newsome 1996; Glickfeld et al. 2013; Znamenskiy and Zador 2013). Finally, because gene correlates of projections are most likely defined by the coexpression of many genes rather than the binary expression of single markers, intersectional genetic tools (Fenno et al. 2014, 2020; Madisen et al. 2015; Ren et al. 2019; Sabatini et al. 2021; Chen et al. 2022a; Pouchelon et al. 2022; Hughes et al. 2023) may be helpful in gaining genetic access to neuronal subpopulations that are defined by projections.

CONCLUDING REMARKS

Understanding the logic of projections is key to establishing comprehensive connectivity maps of brain circuits. Current technology is rapidly advancing to enable the interrogation of gene expression profiles and projections at cellular resolution on a brain-wide scale in mice. Moving forward, our understanding of the logic of projections will likely be limited less by how much data are available, but more by how the gene–projection relationship is conceptualized. Essentially, to effectively use gene expression profiles as a "Rosetta Stone" to associate the "text" of projections with the "text" of other neuronal properties, we need to first understand how the "text" is organized on the Rosetta Stone.

I proposed to describe each neuron as a combination of gene expression features and projection features, that is, a combination of "building blocks" rather than a member of a category (i.e., transcriptomic type). Because coexpressed genes may be associated with underlying cellular processes that constrain the collateralization patterns of projections, finding these building blocks (i.e., coexpressed genes and projection collateralization) makes it straightforward to unravel their relationship. Furthermore, building blocks can also describe transient interactions during development, a major challenge in understanding the gene–projection relationship. Similar decomposition-based approaches are also widely and successfully used in understanding other complex biological data, such as neuronal activity. Although generating a building block list is different from assigning each cell to a type, cell types themselves are a means to decompose complex biological systems (e.g., the brain) into basic building blocks (Zeng 2022).

A developmental perspective cannot only uncover transient relationships that are difficult to observe in adult neurons, but also provide insights into how developmental constraints shape circuit evolution. This latter question is the foundation of modern developmental evolutionary biology (Wagner 2014) and has further implications for understanding how variability in connectivity is generated and the range of expected variabilities across individuals. A developmental approach can benefit from multiple strategies, including systematic cell atlasing using high-throughput approaches, mechanistic investigation through perturbations, developing new multimodal techniques, and designing genetic tools to enable future studies. The key to success relies not only on new technology development and innovative experiments, but also on open data access and effective knowledge sharing among laboratories with complementary expertise. This exciting collaborative effort will hopefully not only construct brain-wide connectivity maps, but also pave the way for achieving a mechanistic understanding of circuit functions by both generating mechanistic circuit models based on the connectivity and providing tools for accessing and manipulating neuronal subpopulations to test such models.

ACKNOWLEDGMENTS

Many ideas in this paper emerged through discussions with Anthony Zador and members of the Zador laboratory. I also thank Jesse Gillis, Gregory Horwitz, and Yu-Chi Sun for valuable comments on the paper, and thank Mara C.P. Rue, Maryam Majeed, Aixin Zhang, Hongkui Zeng, and others at the Allen Institute for discussion.

REFERENCES

Aggoun-Zouaoui D, Innocenti GM. 1994. Juvenile visual callosal axons in kittens display origin- and fate-related morphology and distribution of arbors. *Eur J Neurosci* **6:** 1846–1863. doi:10.1111/j.1460-9568.1994.tb00577.x

Árnadóttir J, Chalfie M. 2010. Eukaryotic mechanosensitive channels. *Annu Rev Biophys* **39:** 111–137. doi:10.1146/annurev.biophys.37.032807.125836

Assal F, Innocenti GM. 1993. Transient intra-areal axons in developing cat visual cortex. *Cereb Cortex* **3:** 290–303. doi:10.1093/cercor/3.4.290

Bandler RC, Vitali I, Delgado RN, Ho MC, Dvoretskova E, Ibarra Molinas JS, Frazel PW, Mohammadkhani M, Machold R, Maedler S, et al. 2022. Single-cell delineation of lineage and genetic identity in the mouse brain. *Nature* **601:** 404–409. doi:10.1038/s41586-021-04237-0

Berg J, Sorensen SA, Ting JT, Miller JA, Chartrand T, Buchin A, Bakken TE, Budzillo A, Dee N, Ding SL, et al. 2021. Human neocortical expansion involves glutamatergic neuron diversification. *Nature* **598:** 151–158. doi:10.1038/s41586-021-03813-8

Borm LE, Mossi Albiach A, Mannens CCA, Janusauskas J, Ozgun C, Fernandez-Garcia D, Hodge R, Castillo F, Hedin CRH, Villablanca EJ, et al. 2023. Scalable in situ single-cell profiling by electrophoretic capture of mRNA using EEL FISH. *Nat Biotechnol* **41:** 222–231.

Bugeon S, Duffield J, Dipoppa M, Ritoux A, Prankerd I, Nicoloutsopoulos D, Orme D, Shinn M, Peng H, Forrest H, et al. 2022. A transcriptomic axis predicts state modulation of cortical interneurons. *Nature* **607:** 330–338. doi:10.1038/s41586-022-04915-7

Cadwell CR, Palasantza A, Jiang X, Berens P, Deng Q, Yilmaz M, Reimer J, Shen S, Bethge M, Tolias KF, et al. 2016. Electrophysiological, transcriptomic and morphologic profiling of single neurons using Patch-seq. *Nat Biotechnol* **34:** 199–203. doi:10.1038/nbt.3445

Callaway EM. 1998. Prenatal development of layer-specific local circuits in primary visual cortex of the macaque monkey. *J Neurosci* **18:** 1505–1527. doi:10.1523/JNEUROSCI.18-04-01505.1998

Capra JA, Erwin GD, McKinsey G, Rubenstein JL, Pollard KS. 2013. Many human accelerated regions are developmental enhancers. *Philos Trans R Soc Lond B Biol Sci* **368:** 20130025. doi:10.1098/rstb.2013.0025

Chalfie M, Horvitz HR, Sulston JE. 1981. Mutations that lead to reiterations in the cell lineages of *C. elegans*. *Cell* **24:** 59–69. doi:10.1016/0092-8674(81)90501-8

Chalfie M, Sulston JE, White JG, Southgate E, Thomson JN, Brenner S. 1985. The neural circuit for touch sensitivity in *Caenorhabditis elegans*. *J Neurosci* **5:** 956–964. doi:10.1523/JNEUROSCI.05-04-00956.1985

Chen X, Sun Y, Zhan H, Kebschull JM, Fischer S, Matho K, Huang ZJ, Gillis J, Zador AM. 2019. High-throughput mapping of long-range neuronal projection using in situ sequencing. *Cell* **179:** 772–786.e19. doi:10.1016/j.cell.2019.09.023

Chen HS, Zhang XL, Yang RR, Wang GL, Zhu XY, Xu YF, Wang DY, Zhang N, Qiu S, Zhan LJ, et al. 2022a. An intein-split transactivator for intersectional neural imaging and optogenetic manipulation. *Nat Commun* **13:** 3605. doi:10.1038/s41467-022-31255-x

Chen X, Fischer S, Zhang A, Gillis J, Zador AM. 2022b. Modular cell type organization of cortical areas revealed by in situ sequencing. bioRxiv doi:10.1101/2022.11.06.515380

Chen Y, Chen X, Baserdem B, Zhan H, Li Y, Davis MB, Kebschull JM, Zador AM, Koulakov AA, Albeanu DF. 2022c. High-throughput sequencing of single neuron projections reveals spatial organization in the olfactory cortex. *Cell* **185:** 4117–4134.e28. doi:10.1016/j.cell.2022.09.038

Chevée M, Robertson JJ, Cannon GH, Brown SP, Goff LA. 2018. Variation in activity state, axonal projection, and position define the transcriptional identity of individual neocortical projection neurons. *Cell Rep* **22:** 441–455. doi:10.1016/j.celrep.2017.12.046

Cho KKA, Shi J, Phensy AJ, Turner ML, Sohal VS. 2023. Long-range inhibition synchronizes and updates prefrontal task activity. *Nature* **617:** 548–554. doi:10.1038/s41586-023-06012-9

Chun JJ, Shatz CJ. 1989. Interstitial cells of the adult neocortical white matter are the remnant of the early generated subplate neuron population. *J Comp Neurol* **282:** 555–569. doi:10.1002/cne.902820407

Clark IC, Gutiérrez-Vázquez C, Wheeler MA, Li Z, Rothhammer V, Linnerbauer M, Sanmarco LM, Guo L, Blain M, Zandee SEJ, et al. 2021. Barcoded viral tracing of single-cell interactions in central nervous system inflammation. *Science* **372:** eabf1230. doi:10.1126/science.abf1230

Condylis C, Ghanbari A, Manjrekar N, Bistrong K, Yao S, Yao Z, Nguyen TN, Zeng H, Tasic B, Chen JL. 2022. Dense functional and molecular readout of a circuit hub in sensory cortex. *Science* **375:** eabl5981. doi:10.1126/science.abl5981

Crow M, Suresh H, Lee J, Gillis J. 2022. Coexpression reveals conserved gene programs that co-vary with cell type across kingdoms. *Nucleic Acids Res* **50:** 4302–4314. doi:10.1093/nar/gkac276

Custo Greig LF, Woodworth MB, Galazo MJ, Padmanabhan H, Macklis JD. 2013. Molecular logic of neocortical projection neuron specification, development and diversity. *Nat Rev Neurosci* **14:** 755–769. doi:10.1038/nrn3586

Daigle TL, Madisen L, Hage TA, Valley MT, Knoblich U, Larsen RS, Takeno MM, Huang L, Gu H, Larsen R, et al. 2018. A suite of transgenic driver and reporter mouse lines with enhanced brain-cell-type targeting and functionality. *Cell* **174:** 465–480.e22. doi:10.1016/j.cell.2018.06.035

Di Bella DJ, Habibi E, Stickels RR, Scalia G, Brown J, Yadollahpour P, Yang SM, Abbate C, Biancalani T, Macosko EZ, et al. 2021. Molecular logic of cellular diversification in the mouse cerebral cortex. *Nature* **595:** 554–559. doi:10.1038/s41586-021-03670-5

Doitsidou M, Flames N, Topalidou I, Abe N, Felton T, Remesal L, Popovitchenko T, Mann R, Chalfie M, Hobert O. 2013. A combinatorial regulatory signature controls terminal differentiation of the dopaminergic nervous system in *C. elegans*. *Genes Dev* **27:** 1391–1405. doi:10.1101/gad.217224.113

Economo MN, Viswanathan S, Tasic B, Bas E, Winnubst J, Menon V, Graybuck LT, Nguyen TN, Smith KA, Yao Z, et al. 2018. Distinct descending motor cortex pathways and their roles in movement. *Nature* **563:** 79–84. doi:10.1038/s41586-018-0642-9

Etchberger JF, Lorch A, Sleumer MC, Zapf R, Jones SJ, Marra MA, Holt RA, Moerman DG, Hobert O. 2007. The molecular signature and *cis*-regulatory architecture of a *C. elegans* gustatory neuron. *Genes Dev* **21:** 1653–1674. doi:10.1101/gad.1560107

Fang W, Bell CM, Sapirstein A, Asami S, Leeper K, Zack DJ, Ji H, Kalhor R. 2022. Quantitative fate mapping: a general framework for analyzing progenitor state dynamics via retrospective lineage barcoding. *Cell* **185:** 4604–4620. e32. doi:10.1016/j.cell.2022.10.028

Felleman DJ, Van Essen DC. 1991. Distributed hierarchical processing in the primate cerebral cortex. *Cereb Cortex* **1:** 1–47. doi:10.1093/cercor/1.1.1

Fenno LE, Mattis J, Ramakrishnan C, Hyun M, Lee SY, He M, Tucciarone J, Selimbeyoglu A, Berndt A, Grosenick L, et al. 2014. Targeting cells with single vectors using multiple-feature Boolean logic. *Nat Methods* **11:** 763–772. doi:10.1038/nmeth.2996

Fenno LE, Ramakrishnan C, Kim YS, Evans KE, Lo M, Vesuna S, Inoue M, Cheung KYM, Yuen E, Pichamoorthy N, et al. 2020. Comprehensive dual- and triple-feature intersectional single-vector delivery of diverse functional payloads to cells of behaving mammals. *Neuron* **107:** 836–853. e11. doi:10.1016/j.neuron.2020.06.003

Finney M, Ruvkun G. 1990. The unc-86 gene product couples cell lineage and cell identity in *C. elegans*. *Cell* **63:** 895–905. doi:10.1016/0092-8674(90)90493-X

Frieda KL, Linton JM, Hormoz S, Choi J, Chow KK, Singer ZS, Budde MW, Elowitz MB, Cai L. 2017. Synthetic recording and in situ readout of lineage information in single cells. *Nature* **541:** 107–111. doi:10.1038/nature20777

Galazo MJ, Emsley JG, Macklis JD. 2016. Corticothalamic projection neuron development beyond subtype specification: Fog2 and intersectional controls regulate interclass neuronal diversity. *Neuron* **91:** 90–106. doi:10.1016/j.neuron.2016.05.024

Galazo MJ, Sweetser D, Macklis JD. 2022. *Tle4* controls both developmental acquisition and postnatal maintenance of corticothalamic projection neuron identity. bioRxiv doi:10.1101/2022.05.09.491192

Gao L, Liu S, Gou L, Hu Y, Liu Y, Deng L, Ma D, Wang H, Yang Q, Chen Z, et al. 2022. Single-neuron projectome of mouse prefrontal cortex. *Nat Neurosci* **25:** 515–529. doi:10.1038/s41593-022-01041-5

Glickfeld LL, Andermann ML, Bonin V, Reid RC. 2013. Cortico-cortical projections in mouse visual cortex are functionally target specific. *Nat Neurosci* **16:** 219–226. doi:10.1038/nn.3300

Gordon PM, Hobert O. 2015. A competition mechanism for a homeotic neuron identity transformation in *C. elegans*. *Dev Cell* **34:** 206–219. doi:10.1016/j.devcel.2015.04.023

Gouwens NW, Sorensen SA, Baftizadeh F, Budzillo A, Lee BR, Jarsky T, Alfiler L, Baker K, Barkan E, Berry K, et al. 2020. Integrated morphoelectric and transcriptomic classification of cortical GABAergic cells. *Cell* **183:** 935–953.e19. doi:10.1016/j.cell.2020.09.057

Guajardo R, Luginbuhl DJ, Han S, Luo L, Li J. 2019. Functional divergence of Plexin B structural motifs in distinct steps of *Drosophila* olfactory circuit assembly. *eLife* **8:** e48594. doi:10.7554/eLife.48594

Hagemann-Jensen M, Ziegenhain C, Sandberg R. 2022. Scalable single-cell RNA sequencing from full transcripts with Smart-seq3xpress. *Nat Biotechnol* **40:** 1452–1457. doi:10.1038/s41587-022-01311-4

Hain D, Gallego-Flores T, Klinkmann M, Macias A, Ciirdaeva E, Arends A, Thum C, Tushev G, Kretschmer F, Tosches MA, et al. 2022. Molecular diversity and evolution of neuron types in the amniote brain. *Science* **377:** eabp8202. doi:10.1126/science.abp8202

Han Y, Kebschull JM, Campbell RAA, Cowan D, Imhof F, Zador AM, Mrsic-Flogel TD. 2018. The logic of single-cell projections from visual cortex. *Nature* **556:** 51–56. doi:10.1038/nature26159

Harris KD, Shepherd GM. 2015. The neocortical circuit: themes and variations. *Nat Neurosci* **18:** 170–181. doi:10.1038/nn.3917

Harris JA, Mihalas S, Hirokawa KE, Whitesell JD, Choi H, Bernard A, Bohn P, Caldejon S, Casal L, Cho A, et al. 2019. Hierarchical organization of cortical and thalamic connectivity. *Nature* **575:** 195–202. doi:10.1038/s41586-019-1716-z

Harris BD, Crow M, Fischer S, Gillis J. 2021. Single-cell co-expression analysis reveals that transcriptional modules are shared across cell types in the brain. *Cell Syst* **12:** 748–756.e3. doi:10.1016/j.cels.2021.04.010

Hayano Y, Zhao H, Kobayashi H, Takeuchi K, Norioka S, Yamamoto N. 2014. The role of T-cadherin in axonal pathway formation in neocortical circuits. *Development* **141:** 4784–4793. doi:10.1242/dev.108290

He M, Tucciarone J, Lee S, Nigro MJ, Kim Y, Levine JM, Kelly SM, Krugikov I, Wu P, Chen Y, et al. 2016. Strategies and tools for combinatorial targeting of GABAergic neurons in mouse cerebral cortex. *Neuron* **91:** 1228–1243. doi:10.1016/j.neuron.2016.08.021

Higo S, Akashi K, Sakimura K, Tamamaki N. 2009. Subtypes of GABAergic neurons project axons in the neocortex. *Front Neuroanat* **3:** 25. doi:10.3389/neuro.05.025.2009

Hobert O. 2008. Regulatory logic of neuronal diversity: terminal selector genes and selector motifs. *Proc Natl Acad Sci* **105:** 20067–20071. doi:10.1073/pnas.0806070105

Houzel JC, Milleret C, Innocenti G. 1994. Morphology of callosal axons interconnecting areas 17 and 18 of the cat. *Eur J Neurosci* **6:** 898–917. doi:10.1111/j.1460-9568.1994.tb00585.x

Huang L, Kebschull JM, Fürth D, Musall S, Kaufman MT, Churchland AK, Zador AM. 2020. BRICseq bridges brain-wide interregional connectivity to neural activity and gene expression in single animals. *Cell* **182:** 177–188.e27. doi:10.1016/j.cell.2020.05.029

Cite this article as *Cold Spring Harb Perspect Biol* doi: 10.1101/cshperspect.a041509

Hughes AC, Pollard BG, Xu B, Gammons JW, Chapman P, Bikoff JB, Schwarz LA. 2023. A novel single vector intersectional AAV strategy for interrogating cellular diversity and brain function. bioRxiv doi:10.1101/2023.02.07.527312

Hunnicutt BJ, Long BR, Kusefoglu D, Gertz KJ, Zhong H, Mao T. 2014. A comprehensive thalamocortical projection map at the mesoscopic level. *Nat Neurosci* **17:** 1276–1285. doi:10.1038/nn.3780

Innocenti GM. 1981. Growth and reshaping of axons in the establishment of visual callosal connections. *Science* **212:** 824–827. doi:10.1126/science.7221566

Innocenti GM, Price DJ. 2005. Exuberance in the development of cortical networks. *Nat Rev Neurosci* **6:** 955–965. doi:10.1038/nrn1790

Jin X, Simmons SK, Guo A, Shetty AS, Ko M, Nguyen L, Jokhi V, Robinson E, Oyler P, Curry N, et al. 2020. In vivo Perturb-seq reveals neuronal and glial abnormalities associated with autism risk genes. *Science* **370:** eaaz6063. doi:10.1126/science.aaz6063

Kalhor R, Kalhor K, Mejia L, Leeper K, Graveline A, Mali P, Church GM. 2018. Developmental barcoding of whole mouse via homing CRISPR. *Science* **361:** eaat9804. doi:10.1126/science.aat9804

Kalmbach BE, Hodge RD, Jorstad NL, Owen S, de Frates R, Yanny AM, Dalley R, Mallory M, Graybuck LT, Radaelli C, et al. 2021. Signature morpho-electric, transcriptomic, and dendritic properties of human layer 5 neocortical pyramidal neurons. *Neuron* **109:** 2914–2927.e5. doi:10.1016/j.neuron.2021.08.030

Kanold PO, Luhmann HJ. 2010. The subplate and early cortical circuits. *Annu Rev Neurosci* **33:** 23–48. doi:10.1146/annurev-neuro-060909-153244

Kebschull JM, Garcia da Silva P, Reid AP, Peikon ID, Albeanu DF, Zador AM. 2016. High-throughput mapping of single-neuron projections by sequencing of barcoded RNA. *Neuron* **91:** 975–987. doi:10.1016/j.neuron.2016.07.036

Killen AC, Barber M, Paulin JJW, Ranscht B, Parnavelas JG, Andrews WD. 2017. Protective role of Cadherin 13 in interneuron development. *Brain Struct Funct* **222:** 3567–3585. doi:10.1007/s00429-017-1418-y

Kim EJ, Zhang Z, Huang L, Ito-Cole T, Jacobs MW, Juavinett AL, Senturk G, Hu M, Ku M, Ecker JR, et al. 2020. Extraction of distinct neuronal cell types from within a genetically continuous population. *Neuron* **107:** 274–282.e6. doi:10.1016/j.neuron.2020.04.018

Klingler E, Tomasello U, Prados J, Kebschull JM, Contestabile A, Galiñanes GL, Fièvre S, Santinha A, Platt R, Huber D, et al. 2021. Temporal controls over inter-areal cortical projection neuron fate diversity. *Nature* **599:** 453–457. doi:10.1038/s41586-021-04048-3

Komiya Y, Habas R. 2008. Wnt signal transduction pathways. *Organogenesis* **4:** 68–75. doi:10.4161/org.4.2.5851

Langlieb J, Sachdev NS, Balderrama KS, Nadaf NM, Raj M, Murray E, Webber JT, Vanderburg C, Gazestani V, Tward D, et al. 2023. The molecular cytoarchitecture of the adult mouse brain. *Nature* **624:** 333–342.

Lee AT, Vogt D, Rubenstein JL, Sohal VS. 2014. A class of GABAergic neurons in the prefrontal cortex sends long-range projections to the nucleus accumbens and elicits acute avoidance behavior. *J Neurosci* **34:** 11519–11525. doi:10.1523/JNEUROSCI.1157-14.2014

Li H, Horns F, Wu B, Xie Q, Li J, Li T, Luginbuhl DJ, Quake SR, Luo L. 2017. Classifying *Drosophila* olfactory projection neuron subtypes by single-cell RNA sequencing. *Cell* **171:** 1206–1220.e22. doi:10.1016/j.cell.2017.10.019

Li J, Guajardo R, Xu C, Wu B, Li H, Li T, Luginbuhl DJ, Xie X, Luo L. 2018a. Stepwise wiring of the *Drosophila* olfactory map requires specific Plexin B levels. *eLife* **7:** e39088. doi:10.7554/eLife.39088

Li M, Santpere G, Imamura Kawasawa Y, Evgrafov OV, Gulden FO, Pochareddy S, Sunkin SM, Li Z, Shin Y, Zhu Y, et al. 2018b. Integrative functional genomic analysis of human brain development and neuropsychiatric risks. *Science* **362:** eaat7615. doi:10.1126/science.aat7615

Lister R, Mukamel EA, Nery JR, Urich M, Puddifoot CA, Johnson ND, Lucero J, Huang Y, Dwork AJ, Schultz MD, et al. 2013. Global epigenomic reconfiguration during mammalian brain development. *Science* **341:** 1237905. doi:10.1126/science.1237905

Lodato S, Arlotta P. 2015. Generating neuronal diversity in the mammalian cerebral cortex. *Annu Rev Cell Dev Biol* **31:** 699–720. doi:10.1146/annurev-cellbio-100814-125353

Luo C, Keown CL, Kurihara L, Zhou J, He Y, Li J, Castanon R, Lucero J, Nery JR, Sandoval JP, et al. 2017. Single-cell methylomes identify neuronal subtypes and regulatory elements in mammalian cortex. *Science* **357:** 600–604. doi:10.1126/science.aan3351

Lust K, Maynard A, Gomes T, Fleck JS, Camp JG, Tanaka EM, Treutlein B. 2022. Single-cell analyses of axolotl telencephalon organization, neurogenesis, and regeneration. *Science* **377:** eabp9262. doi:10.1126/science.abp9262

Madisen L, Garner AR, Shimaoka D, Chuong AS, Klapoetke NC, Li L, van der Bourg A, Niino Y, Egolf L, Monetti C, et al. 2015. Transgenic mice for intersectional targeting of neural sensors and effectors with high specificity and performance. *Neuron* **85:** 942–958. doi:10.1016/j.neuron.2015.02.022

Matho KS, Huilgol D, Galbavy W, He M, Kim G, An X, Lu J, Wu P, Di Bella DJ, Shetty AS, et al. 2021. Genetic dissection of the glutamatergic neuron system in cerebral cortex. *Nature* **598:** 182–187. doi:10.1038/s41586-021-03955-9

Mich JK, Graybuck LT, Hess EE, Mahoney JT, Kojima Y, Ding Y, Somasundaram S, Miller JA, Kalmbach BE, Radaelli C, et al. 2021. Functional enhancer elements drive subclass-selective expression from mouse to primate neocortex. *Cell Rep* **34:** 108754. doi:10.1016/j.celrep.2021.108754

Mitani S, Du H, Hall DH, Driscoll M, Chalfie M. 1993. Combinatorial control of touch receptor neuron expression in *Caenorhabditis elegans*. *Development* **119:** 773–783. doi:10.1242/dev.119.3.773

Mo A, Mukamel EA, Davis FP, Luo C, Henry GL, Picard S, Urich MA, Nery JR, Sejnowski TJ, Lister R, et al. 2015. Epigenomic signatures of neuronal diversity in the mammalian brain. *Neuron* **86:** 1369–1384. doi:10.1016/j.neuron.2015.05.018

Molyneaux BJ, Arlotta P, Menezes JR, Macklis JD. 2007. Neuronal subtype specification in the cerebral cortex. *Nat Rev Neurosci* **8:** 427–437. doi:10.1038/nrn2151

Movshon JA, Newsome WT. 1996. Visual response properties of striate cortical neurons projecting to area MT in macaque monkeys. *J Neurosci* **16:** 7733–7741. doi:10.1523/JNEUROSCI.16-23-07733.1996

Muñoz-Castañeda R, Zingg B, Matho KS, Chen X, Wang Q, Foster NN, Li A, Narasimhan A, Hirokawa KE, Huo B, et al. 2021. Cellular anatomy of the mouse primary motor cortex. *Nature* **598:** 159–166. doi:10.1038/s41586-021-03970-w

Oh SW, Harris JA, Ng L, Winslow B, Cain N, Mihalas S, Wang Q, Lau C, Kuan L, Henry AM, et al. 2014. A mesoscale connectome of the mouse brain. *Nature* **508:** 207–214. doi:10.1038/nature13186

O'Leary DD. 1987. Remodelling of early axonal projections through the selective elimination of neurons and long axon collaterals. *Ciba Found Symp* **126:** 113–142. doi:10.1002/9780470513422.ch8

O'Leary DD, Stanfield BB. 1985. Occipital cortical neurons with transient pyramidal tract axons extend and maintain collaterals to subcortical but not intracortical targets. *Brain Res* **336:** 326–333. doi:10.1016/0006-8993(85)90661-4

O'Leary DD, Stanfield BB. 1986. A transient pyramidal tract projection from the visual cortex in the hamster and its removal by selective collateral elimination. *Brain Res* **392:** 87–99. doi:10.1016/0165-3806(86)90235-X

Özel MN, Gibbs CS, Holguera I, Soliman M, Bonneau R, Desplan C. 2022. Coordinated control of neuronal differentiation and wiring by sustained transcription factors. *Science* **378:** eadd1884. doi:10.1126/science.add1884

Paul A, Crow M, Raudales R, He M, Gillis J, Huang ZJ. 2017. Transcriptional architecture of synaptic communication delineates GABAergic neuron identity. *Cell* **171:** 522–539.e20. doi:10.1016/j.cell.2017.08.032

Peng H, Xie P, Liu L, Kuang X, Wang Y, Qu L, Gong H, Jiang S, Li A, Ruan Z, et al. 2021. Morphological diversity of single neurons in molecularly defined cell types. *Nature* **598:** 174–181. doi:10.1038/s41586-021-03941-1

Pollard KS, Salama SR, Lambert N, Lambot MA, Coppens S, Pedersen JS, Katzman S, King B, Onodera J, Siepel A, et al. 2006. An RNA gene expressed during cortical development evolved rapidly in humans. *Nature* **443:** 167–172. doi:10.1038/nature05113

Pouchelon G, Vergara J, McMahon J, Gorissen BL, Lin JD, Vormstein-Schneider D, Niehaus JL, Burbridge TJ, Wester JC, Sherer M, et al. 2022. A versatile viral toolkit for functional discovery in the nervous system. *Cell Rep Methods* **2:** 100225. doi:10.1016/j.crmeth.2022.100225

Prabhakar S, Noonan JP, Pääbo S, Rubin EM. 2006. Accelerated evolution of conserved noncoding sequences in humans. *Science* **314:** 786. doi:10.1126/science.1130738

Price DJ, Blakemore C. 1985. Regressive events in the postnatal development of association projections in the visual cortex. *Nature* **316:** 721–724. doi:10.1038/316721a0

Qian Y, Li J, Zhao S, Matthews EA, Adoff M, Zhong W, An X, Yeo M, Park C, Yang X, et al. 2022. Programmable RNA sensing for cell monitoring and manipulation. *Nature* **610:** 713–721. doi:10.1038/s41586-022-05280-1

Raj B, Wagner DE, McKenna A, Pandey S, Klein AM, Shendure J, Gagnon JA, Schier AF. 2018. Simultaneous single-cell profiling of lineages and cell types in the vertebrate brain. *Nat Biotechnol* **36:** 442–450. doi:10.1038/nbt.4103

Reilly MB, Cros C, Varol E, Yemini E, Hobert O. 2020. Unique homeobox codes delineate all the neuron classes of *C. elegans*. *Nature* **584:** 595–601. doi:10.1038/s41586-020-2618-9

Ren J, Isakova A, Friedmann D, Zeng J, Grutzner SM, Pun A, Zhao GQ, Kolluru SS, Wang R, Lin R, et al. 2019. Single-cell transcriptomes and whole-brain projections of serotonin neurons in the mouse dorsal and median raphe nuclei. *eLife* **8:** e49424. doi:10.7554/eLife.4942

Rosenberg AB, Roco CM, Muscat RA, Kuchina A, Sample P, Yao Z, Graybuck LT, Peeler DJ, Mukherjee S, Chen W, et al. 2018. Single-cell profiling of the developing mouse brain and spinal cord with split-pool barcoding. *Science* **360:** 176–182. doi:10.1126/science.aam8999

Sabatini PV, Wang J, Rupp AC, Affinati AH, Flak JN, Li C, Olson DP, Myers MG. 2021. tTARGIT AAVs mediate the sensitive and flexible manipulation of intersectional neuronal populations in mice. *eLife* **10:** e66835. doi:10.7554/eLife.66835

Saunders A, Huang KW, Vondrak C, Hughes C, Smolyar K, Sen H, Philson AC, Nemesh J, Wysoker A, Kashin S, et al. 2022. Ascertaining cells' synaptic connections and RNA expression simultaneously with barcoded rabies virus libraries. *Nat Commun* **13:** 6993. doi:10.1038/s41467-022-34334-1

Scala F, Kobak D, Bernabucci M, Bernaerts Y, Cadwell CR, Castro JR, Hartmanis L, Jiang X, Laturnus S, Miranda E, et al. 2021. Phenotypic variation of transcriptomic cell types in mouse motor cortex. *Nature* **598:** 144–150. doi:10.1038/s41586-020-2907-3

Shepherd GM. 2013. Corticostriatal connectivity and its role in disease. *Nat Rev Neurosci* **14:** 278–291. doi:10.1038/nrn3469

Shi H, He Y, Zhou Y, Huang J, Maher K, Wang B, Tang Z, Luo S, Tan P, Wu M, et al. 2023. Spatial atlas of the mouse central nervous system at molecular resolution. *Nature* **622:** 552–561.

Stanfield BB, O'Leary DD. 1985a. Fetal occipital cortical neurones transplanted to the rostral cortex can extend and maintain a pyramidal tract axon. *Nature* **313:** 135–137. doi:10.1038/313135a0

Stanfield BB, O'Leary DD. 1985b. The transient corticospinal projection from the occipital cortex during the postnatal development of the rat. *J Comp Neurol* **238:** 236–248. doi:10.1002/cne.902380210

Stanfield BB, O'Leary DD, Fricks C. 1982. Selective collateral elimination in early postnatal development restricts cortical distribution of rat pyramidal tract neurones. *Nature* **298:** 371–373. doi:10.1038/298371a0

Sun YC, Chen X, Fischer S, Lu S, Zhan H, Gillis J, Zador AM. 2021. Integrating barcoded neuroanatomy with spatial transcriptional profiling enables identification of gene correlates of projections. *Nat Neurosci* **24:** 873–885. doi:10.1038/s41593-021-00842-4

Tasic B, Menon V, Nguyen TN, Kim TK, Jarsky T, Yao Z, Levi B, Gray LT, Sorensen SA, Dolbeare T, et al. 2016. Adult mouse cortical cell taxonomy revealed by single cell transcriptomics. *Nat Neurosci* **19:** 335–346. doi:10.1038/nn.4216

Tasic B, Yao Z, Graybuck LT, Smith KA, Nguyen TN, Bertagnolli D, Goldy J, Garren E, Economo MN, Viswanathan S, et al. 2018. Shared and distinct transcriptomic cell types across neocortical areas. *Nature* **563:** 72–78. doi:10.1038/s41586-018-0654-5

Tomioka R, Rockland KS. 2007. Long-distance corticocortical GABAergic neurons in the adult monkey white and

Cite this article as *Cold Spring Harb Perspect Biol* doi: 10.1101/cshperspect.a041509

gray matter. *J Comp Neurol* **505:** 526–538. doi:10.1002/cne
.21504

Topalidou I, Chalfie M. 2011. Shared gene expression in distinct neurons expressing common selector genes. *Proc Natl Acad Sci* **108:** 19258–19263. doi:10.1073/pnas
.1111684108

Tosches MA, Yamawaki TM, Naumann RK, Jacobi AA, Tushev G, Laurent G. 2018. Evolution of pallium, hippocampus, and cortical cell types revealed by single-cell transcriptomics in reptiles. *Science* **360:** 881–888. doi:10.1126/science.aar4237

Tsyporin J, Tastad D, Ma X, Nehme A, Finn T, Huebner L, Liu G, Gallardo D, Makhamreh A, Roberts JM, et al. 2021. Transcriptional repression by FEZF2 restricts alternative identities of cortical projection neurons. *Cell Rep* **35:** 109269. doi:10.1016/j.celrep.2021.109269

Turner-Evans DB, Jensen KT, Ali S, Paterson T, Sheridan A, Ray RP, Wolff T, Lauritzen JS, Rubin GM, Bock DD, et al. 2020. The neuroanatomical ultrastructure and function of a biological ring attractor. *Neuron* **108:** 145–163.e10. doi:10.1016/j.neuron.2020.08.006

Wagner GP. 2014. *Homology, genes, and evolutionary innovation.* Princeton University Press, Princeton, NJ.

Way JC, Chalfie M. 1988. mec-3, a homeobox-containing gene that specifies differentiation of the touch receptor neurons in *C. elegans. Cell* **54:** 5–16. doi:10.1016/0092-8674(88)90174-2

Way JC, Chalfie M. 1989. The mec-3 gene of *Caenorhabditis elegans* requires its own product for maintained expression and is expressed in three neuronal cell types. *Genes Dev* **3:** 1823–1833. doi:10.1101/gad.3.12a.1823

Webster MJ, Ungerleider LG, Bachevalier J. 1991. Connections of inferior temporal areas TE and TEO with medial temporal-lobe structures in infant and adult monkeys. *J Neurosci* **11:** 1095–1116. doi:10.1523/JNEUROSCI.11-04-01095.1991

Wei X, Fu S, Li H, Liu Y, Wang S, Feng W, Yang Y, Liu X, Zeng YY, Cheng M, et al. 2022. Single-cell stereo-seq reveals induced progenitor cells involved in axolotl brain regeneration. *Science* **377:** eabp9444. doi:10.1126/science.abp9444

Wenick AS, Hobert O. 2004. Genomic *cis*-regulatory architecture and *trans*-acting regulators of a single interneuron-specific gene battery in *C. elegans. Dev Cell* **6:** 757–770. doi:10.1016/j.devcel.2004.05.004

Winnubst J, Bas E, Ferreira TA, Wu Z, Economo MN, Edson P, Arthur BJ, Bruns C, Rokicki K, Schauder D, et al. 2019. Reconstruction of 1,000 projection neurons reveals new cell types and organization of long-range connectivity in the mouse brain. *Cell* **179:** 268–281.e13. doi:10.1016/j.cell.2019.07.042

Woych J, Ortega Gurrola A, Deryckere A, Jaeger ECB, Gumnit E, Merello G, Gu J, Joven Araus A, Leigh ND, Yun M, et al. 2022. Cell-type profiling in salamanders identifies innovations in vertebrate forebrain evolution. *Science* **377:** eabp9186. doi:10.1126/science.abp9186

Wu J, Duggan A, Chalfie M. 2001. Inhibition of touch cell fate by *egl-44* and *egl-46* in *C. elegans. Genes Dev* **15:** 789–802. doi:10.1101/gad.857401

Yao Z, Liu H, Xie F, Fischer S, Adkins RS, Aldridge AI, Ament SA, Bartlett A, Behrens MM, Van den Berge K, et al. 2021a. A transcriptomic and epigenomic cell atlas of the mouse

primary motor cortex. *Nature* **598:** 103–110. doi:10.1038/s41586-021-03500-8

Yao Z, van Velthoven CTJ, Nguyen TN, Goldy J, Sedeno-Cortes AE, Baftizadeh F, Bertagnolli D, Casper T, Chiang M, Crichton K, et al. 2021b. A taxonomy of transcriptomic cell types across the isocortex and hippocampal formation. *Cell* **184:** 3222–3241.e26. doi:10.1016/j.cell.2021.04.021

Yao S, Wang Q, Hirokawa KE, Ouellette B, Ahmed R, Bomben J, Brouner K, Casal L, Caldejon S, Cho A, et al. 2023a. A whole-brain monosynaptic input connectome to neuron classes in mouse visual cortex. *Nat Neurosci* **26:** 350–364. doi:10.1038/s41593-022-01219-x

Yao Z, van Velthoven CTJ, Kunst M, Zhang M, McMillen D, Lee C, Jung W, Goldy J, Abdelhak A, Aitken M, et al. 2023b. A high-resolution transcriptomic and spatial atlas of cell types in the whole mouse brain. *Nature* **624:** 317–332.

Yuan L, Chen X, Zhan H, Gilbert HL, Zador AM. 2023. Massive multiplexing of spatially resolved single neuron projections with axonal BARseq. bioRxiv doi:10.1101/2023.02.18.528865

Yuste R, Hawrylycz M, Aalling N, Aguilar-Valles A, Arendt D, Armañanzas R, Ascoli GA, Bielza C, Bokharaie V, Bergmann TB, et al. 2020. A community-based transcriptomics classification and nomenclature of neocortical cell types. *Nat Neurosci* **23:** 1456–1468. doi:10.1038/s41593-020-0685-8

Zador AM, Dubnau J, Oyibo HK, Zhan H, Cao G, Peikon ID. 2012. Sequencing the connectome. *PLoS Biol* **10:** e1001411. doi:10.1371/journal.pbio.1001411

Zeisel A, Muñoz-Manchado AB, Codeluppi S, Lönnerberg P, La Manno G, Juréus A, Marques S, Munguba H, He L, Betsholtz C, et al. 2015. Brain structure. Cell types in the mouse cortex and hippocampus revealed by single-cell RNA-seq. *Science* **347:** 1138–1142. doi:10.1126/science.aaa1934

Zeisel A, Hochgerner H, Lonnerberg P, Johnsson A, Memic F, van der Zwan J, Haring M, Braun E, Borm LE, La Manno G, et al. 2018. Molecular architecture of the mouse nervous system. *Cell* **174:** 999–1014.e22. doi:10.1016/j.cell.2018.06.021

Zeng H. 2022. What is a cell type and how to define it? *Cell* **185:** 2739–2755. doi:10.1016/j.cell.2022.06.031

Zeng H, Sanes JR. 2017. Neuronal cell-type classification: challenges, opportunities and the path forward. *Nat Rev Neurosci* **18:** 530–546. doi:10.1038/nrn.2017.85

Zhang M, Eichhorn SW, Zingg B, Yao Z, Cotter K, Zeng H, Dong H, Zhuang X. 2021a. Spatially resolved cell atlas of the mouse primary motor cortex by MERFISH. *Nature* **598:** 137–143. doi:10.1038/s41586-021-03705-x

Zhang Z, Zhou J, Tan P, Pang Y, Rivkin AC, Kirchgessner MA, Williams E, Lee CT, Liu H, Franklin AD, et al. 2021b. Epigenomic diversity of cortical projection neurons in the mouse brain. *Nature* **598:** 167–173. doi:10.1038/s41586-021-03223-w

Zhang A, Jin L, Yao S, Matsuyama M, van Velthoven C, Sullivan H, Sun N, Kellis M, Tasic B, Tasic B, et al. 2023a. Rabies virus-based barcoded neuroanatomy resolved by single-cell RNA and *in situ* sequencing. *eLife* **12:** RP87866.

Zhang M, Pan X, Jung W, Halpern AR, Eichhorn SW, Lei Z, Cohen L, Smith KA, Tasic B, Yao Z, et al. 2023b. Molecularly defined and spatially resolved cell atlas of the whole mouse brain. *Nature* **624:** 343–354.

Zhao Q, Yu CD, Wang R, Xu QJ, Dai Pra R, Zhang L, Chang RB. 2022. A multidimensional coding architecture of the vagal interoceptive system. *Nature* **603:** 878–884. doi:10.1038/s41586-022-04515-5

Zheng C, Chalfie M. 2016. Securing neuronal cell fate in *C. elegans. Curr Top Dev Biol* **116:** 167–180. doi:10.1016/bs.ctdb.2015.11.011

Zheng C, Jin FQ, Chalfie M. 2015. Hox proteins act as transcriptional guarantors to ensure terminal differentiation. *Cell Rep* **13:** 1343–1352. doi:10.1016/j.celrep.2015.10.044

Zhu Y, Sousa AMM, Gao T, Skarica M, Li M, Santpere G, Esteller-Cucala P, Juan D, Ferrández-Peral L, Gulden FO, et al. 2018. Spatiotemporal transcriptomic divergence across human and macaque brain development. *Science* **362:** eaat8077. doi:10.1126/science.aat8077

Zingg B, Hintiryan H, Gou L, Song MY, Bay M, Bienkowski MS, Foster NN, Yamashita S, Bowman I, Toga AW, et al. 2014. Neural networks of the mouse neocortex. *Cell* **156:** 1096–1111. doi:10.1016/j.cell.2014.02.023

Znamenskiy P, Zador AM. 2013. Corticostriatal neurons in auditory cortex drive decisions during auditory discrimination. *Nature* **497:** 482–485. doi:10.1038/nature12077

Cite this article as *Cold Spring Harb Perspect Biol* doi: 10.1101/cshperspect.a041509

The Unusual Effectiveness of Evolution in Systems Neuroscience

Arkarup Banerjee,[1] Steven M. Phelps,[2] and Justus M. Kebschull[3,4,5,6]

[1]School of Biological Sciences, Cold Spring Harbor Laboratory, Cold Spring Harbor, New York 11724, USA

[2]Department of Integrative Biology, University of Texas at Austin, Austin, Texas 78712, USA

[3]Department of Biomedical Engineering, Johns Hopkins University, Baltimore, Maryland 21218, USA

[4]Kavli Neuroscience Discovery Institute, Johns Hopkins University, Baltimore, Maryland 21218, USA

[5]Department of Neuroscience, Johns Hopkins University, Baltimore, Maryland 21205, USA

[6]Center for Functional Anatomy and Evolution, Johns Hopkins University, Baltimore, Maryland 21287, USA

Correspondence: abanerjee@cshl.edu

This perspective advocates for "evolutionary systems neuroscience" as a framework combining evolutionary biology with neural circuit analysis. Evolution creates natural circuit modifications that preserve essential functions while enabling new behaviors. Modern technologies now allow researchers to investigate causal connections from genes to circuits to behaviors with unprecedented precision. By studying both convergent and divergent evolution, we can uncover both broad computational principles and specific implementation mechanisms. Across diverse examples—from insect courtship to rodent communication—we explore how targeted circuit changes drive behavioral innovation without disrupting core functions. This framework may reveal "deep homologies" in neural mechanisms, similar to how evolutionary developmental biology (evo-devo) identified conserved genetic toolkits in morphological development. This evolutionary lens promises not just to reveal how brains work, but why they work the way they do—providing insights that extend beyond neuroscience to complex adaptive systems more broadly.

In 1960, Eugene Wigner wrote an influential essay on "the unreasonable effectiveness of mathematics in the natural sciences" (Wigner 1960). He argued that while it did not have to be that way, a vast number of phenomena from buoyancy to black holes can be concisely and precisely described using mathematics. Arguably, that same transformative role in biology is played by the theory of evolution. Across biological disciplines, evolutionary frameworks have repeatedly provided deep insights that would otherwise remain obscure. Evolutionary genomics has reconstructed disease outbreaks like SARS-CoV2 (Markov et al. 2023), revealed functional domains of coding sequences (Alföldi and Lindblad-Toh 2013), clarified the progression of cancer (Merlo et al. 2006; Navin et al. 2011), and identified human-specific adaptations in the genome (Berto et al. 2020). Similarly, the evolutionary reframing of developmental biology (evo-devo) transformed our understanding of animal bodies and the molecular mechanisms that generate body plans (Shubin et al. 2009). In each case, an evolutionary perspective organized fragmented facts into coherent insights about the underlying genetic and developmental mechanisms.

We here argue that systems neuroscience stands at a similar inflection point, where evolutionary approaches can provide unique insights into neural circuit function. Our nervous systems were not constructed by the meticulous efforts of an engineer but by the haphazard puttering of evolution. In fact, the brain and behavior of every species has emerged through this same process of modification and descent. Thus, evolutionary biology offers an essential perspective for systems neuroscience. By examining how neural circuits have been modified through evolution, we can identify design principles that explain why circuits take their particular forms and reveal how "changes" in the brain generate "novel" behavior.

EVOLUTIONARY PERTURBATIONS TO NEURAL CIRCUITS

A principal goal of systems neuroscience is to understand how computations performed by neural circuits generate behavior. A powerful approach is to perturb specific neural circuit elements and determine their causal effects on behavior. From intracortical electrical stimulation (Fritsch and Hitzig 1870) to cell-type-specific optogenetics (Boyden et al. 2005), our tools of neural circuit dissection have become more precise and specific (Boyden et al. 2005; Luo et al. 2018). Such acute perturbations give us unprecedented control over neural activity and have revolutionized the search for mechanistic explanations of behavior. However, our current tools also have limitations. They often push circuit dynamics beyond the bounds of normal function, which can severely disrupt behavior (i.e., loss of function). There is growing appreciation that such "off-manifold" perturbations can mis- or overestimate causal effects on behav-

ior (Otchy et al. 2015; Jazayeri and Afraz 2017; Wolff and Ölveczky 2018; Banerjee et al. 2021). This may be especially problematic for circuit nodes with highly recurrent connectivity such as the cortex (Li et al. 2019). Moreover, if silencing a brain region leads to a severe disruption of behavior, it may preclude a detailed understanding of how moment-to-moment neural computations generate behavior.

To illustrate the distinction between off- and on-manifold perturbations, we can compare two typical strategies for analyzing morphological traits. Random mutagenesis represents a classic off-manifold approach: Complete gene knockouts may demonstrate necessity but reveal little about how its natural variations sculpt phenotypes. In contrast, crossing phenotypically distinct parents (producing F1 and F2 hybrids) creates on-manifold perturbations that often reveal more nuanced genotype-to-phenotype relationships. For instance, studies of appendage evolution in stickleback fish highlight this dichotomy. Knocking out the Pitx1 gene completely eliminates pelvic spines—much like silencing a circuit node wipes out its function (Shapiro et al. 2004)—but tells us little about how graded shifts in Pitx1 expression tune spine length. Population genetic approaches, however, found that natural deletions in a small Pitx1 enhancer region subtly adjust spine size across lake populations (Chan et al. 2010). Likewise, global loss of Shh produces severe limb truncations (Chiang et al. 2001), whereas small, naturally occurring mutations in the limb-specific zone of polarizing activity regulatory sequence (ZRS) enhancer drive extra digits or modest digit-number shifts (Lettice et al. 2003; Kvon et al. 2016). This shows that while random mutagenesis can establish whether a gene is necessary, it alone does not reveal how genes normally shape traits through natural variations.

Similarly, in neuroscience, acute off-manifold manipulations tell us which circuit nodes are essential. In contrast, evolution's own "experiments" offer a complementary approach. Over longer timescales, they also manipulate circuits, necessarily settling on allowable "on-manifold" configurations tailored toward

Cite this article as *Cold Spring Harb Perspect Biol* doi: 10.1101/cshperspect.a041510

natural behaviors (Fig. 1). Evolution may alter the functions of preexisting circuits by changing the intrinsic properties of neurons, the density of neuronal projections, or the number of cells of a given type; most dramatically, it may duplicate entire regions or circuits (Chakraborty and Jarvis 2015; Katz 2016; Katz and Hale 2017; Kebschull et al. 2020; Roberts et al. 2022). Many of these evolutionary changes parallel circuit perturbations one might perform as a systems neuroscientist— by changing the excitability of cells, for example, or altering input from one region to another. Unlike traditional manipulations, however, evolutionary perturbations must retain essential brain functions, even as circuits are modified and repurposed for new uses. This distinction is crucial. While experimental perturbations can reveal whether a circuit element is necessary for a behavior, evolutionary comparisons can reveal how quantitative changes in circuit properties enable new behavioral capabilities without disrupting core functions. These natural experiments complement traditional perturbation approaches by revealing viable circuit configurations that support novel behaviors.

DESIGN PRINCIPLES IN NEURAL CIRCUIT EVOLUTION

Out of all possible evolutionary perturbations to brain circuits, it is probable that some types of changes are easier or more effective to make and that these changes are more likely to be passed on to progeny than others. We argue that understanding the identity as well as physiological and computational consequences of common changes across short and long time scales is critical for understanding how changing behaviors are caused by changing brains. Understanding the nature of common changes will highlight structural "design" principles of brain circuits, narrowing hypothesis space and providing systems neuroscience with a path toward a more generalizable understanding of brain function.

The field is just beginning to uncover such principles. Recent comparative single-cell transcriptomics work has already yielded promising insights. A series of studies in mice, birds, turtles, lizards, and amphibians demonstrated deep conservation of inhibitory cell types in the pallium, which contrasts strongly with the rapid diversification of intermingled excitatory projection neurons (Tosches et al. 2018; Colquitt

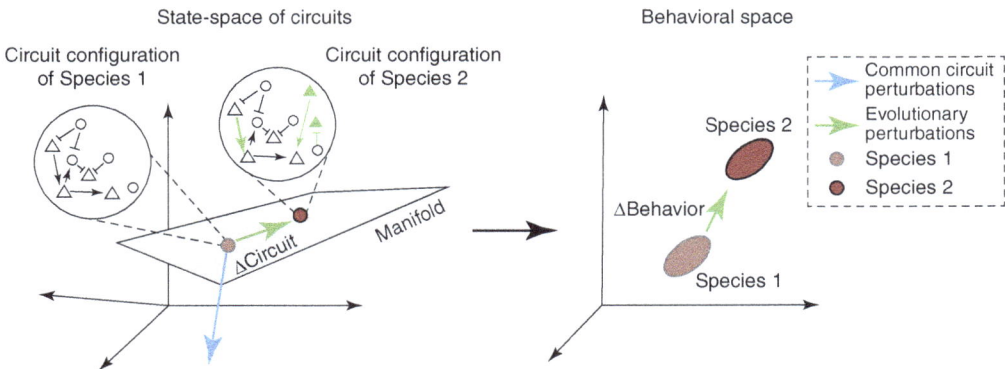

Figure 1. Evolution as method of circuit perturbation. In the space of all possible circuit configurations, those that have been selected over evolution exist on a low-dimensional manifold. Commonly used circuit perturbations (e.g., optogenetics, pharmacological silencing) often disrupt neural activity severely, leading to over-estimation of causal effects. Evolutionary tinkering over longer timescales also manipulates circuits, but it necessarily settles on robust configurations and dynamics tailored toward natural behaviors. By leveraging evolution as a perturbation method, one may be able to discover circuit configurations that allow a new behavior to emerge in recently diverged species, providing insights into how neural circuits produce behavior more generally.

et al. 2021; Hain et al. 2022; Lust et al. 2022; Woych et al. 2022; Zaremba et al. 2025). Intriguingly, the same holds true for the inhibitory and excitatory neurons of the cerebellar nuclei (Kebschull et al. 2020), despite completely different developmental origins of the pallium and cerebellum. At the same time, these studies highlight that essentially all brain regions are constantly evolving and that new cell types appear both in "old" and "new" brain structures (Kebschull et al. 2020; Faltine-Gonzalez and Kebschull 2022; Hain et al. 2022). Over shorter evolutionary timescales, such as across mammals, coarse transcriptomic identities within brain regions are generally conserved, but cell type abundances change, as do the transcriptomic patterns of these cell types (Hodge et al. 2019; Krienen et al. 2020, 2023; Bakken et al. 2021; Wei et al. 2022; Chen et al. 2023), suggesting more subtly altered functional properties of conserved cell types. For a more in-depth analysis of cell type evolution, we point the reader to a series of excellent reviews dedicated to the topic (Arendt 2008; Arendt et al. 2016; Tosches 2017, 2021; Sachkova and Burkhardt 2019).

Comparative single-cell and spatial transcriptomics is also beginning to shed light on evolutionary changes to brain structure beyond individual cell types. The cerebellar nuclei vary in number across vertebrates (Kebschull et al. 2024). A comparison of the cell types present in these different regions in chickens, mice, and humans revealed an archetypal cerebellar nucleus composed of a deeply conserved cell type set (Kebschull et al. 2020). This pattern suggests a mechanism for the evolution of new brain regions by duplication and divergence of ancestral regions.

Brain structure, of course, is more than just transcriptomic cell types and brain regions. With the emergence of more accessible high-throughput neuroanatomical methods, including electron microscopy for small brains and viral mapping tools for larger brains, we expect that wiring design principles will be discovered for connectivity in due course and that these principles might explain commonly observed neuronal circuit structures (Luo 2021). The first results are tantalizing, as summarized in Roberts

et al. (2022). In *Caenorhabditis elegans*, wiring optimization explains the arrangement of the pharyngeal circuit (Chen et al. 2006). In drosophilids, subtle central and peripheral wiring changes seem to underlie varying courtship behaviors (Seeholzer et al. 2018; Ding et al. 2019). In songbirds, the duplication of an entire motor control circuit has been proposed to give rise to the song circuitry (Chakraborty and Jarvis 2015; Moll et al. 2025). And in mammals, a recent study points to specific changes in projection strength between laboratory mice and Alston's singing mice (Isko et al. 2024). Which of these changes might be rules and which are exceptions remains an open question. Resolution of this question will require more comparative connectomics data sets in both closely related and widely divergent species.

EVOLUTIONARY APPROACHES TO UNDERSTANDING THE NEURAL BASIS OF BEHAVIORS

Leveraging natural diversity and evolutionary thinking can also help us understand the circuit mechanisms that underlie behavior (Cisek 2019; Cisek and Hayden 2022). One useful "intuition-pump" (Dennett 2014) in systems neuroscience is to consider a behavioral process at three levels as proposed by Marr: the computational level (what problem is being solved and why), the algorithmic level (what strategy achieves the solution), and the implementation level (how it is physically realized) (Marr and Poggio 1976). In this section, we highlight how different evolutionary scenarios may offer complementary insights into Marr's levels of explanation.

Convergent evolution—when distantly related species independently evolve similar capabilities—reveals crucial algorithmic principles that transcend specific neural implementations. Consider echolocation in bats and dolphins (Thomas et al. 2002), or tool use in primates and corvids (Van Lawick-Goodall 1971). These distantly related lineages evolved "similar" capabilities using vastly "different" neural architectures. These natural experiments identify robust algorithmic solutions that emerge repeatedly when similar computational problems are en-

Cite this article as *Cold Spring Harb Perspect Biol* doi: 10.1101/cshperspect.a041510

countered, even when the neural substrates differ dramatically. For example, recent work showed evidence for an evolutionarily conserved scheme for gain-control that uses divisive normalization for reformatting odor concentration in early olfactory circuits of insects, fish, and mice (Olsen et al. 2010; Hong and Wilson 2013; Zhu et al. 2013; Banerjee et al. 2015; Shen et al. 2025). Since vertebrate and invertebrate olfactory systems likely evolved independently, this finding suggests that evolution converged on similar algorithmic solutions (i.e., divisive normalization) despite stark differences in neural circuit architectures implementing this computation. In another striking convergence, recent studies have demonstrated that neural circuits encoding head-direction information are surprisingly similar between insects and fish (Seelig and Jayaraman 2015; Heinze 2023; Petrucco et al. 2023). In addition to these examples, we point the reader to other excellent articles that highlight how functional convergence can lead to deep algorithmic insights (Laurent and Gabbiani 1998; Newcomb et al. 2012; Seelig and Jayaraman 2015; Hemberger et al. 2016; Jarvis 2019; Petrucco et al. 2023; Wirthlin et al. 2024).

However, efforts to understand the neural circuit implementation underlying these algorithms at various levels—gene expression, cell types, cytoarchitecture, and connectivity—across vastly different species face technical and conceptual challenges. Disagreements about assessing homology often exist across these various nested hierarchical levels. In addition to these technical difficulties, conceptual challenges exist as well. For example, Katz and colleagues examined the central pattern generators that underlie homologous swimming behaviors in distantly related sea slugs (Sakurai and Katz 2017). They manipulated synapse function and electrical connectivity to show that very different circuit configurations are used to generate the left–right alternation needed for swimming. The implementation of the behavior has changed over time, even as the behavior itself has remained static. Such "circuit drifts" reveal the limits of comparing distantly related species. Even when a behavior is constant, changes accrue that obscure the relationship between nervous system and behavior. These examples highlight the technical and conceptual issues that may limit our ability to understand how specific circuit elements implement key aspects of the behavior under study.

Complementarily, divergent evolution that leads to stark behavioral differences between closely related species can illuminate all three of Marr's levels, with distinctive advantages at the implementation level that are largely inaccessible when comparing distant lineages (Jourjine and Hoekstra 2021; Ding 2025). Focusing on recently diverged species with dramatic behavioral differences may minimize the effect of drift and allow us to discover how neural differences drive behavioral change. When examining species that diverged relatively recently, one may assume that most circuit elements remain homologous and directly comparable. This conservation of neural architecture creates a natural control condition, allowing us to pinpoint the specific implementation-level changes that enable new computations. We can trace precise causal pathways from genetic modifications to circuit-level changes to behavioral innovations, revealing exactly how neural hardware implements computational functions (Fig. 2A–C). This granular implementation-level analysis is typically impossible across distantly related species, where differences in overall brain organization make direct circuit comparisons challenging. Therefore, comparisons between carefully chosen closely related species provide a unique window into the mechanistic bases for behavior.

By strategically combining these approaches—studying convergent evolution across distant lineages to identify algorithmic principles and divergent evolution between close relatives to understand implementation mechanisms—we can develop a more complete understanding of how neural circuits generate behavior. This dual approach leverages evolution's natural experiments to reveal both the computational strategies that recur across animal lineages and the specific neural mechanisms that implement these strategies in particular species.

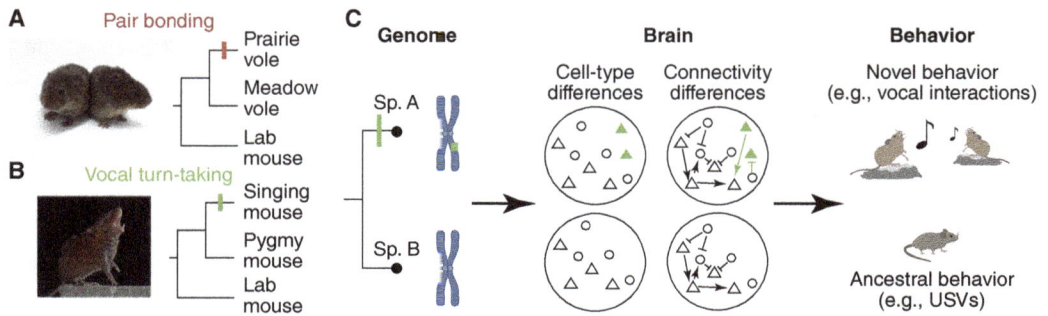

Figure 2. (*A,B*) Examples of drastic behavioral changes among recently diverged species. Prairie voles are monogamous compared to the closely related meadow voles and laboratory mice, which are both polygamous. Similarly, Alston's singing mouse has evolved a series of vocal innovations that include human-audible, stereotyped songs used in vocal turn-taking with other conspecifics. (*C*) Leveraging drastic behavioral divergences in closely related species is ideally suited to understand the genetic, cellular, and circuit-level mechanisms underlying naturalistic behavior. (USVs) Ultrasonic vocalizations.

EXAMPLES OF CIRCUIT ADAPTATIONS IN CLOSELY RELATED SPECIES

Across many phylogenetic lineages, evolutionary divergence has generated remarkable behavioral innovations through targeted modifications of neural circuits. These natural experiments reveal mechanistic principles that might otherwise remain hidden. By examining circuit adaptations across species, we can identify conserved principles of neural circuit evolution and gain insights into how even the most sophisticated behaviors arise.

The relatively compact nervous systems of insects provide elegant examples of how discrete circuit changes can yield dramatic behavioral differences. Closely related *Drosophila* species exhibit diverse courtship rituals that depend upon sexually dimorphic neural circuits (Demir and Dickson 2005; Yamamoto and Koganezawa 2013; Kallman et al. 2015; Coen and Murthy 2016; Baker et al. 2024). For example, researchers have identified changes in excitation–inhibition balance of inputs to P1 neurons that can explain the behavioral divergence between species (Seeholzer et al. 2018; Ding et al. 2019; Coleman et al. 2024). The power of this example lies in its specificity—a localized change in a defined circuit motif drives species-specific behavior. Even in these relatively simple nervous systems, evolution appears to target specific circuit nodes

that maximize behavioral plasticity while preserving essential functions.

We find similar principles at work in the evolution of sensorimotor behaviors in rodents. Two closely related deer mouse species (*Peromyscus maniculatus* and *Peromyscus polionotus*) have evolved different escape responses to threat stimuli that match their distinct habitats: The forest-dwelling species darts away, while the open-field species tends to freeze (Baier et al. 2025). Despite the greater complexity of mammalian brains, this behavioral difference has been traced to specific neural circuit modifications in the dorsal periaqueductal gray (dPAG). dPAG neurons are actively engaged in the species that darts to escape (*P. maniculatus*) but remains largely silent in the sister species that prefers to freeze (*P. polionotus*). Strikingly, optogenetic activation of excitatory dPAG neurons reliably elicit escape in *P. maniculatus* but not in *P. polionotus*, confirming that this circuit difference is causally related to the behavioral divergence (Baier et al. 2025). This example demonstrates that even in more complex vertebrate brains, modifications of specific circuit nodes can lead to rapid behavioral evolution.

The neural substrates of social behavior highlight how evolution modifies increasingly complex circuits to generate sophisticated behavioral adaptations. The monogamous prairie vole (*Microtus ochrogaster*) and promiscuous

 Cite this article as *Cold Spring Harb Perspect Biol* doi: 10.1101/cshperspect.a041510

meadow vole (*Microtus pennsylvanicus*) exhibit dramatically different mating systems and social organizations (Carter and Getz 1993; McGraw and Young 2010). This behavioral divergence (Fig. 2A) correlates with specific changes in the distribution of oxytocin and vasopressin receptors within reward-related brain regions, including the nucleus accumbens and prefrontal cortex (Insel et al. 1994; Walum and Young 2018). These forebrain regions represent a significant increase in circuit complexity compared to midbrain structures, yet evolution has again targeted specific circuit elements—in this case, the expression patterns of neuromodulator receptors—to reconfigure neural function (Walum and Young 2018; Froemke and Young 2021). Prefrontal cortical neurons that project to nucleus accumbens may promote partner-associated reward, effectively transforming a previously neutral stimulus into a rewarding one (Amadei et al. 2017; Pierce et al. 2024). These studies demonstrate how relatively subtle molecular changes in complex forebrain circuits can generate profound differences in social organization.

Mammalian vocal communication systems offer another compelling example of how circuit evolution may drive behavioral innovation across species. Alston's singing mouse (*Scotinomys teguina*) produces long, stereotyped, human-audible songs that facilitate antiphonal (call-and-response) interactions with some parallels to human conversational turn-taking—a behavior absent in laboratory mice (*Mus musculus*), which diverged from singing mice ~20 million years ago (Campbell et al. 2010; Pasch et al. 2013; Scott and Steppan 2017; Banerjee et al. 2019; Okobi et al. 2019). This complex vocal behavior requires coordination between cortical and subcortical systems (Okobi et al. 2019; Banerjee et al. 2024). Recent neurobiological investigations reveal that singing mice have evolved this novel vocal mode while retaining the ancestral ultrasonic vocalizations (USVs) common to rodents (Isko et al. 2024; Zheng et al. 2025). Songs, which are loud and have a stereotyped rhythm, are used for long-range communication, whereas USVs, which are softer and do not possess a consistent

rhythm, are used for short-range communication. Despite these drastic acoustic and usage differences, both vocal modes are produced by similar peripheral sound production mechanisms, exhibit identical phonetic coupling with respiration, and share vocal gating by the midbrain periaqueductal gray (PAG; Zheng et al. 2025). In contrast to these similarities at the subcortical level, a recent comparative barcoded neuroanatomy study has identified a selective expansion of projections from the orofacial motor cortex (OMC) to two specific downstream targets: an auditory cortical region and the PAG (Isko et al. 2024). Moreover, this cortical region has been shown to be causally involved in this singing behavior using electrical stimulation, focal cooling, pharmacology, and chronic electrophysiology (Okobi et al. 2019; Banerjee et al. 2024). This finding suggests that significant behavioral divergences over relatively brief evolutionary timescales may not require wholesale reorganization of neural architecture but rather proceed by targeted modifications to long-range projection patterns of cortical neurons. The singing mouse example suggests how evolution can generate complex, temporally structured behaviors through quantitative changes in central circuits rather than dramatic neural reorganization of the motor periphery.

Are there any common themes that emerge from these examples? Or are evolutionary solutions so specific for individual species that it is hard to imagine common principles that apply to most? One might expect large behavioral differences in recently diverged species to require drastic neural circuit changes. However, an emerging theme is that a relatively small number of quantitative modifications—in both genomes and central circuits—may be sufficient to produce significant behavioral divergence in closely related species. At short evolutionary timescales, anatomical structures and cell types are likely to remain conserved, with targeted modifications in circuit connectivity driving rapid behavioral diversification. This is an essential insight. It also bodes well for our ability to use a comparative approach to identify structural, functional, and molecular mechanisms that underlie species differences in behavior.

THE TECHNICAL REVOLUTION ENABLING EVOLUTIONARY SYSTEMS NEUROSCIENCE

Recent technological breakthroughs have made this the ideal moment for synthesizing evolutionary and systems approaches. A suite of powerful new tools at the genetic, circuit, and behavioral levels enable unprecedented analysis of the structure and function of neural circuits across species. Efficient long-read sequencing technologies make the assembly of genomes an increasingly routine task (Kirsche and Schatz 2021; Marx 2023), while single-cell and spatial transcriptomics and related omics approaches are readily accessible and are applicable across species, allowing the in-depth analysis of transcriptomic cell types and their distribution across vast evolutionary timescales (Tosches et al. 2018; Hodge et al. 2019; Kebschull et al. 2020; Bakken et al. 2021; Hain et al. 2022; Wei et al. 2022; Woych et al. 2022; Chen et al. 2023; Krienen et al. 2023; Zaremba et al. 2025).

The accessibility of electron microscopy has increased dramatically, yielding complete connectomes of model organisms like *Drosophila* (Zheng et al. 2018; Schlegel et al. 2024) while making it feasible to trace circuits in nontraditional species with small brains (Sayre et al. 2021; Winding et al. 2023; Cook et al. 2025). The development of viral tracers (Jaeger et al. 2025; Lust and Tanaka 2025), brain clearing, and standard reference atlases is making mesoscale connectomics feasible for a broad range of species with larger brains. Meanwhile, high-throughput molecular techniques like MAPseq (Kebschull et al. 2016; Han et al. 2018; Huang et al. 2020) and BARseq (Chen et al. 2019; Sun et al. 2021) provide single-cell resolution maps of connectivity, allowing detailed comparison of circuit architecture between species (Isko et al. 2024).

Advanced viral tools are bringing opto- and chemogenetic tools and sensors to a broad range of vertebrates (Nectow and Nestler 2020; Graybuck et al. 2021), while CRISPR is giving us unprecedented genetic access across animals (Heidenreich and Zhang 2016). Advanced electrophysiology and multiphoton imaging methods enable monitoring of neural activity during naturalistic behaviors across diverse species. Machine learning–based pose estimation tools like DeepLabCut (Mathis et al. 2018) and SLEAP (Pereira et al. 2022) allow automated quantification of natural behaviors in any species, making it possible to precisely characterize behavioral variations across closely related organisms.

Together, these technical advances make it possible to connect genetic changes to circuit modifications to behavioral differences at a mechanistic level that was previously unattainable. The time is ripe for systematic comparison of neural circuits across carefully chosen species to reveal both deep homologies and evolutionary innovations.

CONCLUSION AND OUTLOOK: TOWARD EVOLUTIONARY SYSTEMS NEUROSCIENCE

We propose that incorporating evolutionary thinking into systems neuroscience—forming what we might call "evolutionary systems neuroscience"—offers a transformative approach to understanding brain function. Just as evo-devo revolutionized our understanding of body plans by revealing the genetic mechanisms underlying form (Müller 2007; Mallarino and Abzhanov 2012), evolutionary systems neuroscience can reveal the circuit-level mechanisms underlying behavior. The key insight is that evolution provides us with natural experiments—variations on neural circuitry that maintain function while enabling behavioral innovation.

A particularly promising outcome of this synthesis may be the discovery of "deep homologies" in neural circuit mechanisms across different behavioral adaptations (Shubin et al. 2009). Akin to conserved genetic toolkits deployed repeatedly during morphological evolution, evolutionary systems neuroscience is beginning to reveal conserved circuit-level modifications that repeatedly drive behavioral innovation. One intriguing example may be found in the unique properties of the neocortex. Its relatively recent evolutionary origin, late developmental maturation, and distinctive architectural features—including elaborate recurrent connections and hierarchical control of subcortical circuits—uniquely position it as a substrate for behavioral innovation without disrupting core functions (Krubitzer and Prescott 2018; Sherman and Usrey 2021; Halley et al. 2022). Therefore, cortical control may play a disproportionately important role in many independently evolved behavioral innovations, providing a compelling example of deep homology in systems neuroscience.

In general, evolution may work with a relatively limited toolkit of circuit-level mechanisms that are modified and repurposed rather than reinvented. By identifying these common mechanisms, evolutionary systems neuroscience promises not just to reveal how brains work, but why they work the way they do—providing a more complete understanding of the neural basis of behavior than either evolutionary or mechanistic approaches alone could achieve. As this synthesis continues to develop, it will yield insights that extend beyond neuroscience to inform our understanding of complex adaptive systems more broadly, from biological development to artificial intelligence. The application of evolutionary thinking to neural circuit analysis represents not just a methodological advance, but a conceptual framework that may fundamentally reshape how we understand the relationship between brains, behavior, and the environments that shaped them.

ACKNOWLEDGMENTS

This work was supported by Klingenstein Simons Fellowships to A.B. and J.M.K., Searle Scholars Fellowship to A.B., and a Packard Fellowship and Sloan Research Fellowship to J.M.K. We thank Clifford E. Harpole, Xiaoyue Mike Zheng, and the anonymous editor for constructive comments on earlier versions of this paper.

REFERENCES

Alföldi J, Lindblad-Toh K. 2013. Comparative genomics as a tool to understand evolution and disease. *Genome Res* **23:** 1063–1068. doi:10.1101/gr.157503.113

Amadei EA, Johnson ZV, Jun Kwon Y, Shpiner AC, Saravanan V, Mays WD, Ryan SJ, Walum H, Rainnie DG, Young LJ, et al. 2017. Dynamic corticostriatal activity biases social bonding in monogamous female prairie voles. *Nature* **546:** 297–301. doi:10.1038/nature22381

Arendt D. 2008. The evolution of cell types in animals: emerging principles from molecular studies. *Nat Rev Genet* **9:** 868–882. doi:10.1038/nrg2416

Arendt D, Musser JM, Baker CVH, Bergman A, Cepko C, Erwin DH, Pavlicev M, Schlosser G, Widder S, Laubichler MD, et al. 2016. The origin and evolution of cell types. *Nat Rev Genet* **17:** 744–757. doi:10.1038/nrg.2016.127

Baier F, Reinhard K, Nuttin B, Sans-Dublanc A, Liu C, Tong V, Murmann JS, Wierda K, Farrow K, Hoekstra HE. 2025. The neural basis of species-specific defensive behaviour in *Peromyscus* mice. *Nature* doi:10.1038/s41586-025-09241-2

Baker CA, Guan XJ, Choi M, Murthy M. 2024. The role of *fruitless* in specifying courtship behaviors across divergent *Drosophila* species. *Sci Adv* **10:** eadk1273. doi:10.1126/sciadv.adk1273

Bakken TE, Jorstad NL, Hu Q, Lake BB, Tian W, Kalmbach BE, Crow M, Hodge RD, Krienen FM, Sorensen SA, et al. 2021. Comparative cellular analysis of motor cortex in human, marmoset and mouse. *Nature* **598:** 111–119. doi:10.1038/s41586-021-03465-8

Banerjee A, Marbach F, Anselmi F, Koh MS, Davis MB, Garcia da Silva P, Delevich K, Oyibo HK, Gupta P, Li B, et al. 2015. An interglomerular circuit gates glomerular output and implements gain control in the mouse olfactory bulb. *Neuron* **87:** 193–207. doi:10.1016/j.neuron.2015.06.019

Banerjee A, Phelps SM, Long MA. 2019. Singing mice. *Curr Biol* **29:** R190–R191. doi:10.1016/j.cub.2018.11.048

Banerjee A, Egger R, Long MA. 2021. Using focal cooling to link neural dynamics and behavior. *Neuron* **109:** 2508–2518. doi:10.1016/j.neuron.2021.05.029

Banerjee A, Chen F, Druckmann S, Long MA. 2024. Temporal scaling of motor cortical dynamics reveals hierarchical control of vocal production. *Nat Neurosci* **27:** 527–535. doi:10.1038/s41593-023-01556-5

Berto S, Liu Y, Konopka G. 2020. Genomics at cellular resolution: insights into cognitive disorders and their evolution. *Hum Mol Genet* **29:** R1–R9. doi:10.1093/hmg/ddaa117

Boyden ES, Zhang F, Bamberg E, Nagel G, Deisseroth K. 2005. Millisecond-timescale, genetically targeted optical control of neural activity. *Nat Neurosci* **8:** 1263–1268. doi:10.1038/nn1525

Campbell P, Pasch B, Pino JL, Crino OL, Phillips M, Phelps SM. 2010. Geographic variation in the songs of neotropical singing mice: testing the relative importance of drift and local adaptation. *Evolution* **64:** 1955–1972. doi:10.1111/j.1558-5646.2010.00962.x

Carter CS, Getz LL. 1993. Monogamy and the prairie vole. *Sci Am* **268:** 100–106. doi:10.1038/scientificamerican0693-100

Chakraborty M, Jarvis ED. 2015. Brain evolution by brain pathway duplication. *Philos Trans R Soc Lond B Biol Sci* **370:** 20150056. doi:10.1098/rstb.2015.0056

Chan YF, Marks ME, Jones FC, Villarreal G Jr, Shapiro MD, Brady SD, Southwick AM, Absher DM, Grimwood J, Schmutz J, et al. 2010. Adaptive evolution of pelvic reduction in sticklebacks by recurrent deletion of a *Pitx1* enhancer. *Science* **327:** 302–305. doi:10.1126/science.1182213

Chen BL, Hall DH, Chklovskii DB. 2006. Wiring optimization can relate neuronal structure and function. *Proc Natl Acad Sci* **103:** 4723–4728. doi:10.1073/pnas.0506806103

Chen X, Sun YC, Zhan H, Kebschull JM, Fischer S, Matho K, Huang ZJ, Gillis J, Zador AM. 2019. High-throughput mapping of long-range neuronal projection using in situ sequencing. *Cell* **179:** 772–786.e19. doi:10.1016/j.cell.2019.09.023

Chen A, Sun Y, Lei Y, Li C, Liao S, Meng J, Bai Y, Liu Z, Liang Z, Zhu Z, et al. 2023. Single-cell spatial transcriptome reveals cell-type organization in the macaque cortex. *Cell* **186:** 3726–3743.e24. doi:10.1016/j.cell.2023.06.009

Chiang C, Litingtung Y, Harris MP, Simandl BK, Li Y, Beachy PA, Fallon JF. 2001. Manifestation of the limb prepattern: limb development in the absence of sonic hedgehog function. *Dev Biol* **236:** 421–435. doi:10.1006/dbio.2001.0346

Cisek P. 2019. Resynthesizing behavior through phylogenetic refinement. *Atten Percept Psychophys* **81:** 2265–2287.

Cisek P, Hayden BY. 2022. Neuroscience needs evolution. *Philos Trans R Soc Lond B Biol Sci* **377:** 20200518.

Coen P, Murthy M. 2016. Singing on the fly: sensorimotor integration and acoustic communication in *Drosophila*. *Curr Opin Neurobiol* **38:** 38–45. doi:10.1016/j.conb.2016.01.013

Coleman RT, Morantte I, Koreman GT, Cheng ML, Ding Y, Ruta V. 2024. A modular circuit coordinates the diversification of courtship strategies. *Nature* **635:** 142–150. doi:10.1038/s41586-024-08028-1

Colquitt BM, Merullo DP, Konopka G, Roberts TF, Brainard MS. 2021. Cellular transcriptomics reveals evolutionary identities of songbird vocal circuits. *Science* **371:** eabd9704. doi: 10.1126/science.abd9704

Cook SJ, Kalinski CA, Loer CM, Memar N, Majeed M, Stephen SR, Bumbarger DJ, Riebesell M, Conradt B, Schnabel R, et al. 2025. Comparative connectomics of two distantly related nematode species reveals patterns of nervous system evolution. *Science* **389:** eadx2143.

Demir E, Dickson BJ. 2005. Fruitless splicing specifies male courtship behavior in *Drosophila*. *Cell* **121:** 785–794. doi:10.1016/j.cell.2005.04.027

Dennett DC. 2014. *Intuition pumps and other tools for thinking*. W.W. Norton, New York.

Ding Y. 2025. Evolution of neural circuits in the origin of behavioral novelty. *Curr Opin Behav Sci* **63:** 101520.

Ding Y, Lillvis JL, Cande J, Berman GJ, Arthur BJ, Long X, Xu M, Dickson BJ, Stern DL. 2019. Neural evolution of context-dependent fly song. *Curr Biol* **29:** 1089–1099.e7. doi:10.1016/j.cub.2019.02.019

Faltine-Gonzalez DZ, Kebschull JM. 2022. A mosaic of new and old cell types. *Science* **377:** 1043–1044. doi:10.1126/science.add9465

Fritsch G, Hitzig E. 1870. Uber die elektrishe erregbarkeit des grosshirns [Electric excitability of the cerebrum]. *Arch Anat Physiol Wissen* **37:** 300–332.

Froemke RC, Young LJ. 2021. Oxytocin, neural plasticity, and social behavior. *Annu Rev Neurosci* **44:** 359–381. doi:10.1146/annurev-neuro-102320-102847

Graybuck LT, Daigle TL, Sedeño-Cortés AE, Walker M, Kalmbach B, Lenz GH, Morin E, Nguyen TN, Garren E, Bendrick JL, et al. 2021. Enhancer viruses for combinatorial cell-subclass-specific labeling. *Neuron* **109:** 1449–1464.e13. doi:10.1016/j.neuron.2021.03.011

Hain D, Gallego-Flores T, Klinkmann M, Macias A, Ciirdaeva E, Arends A, Thum C, Tushev G, Kretschmer F, Tosches MA, et al. 2022. Molecular diversity and evolution of neuron types in the amniote brain. *Science* **377:** eabp8202. doi:10.1126/science.abp8202

Halley AC, Baldwin MKL, Cooke DF, Englund M, Pineda CR, Schmid T, Yartsev MM, Krubitzer L. 2022. Coevolution of motor cortex and behavioral specializations associated with flight and echolocation in bats. *Curr Biol* **32:** 2935–2941.e3. doi:10.1016/j.cub.2022.04.094

Han Y, Kebschull JM, Campbell RAA, Cowan D, Imhof F, Zador AM, Mrsic-Flogel TD. 2018. The logic of single-cell projections from visual cortex. *Nature* **556:** 51–56. doi:10.1038/nature26159

Heidenreich M, Zhang F. 2016. Applications of CRISPR-Cas systems in neuroscience. *Nat Rev Neurosci* **17:** 36–44. doi:10.1038/nrn.2015.2

Heinze S. 2023. Neuroscience: fish and fly headed in the same direction. *Curr Biol* **33:** R677–R679. doi:10.1016/j.cub.2023.05.006

Hemberger M, Pammer L, Laurent G. 2016. Comparative approaches to cortical microcircuits. *Curr Opin Neurobiol* **41:** 24–30. doi:10.1016/j.conb.2016.07.009

Hodge RD, Bakken TE, Miller JA, Smith KA, Barkan ER, Graybuck LT, Close JL, Long B, Johansen N, Penn O, et al. 2019. Conserved cell types with divergent features in human versus mouse cortex. *Nature* **573:** 61–68. doi:10.1038/s41586-019-1506-7

Hong EJ, Wilson RI. 2013. Olfactory neuroscience: normalization is the norm. *Curr Biol* **23:** R1091–R1093. doi:10.1016/j.cub.2013.10.056

Huang L, Kebschull JM, Fürth D, Musall S, Kaufman MT, Churchland AK, Zador AM. 2020. BRICseq bridges brain-wide interregional connectivity to neural activity and gene expression in single animals. *Cell* **182:** 177–188.e27. doi:10.1016/j.cell.2020.05.029

Insel TR, Wang ZX, Ferris CF. 1994. Patterns of brain vasopressin receptor distribution associated with social organization in microtine rodents. *J Neurosci* **14:** 5381–5392. doi:10.1523/JNEUROSCI.14-09-05381.1994

Isko EC, Harpole CE, Zheng XM, Zhan H, Davis MB, Zador AM, Banerjee A. 2024. Selective expansion of motor cortical projections in the evolution of vocal novelty. bioRxiv doi:10.1101/2024.09.13.612752

Jaeger ECB, Vijatovic D, Deryckere A, Zorin N, Nguyen AL, Ivanian G, Woych J, Arnold RC, Gurrola AO, Shvartsman A, et al. 2025. Adeno-associated viral tools to trace neural development and connectivity across amphibians. *Dev Cell* **60:** 794–812.e6. doi:10.1016/j.devcel.2024.10.025

Jarvis ED. 2019. Evolution of vocal learning and spoken language. *Science* **366:** 50–54. doi:10.1126/science.aax0287

Jazayeri M, Afraz A. 2017. Navigating the neural space in search of the neural code. *Neuron* **93:** 1003–1014. doi:10 .1016/j.neuron.2017.02.019

Jourjine N, Hoekstra HE. 2021. Expanding evolutionary neuroscience: insights from comparing variation in behavior. *Neuron* **109:** 1084–1099.

Kallman BR, Kim H, Scott K. 2015. Excitation and inhibition onto central courtship neurons biases *Drosophila* mate choice. *eLife* **4:** e11188. doi:10.7554/eLife.11188

Katz PS. 2016. Evolution of central pattern generators and rhythmic behaviours. *Philos Trans R Soc Lond B Biol Sci* **371:** 20150057. doi:10.1098/rstb.2015.0057

Katz PS, Hale ME. 2017. Evolution of motor systems. In *Neurobiology of motor control*, pp. 135–176. Wiley, Hoboken, NJ.

Kebschull JM, Garcia da Silva P, Reid AP, Peikon ID, Albeanu DF, Zador AM. 2016. High-throughput mapping of single-neuron projections by sequencing of barcoded RNA. *Neuron* **91:** 975–987. doi:10.1016/j.neuron.2016.07 .036

Kebschull JM, Richman EB, Ringach N, Friedmann D, Albarran E, Kolluru SS, Jones RC, Allen WE, Wang Y, Cho SW, et al. 2020. Cerebellar nuclei evolved by repeatedly duplicating a conserved cell-type set. *Science* **370:** eabd5059. doi:10.1126/science.abd5059

Kebschull JM, Casoni F, Consalez GG, Goldowitz D, Hawkes R, Ruigrok TJH, Schilling K, Wingate R, Wu J, Yeung J, et al. 2024. Cerebellum lecture: the cerebellar nuclei—core of the cerebellum. *Cerebellum* **23:** 620–677. doi:10.1007/ s12311-022-01506-0

Kirsche M, Schatz MC. 2021. Democratizing long-read genome assembly. *Cell Syst* **12:** 945–947. doi:10.1016/j.cels .2021.09.010

Krienen FM, Goldman M, Zhang Q, C H Del Rosario R, Florio M, Machold R, Saunders A, Levandowski K, Zaniewski H, Schuman B, et al. 2020. Innovations present in the primate interneuron repertoire. *Nature* **586:** 262–269. doi:10.1038/s41586-020-2781-z

Krienen FM, Levandowski KM, Zaniewski H, del Rosario RCH, Schroeder ME, Goldman M, Wienisch M, Lutservitz A, Beja-Glasser VF, Chen C, et al. 2023. A marmoset brain cell census reveals regional specialization of cellular identities. *Sci Adv* **9:** eadk3986. doi:10.1126/sciadv .adk3986

Krubitzer LA, Prescott TJ. 2018. The combinatorial creature: cortical phenotypes within and across lifetimes. *Trends Neurosci* **41:** 744–762. doi:10.1016/j.tins.2018.08.002

Kvon EZ, Kamneva OK, Melo US, Barozzi I, Osterwalder M, Mannion BJ, Tissières V, Pickle CS, Plajzer-Frick I, Lee EA, et al. 2016. Progressive loss of function in a limb enhancer during snake evolution. *Cell* **167:** 633–642. e11. doi:10.1016/j.cell.2016.09.028

Laurent G, Gabbiani F. 1998. Collision-avoidance: nature's many solutions. *Nat Neurosci* **1:** 261–263. doi:10.1038/ 1071

Lettice LA, Heaney SJH, Purdie LA, Li L, de Beer P, Oostra BA, Goode D, Elgar G, Hill RE, de Graaff E. 2003. A long-range Shh enhancer regulates expression in the developing limb and fin and is associated with preaxial polydactyly. *Hum Mol Genet* **12:** 1725–1735. doi:10.1093/hmg/ ddg180

Li N, Chen S, Guo ZV, Chen H, Huo Y, Inagaki HK, Chen G, Davis C, Hansel D, Guo C, et al. 2019. Spatiotemporal constraints on optogenetic inactivation in cortical circuits. *eLife* **8:** e48622. doi:10.7554/eLife.48622

Luo L. 2021. Architectures of neuronal circuits. *Science* **373:** eabg7285. doi:10.1126/science.abg7285

Luo L, Callaway EM, Svoboda K. 2018. Genetic dissection of neural circuits: a decade of progress. *Neuron* **98:** 256–281. doi:10.1016/j.neuron.2018.03.040

Lust K, Tanaka EM. 2025. Adeno-associated viruses for efficient gene expression in the axolotl nervous system. *Proc Natl Acad Sci* **122:** e2421373122. doi:10.1073/pnas .2421373122

Lust K, Maynard A, Gomes T, Fleck JS, Camp JG, Tanaka EM, Treutlein B. 2022. Single-cell analyses of axolotl telencephalon organization, neurogenesis, and regeneration. *Science* **377:** eabp9262. doi:10.1126/science.abp9262

Mallarino R, Abzhanov A. 2012. Paths less traveled: evo-devo approaches to investigating animal morphological evolution. *Annu Rev Cell Dev Biol* **28:** 743–763. doi:10 .1146/annurev-cellbio-101011-155732

Markov PV, Ghafari M, Beer M, Lythgoe K, Simmonds P, Stilianakis NI, Katzourakis A. 2023. The evolution of SARS-CoV-2. *Nat Rev Microbiol* **21:** 361–379. doi:10 .1038/s41579-023-00878-2

Marr D, Poggio TA. 1976. From understanding computation to understanding neural circuitry. *MIT Libraries.* https ://dspace.mit.edu/handle/1721.1/5782

Marx V. 2023. Method of the year: long-read sequencing. *Nat Methods* **20:** 6–11. doi:10.1038/s41592-022-01730-w

Mathis A, Mamidanna P, Cury KM, Abe T, Murthy VN, Mathis MW, Bethge M. 2018. DeepLabCut: markerless pose estimation of user-defined body parts with deep learning. *Nat Neurosci* **21:** 1281–1289. doi:10.1038/ s41593-018-0209-y

McGraw LA, Young LJ. 2010. The prairie vole: an emerging model organism for understanding the social brain. *Trends Neurosci* **33:** 103–109. doi:10.1016/j.tins.2009.11 .006

Merlo LMF, Pepper JW, Reid BJ, Maley CC. 2006. Cancer as an evolutionary and ecological process. *Nat Rev Cancer* **6:** 924–935. doi:10.1038/nrc2013

Moll FW, Kersten Y, Erdle S, Nieder A. 2025. Exploring anatomical links between the crow's nidopallium caudolaterale and its song system. *J Comp Neurol* **533:** e70028. doi:10.1002/cne.70028

Müller GB. 2007. Evo-devo: extending the evolutionary synthesis. *Nat Rev Genet* **8:** 943–949. doi:10.1038/nrg2219

Navin N, Kendall J, Troge J, Andrews P, Rodgers L, McIndoo J, Cook K, Stepansky A, Levy D, Esposito D, et al. 2011. Tumour evolution inferred by single-cell sequencing. *Nature* **472:** 90–94. doi:10.1038/nature09807

Nectow AR, Nestler EJ. 2020. Viral tools for neuroscience. *Nat Rev Neurosci* **21:** 669–681. doi:10.1038/s41583-020- 00382-z

Newcomb JM, Sakurai A, Lillvis JL, Gunaratne CA, Katz PS. 2012. Homology and homoplasy of swimming behaviors and neural circuits in the Nudipleura (Mollusca, Gastropoda, Opisthobranchia). *Proc Natl Acad Sci* **109** (Suppl 1)**:** 10669–10676. doi:10.1073/pnas.1201877109

Okobi DE Jr, Banerjee A, Matheson AMM, Phelps SM. Long MA. 2019. Motor cortical control of vocal interaction in neotropical singing mice. *Science* **363**: 983–988. doi:10.1126/science.aau9480

Olsen SR, Bhandawat V, Wilson RI. 2010. Divisive normalization in olfactory population codes. *Neuron* **66**: 287–299. doi:10.1016/j.neuron.2010.04.009

Otchy TM, Wolff SBE, Rhee JY, Pehlevan C, Kawai R, Kempf A, Gobes SMH, Ölveczky BP. 2015. Acute off-target effects of neural circuit manipulations. *Nature* **528**: 358–363. doi:10.1038/nature16442

Pasch B, Bolker BM, Phelps SM. 2013. Interspecific dominance via vocal interactions mediates altitudinal zonation in neotropical singing mice. *Am Nat* **182**: E161–E173. doi:10.1086/673263

Pereira TD, Tabris N, Matsliah A, Turner DM, Li J, Ravindranath S, Papadoyannis ES, Normand E, Deutsch DS, Wang ZY, et al. 2022. SLEAP: a deep learning system for multi-animal pose tracking. *Nat Methods* **19**: 486–495. doi:10.1038/s41592-022-01426-1

Petrucco L, Lavian H, Wu YK, Svara F, Štih V, Portugues R. 2023. Neural dynamics and architecture of the heading direction circuit in zebrafish. *Nat Neurosci* **26**: 765–773. doi:10.1038/s41593-023-01308-5

Pierce AF, Protter DSW, Watanabe YL, Chapel GD, Cameron RT, Donaldson ZR. 2024. Nucleus accumbens dopamine release reflects the selective nature of pair bonds. *Curr Biol* **34**: 519–530.e5. doi:10.1016/j.cub.2023.12.041

Roberts RJV, Pop S, Prieto-Godino LL. 2022. Evolution of central neural circuits: state of the art and perspectives. *Nat Rev Neurosci* **23**: 725–743. doi:10.1038/s41583-022-00644-y

Sachkova M, Burkhardt P. 2019. Exciting times to study the identity and evolution of cell types. *Development* **146**: dev178996. doi:10.1242/dev.178996

Sakurai A, Katz PS. 2017. Artificial synaptic rewiring demonstrates that distinct neural circuit configurations underlie homologous behaviors. *Curr Biol* **27**: 1721–1734.e3. doi:10.1016/j.cub.2017.05.016

Sayre ME, Templin R, Chavez J, Kempenaers J, Heinze S. 2021. A projectome of the bumblebee central complex. *eLife* **10**: e68911. doi:10.7554/eLife.68911

Schlegel P, Yin Y, Bates AS, Dorkenwald S, Eichler K, Brooks P, Han DS, Gkantia M, Dos Santos M, Munnelly E, et al. 2024. Whole-brain annotation and multi-connectome cell typing of *Drosophila*. *Nature* **634**: 139–152. doi:10.1038/s41586-024-07686-5

Scott J, Steppan JJ. 2017. Muroid rodent phylogenetics: 900-species tree reveals increasing diversification rates *PLoS One* **12**: e0183070. doi:10.1371/journal.pone.0183070

Seeholzer LF, Seppo M, Stern DL, Ruta V. 2018. Evolution of a central neural circuit underlies *Drosophila* mate preferences. *Nature* **559**: 564–569. doi:10.1038/s41586-018-0322-9

Seelig JD, Jayaraman V. 2015. Neural dynamics for landmark orientation and angular path integration. *Nature* **521**: 186–191. doi:10.1038/nature14446

Shapiro MD, Marks ME, Peichel CL, Blackman BK, Nereng KS, Jónsson B, Schluter D, Kingsley DM. 2004. Genetic and developmental basis of evolutionary pelvic reduction

in threespine sticklebacks. *Nature* **428**: 717–723. doi:10.1038/nature02415

Shen Y, Banerjee A, Albeanu DF, Navlakha S. 2025. An evolutionarily conserved scheme for reformatting odor concentration in early olfactory circuits. bioRxiv doi:10.1101/2025.01.23.634259

Sherman SM, Usrey WM. 2021. Cortical control of behavior and attention from an evolutionary perspective. *Neuron* **109**: 3048–3054. doi:10.1016/j.neuron.2021.06.021

Shubin N, Tabin C, Carroll S. 2009. Deep homology and the origins of evolutionary novelty. *Nature* **457**: 818–823. doi:10.1038/nature07891

Sun YC, Chen X, Fischer S, Lu S, Zhan H, Gillis J, Zador AM. 2021. Integrating barcoded neuroanatomy with spatial transcriptomic profiling enables identification of gene correlates of projections. *Nat Neurosci* **24**: 873–885. doi:10.1038/s41593-021-00842-4

Tosches MA. 2017. Developmental and genetic mechanisms of neural circuit evolution. *Dev Biol* **431**: 16–25. doi:10.1016/j.ydbio.2017.06.016

Tosches MA. 2021. From cell types to an integrated understanding of brain evolution: the case of the cerebral cortex. *Annu Rev Cell Dev Biol* **37**: 495–517. doi:10.1146/annurev-cellbio-120319-112654

Tosches MA, Yamawaki TM, Naumann RK, Jacobi AA, Tushev G, Laurent G. 2018. Evolution of pallium, hippocampus, and cortical cell types revealed by single-cell transcriptomics in reptiles. *Science* **360**: 881–888. doi:10.1126/science.aar4237

Van Lawick-Goodall J. 1971. Tool-using in primates and other vertebrates. *Adv Stud Behav* **3**: 195–249. doi:10.1016/S0065-3454(08)60157-6

Thomas JA, Moss CF, Vater M. 2002. *Echolocation in bats and dolphins*. University of Chicago Press, Chicago.

Walum H, Young LJ. 2018. The neural mechanisms and circuitry of the pair bond. *Nat Rev Neurosci* **19**: 643–654. doi:10.1038/s41583-018-0072-6

Wei JR, Hao ZZ, Xu C, Huang M, Tang L, Xu N, Liu R, Shen Y, Teichmann SA, Miao Z, et al. 2022. Identification of visual cortex cell types and species differences using single-cell RNA sequencing. *Nat Commun* **13**: 6902. doi:10.1038/s41467-022-34590-1

Wigner EP. 1960. The unreasonable effectiveness of mathematics in the natural sciences. Richard Courant lecture in mathematical sciences delivered at New York University, May 11, 1959. *Commun Pure Appl Math* **13**: 1–14. doi:10.1002/cpa.3160130102

Winding M, Pedigo BD, Barnes CL, Patsolic HG, Park Y, Kazimiers T, Fushiki A, Andrade IV, Khandelwal A, Valdes-Aleman J, et al. 2023. The connectome of an insect brain. *Science* **379**: eadd9330. doi:10.1126/science.add9330

Wirthlin ME, Schmid TA, Elie JE, Zhang X, Kowalczyk A, Redlich R, Shvareva VA, Rakuljic A, Ji MB, Bhat NS, et al. 2024. Vocal learning-associated convergent evolution in mammalian proteins and regulatory elements. *Science* **383**: eabn3263. doi:10.1126/science.abn3263

Wolff SB, Ölveczky BP. 2018. The promise and perils of causal circuit manipulations. *Curr Opin Neurobiol* **49**: 84–94. doi:10.1016/j.conb.2018.01.004

Cite this article as *Cold Spring Harb Perspect Biol* doi: 10.1101/cshperspect.a041510

Woych J, Ortega Gurrola A, Deryckere A, Jaeger ECB, Gumnit E, Merello G, Gu J, Joven Araus A, Leigh ND, Yun M, et al. 2022. Cell-type profiling in salamanders identifies innovations in vertebrate forebrain evolution. *Science* **377**: eabp9186. doi:10.1126/science.abp9186

Yamamoto D, Koganezawa M. 2013. Genes and circuits of courtship behaviour in *Drosophila* males. *Nat Rev Neurosci* **14**: 681–692. doi:10.1038/nrn3567

Zaremba B, Fallahshahroudi A, Schneider C, Schmidt J, Sarropoulos I, Leushkin E, Berki B, Van Poucke E, Jensen P, Senovilla-Ganzo R, et al. 2025. Developmental origins and evolution of pallial cell types and structures in birds. *Science* **387**: eadp5182. doi:10.1126/science.adp5182

Zheng Z, Lauritzen JS, Perlman E, Robinson CG, Nichols M, Milkie D, Torrens O, Price J, Fisher CB, Sharifi N, et al. 2018. A complete electron microscopy volume of the brain of adult *Drosophila melanogaster*. *Cell* **174**: 730–743.e22. doi:10.1016/j.cell.2018.06.019

Zheng XM, Harpole CE, Davis MB, Banerjee A. 2025. Parametric modulation of a shared midbrain circuit drives distinct vocal modes in a singing mouse. bioRxiv doi:10.1101/2025.04.04.647309

Zhu P, Frank T, Friedrich RW. 2013. Equalization of odor representations by a network of electrically coupled inhibitory interneurons. *Nat Neurosci* **16**: 1678–1686. doi:10.1038/nn.3528

Modeling the Emergence of Circuit Organization and Function during Development

Shreya Lakhera,[1] Elizabeth Herbert,[1] and Julijana Gjorgjieva

School of Life Sciences, Technical University of Munich, 85354 Freising, Germany

Correspondence: gjorgjieva@tum.de

Developing neural circuits show unique patterns of spontaneous activity and structured network connectivity shaped by diverse activity-dependent plasticity mechanisms. Based on extensive experimental work characterizing patterns of spontaneous activity in different brain regions over development, theoretical and computational models have played an important role in delineating the generation and function of individual features of spontaneous activity and their role in the plasticity-driven formation of circuit connectivity. Here, we review recent modeling efforts that explore how the developing cortex and hippocampus generate spontaneous activity, focusing on specific connectivity profiles and the gradual strengthening of inhibition as the key drivers behind the observed developmental changes in spontaneous activity. We then discuss computational models that mechanistically explore how different plasticity mechanisms use this spontaneous activity to instruct the formation and refinement of circuit connectivity, from the formation of single neuron receptive fields to sensory feature maps and recurrent architectures. We end by highlighting several open challenges regarding the functional implications of the discussed circuit changes, wherein models could provide the missing step linking immature developmental and mature adult information processing capabilities.

Development is a complex, protracted, and dynamic process over which multiple mechanisms interact to establish the organization and function of adult neural circuits. Although early development is primarily driven by genetic programs, activity-dependent processes play a crucial role in refining network connectivity and shaping the emergence of functional properties. Before the onset of sensory experience, many developing neural circuits generate spontaneous activity with nonrandom spatiotemporal correlations that change as the animal matures. An emerging viewpoint is that this activity, via activity-dependent synaptic plasticity mechanisms, drives refinements underlying the emergence of circuit function. With the recent advent of new technologies to simultaneously record the activity of many neurons over a prolonged time period, we have begun to characterize how synaptic, cellular, and population activity features evolve over development. A key challenge in understanding circuit development lies in deciphering how such spontaneous activity is generated in early networks and how it could shape and refine network connectivity by activity-dependent synaptic plasticity. Theoretical frameworks and computational models offer a

---

[1]These authors contributed equally to this work.

Cite this article as *Cold Spring Harb Perspect Biol* doi: 10.1101/cshperspect.a041511

powerful means to investigate this process by proposing and exploring specific hypotheses and to provide unique insights into the mechanisms shaping neural circuitry and function.

Here, we highlight recent contributions of theoretical and computational studies to our understanding of activity-dependent developmental processes, focusing mainly on sensory systems and the hippocampus in rodents. We first discuss models that shed light on the generation of early spontaneous activity characterized by synchronous bursts or propagating waves at or shortly after birth, and the progressive decorrelation and sparsification of such activity over subsequent stages of development. We then consider how in silico simulated scenarios can inform how spontaneous activity patterns shape circuit refinements via activity-dependent plasticity mechanisms, leading to the emergence of input selectivity and refined processing capabilities. Beyond establishing feedforward receptive fields and tuning properties between the sensory periphery and downstream regions, we also focus on how recurrent network structures are established downstream, and the role of inhibitory synaptic plasticity in this refinement process.

SPONTANEOUS ACTIVITY IN DEVELOPING CIRCUITS

Before the onset of sensory experience, neuronal networks in the developing brain are not merely silent. Instead, they spontaneously generate patterns of electrical activity that play a crucial role in establishing and refining neural circuits. Early in development, this activity typically takes the form of synchronous bursts, oscillations, or traveling waves, which punctuate periods of baseline quiescence and transiently activate large numbers of neurons at the same time (Khazipov et al. 2004; Yang et al. 2009; Seelke and Blumberg 2010). As development progresses, neurons can sustain higher baseline firing rates but display less correlated firing (Golshani et al. 2009; Colonnese et al. 2010; Chini et al. 2022). Spontaneous network activity also becomes sparser, activating fewer neurons at any given time (Kerr et al. 2005; Rochefort et al. 2009; Frye and MacLean 2016; Wosniack et al. 2021). This transition from intermittent synchronous bursting to sustained and decorrelated spontaneous activity reflects the maturation of neural circuits toward a system capable of precise, reliable, and efficient stimulus encoding in adulthood (Gjorgjieva et al. 2014; Avitan et al. 2021; Glanz et al. 2021; Jia et al. 2022; Trägenap et al. 2023). Experiments have identified multiple factors that influence this shift, including changes in intrinsic neuronal properties, synaptic transmission, and connectivity motifs (Mease et al. 2013; Lohmann and Kessels 2014; Pouchelon et al. 2021; Kalemaki et al. 2022). Theoretical models can help to disentangle the contribution of these factors and propose mechanisms behind the generation of spontaneous activity observed experimentally. Here, we first highlight recent theoretical work that addresses the generation of early synchronous events and the developmental shift toward progressively sparser, decorrelated, and persistent activity.

GENERATION OF SYNCHRONOUS NETWORK EVENTS IN EARLY DEVELOPMENT

Transient bursts or waves of neural activity that synchronize large networks of neurons are a key signature of early brain development and have been observed across regions and developmental stages (Hanganu et al. 2006; Golshani et al. 2009; Ackman et al. 2012; Kirmse et al. 2015; Smith et al. 2018; for review, see Martini et al. 2021). Generated spontaneously before the onset of mature sensory transduction or refined motor abilities, these events lack any obvious correlate or trigger in the external world. In peripheral sensory areas such as the retina and cochlea, specific intrinsic neuronal mechanisms enable neurons to spontaneously burst (Blankenship and Feller 2010; Ackman et al. 2012; Wang and Bergles 2015). For example, before eye opening (~P15 in rodents), the retina generates bursts of action potentials that exhibit strong spatiotemporal correlations and propagate as waves, known as retinal waves. Experimental work has classified retinal waves into three different developmental stages based on their spatiotemporal properties and the mechanisms of generation (Blankenship and Feller 2010). Theoretical models have been central to understanding retinal waves, including the mechanisms for their initiation, propagation, and termination, focusing on the interaction between intrinsic neuronal properties and electrical or synaptic transmission, reviewed in Godfrey and Eglen (2009) and Gjorgjieva and Eglen (2011).

The early developmental spontaneous events in the sensory periphery propagate to the cortex, where they can trigger the synchronous activation of local networks. In rodents, in vivo electrophysiology has identified spindle bursts, transient oscillations of the local field potential (LFP) present during the first postnatal week in the somatosensory (Khazipov et al. 2004; Minlebaev et al. 2007), visual (Hanganu et al. 2006; Colonnese et al. 2010), motor (An et al. 2014), and prefrontal cortex (Fig. 1A; Brockmann et al. 2011). In human preterm babies, these events are observed at a stage

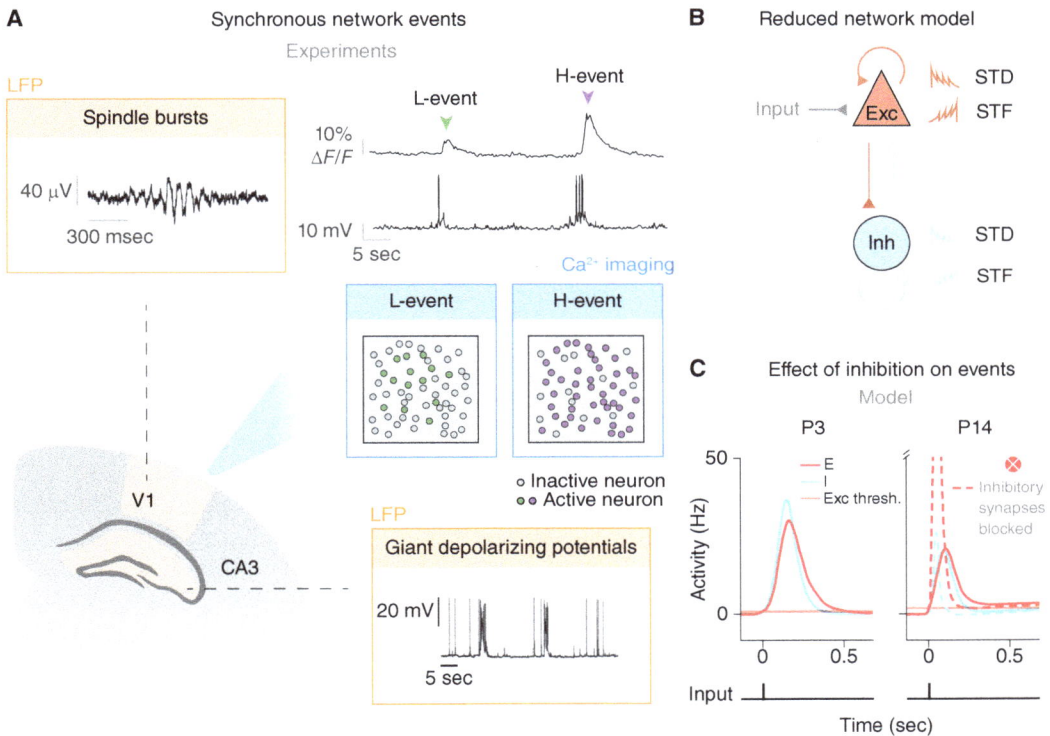

Figure 1. Spontaneous activity in the cortex: experiments and models. (*A*) Developmental spontaneous activity is most commonly observed in the cortex and hippocampus through electrophysiological local field potential (LFP) recordings or calcium imaging. Shown are representative traces of spindle bursts (LFP recordings in V1; based on data in Hanganu et al. 2006), low-synchronicity L-events, and high-synchronicity H-events (calcium imaging in V1; based on data in Siegel et al. 2012 and Leighton et al. 2021), and giant depolarizing potentials (GDPs; whole-cell patch-clamp recordings in CA3; based on data in Lombardi et al. 2018). (*B*) Reduced mean-field model of a recurrent EI network with short-term plastic connections (based on data in Rahmati et al. 2017). All excitatory (red) and inhibitory (blue) synapses are subject to short-term depression (STD) and short-term facilitation (STF). The model generates bursts of spontaneous activity reflective of spindle bursts or GDPs in response to input perturbations. (*C*) Providing a pulse input to the excitatory population leads to a large network event (*left*), which is terminated by STD in the excitatory connections. The model captures several developmental trends observed experimentally, including the reduction in amplitude of network events in the second postnatal week when inhibitory synaptic strength is stronger, and accordingly larger events when inhibition is blocked (*right*). (Simulations based on Rahmati et al. 2017.)

before eye opening in rodents and correspond to delta brush oscillations in noninvasive electroencephalogram (EEG) imaging (Milh et al. 2007). Spindle bursts in sensory cortices are elicited by activity occurring in the respective sense organs; for example, spindle bursts occurring in somatosensory and motor cortices are driven by spontaneous movements and myoclonic twitches (Kreider and Blumberg 2000; Khazipov et al. 2004; Blumberg et al. 2013), and those in the visual cortex are triggered by retinal waves (Hanganu et al. 2006). During the second postnatal week, two-photon Ca^{2+} imaging in the mouse primary visual cortex has found that the spindle bursts triggered by retinal waves manifest as low-synchronicity events known as L-events, which have low cell participation (20%–80% in a recorded population of 50–100 cells) and low amplitude (Siegel et al. 2012; Leighton et al. 2021; Tezuka et al. 2022). Such events occur alongside high-synchronicity events with a much higher cell participation (>80%) known as H-events, which have a high amplitude and are largely independent of retinal drive (Fig. 1A; Hanganu et al. 2006; Siegel et al. 2012; Leighton et al. 2021; Tezuka et al. 2022). Wide-field Ca^{2+} imaging of spontaneous activity during the same time period also identified two types of events with local and global spatial spread, and linked them, respectively, to the putative L- and H-events measured with two-photon imaging based on similar amplitudes and frequencies (Ackman et al. 2012; Gribizis et al. 2019; Leighton et al. 2021).

A class of theoretical models has provided important insight into the generation and spatiotemporal evolution of these periphery-driven spontaneous events in the sensory cortex from a dynamical systems perspective. Known as mean-field models, these network models consist of coupled excitatory and inhibitory units that represent the mean activity of a cortical network by averaging entire excitatory and inhibitory populations and individual synaptic properties (Fig. 1B). Constrained with neuronal and synaptic parameters from early development, Rahmati et al. (2017) used a mean-field model to demonstrate that events such as spindle bursts can be initiated when peripheral inputs drive the cortical excitatory population above an intrinsic threshold of

instability (Fig. 1C; Rahmati et al. 2017). Specifically, when using parameters corresponding to early developmental stages (P3–P10, rodent visual cortex), the model network remains silent in the absence of input, consistent with postnatal circuits being largely quiescent at rest. In the framework of dynamical systems, this corresponds to having a single stable fixed point at a firing rate of zero, to which the network returns after a small perturbation triggered by the peripheral drive. However, introducing short-term plasticity (STP) to the synaptic weights allows supra-threshold inputs to the excitatory population to be first amplified and then attenuated, producing a transient network event that strongly activates both excitatory and inhibitory populations. This event is reminiscent of experimentally observed spindle bursts that recruit both glutamatergic pyramidal neurons and GABAergic interneurons (Khazipov et al. 2004; Hanganu et al. 2006). Mathematical analysis revealed that this network event arises from a second, transient fixed point embedded in the network's fast-firing dynamics, which is induced by the external input and vanishes because of short-term depression at excitatory synapses after the network event reaches peak amplitude.

Inhibitory activity has been difficult to characterize experimentally in early development because parvalbumin (PV), a protein expressed by the main subtype of inhibitory interneurons, is expressed late, hence prohibiting calcium imaging (see, e.g., Leighton et al. 2021; Pouchelon et al. 2021; Baruchin et al. 2022), and the onset of inhibitory influence has been the subject of frequent debate. GABAergic interneurons are slow to acquire their adult inhibitory properties, but the exact timescale over which inhibition develops in the in vivo sensory cortex and its role in regulating cortical dynamics at this stage remain unclear (Kirmse et al. 2015; Valeeva et al. 2016; Murata and Colonnese 2020; Peerboom and Wierenga 2021). In addition to being amenable to mathematical analysis, the mean-field model discussed above was also used to study the role of inhibition in generating spontaneous activity, specifically in modulating the initiation and spatiotemporal properties of network events (Rahmati et al. 2017). First, when using parameters based on the first postnatal week when inhibitory synaptic strength is weak, network events can be easily triggered. However, when using parameters based on the second postnatal week when inhibitory synaptic strength is stronger, network events become harder to generate. Simulating blocked inhibition at this later stage results in network events with much larger amplitude, implying that the developmental strengthening of inhibition suppresses network events (Fig. 1C). This supports recent experimental findings that GABAergic interneurons already exert an inhibitory effect in the cortical network by the second postnatal week (Kirmse et al. 2015; Valeeva et al. 2016; Che et al. 2018; Murata and Colonnese 2020; Leighton et al. 2021) and suppress synchronous events occurring in sensory cortices at this time (Duan et al. 2020). Indeed, recent in vivo Ca^{2+} imaging and electrophysiological recordings have identified strong inhibition from somatostatin-expressing (SST) interneurons as the main factor controlling the ratio between H-events and spindle-burst-like L-events (Leighton et al. 2021) and restricting the recruitment of excitatory neurons into L-events.

In the developing hippocampus, certain patterns of spontaneous activity are believed to rely on mechanisms endogenous to the local circuit, in contrast to the periphery-driven events in the sensory cortex. Electrophysiological recordings in slices throughout the first postnatal week have characterized one of the most prominent activity patterns in hippocampal areas CA1 and CA3, giant depolarizing potentials (GDPs). Spontaneously active excitatory pyramidal cells with strong recurrent connections are commonly assumed responsible for the initiation of GDPs (Griguoli and Cherubini 2017). As in the cortex (Rahmati et al. 2017), the same class of theoretical mean-field models has also been used to understand the generation of in vivo events likely corresponding to GDPs in the hippocampus (Graf et al. 2022). Dynamical systems analysis of the model revealed an input-dependent transition between two states in the network: first, a silent state with a fixed point at zero firing rate in which input perturbations to the excitatory population trigger GDPs via intrinsic instability-driven dynamics, and, second, an active state

with an additional fixed point at a nonzero firing rate in which activity is continuous and GDPs cannot be generated. Simulating the developmental strengthening of inhibitory synapses allows the network to spend more time in the active state, which the authors relate to an experimentally observed transition from discontinuous to continuous activity in mouse hippocampus during the second postnatal week (Graf et al. 2022).

As in the sensory cortex, GABAergic interneurons have also been found to play an important role in shaping the properties of GDPs and network events at early postnatal stages in the hippocampus, both in vitro (Flossmann et al. 2019) and in vivo (Dard et al. 2022). Ca^{2+} imaging, electrophysiology, and optogenetic manipulation in vitro have revealed an excitatory role of GABA during the first postnatal week, finding that the activation of GABAergic SST interneurons can strongly promote GDPs during this time (Flossmann et al. 2019). Further analysis of the aforementioned mean-field model revealed that an excitatory action of GABA may support GDP initiation by modulating the intrinsic threshold of instability of the excitatory population (Flossmann et al. 2019). These studies exemplify the effectiveness of reduced models in identifying key features of network dynamics conducive to synchronous network events, such as instability thresholds and STP mechanisms.

The distribution of synaptic connections and single-neuron properties within a network—typically averaged out in the mean-field models discussed earlier—may also be central to generating synchronous network events. Indeed, electrophysiological and imaging experiments have identified GABAergic interneurons (putative SST neurons identified in Flossmann et al. 2019) in both CA1 and CA3 of the hippocampus to be "functional hubs," whose single-neuron stimulation can influence the activity of the entire population (Bonifazi et al. 2009; Picardo et al. 2011). These functional hub neurons have been shown in vitro to synchronize network-wide activity within CA1 and CA3 and thus orchestrate GDPs at the single-neuron level in the first postnatal week (Bonifazi et al. 2009), consistent with a potentially excitatory action of GABA at this stage. Modeling work suggests that such functional hubs may arise because of the unique combination of network topology and intrinsic cellular excitability present in the hippocampus during early development. In developing networks, both early-born, "mature" neurons and later-born, "young" neurons coexist simultaneously, because different neuronal populations mature at different rates. Assuming that younger neurons are more excitable but more sparsely connected, whereas more mature neurons are less excitable but more densely connected (Doetsch and Hen 2005; Ge et al. 2006; Picardo et al. 2011), Luccioli et al. (2014) showed in a spiking network model of excitatory neurons that this developmental heterogeneity in neuronal maturity could give rise to the existence of functional hubs. In the model, functional hubs were identified as highly excitable neurons that were the first to activate upon the onset of a synchronous network burst and could strongly impact network dynamics upon single-neuron stimulation. Because activity in the entorhinal cortex drives the maturation of the hippocampus, Donato et al. (2017) and Mòdol et al. (2017) investigated functional connectivity in the entorhinal cortex and found similar hub neurons. They further identified a second class of interneurons called low functional connected drivers, capable of impacting network dynamics when stimulated but not reliably activated at the onset of network bursts. The earlier maturation of the entorhinal cortex compared to CA3 and CA1 (Donato et al. 2017) suggests a higher probability of finding interneurons with more adult-like, inhibitory neurotransmission. Indeed, by including inhibitory neurons in the model of excitatory neurons with developmental heterogeneity, Luccioli et al. (2018) could explain the presence of low functionally connected driver cells. Together, these models highlight how the ability of single neurons to orchestrate diverse network-wide dynamics can emerge from the specific interaction between connectivity profiles, excitability, and neurotransmission present in different regions of the neonatal brain.

The same factors may also influence the spatial structure of spontaneous activity patterns across developmental ages and species. In the ferret, multiple cortices (visual, auditory, and parietal cortex) show modular patterns of spontaneous activity with long-range spatial correlations (Smith et al. 2018; Powell et al. 2022), different from the spindle bursts, L-events, or H-events in the mouse visual cortex. In the visual cortex, these modular patterns are present before eye opening and are independent of the retinal drive, suggesting their generation relies on recurrent connectivity within the cortex. However, at this age, long-range horizontal connections between orientation-selective neurons have not yet been established. Long-range activity correlations can nonetheless arise from heterogeneously oriented, elliptical connectivity profiles that induce directions of local facilitation and lateral suppression and allow activity to spread through neighboring neurons (Smith et al. 2018; Dahmen et al. 2022). This implies that long-range recurrent connections are not necessary for generating highly synchronous activity in distant neurons and suggests moreover that heterogeneity in lateral connectivity may provide a backbone for the subsequent organization of long-range connections via activity-dependent Hebbian plasticity mechanisms.

DESYNCHRONIZATION AND SPARSIFICATION OF NEURAL ACTIVITY THROUGHOUT DEVELOPMENT

Spontaneous activity in the neocortex becomes progressively more desynchronized during development, giving way to an asynchronous activity state that forms the basis of information processing in adult circuits. In this state, individual neurons sustain higher firing rates on average (as opposed to the quiescence of early networks punctuated by infrequent bursts), but population activity is sparse, with fewer neurons active at any given time (Fig. 2A). Continuous low-amplitude oscillations replace network bursts observed in the LFP (Colonnese et al. 2010; Harris and Thiele 2011; Shen and Colonnese 2016), and the activity of neighboring neurons observed in Ca^{2+} imaging and electrophysiology strongly decorrelates (Golshani et al. 2009; Rochefort et al. 2009; Che et al. 2018; Chini et al. 2022). The gradual developmental strengthening of recurrent excitatory connections between pyramidal

Cite this article as *Cold Spring Harb Perspect Biol* doi: 10.1101/cshperspect.a041511

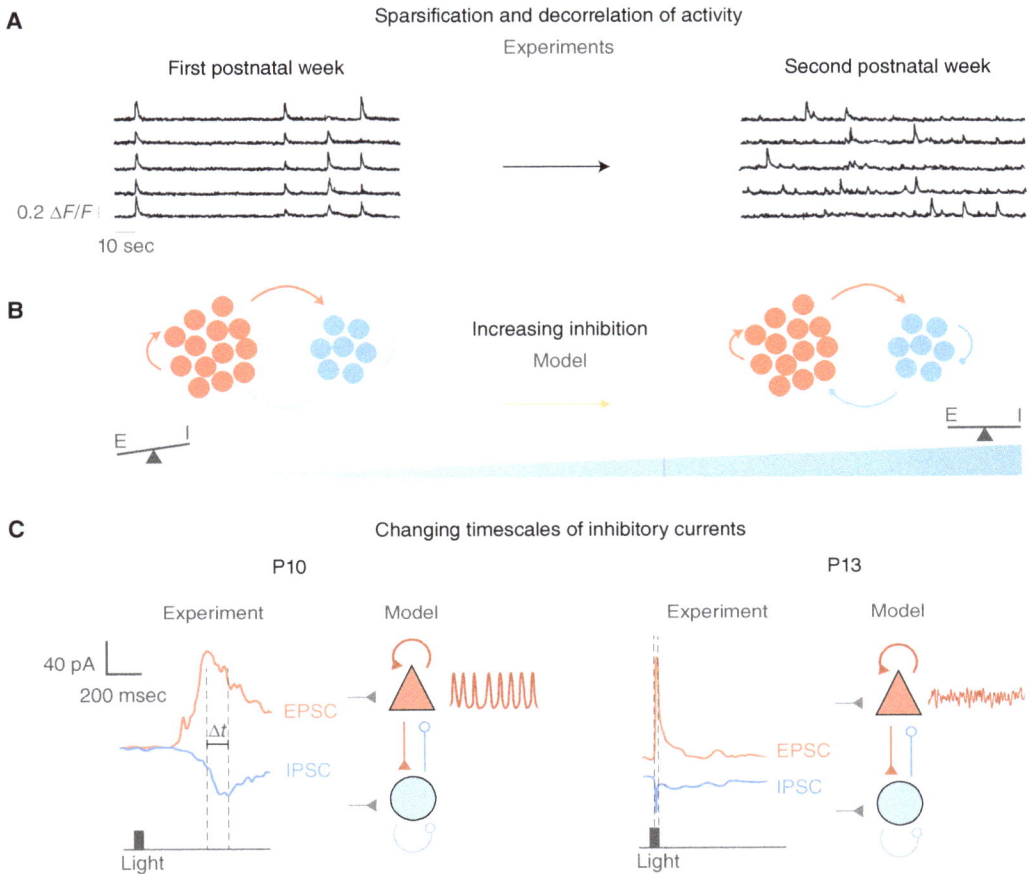

Figure 2. Mechanisms for sparsification of neural activity. (*A*) Network activity transitions from silent activity punctuated by synchronous network bursts to sparse, continuous, and decorrelated activity over the second postnatal week. (Based on data from Rochefort et al. 2009.) (*B*) A relative increase in inhibition through the strengthening of inhibitory synapses (Chini et al. 2022) or the increase in excitability of inhibitory interneurons (Maldonado et al. 2021) are plausible factors driving this transition. (*C*) Faster kinetics of inhibitory synaptic currents reduce the time delay between excitatory and inhibitory responses to perturbations, causing large network events to disappear. (Based on data from Romagnoni et al. 2020.)

neurons (Ko et al. 2013) likely facilitates the ability of cortical circuits to sustain persistent activity (Barak and Tsodyks 2007; Rahmati et al. 2017). The concurrent desynchronization and sparsification of such activity arise from multiple developmental changes occurring in parallel at the cellular and circuit levels. Recent modeling work has dissected the mechanisms underlying these developmental changes in spontaneous activity, including changes in the relative balance between excitation and inhibition, in single-neuron properties, and in connectivity patterns.

One factor driving the developmental desynchronization and sparsification of activity is the strong increase in the inhibitory influence of GABAergic interneurons. Across brain regions, the weak impact of inhibition at the cellular and network level at birth progressively increases over development (Fig. 2B). Throughout development, neurons receive a decreasing ratio of excitatory to inhibitory conductances across stimulus conditions until an approximate balance is reached by adulthood (Dorrn et al. 2010; Zhang et al. 2011). This strengthening of inhibition over development is driven by multiple factors, including the progressive migration and integration of interneurons into the cortex (Daw et al. 2007), an overall increase in the number and density of inhibitory synapses (Etherington and Williams 2011), the gradual shift in chloride reversal potential (Lazarus and Huang 2011), changes in active and passive membrane properties (Le Magueresse and Monyer 2013), and an acceleration of the kinetics of postsynaptic currents due to changing receptor subunit compositions (Pangratz-Fuehrer and Hestrin 2011).

Computational models elucidate how these diverse factors that increase inhibitory influence interact to produce developmental changes at the network level. For example, Rahmati et al. (2017) observed two notable developmental changes when mean-field models for spontaneous activity generation in the visual cortex were parameterized for later developmental stages (P10 and P20). First, synchronous network bursts became harder to generate, consistent with the experimentally observed disappearance of retinal wave–triggered spindle bursts by P14 (Murata and Colonnese 2016; Shen and Colonnese 2016). Second, new stable fixed points in the model arose at nonzero firing rates, indicating the capability of the network to sustain persistent activity states that can be metastable and have been implicated in different computational regimes of adult networks (Mazzucato et al. 2019; Stringer et al. 2019; Modol et al. 2020; Nakazawa et al. 2020). In the model, two factors drove the disappearance of network bursts: a strong increase in absolute inhibitory synaptic strength and a decrease in the efficacy of synaptic transmission by a change in excitatory short-term facilitation. Several other factors support the emergence of persistent activity states, including an increase in the activation thresholds in both excitatory and inhibitory populations and a speeding up of inhibitory short-term depression. Notably, many of the model parameters driving developmental changes in spontaneous activity boost the strength of inhibitory synaptic connections relative to excitatory. Hence, the model makes specific predictions for the role of each parameter, which could subsequently be tested in perturbation experiments that selectively manipulate each biological correlate (Rahmati et al. 2017).

In addition to inhibitory strengths, the timing of inhibition relative to excitation could also be an important factor behind the developmental sparsification in network dynamics. In the rodent visual cortex, the delay between the peaks of light-evoked inhibitory and excitatory postsynaptic currents decreases substantially between P10 and P13 shortly before eye opening (Fig. 2C), suggesting that inhibition speeds up fractionally faster than excitation (Colonnese 2014). In a different mean-field excitatory–inhibitory model, this change in delay was found to be the key parameter driving a switch from oscillatory to stable dynamics that matches activity changes observed in vivo during development (Romagnoni et al. 2020). Several biophysical factors could underlie the faster kinetics of inhibitory currents, including the progressive integration of fast-spiking PV-expressing interneurons into the cortex (Colonnese 2014; Modol et al. 2020) or the rearrangements in thalamocortical input to interneuron subtypes (Daw et al. 2007; Marques-Smith et al. 2016; Tuncdemir et al. 2016; Guan et al. 2017). In the model, the transition could occur even in the absence of changes in feedforward input, suggesting that increased thalamic input is a complementary but not necessary component of the switch (Romagnoni et al. 2020).

Beyond the mean-field models that average activity across individual neurons, increasing the net strength of inhibition in large spiking networks of excitatory and inhibitory neurons also leads to sparser firing with reduced pairwise spike correlations between neurons (Chini et al. 2022). This result recapitulates the sparsification occurring in the first two postnatal weeks and also when optogenetically stimulating inhibitory neurons in the prefrontal cortex in vivo (Chini et al. 2022). Decorrelation of neuronal activity due to increased inhibition has also been observed in the mouse primary visual cortex upon application of the neuromodulator oxytocin, which increases the excitability of SST interneurons, and was also captured in a large spiking network (Maldonado et al. 2021). Therefore, despite the distinct manner of increasing net inhibition in each case—by increasing the ratio of inhibitory and excitatory conductances (Chini et al. 2022) or by increasing the resting membrane potential of a subset of inhibitory neurons (Maldonado et al. 2021)—the similar influence on network decorrelation and sparsification points to the robustness of the proposed mechanism.

Interestingly, although the period of sparsification coincides with the onset of sensory experience, the sparsification process is largely independent of sensory inputs, and sensory deprivation has little effect on the sparsification of neural activity (Golshani et al. 2009; Rochefort et al. 2009). In line with this, spontaneous activity drives activity-dependent changes in network connectivity that may also support the sparsification of spontaneous activity observed in the cortex. Although feedforward thalamocortical connections in sensory pathways are initially diffuse and imprecise, activity-dependent synaptic plasticity can use developmental spontaneous activity to refine them into specific topographic projections over the first postnatal weeks of rodent life (Ackman and Crair 2014; Richter and Gjorgjieva 2017; Thompson et al. 2017). A network model for this type of refinement can lead to cortical activity in which network bursts downstream from the sensory periphery occur more rarely and recruit fewer neurons, reminiscent of sparsification (Wosniack et al. 2021).

Alongside the developmental increase in inhibition and refinement of network connectivity, changes in biophysical single-neuron properties may also contribute to changes in synchronization between neuronal populations. A developmental increase in the ratio of sodium to potassium conductances can endow single neurons with the property of gain scaling as they mature, which allows them to respond adaptively to the amplitude of stimuli encountered (Mease et al. 2013). Modeling work has shown that the absence of gain scaling at birth can allow peripheral spontaneous activity to be propagated across many downstream layers in the form of synchronous bursts. As single-neuron gain scaling emerges over the first postnatal week, network responses become graded and less synchronous and recruit fewer cells at any given time, also

reminiscent of sparsification (Gjorgjieva et al. 2014).

In summary, computational models have provided unique insights into both the generation of synchronous network bursts prevalent in the early postnatal brain and the transition to sustained, continuous, and decorrelated activity that takes place over subsequent phases of development. By capturing transitions between dynamical regimes, models have been especially informative in understanding the impact of cellular and synaptic parameters, either by extracting such parameters from experiments (Rahmati et al. 2017) or by deriving them based on target features of activity (Romagnoni et al. 2020). The latter approach can be used to ultimately derive a specific developmental parameter trajectory that recapitulates in vivo activity patterns, thus forming predictions that can be tested experimentally.

ACTIVITY-DEPENDENT REFINEMENT OF CONNECTIVITY THROUGH SYNAPTIC PLASTICITY

As spontaneous activity patterns evolve during development, they play a crucial role in setting up structured connectivity in developing neural circuits. Emerging evidence suggests that activity-dependent synaptic plasticity mechanisms use the synchronized and wave-like activity patterns generated in developing circuits to establish the connectivity required for sensory processing and behavior in mature circuits. In sensory cortices in particular, two types of connectivity have typically been the subject of experimental and theoretical work. The first is the establishment of receptive fields at the single-neuron level, defined as the set of inputs to the neuron that gives rise to its response properties (Fig. 3A). These receptive fields refine during development by reducing the number of inputs (Chen and Regehr 2000; Thompson et al. 2017) and often (but not always) represent the neuron's tuning to a particular feature of the sensory input—for example, the orientation of an object in the visual cortex or tone frequency in the auditory cortex (Hubel and Wiesel 1962; Knudsen and Konishi 1978; Clopath et al. 2010; Gjorgjieva et al. 2011; Vogels et al. 2011). The second is the emergence of

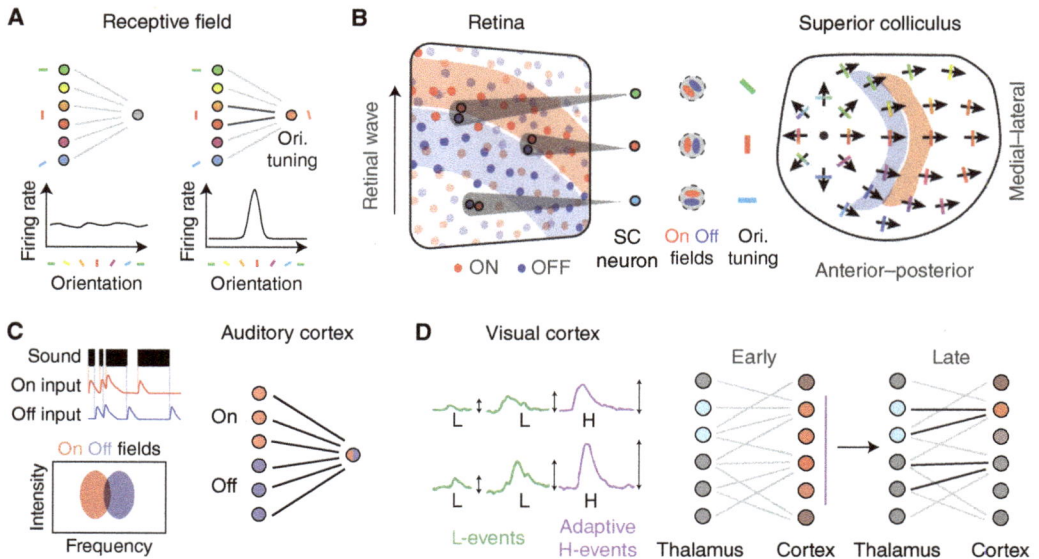

Figure 3. Spontaneous activity-driven refinement of feedforward connections. (*A*) The emergence of a receptive field in a postsynaptic neuron by the selective refinement of multiple presynaptic inputs. Line thickness indicates synaptic strength. (*B*) Stage III waves in the retina that asynchronously activate ON (red) and OFF (blue) retinal ganglion cells (Kerschensteiner and Wong 2008) can generate receptive fields in the superior colliculus consisting of separate but adjacent ON and OFF subfields, leading to orientation selectivity. Note that here orientation selectivity is directly represented by the composition of the receptive field and not only by the firing rate of the neuron due to the inputs it receives, as in *A*. The directional bias of stage III waves generates the concentric orientation map in the superior colliculus. (Schematic adapted from Teh et al. 2023.) (*C*) Natural auditory stimuli invoke alternating responses to sound onset (ON) and offset (OFF) (*upper left*). In a model of ON and OFF inputs to an auditory cortex neuron (*right*), this leads to a separation of ON and OFF subfields at adjacent frequencies (*lower left*) as seen in the adult. (Schematic adapted from Sollini et al. 2018.) (*D*) Modeling the role of spontaneous low-synchronicity (L) events and high-synchronicity (H) events in the visual cortex on the refinement of thalamocortical connectivity. L-events refine receptive fields in the cortex by strengthening the connections to fewer thalamocortical inputs, generating topographic connectivity, which preserves neighboring relations between the thalamus and the cortex. H-events adaptively regulate their amplitude to homeostatically regulate synaptic strength. (Schematic adapted from Wosniack et al. 2021.)

organized feedforward connectivity (typically called a map) between neuronal populations, one defined as an input and the other as a target projection area. In most sensory systems, these connectivity maps are topographically organized, whereby stimulus features in the input area (e.g., the sensory periphery) are represented in downstream areas (e.g., the cortex) in a manner that preserves neighboring relations. For instance, the map between the retina and downstream visual areas—such as the dorsal lateral geniculate nucleus (dLGN) of the thalamus, the superior colliculus, and the primary visual cortex—is known as a retinotopic map (Schuett et

al. 2002; Mrsic-Flogel et al. 2005; Piscopo et al. 2013). Similar topographic maps exist in the somatosensory and auditory systems (Garrett et al. 2014; Mizuno et al. 2018; Antón-Bolaños et al. 2019; Kersbergen et al. 2022). Here, we review theoretical and modeling frameworks inspired by extensive experimental evidence of the multiple ways in which spatiotemporal correlations in early peripheral activity can set up receptive fields and establish topographic connectivity (Cang and Feldheim 2013; Kirkby et al. 2013; Richter and Gjorgjieva 2017). Refined feedforward connectivity influences and interacts with downstream network activity patterns in a manner that ultimately shapes local recurrent connectivity and, indirectly, top-down modulatory feedback (Murakami et al. 2022). We also review recent theoretical work studying how local network dynamics instruct the formation of appropriate recurrent connections (Ko et al. 2013; Kim et al. 2020). The emerging picture is that of feedforward and recurrent connections becoming gradually aligned with each other, which is beneficial for the amplification and stabilization of sensory-evoked responses following early development (Lempel and Fitzpatrick 2023).

The phenomenological description of activity-dependent plasticity mechanisms that are used to translate correlated activity patterns into synaptic connectivity changes are known as learning rules. Most of these learning rules are based on the Hebbian principle that coincident pre- and postsynaptic activity potentiates synaptic weights (Hebb 1949) but can depend on different aspects of neural activity, including firing rates, spike timing, voltage, and calcium (Bi and Poo 1998; Abbott and Nelson 2000; Song et al. 2000; Clopath et al. 2010; Graupner and Brunel 2012; Magee and Grienberger 2020). Their exact dependence varies across brain regions, species, neuron type, and developmental age (Turrigiano and Nelson 2004; Lohmann and Kessels 2014). Most experimental studies characterizing spontaneous activity in development have found much slower correlation timescales than the corresponding correlation timescales in adult activity, raising questions about the nature of the activity-dependent learning rules using this activity (Butts et al. 2007; Colonnese and Khazipov 2010; Winnubst et al. 2015). For instance, retinal waves have burst durations of ~1 sec (Meister et al. 1991; Wong et al. 1993; Ackman et al. 2012) and transmit information at a timescale of 1–2 sec (Butts and Rokhsar 2001). Therefore, several models have demonstrated that the activity-dependent plasticity rules that interpret developmental spontaneous activity patterns into connectivity must be operating on similar timescales (Butts et al. 2007; Gjorgjieva et al. 2009; Bennett and Bair 2015; Winnubst et al. 2015). Below we review these models according to the type of connectivity they investigate.

EMERGENCE AND REFINEMENT OF STRUCTURED FEEDFORWARD CONNECTIVITY

Locally correlated activity patterns that activate a few neighboring neurons are ideally poised to establish topographic feedforward connectivity and refine sensory receptive fields. Classical theoretical studies explained how Hebbian plasticity in rate-based models can instruct topographic map formation in feedforward connections using locally correlated activity (Hebb 1949; Miller and MacKay 1994; Richter and Gjorgjieva 2017). However, more recent work has provided extensive new evidence about the nature of correlated activity and the type of activity-dependent plasticity rules adapted to this activity.

Activity-based connectivity refinement is best understood in the visual system, where the structure of the driving input from the sensory periphery has been extensively characterized. For instance, retinal waves provide a source of correlated activity that drives the segregation of retinotopically organized inputs into distinct regions in the thalamus and visual cortex, leading to the formation of the retinotopic map between the retina and thalamus, as well as thalamus and visual cortex (Kirkby et al. 2013; Ko et al. 2013; Ackman and Crair 2014; Thompson et al. 2017). Because retinal waves consist of bursts, phenomenological learning rules that depend on bursts have been proposed to instruct receptive field formation and refinement (Butts et al. 2007; Gjorgjieva et al. 2009). Even when using spike-timing-dependent plasticity (STDP), modeling work has shown that the time constant of STDP should be matched to retinal wave speed to set the spatial scale of emergent connectivity and receptive field size (Bennett and Bair 2015). Biologically, this matching of timescales between plasticity rules and spontaneous activity correlations could be attributed to development-specific synaptic properties, such as a shift in the receptor composition of N-methyl-D-aspartate (NMDA) receptors (Liu and Chen 2008) or the dominance of slower NMDA over faster α-amino-3-hydroxy-5-methyl-4-isoxazolepropionic acid (AMPA) receptors (Tikidji-Hamburyan et al. 2023). Recent modeling work found that such dominance is necessary to prevent initial unrefined connections from the retina to thalamic neurons from generating misinformative, so-called "parasitic" correlations at fast timescales, which could misguide connectivity refinement driven by plasticity and limit the transmission of spatially informative input correlations at slower timescales (Tikidji-Hamburyan et al. 2023).

With the developmental modification of spontaneous activity properties, the properties of the receptive fields and maps driven by this activity through synaptic plasticity also evolve. As developmental spontaneous activity is localized in space and time, more precise temporal correlations begin refining finer receptive field aspects, including their segregation into separate subfields. By providing simulated stage III retinal waves as input to a feedforward rate network with Hebbian plasticity, Teh et al. (2023) studied how specific temporal correlations in these waves affect the organization of receptive fields in the superior colliculus. In particular, the unique firing during stage III retinal waves where ON retinal ganglion cells precede the firing of OFF retinal ganglion cells (Kerschensteiner and Wong 2008) was found to guide the emergence of orientation tuning of individual collicular cells by generating aligned ON and OFF subfields (Fig. 3B, left). Additionally, modeling spatiotemporal correlations of retinal waves supported the formation of the orientation preference map as in the mouse superior colliculus, whereas introducing a propagation bias as found at the transition of stage II to stage III waves induced a concentric organization of orientation tuning and aligned the center of the orientation map to the center of vision (Fig. 3B). This extends previous modeling work that has found that temporally delayed inputs can generate direction selectivity in feedforward networks, either through lagged input channels (Blais et al. 2000) or delayed inhibitory activity (Gjorgjieva et al. 2011) in single neurons by considering neuronal populations organized in space.

In some sensory systems, as the activity transitions from spontaneous to sensory-evoked, the temporal correlations of evoked patterns acquire properties appropriate to induce subfield segregation. In the auditory system, Sollini et al. (2018)

showed that the natural alteration of sound onset and offset present in environmental sounds, combined with Hebbian plasticity, is sufficient to gradually segregate ON and OFF receptive fields of neurons in the mouse auditory cortex, which are located at adjacent but nonoverlapping frequencies by adulthood (Fig. 3C; Sollini et al. 2018). This result implies that the statistical structure of natural stimuli in the external world can be well-suited to continue the process of receptive field refinement initialized by spontaneous activity. This is supported by experiments in the mouse and ferret visual cortex showing that sensory experience after eye opening further refines connectivity (Ko et al. 2013) and sets up direction tuning (Li et al. 2006; Chang et al. 2020) following the emergence of orientation-tuned receptive fields without evoked visual input (White et al. 2001).

Once retinal waves propagate to the visual cortex, they invoke spontaneous LFP spindle bursts and local calcium events, which continue to shape connectivity in local circuits (Hanganu et al. 2006; Colonnese and Khazipov 2010). Modeling work has shown that the local, low-synchronicity calcium events triggered by retinal waves are sufficient to establish thalamocortical receptive fields in a feedforward rate-based network via Hebbian plasticity (Fig. 3D; Wosniack et al. 2021). In contrast, the global, high-synchronicity events that simultaneously activate larger neural populations were found to adapt to ongoing activity in the network and thus homeostatically regulate connection strength (Wosniack et al. 2021).

Simultaneous to the development of experimental approaches to study activity and plasticity at the cellular and network level, methods now allow us to investigate the subcellular or synaptic organization during early postnatal development (Kleindienst et al. 2011; Takahashi et al. 2012). Multiple studies have revealed that already early in development and before the onset of sensory experience, synapses organize on individual dendritic branches in synaptic clusters, highly correlated groups of synapses near each other in dendritic space (Kleindienst et al. 2011; Takahashi et al. 2012; Leighton et al. 2023). Kirchner and Gjorgjieva built a model for synaptic cluster formation based on experimentally identified interacting neurotrophic molecules directly implicated in the developmental emergence of functional clusters in the visual cortex before eye opening (Niculescu et al. 2018; Kirchner and Gjorgjieva 2021). Beyond capturing the emergence of functional clusters driven by spontaneous activity during development, the model could also relate them to the synaptic clusters in the adult based on their representation of specific stimulus features. Interestingly, by including species-specific differences such as the amplification factor of visual space (modeled by the receptive field spread), the model could infer clustering in different species with respect to different sensory features, including orientation selectivity in ferrets (Wilson et al. 2016) and receptive field overlap in mice (Iacaruso et al. 2017), shedding light on how interspecies differences emerge during development.

Taken together, these computational models at the subcellular, cellular, and network levels make concrete proposals for the role of developmental patterns of spontaneous activity in shaping organization at different scales. A common emerging theme is the gradual refinement occurring simultaneously with the evolution of activity patterns, with coarser receptive fields and maps forming early on, and more refined features emerging later in development with the onset of sensory experience. Most models rely on forms of Hebbian synaptic plasticity that use slower developmental timescales to instruct this refinement.

EMERGENCE AND REFINEMENT OF STRUCTURED RECURRENT CONNECTIVITY

Many adult computations rely on the structured recurrent connectivity that emerges in development concurrently with feedforward connectivity through activity-dependent plasticity to ultimately shape the highly structured microcircuits and networks that underlie adult sensory processing. In the sensory cortex, a key connectivity feature that emerges shortly after eye opening is strong bidirectional connections between neurons with similar tuning properties (Ko et al. 2013; Kim et al. 2020; Crodelle and McLaughlin 2021). Using multistage spiking network simu-

lations, Ko et al. (2013) investigated the sequential development of connectivity in the visual cortex. Their model showed that first, early gap-junction coupling facilitates the activity-dependent refinement of structured feedforward synaptic connections. Next, strong bidirectional connections develop among similarly tuned cortical neurons, forming functional subnetworks organized into so-called assemblies (Fig. 4A). Notably, allowing recurrent and feedforward connections to be simultaneously plastic produced short-lived biases in recurrent connectivity and corrupted the establishment of feedforward receptive fields. These modeling results support the notion that recurrent connections refine after feedforward connections are appropriately formed. In contrast, however, driving recurrent neural network models with inputs emulating natural image statistics or allowing the models to implement a richer set of biologically plausible activity-dependent synaptic plasticity can allow feedforward receptive fields and recurrent assembly structure to emerge through synaptic plasticity simultaneously (Miconi et al. 2016; Eckmann et al. 2023). These modeling efforts lead to the conclusion that input correlations in natural images or other visually structured input may provide a powerful instruction for the simultaneous feedforward or recurrent connectivity organization, or that an appropriate synergy of synaptic plasticity mechanisms may operate at each pathway. Although these models involve more parameters than previously discussed mean-field models, they also allow analytical treatment under certain assumptions (Eckmann et al. 2023). A recent modeling study suggested that bidirectional recurrent connectivity could also arise through spontaneous activity before eye opening in species with orientation maps (Kim et al. 2020). Modeling stage III retinal waves using retinal ganglion cell mosaics measured in cats and monkeys produced the coactivation of V1 network neurons with similar tuning and induced horizontal connections between neurons with shared orientation preference (Fig. 4B; Kim et al. 2020). After the establishment of recurrent connections, the authors observed that the spontaneous network activity correlates with the underlying orientation map, as experimen-

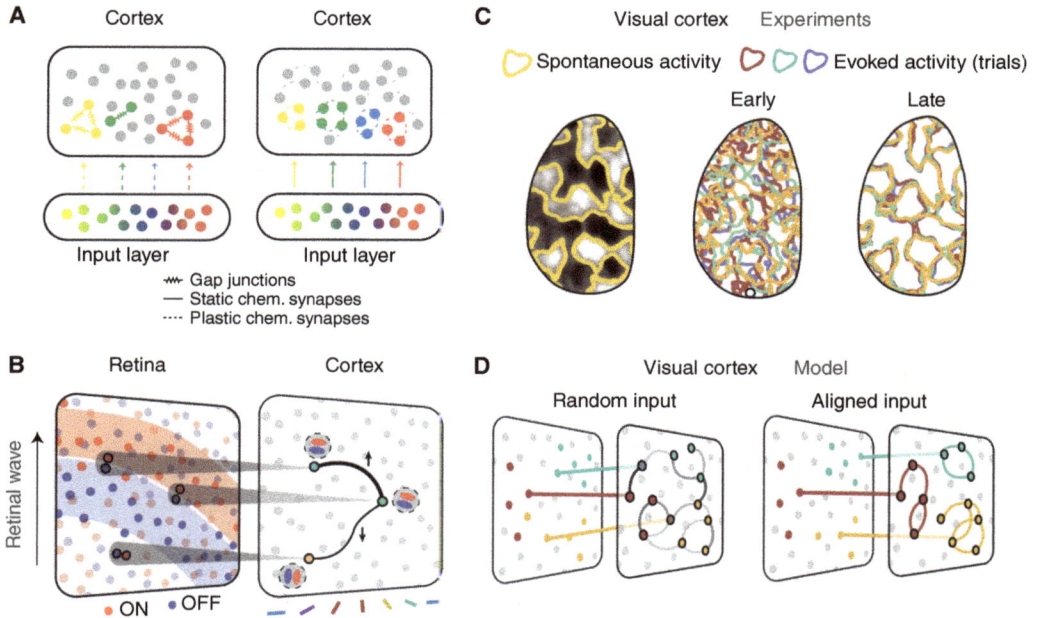

Figure 4. The interaction between the plasticity of feedforward and recurrent connectivity. (*A*) Gap-junction-connected neurons acquire similar feedforward receptive fields (*left*). Feedforward input can induce bidirectional recurrent connections between neurons with similar tuning (*right*). (Schematic adapted from Ko et al. 2013.) (*B*) Stage III retinal waves can also induce bidirectional recurrent connections between similarly tuned neurons in the visual cortex. (Schematic adapted from Kim et al. 2020.) (*C*) Evoked and spontaneous activity in the ferret visual cortex change over sensory experience; both become more similar to each other, and evoked activity becomes more stable over trials. (Schematic adapted from Trägenap et al. 2023.) (*D*) Alignment between feedforward inputs and recurrent connectivity is proposed to cause the changes in network activity over sensory experience shown in *C*. (Schematic based on data in Trägenap et al. 2023.)

tally observed in cats and ferrets (Kenet et al. 2003; Smith et al. 2018).

Such organization of recurrent connections has been framed as a process of feedforward-recurrent alignment, in which activity generated by feedforward connections becomes progressively more aligned to the activity generated spontaneously by the recurrent cortical network (Fig. 4D; Trägenap et al. 2023; see also Fiser et al. 2010 for a related proposal). Experimental work in the ferret visual cortex has shown that, although visually evoked activity before eye opening (by temporarily forcing the eyes open) is highly variable, it gradually becomes more reliable with visual experience (Fig. 4C; Trägenap et al. 2023; see also Avitan et al. 2021; Avitan and Stringer 2022). The authors confirmed with a phenomenological model that the progressive alignment of feedforward and recur-

rent connections underlies the observed change in network activity. In a rate-based recurrent network model, the authors generated feedforward input patterns with varying degrees of alignment to the recurrent connectivity. They showed that as alignment increases, visually evoked responses and spontaneous activity change and become more similar to each other as the network selectively amplifies the aligned components of the feedforward input patterns (Fig. 4D). These results highlight that feedforward and recurrent connections may become more aligned throughout development and lead to more reliable sensory representations, thus preparing the cortical circuit for mature sensory processing. Recent experimental work in the ferret visual cortex has confirmed this hypothesis by focusing on the connectivity supporting orientation selectivity from layer 4 to layer 2/3 as well as recurrent connectivity within layer 2/3 (Lempel and Fitzpatrick 2023). Future theoretical and experimental work could examine whether refinement through plasticity and sensory inputs indeed causes this gradual alignment between feedforward and recurrent connections.

INHIBITORY CONTROL OF PLASTICITY AND EMERGENT CONNECTIVITY

As neural circuits in the developing brain generate and are refined by spontaneous and early sensory activity, there must be safety checks in place to prevent excessive excitation and ensure stable activity levels. This control is mainly achieved by inhibition, which must be appropriately incorporated to prevent epileptic activity as developing neurons and synapses acquire mature ion channel composition and synaptic transmission abilities (as discussed in the section Desynchronization and Sparsification of Neural Activity throughout Development). Synaptic plasticity at inhibitory synapses could be an additional factor underlying the developmental strengthening of inhibition. Alongside promoting spontaneous activity sparsification (Golshani et al. 2009; Larisch et al. 2021; Chini et al. 2022), inhibition and inhibitory plasticity affect multiple properties of developing networks, including the gradual emergence of excitation/inhibition (E/I) balance (Froemke 2015), controlling neuronal firing properties (Gjorgjieva et al. 2014; Avitan and Goodhill 2018; Glanz et al. 2021; Jia et al. 2022; Trägenap et al. 2023), and regulating excitatory plasticity (Agnes et al. 2020; Miehl and Gjorgjieva 2022). E/I balance typically refers to the coregulation of excitation and inhibition as measured by the ratio of excitatory and inhibitory neuronal inputs (Froemke 2015; Hennequin et al. 2017). E/I balance can occur in multiple forms at the population, single-neuron, and dendritic level and at a range of timescales, with different computational implications for circuit processing (Dorrn et al. 2010; House et al. 2011; Barnes et al. 2015; Field et al. 2020; Wu et al. 2022). Various inhibitory plasticity rules have been proposed to regulate E/I balance in computational models, often in conjunction with excitatory plasticity (Vogels et al. 2011; Luz and Sha-

mir 2012; Kleberg et al. 2014; Wu et al. 2022). As with the activity-dependent refinement of excitatory connectivity, the coordinated interaction of excitatory and inhibitory plasticity is typically studied in models with feedforward connectivity, which include the emergence of sensory receptive fields (Vogels et al. 2011; Clopath et al. 2016; Miehl and Gjorgjieva 2022), place fields (Weber and Sprekeler 2018), and grid fields (Weber and Sprekeler 2018) in a single postsynaptic neuron based on input statistics (Luz and Shamir 2012; Kleberg et al. 2014; Agnes et al. 2020). In recurrent circuits, inhibitory plasticity also shapes strongly connected cells such as neuronal assemblies (Litwin-Kumar et al. 2017; Miehl and Gjorgjieva 2022) and chain-like structures (Zhang et al. 2014; Maes et al. 2020), as well as ensures tuning diversity and efficient sensory representation (Larisch et al. 2021).

However, the delayed emergence of inhibitory long-term plasticity may be insufficient to establish E/I balance early on in development (Dorrn et al. 2010). One proposal for achieving stable firing in the absence of early developmental E/I balance is STP (Jia et al. 2022), which was already shown to play a role in generating spontaneous activity (Rahmati et al. 2017). In a feedforward model, Jia et al. (2022) proposed that short-term depression at excitatory synapses can stabilize neural activity in very young animals when inhibition is immature. As inhibition strengthens, the model supports the switch of STP at excitatory synapses from depression to facilitation (Reyes and Sakmann 1999). This switch stabilizes activity and supports E/I cotuning and the emergence of temporally precise spiking in the model. This is similar to the finding by Rahmati et al. (2017) that faster recovery from excitatory short-term depression supports the transition from synchronous to decorrelated spontaneous activity. These results underscore the need to simultaneously consider different forms of synaptic plasticity that underlie activity-dependent changes in connectivity.

Inhibitory neurons can be divided into multiple distinct subtypes based on their electrophysiological, morphological, and transcriptomic properties (Tremblay et al. 2016). Computational models have benefited from multiple experimental studies of interneuron-specific plasticity and explored its role in feedforward and recurrent networks (Wilmes et al. 2016; Agnes et al. 2020; Wu et al. 2022; Agnes and Vogels 2024; Lagzi and Fairhall 2024). As understanding how inhibition matures developmentally is proving crucial for understanding circuit development, new models need to consider this inhibitory diversity in setting up mature circuits. This includes, for instance, the study of transient circuits where SST interneurons are integrated before PV (Daw et al. 2007; Marques-Smith et al. 2016; Tuncdemir et al. 2016; Guan et al. 2017). Moreover, increasing knowledge of the different short-term dynamics of interneuron subtypes (Campagnola et al. 2022) and their different targeting biases onto pyramidal cells (Jiang et al. 2015) all point toward diverse and understudied influences on developing network dynamics and plasticity.

CONCLUSIONS AND OUTLOOK

Development is a dynamic process in which multiple mechanisms coordinate to establish the single neuron, synaptic, and network properties of adult neuronal circuits. Here, we reviewed the types of early developmental spontaneous and sensory-evoked activity and their interaction with synaptic plasticity to shape network connectivity and function. We focused on mechanistic models at different levels of abstraction, constrained by experimental data, which explore the components underlying spontaneous activity generation and activity-dependent organization. These models reveal nontrivial interactions between excitation, inhibition, changing connectivity, and ongoing activity that weave together to guide brain development.

Despite this progress, many open challenges remain. One question concerns the dynamical regime of immature networks. In the adult, experimental evidence suggests that cortical regions operate in the inhibition-stabilized regime, where strong recurrent excitatory connectivity requires strong recurrent inhibition to stabilize activity, with ample implications for information processing (Ozeki et al. 2009; Adesnik 2017; Kato et al. 2017; Sanzeni et al. 2020; Sadeh and Clopath 2021). However, at what stage networks become

inhibition-stabilized during development is unclear. On one hand, low connection density between excitatory pyramidal cells, the delayed integration of inhibitory interneurons, and the gradual strengthening of inhibition measured in the rodent neocortex suggest that cortical networks may not yet operate as an inhibition-stabilized network during the first postnatal weeks. Certain theoretical work presented in this review supports this idea, finding that networks modeled early on (∼P3–P10) are intrinsically stable and that the inhibition-stabilized regime emerges and becomes accessible as development progresses (Rahmati et al. 2017; Graf et al. 2022; Kirmse and Zhang 2022). However, we also reported work supporting an inhibitory role of GABA already by P2 that, combined with theoretical modeling, suggests that recurrent networks in the PFC may operate in the inhibition-stabilized regime shortly after birth (Chini et al. 2022).

A second challenge concerns the extent to which simultaneously developing neural circuits interact. The developing brain assembles into a hierarchical circuit with extensive inter- and intraregion connections (Murakami et al. 2022). In recent years, experiments have begun to understand the crucial role of feedback connectivity in adult animals for various computations such as predictive coding and context modulation (Marques et al. 2018; Morimoto et al. 2021). This leads to the question of how feedback connectivity affects spontaneous activity and connectivity refinement in postnatal development. Some experimental work has implicated corticothalamic feedback in the generation of spontaneous spindle bursts in the cortex in the first two postnatal weeks (Murata and Colonnese 2016). A computational study has further proposed that, in the second postnatal week, feedback connections are optimized to prevent misinformative correlations (Tikidji-Hamburyan et al. 2023). However, a complete understanding of the emergence of structured feedback connectivity and its interaction with spontaneous activity needs further experimental and computational research.

A third challenge concerns the generalizability of models constrained by experimental data from sensory cortices to other brain regions. Several kinds of spontaneous activity patterns are observed in the hippocampus, but their role in the generation of different hippocampal representations of space (head direction cells, grid cells, place cells) that emerge at different developmental time points remains unclear (Langston et al. 2010; Wills et al. 2010; Wills and Cacucci 2014; Tan et al. 2017). Similarly, the propagation of activity and refinement of connectivity between the hippocampus, olfactory, and prefrontal cortex (Gretenkord et al. 2019; Chini and Hanganu-Opatz 2021; Xu et al. 2021) remains a rich avenue for further theoretical and experimental research.

ACKNOWLEDGMENTS

We thank Mattia Chini, Christian Lohmann, Deyue Kong, Irina Pochinok, Sigrid Trägenap, and Yue (Kris) Wu for feedback on the manuscript and the entire "Computation in Neural Circuits Group" for discussions. This project has received funding from the European Union's Horizon 2020 research and innovation program under the Marie Skłodowska-Curie grant agreement no. 860949 and the European Research Council grant agreement no. 804824 to J.G.

REFERENCES

Abbott LF, Nelson SB. 2000. Synaptic plasticity: taming the beast. *Nat Neurosci* **3**: 1178–1183. doi:10.1038/81453

Ackman JB, Crair MC. 2014. Role of emergent neural activity in visual map development. *Curr Opin Neurobiol* **24**: 166–175. doi:10.1016/j.conb.2013.11.011

Ackman JB, Burbridge TJ, Crair MC. 2012. Retinal waves coordinate patterned activity throughout the developing visual system. *Nature* **490**: 219–225. doi:10.1038/nature11529

Adesnik H. 2017. Synaptic mechanisms of feature coding in the visual cortex of awake mice. *Neuron* **95**: 1147–1159.e4. doi:10.1016/j.neuron.2017.08.014

Agnes EJ, Vogels TP. 2024. Co-dependent excitatory and inhibitory plasticity accounts for quick, stable and long-lasting memories in biological networks. *Nat Neurosci* doi:10.1038/s41593-024-01597-4

Agnes EJ, Luppi AI, Vogels TP. 2020. Complementary inhibitory weight profiles emerge from plasticity and allow flexible switching of receptive fields. *J Neurosci* **40**: 9634–9649. doi:10.1523/JNEUROSCI.0276-20.2020

An S, Kilb W, Luhmann HJ. 2014. Sensory-evoked and spontaneous γ and spindle bursts in neonatal rat motor cortex.

J Neurosci **34**: 10870–10883. doi:10.1523/JNEUROSCI.4539-13.2014

Antón-Bolaños N, Sempere-Ferràndez A, Guillamón-Vivancos T, Martini FJ, Pérez-Saiz L, Gezelius H, Filipchuk A, Valdeolmillos M, López-Bendito G. 2019. Prenatal activity from thalamic neurons governs the emergence of functional cortical maps in mice. *Science* **364**: 987–990. doi:10.1126/science.aav7617

Avitan L, Goodhill GJ. 2018. Code under construction: neural coding over development. *Trends Neurosci* **41**: 599–609. doi:10.1016/j.tins.2018.05.011

Avitan L, Stringer C. 2022. Not so spontaneous: multi-dimensional representations of behaviors and context in sensory areas. *Neuron* **110**: 3064–3075. doi:10.1016/j.neuron.2022.06.019

Avitan L, Pujic Z, Mölter J, Zhu S, Sun B, Goodhill GJ. 2021. Spontaneous and evoked activity patterns diverge over development. *eLife* **10**: e61942. doi:10.7554/eLife.61942

Barak O, Tsodyks M. 2007. Persistent activity in neural networks with dynamic synapses. *PLoS Comput Biol* **3**: e35. doi:10.1371/journal.pcbi.0030035

Barnes SJ, Sammons RP, Jacobsen RI, Mackie J, Keller GB, Keck T. 2015. Subnetwork-specific homeostatic plasticity in mouse visual cortex in vivo. *Neuron* **86**: 1290–1303. doi:10.1016/j.neuron.2015.05.010

Baruchin LJ, Ghezzi F, Kohl MM, Butt SJB. 2022. Contribution of interneuron subtype-specific GABAergic signaling to emergent sensory processing in mouse somatosensory whisker barrel cortex. *Cereb Cortex* **32**: 2538–2554. doi:10.1093/cercor/bhab363

Bennett JEM, Bair W. 2015. Refinement and pattern formation in neural circuits by the interaction of traveling waves with spike-timing dependent plasticity. *PLoS Comput Biol* **11**: e1004422. doi:10.1371/journal.pcbi.1004422

Bi G, Poo M. 1998. Synaptic modifications in cultured hippocampal neurons: dependence on spike timing, synaptic strength, and postsynaptic cell type. *J Neurosci* **18**: 10464–10472. doi:10.1523/JNEUROSCI.18-24-10464.1998

Blais B, Cooper LN, Shouval H. 2000. Formation of direction selectivity in natural scene environments. *Neural Comput* **12**: 1057–1066. doi:10.1162/089976600300015501

Blankenship AG, Feller MB. 2010. Mechanisms underlying spontaneous patterned activity in developing neural circuits. *Nat Rev Neurosci* **11**: 18–29. doi:10.1038/nrn2759

Blumberg MS, Coleman CM, Gerth AI, McMurray B. 2013. Spatiotemporal structure of REM sleep twitching reveals developmental origins of motor synergies. *Curr Biol* **23**: 2100–2109. doi:10.1016/j.cub.2013.08.055

Bonifazi P, Goldin M, Picardo MA, Jorquera I, Cattani A, Bianconi G, Represa A, Ben-Ari Y, Cossart R. 2009. GABAergic hub neurons orchestrate synchrony in developing hippocampal networks. *Science* **326**: 1419–1424. doi:10.1126/science.1175509

Brockmann MD, Pöschel B, Cichon N, Hanganu-Opatz IL. 2011. Coupled oscillations mediate directed interactions between prefrontal cortex and hippocampus of the neonatal rat. *Neuron* **71**: 332–347. doi:10.1016/j.neuron.2011.05.041

Butts DA, Rokhsar DS. 2001. The information content of spontaneous retinal waves. *J Neurosci* **21**: 961–973. doi:10.1523/JNEUROSCI.21-03-00961.2001

Butts DA, Kanold PO, Shatz CJ. 2007. A burst-based "Hebbian" learning rule at retinogeniculate synapses links retinal waves to activity-dependent refinement. *PLoS Biol* **5**: e61. doi:10.1371/journal.pbio.0050061

Campagnola L, Seeman SC, Chartrand T, Kim L, Hoggarth A, Gamlin C, Ito S, Trinh J, Davoudian P, Radaelli C, et al. 2022. Local connectivity and synaptic dynamics in mouse and human neocortex. *Science* **375**: eabj5861. doi:10.1126/science.abj5861

Cang J, Feldheim DA. 2013. Developmental mechanisms of topographic map formation and alignment. *Annu Rev Neurosci* **36**: 51–77. doi:10.1146/annurev-neuro-062012-170341

Chang JT, Whitney D, Fitzpatrick D. 2020. Experience-dependent reorganization drives development of a binocularly unified cortical representation of orientation. *Neuron* **107**: 338–350.e5. doi:10.1016/j.neuron.2020.04.022

Che A, Babij R, Iannone AF, Fetcho RN, Ferrer M, Liston C, Fishell G, De Marco García NV. 2018. Layer I interneurons sharpen sensory maps during neonatal development. *Neuron* **99**: 98–116.e7. doi:10.1016/j.neuron.2018.06.002

Chen C, Regehr WG. 2000. Developmental remodeling of the retinogeniculate synapse. *Neuron* **28**: 955–966. doi:10.1016/S0896-6273(00)00166-5

Chini M, Hanganu-Opatz IL. 2021. Prefrontal cortex development in health and disease: lessons from rodents and humans. *Trends Neurosci* **44**: 227–240. doi:10.1016/j.tins.2020.10.017

Chini M, Pfeffer T, Hanganu-Opatz I. 2022. An increase of inhibition drives the developmental decorrelation of neural activity. *eLife* **11**: e78811. doi:10.7554/eLife.78811

Clopath C, Büsing L, Vasilaki E, Gerstner W. 2010. Connectivity reflects coding: a model of voltage-based STDP with homeostasis. *Nat Neurosci* **13**: 344–352. doi:10.1038/nn.2479

Clopath C, Vogels TP, Froemke RC, Sprekeler H. 2016. Receptive field formation by interacting excitatory and inhibitory synaptic plasticity. bioRxiv doi:10.1101/066589

Colonnese MT. 2014. Rapid developmental emergence of stable depolarization during wakefulness by inhibitory balancing of cortical network excitability. *J Neurosci* **34**: 5477–5485. doi:10.1523/JNEUROSCI.3659-13.2014

Colonnese MT, Khazipov R. 2010. "Slow activity transients" in infant rat visual cortex: a spreading synchronous oscillation patterned by retinal waves. *J Neurosci* **30**: 4325–4337. doi:10.1523/JNEUROSCI.4995-09.2010

Colonnese MT, Kaminska A, Minlebaev M, Milh M, Bloem B, Lescure S, Moriette G, Chiron C, Ben-Ari Y, Khazipov R. 2010. A conserved switch in sensory processing prepares developing neocortex for vision. *Neuron* **67**: 480–498. doi:10.1016/j.neuron.2010.07.015

Crodelle J, McLaughlin DW. 2021. Modeling the role of gap junctions between excitatory neurons in the developing visual cortex. *PLoS Comput Biol* **17**: e1007915. doi:10.1371/journal.pcbi.1007915

Dahmen D, Layer M, Deutz L, Dąbrowska PA, Voges N, von Papen M, Brochier T, Riehle A, Diesmann M, Grün S, et al. 2022. Global organization of neuronal activity only requires unstructured local connectivity. *eLife* **11**: e68422. doi:10.7554/eLife.68422

Dard RF, Leprince E, Denis J, Rao Balappa S, Suchkov D, Boyce R, Lopez C, Giorgi-Kurz M, Szwagier T, Dumont T, et al. 2022. The rapid developmental rise of somatic inhibition disengages hippocampal dynamics from self-motion. *eLife* **11**: e78116. doi:10.7554/eLife.78116

Daw MI, Ashby MC, Isaac JTR. 2007. Coordinated developmental recruitment of latent fast spiking interneurons in layer IV barrel cortex. *Nat Neurosci* **10**: 453–461. doi:10.1038/nn1866

Doetsch F, Hen R. 2005. Young and excitable: the function of new neurons in the adult mammalian brain. *Curr Opin Neurobiol* **15**: 121–128. doi:10.1016/j.conb.2005.01.018

Donato F, Jacobsen RI, Moser MB, Moser EI. 2017. Stellate cells drive maturation of the entorhinal-hippocampal circuit. *Science* **355**: eaai8178. doi:10.1126/science.aai8178

Dorrn AL, Yuan K, Barker AJ, Schreiner CE, Froemke RC. 2010. Developmental sensory experience balances cortical excitation and inhibition. *Nature* **465**: 932–936. doi:10.1038/nature09119

Duan ZRS, Che A, Chu P, Modol L, Bollmann Y, Babij R, Fetcho RN, Otsuka T, Fuccillo MV, Liston C, et al. 2020. GABAergic restriction of network dynamics regulates interneuron survival in the developing cortex. *Neuron* **105**: 75–92.e5. doi:10.1016/j.neuron.2019.10.008

Eckmann S, Young EJ, Gjorgjieva J. 2023. Synapse-type-specific competitive Hebbian learning forms functional recurrent networks. bioRxiv doi:10.1101/2022.03.11.483899v3

Etherington SJ, Williams SR. 2011. Postnatal development of intrinsic and synaptic properties transforms signaling in the layer 5 excitatory neural network of the visual cortex. *J Neurosci* **31**: 9526–9537. doi:10.1523/JNEUROSCI.0458-11.2011

Field RE, D'amour JA, Tremblay R, Miehl C, Rudy B, Gjorgjieva J, Froemke RC. 2020. Heterosynaptic plasticity determines the set point for cortical excitatory-inhibitory balance. *Neuron* **106**: 842–854.e4. doi:10.1016/j.neuron.2020.03.002

Fiser J, Berkes P, Orbán G, Lengyel M. 2010. Statistically optimal perception and learning: from behavior to neural representations. *Trends Cogn Sci* **14**: 119–130. doi:10.1016/j.tics.2010.01.003

Flossmann T, Kaas T, Rahmati V, Kiebel SJ, Witte OW, Holthoff K, Kirmse K. 2019. Somatostatin interneurons promote neuronal synchrony in the neonatal hippocampus. *Cell Rep* **26**: 3173–3182.e5. doi:10.1016/j.celrep.2019.02.061

Froemke RC. 2015. Plasticity of cortical excitatory-inhibitory balance. *Annu Rev Neurosci* **38**: 195–219. doi:10.1146/annurev-neuro-071714-034002

Frye CG, MacLean JN. 2016. Spontaneous activations follow a common developmental course across primary sensory areas in mouse neocortex. *J Neurophysiol* **116**: 431–437. doi:10.1152/jn.00172.2016

Garrett ME, Nauhaus I, Marshel JH, Callaway EM. 2014. Topography and areal organization of mouse visual cortex. *J Neurosci* **34**: 12587–12600. doi:10.1523/JNEUROSCI.1124-14.2014

Ge S, Goh ELK, Sailor KA, Kitabatake Y, Ming G, Song H. 2006. GABA regulates synaptic integration of newly generated neurons in the adult brain. *Nature* **439**: 589–593. doi:10.1038/nature04404

Gjorgjieva J, Eglen SJ. 2011. Modeling developmental patterns of spontaneous activity. *Curr Opin Neurobiol* **21**: 679–684. doi:10.1016/j.conb.2011.05.015

Gjorgjieva J, Toyoizumi T, Eglen SJ. 2009. Burst-time-dependent plasticity robustly guides ON/OFF segregation in the lateral geniculate nucleus. *PLoS Comput Biol* **5**: e1000618. doi:10.1371/journal.pcbi.1000618

Gjorgjieva J, Clopath C, Audet J, Pfister JP. 2011. A triplet spike-timing-dependent plasticity model generalizes the Bienenstock–Cooper–Munro rule to higher-order spatiotemporal correlations. *Proc Natl Acad Sci* **108**: 19383–19388. doi:10.1073/pnas.1105933108

Gjorgjieva J, Mease RA, Moody WJ, Fairhall AL. 2014. Intrinsic neuronal properties switch the mode of information transmission in networks. *PLoS Comput Biol* **10**: e1003962. doi:10.1371/journal.pcbi.1003962

Glanz RM, Dooley JC, Sokoloff G, Blumberg MS. 2021. Sensory coding of limb kinematics in motor cortex across a key developmental transition. *J Neurosci* **41**: 6905–6918. doi:10.1523/JNEUROSCI.0921-21.2021

Godfrey KB, Eglen SJ. 2009. Theoretical models of spontaneous activity generation and propagation in the developing retina. *Mol Biosyst* **5**: 1527–1535. doi:10.1039/b907213f

Golshani P, Gonçalves JT, Khoshkhoo S, Mostany R, Smirnakis S, Portera-Cailliau C. 2009. Internally mediated developmental desynchronization of neocortical network activity. *J Neurosci* **29**: 10890–10899. doi:10.1523/JNEUROSCI.2012-09.2009

Graf J, Rahmati V, Majoros M, Witte OW, Geis C, Kiebel SJ, Holthoff K, Kirmse K. 2022. Network instability dynamics drive a transient bursting period in the developing hippocampus in vivo. *eLife* **11**: e82756. doi:10.7554/eLife.82756

Graupner M, Brunel N. 2012. Calcium-based plasticity model explains sensitivity of synaptic changes to spike pattern, rate, and dendritic location. *Proc Natl Acad Sci* **109**: 3991–3996. doi:10.1073/pnas.1109359109

Gretenkord S, Kostka JK, Hartung H, Watznauer K, Fleck D, Minier-Toribio A, Spehr M, Hanganu-Opatz IL. 2019. Coordinated electrical activity in the olfactory bulb gates the oscillatory entrainment of entorhinal networks in neonatal mice. *PLoS Biol* **17**: e2006994. doi:10.1371/journal.pbio.2006994

Gribizis A, Ge X, Daigle TL, Ackman JB, Zeng H, Lee D, Crair MC. 2019. Visual cortex gains independence from peripheral drive before eye opening. *Neuron* **104**: 711–723.e3. doi:10.1016/j.neuron.2019.08.015

Griguoli M, Cherubini E. 2017. Early correlated network activity in the hippocampus: its putative role in shaping neuronal circuits. *Front Cell Neurosci* **11**: 255. doi:10.3389/fncel.2017.00255

Guan W, Cao JW, Liu LY, Zhao ZH, Fu Y, Yu YC. 2017. Eye opening differentially modulates inhibitory synaptic transmission in the developing visual cortex. *eLife* **6**: e32337. doi:10.7554/eLife.32337

Hanganu IL, Ben-Ari Y, Khazipov R. 2006. Retinal waves trigger spindle bursts in the neonatal rat visual cortex. *J Neurosci* **26**: 6728–6736. doi:10.1523/JNEUROSCI.0752-06.2006

Harris KD, Thiele A. 2011. Cortical state and attention. *Nat Rev Neurosci* **12**: 509–523. doi:10.1038/nrn3084

Hebb DO. 1949. *The organization of behavior; a neuropsychological theory*. Wiley, Oxford.

Hennequin G, Agnes EJ, Vogels TP. 2017. Inhibitory plasticity: balance, control, and codependence. *Annu Rev Neurosci* **40**: 557–579. doi:10.1146/annurev-neuro-072116-031005

House DRC, Elstrott J, Koh E, Chung J, Feldman DE. 2011. Parallel regulation of feedforward inhibition and excitation during whisker map plasticity. *Neuron* **72**: 819–831. doi:10.1016/j.neuron.2011.09.008

Hubel DH, Wiesel TN. 1962. Receptive fields, binocular interaction and functional architecture in the cat's visual cortex. *J Physiol* **160**: 106–154. doi:10.1113/jphysiol.1962.sp006837

Iacaruso MF, Gasler IT, Hofer SB. 2017. Synaptic organization of visual space in primary visual cortex. *Nature* **547**: 449–452. doi:10.1038/nature23019

Jia DW, Vogels TP, Costa RP. 2022. Developmental depression-to-facilitation shift controls excitation-inhibition balance. *Commun Biol* **5**: 1–12. doi:10.1038/s42003-021-02997-z

Jiang X, Shen S, Cadwell CR, Berens P, Sinz F, Ecker AS, Patel S, Tolias AS. 2015. Principles of connectivity among morphologically defined cell types in adult neocortex. *Science* **350**: aac9462. doi:10.1126/science.aac9462

Kalemaki K, Velli A, Christodoulou O, Denaxa M, Karagogeos D, Sidiropoulou K. 2022. The developmental changes in intrinsic and synaptic properties of prefrontal neurons enhance local network activity from the second to the third postnatal weeks in mice. *Cereb Cortex* **32**: 3633–3650. doi:10.1093/cercor/bhab438

Kato HK, Asinof SK, Isaacson JS. 2017. Network-level control of frequency tuning in auditory cortex. *Neuron* **95**: 412–423.e4. doi:10.1016/j.neuron.2017.06.019

Kenet T, Bibitchkov D, Tsodyks M, Grinvald A, Arieli A. 2003. Spontaneously emerging cortical representations of visual attributes. *Nature* **425**: 954–956. doi:10.1038/nature02078

Kerr JND, Greenberg D, Helmchen F. 2005. Imaging input and output of neocortical networks in vivo. *Proc Natl Acad Sci* **102**: 14063–14068. doi:10.1073/pnas.0506029102

Kersbergen CJ, Babola TA, Rock J, Bergles DE. 2022. Developmental spontaneous activity promotes formation of sensory domains, frequency tuning and proper gain in central auditory circuits. *Cell Rep* **41**: 111649. doi:10.1016/j.celrep.2022.111649

Kerschensteiner D, Wong ROL. 2008. A precisely timed asynchronous pattern of ON and OFF retinal ganglion cell activity during propagation of retinal waves. *Neuron* **58**: 851–858. doi:10.1016/j.neuron.2008.04.025

Khazipov R, Sirota A, Leinekugel X, Holmes GL, Ben-Ari Y, Buzsáki G. 2004. Early motor activity drives spindle bursts in the developing somatosensory cortex. *Nature* **432**: 758–761. doi:10.1038/nature03132

Kim J, Song M, Jang J, Paik S-B. 2020. Spontaneous retinal waves can generate long-range horizontal connectivity in visual cortex. *J Neurosci* **40**: 6584–6599. doi:10.1523/JNEUROSCI.0649-20.2020

Kirchner JH, Gjorgjieva J. 2021. Emergence of local and global synaptic organization on cortical dendrites. *Nat Commun* **12**: 4005. doi:10.1038/s41467-021-23557-3

Kirkby LA, Sack GS, Firl A, Feller MB. 2013. A role for correlated spontaneous activity in the assembly of neural circuits. *Neuron* 80: 1129–1144. doi:10.1016/j.neuron.2013.10.030

Kirmse K, Kummer M, Kovalchuk Y, Witte OW, Garaschuk O, Holthoff K. 2015. GABA depolarizes immature neurons and inhibits network activity in the neonatal neocortex in vivo. *Nat Commun* 6: 7750. doi:10.1038/ncomms8750

Kirmse K, Zhang C. 2022. Principles of GABAergic signaling in developing cortical network dynamics. *Cell Rep* 38: 110568. doi:10.1016/j.celrep.2022.110568

Kleberg FI, Fukai T, Gilson M. 2014. Excitatory and inhibitory STDP jointly tune feedforward neural circuits to selectively propagate correlated spiking activity. *Front Comput Neurosci* 8: 53. doi:10.3389/fncom.2014.00053

Kleindienst T, Winnubst J, Roth-Alpermann C, Bonhoeffer T, Lohmann C. 2011. Activity-dependent clustering of functional synaptic inputs on developing hippocampal dendrites. *Neuron* 72: 1012–1024. doi:10.1016/j.neuron.2011.10.015

Knudsen EI, Konishi M. 1978. Space and frequency are represented separately in auditory midbrain of the owl. *J Neurophysiol* 41: 870–884. doi:10.1152/jn.1978.41.4.870

Ko H, Cossell L, Baragli C, Antolik J, Clopath C, Hofer SB, Mrsic-Flogel TD. 2013. The emergence of functional microcircuits in visual cortex. *Nature* 496: 96–100. doi:10.1038/nature12015

Kreider JC, Blumberg MS. 2000. Mesopontine contribution to the expression of active "twitch" sleep in decerebrate week-old rats. *Brain Res* 872: 149–159. doi:10.1016/S0006-8993(00)02518-X

Lagzi F, Fairhall AL. 2024. Emergence of co-tuning in inhibitory neurons as a network phenomenon mediated by randomness, correlations, and homeostatic plasticity. *Sci Adv* 10: eadi4350. doi:10.1126/sciadv.adi4350

Langston RF, Ainge JA, Couey JJ, Canto CB, Bjerknes TL, Witter MP, Moser EI, Moser MB. 2010. Development of the spatial representation system in the rat. *Science* 328: 1576–1580. doi:10.1126/science.1188210

Larisch R, Gönner L, Teichmann M, Hamker FH. 2021. Sensory coding and contrast invariance emerge from the control of plastic inhibition over emergent selectivity. *PLoS Comput Biol* 17: e1009566. doi:10.1371/journal.pcbi.1009566

Lazarus MS, Huang ZJ. 2011. Distinct maturation profiles of perisomatic and dendritic targeting GABAergic interneurons in the mouse primary visual cortex during the critical period of ocular dominance plasticity. *J Neurophysiol* 106: 775–787. doi:10.1152/jn.00729.2010

Leighton AH, Cheyne JE, Houwen GJ, Maldonado PP, De Winter F, Levelt CN, Lohmann C. 2021. Somatostatin interneurons restrict cell recruitment to retinally driven spontaneous activity in the developing cortex. *Cell Rep* 36: 109316. doi:10.1016/j.celrep.2021.109316

Leighton AH, Cheyne JE, Lohmann C. 2023. Clustered synapses develop in distinct dendritic domains in visual cortex before eye opening. *eLife* 12: RP93498. doi:10.7554/eLife.93498.1

Le Magueresse C, Monyer H. 2013. GABAergic interneurons shape the functional maturation of the cortex. *Neuron* 77: 388–405. doi:10.1016/j.neuron.2013.01.011

Lempel AA, Fitzpatrick D. 2023. Developmental alignment of feedforward inputs and recurrent network activity drives increased response selectivity and reliability in primary visual cortex following the onset of visual experience. bioRxiv doi:10.1101/2023.07.09.547747

Li Y, Fitzpatrick D, White LE. 2006. The development of direction selectivity in ferret visual cortex requires early visual experience. *Nat Neurosci* 9: 676–681. doi:10.1038/nn1684

Litwin-Kumar A, Harris KD, Axel R, Sompolinsky H, Abbott LF. 2017. Optimal degrees of synaptic connectivity. *Neuron* 93: 1153–1164.e7. doi:10.1016/j.neuron.2017.01.030

Liu X, Chen C. 2008. Different roles for AMPA and NMDA receptors in transmission at the immature retinogeniculate synapse. *J Neurophysiol* 99: 629–643. doi:10.1152/jn.01171.2007

Lohmann C, Kessels HW. 2014. The developmental stages of synaptic plasticity. *J Physiol* 592: 13–31. doi:10.1113/jphysiol.2012.235119

Lombardi A, Jedlicka P, Luhmann HJ, Kilb W. 2018. Giant depolarizing potentials trigger transient changes in the intracellular Cl^- concentration in CA3 pyramidal neurons of the immature mouse hippocampus. *Front Cell Neurosci* 12: 420. doi:10.3389/fncel.2018.00420

Luccioli S, Ben-Jacob E, Barzilai A, Bonifazi P, Torcini A. 2014. Clique of functional hubs orchestrates population bursts in developmentally regulated neural networks. *PLoS Comput Biol* 10: e1003823. doi:10.1371/journal.pcbi.1003823

Luccioli S, Angulo-Garcia D, Cossart R, Malvache A, Módol L, Sousa VH, Bonifazi P, Torcini A. 2018. Modeling driver cells in developing neuronal networks. *PLoS Comput Biol* 14: e1006551. doi:10.1371/journal.pcbi.1006551

Luz Y, Shamir M. 2012. Balancing feed-forward excitation and inhibition via Hebbian inhibitory synaptic plasticity. *PLoS Comput Biol* 8: e1002334. doi:10.1371/journal.pcbi.1002334

Maes A, Barahona M, Clopath C. 2020. Learning spatiotemporal signals using a recurrent spiking network that discretizes time. *PLoS Comput Biol* 16: e1007606. doi:10.1371/journal.pcbi.1007606

Magee JC, Grienberger C. 2020. Synaptic plasticity forms and functions. *Annu Rev Neurosci* 43: 95–117. doi:10.1146/annurev-neuro-090919-022842

Maldonado PP, Nuno-Perez A, Kirchner JH, Hammock E, Gjorgjieva J, Lohmann C. 2021. Oxytocin shapes spontaneous activity patterns in the developing visual cortex by activating somatostatin interneurons. *Curr Biol* 31: 322–333.e5. doi:10.1016/j.cub.2020.10.028

Marques T, Nguyen J, Fioreze G, Petreanu L. 2018. The functional organization of cortical feedback inputs to primary visual cortex. *Nat Neurosci* 21: 757–764. doi:10.1038/s41593-018-0135-z

Marques-Smith A, Lyngholm D, Kaufmann AK, Stacey JA, Hoerder-Suabedissen A, Becker EBE, Wilson MC, Molnár Z, Butt SJB. 2016. A transient translaminar GABAergic interneuron circuit connects thalamocortical recipient layers in neonatal somatosensory cortex. *Neuron* 89: 536–549. doi:10.1016/j.neuron.2016.01.015

Martini FJ, Guillamón-Vivancos T, Moreno-Juan V, Valdeolmillos M, López-Bendito G. 2021. Spontaneous activity in developing thalamic and cortical sensory networks. *Neuron* **109:** 2519–2534. doi:10.1016/j.neuron.2021.06.026

Mazzucato L, La Camera G, Fontanini A. 2019. Expectation-induced modulation of metastable activity underlies faster coding of sensory stimuli. *Nat Neurosci* **22:** 787–796. doi:10.1038/s41593-019-0364-9

Mease RA, Famulare M, Gjorgjieva J, Moody WJ, Fairhall AL. 2013. Emergence of adaptive computation by single neurons in the developing cortex. *J Neurosci* **33:** 12154–12170. doi:10.1523/JNEUROSCI.3263-12.2013

Meister M, Wong ROL, Baylor DA, Shatz CJ. 1991. Synchronous bursts of action potentials in ganglion cells of the developing mammalian retina. *Science* **252:** 939–943. doi:10.1126/science.2035024

Miconi T, McKinstry JL, Edelman GM. 2016. Spontaneous emergence of fast attractor dynamics in a model of developing primary visual cortex. *Nat Commun* **7:** 13208. doi:10.1038/ncomms13208

Miehl C, Gjorgjieva J. 2022. Stability and learning in excitatory synapses by nonlinear inhibitory plasticity. *PLoS Comput Biol* **18:** e1010682. doi:10.1371/journal.pcbi.1010682

Milh M, Kaminska A, Huon C, Lapillonne A, Ben-Ari Y, Khazipov R. 2007. Rapid cortical oscillations and early motor activity in premature human neonate. *Cereb Cortex* **17:** 1582–1594. doi:10.1093/cercor/bhl069

Miller KD, MacKay DJC. 1994. The role of constraints in Hebbian learning. *Neural Comput* **6:** 100–126. doi:10.1162/neco.1994.6.1.100

Minlebaev M, Ben-Ari Y, Khazipov R. 2007. Network mechanisms of spindle-burst oscillations in the neonatal rat barrel cortex in vivo. *J Neurophysiol* **97:** 692–700. doi:10.1152/jn.00759.2006

Mizuno H, Ikezoe K, Nakazawa S, Sato T, Kitamura K, Iwasato T. 2018. Patchwork-type spontaneous activity in neonatal barrel cortex layer 4 transmitted via thalamocortical projections. *Cell Rep* **22:** 123–135. doi:10.1016/j.celrep.2017.12.012

Mòdol L, Sousa VH, Malvache A, Tressard T, Baude A, Cossart R. 2017. Spatial embryonic origin delineates GABAergic hub neurons driving network dynamics in the developing entorhinal cortex. *Cereb Cortex* **27:** 4649–4661. doi:10.1093/cercor/bhx198

Modol L, Bollmann Y, Tressard T, Baude A, Che A, Duan ZRS, Babij R, De Marco García NV, Cossart R. 2020. Assemblies of perisomatic GABAergic neurons in the developing barrel cortex. *Neuron* **105:** 93–105.e4. doi:10.1016/j.neuron.2019.10.007

Morimoto MM, Uchishiba E, Saleem AB. 2021. Organization of feedback projections to mouse primary visual cortex. *iScience* **24:** 102450. doi:10.1016/j.isci.2021.102450

Mrsic-Flogel TD, Hofer SB, Creutzfeldt C, Cloëz-Tayarani I, Changeux JP, Bonhoeffer T, Hübener M. 2005. Altered map of visual space in the superior colliculus of mice lacking early retinal waves. *J Neurosci* **25:** 6921–6928. doi:10.1523/JNEUROSCI.1555-05.2005

Murakami T, Matsui T, Uemura M, Ohki K. 2022. Modular strategy for development of the hierarchical visual network in mice. *Nature* **608:** 578–585. doi:10.1038/s41586-022-05045-w

Murata Y, Colonnese MT. 2016. An excitatory cortical feedback loop gates retinal wave transmission in rodent thalamus. *eLife* **5:** e18816. doi:10.7554/eLife.18816

Murata Y, Colonnese MT. 2020. GABAergic interneurons excite neonatal hippocampus in vivo. *Sci Adv* **6:** eaba1430. doi:10.1126/sciadv.aba1430

Nakazawa S, Yoshimura Y, Takagi M, Mizuno H, Iwasato T. 2020. Developmental phase transitions in spatial organization of spontaneous activity in postnatal barrel cortex layer 4. *J Neurosci* **40:** 7637–7650. doi:10.1523/JNEUROSCI.1116-20.2020

Niculescu D, Michaelsen-Preusse K, Güner Ü, van Dorland R, Wierenga CJ, Lohmann C. 2018. A BDNF-mediated push-pull plasticity mechanism for synaptic clustering. *Cell Rep* **24:** 2063–2074. doi:10.1016/j.celrep.2018.07.073

Ozeki H, Finn IM, Schaffer ES, Miller KD, Ferster D. 2009. Inhibitory stabilization of the cortical network underlies visual surround suppression. *Neuron* **62:** 578–592. doi:10.1016/j.neuron.2009.03.028

Pangratz-Fuehrer S, Hestrin S. 2011. Synaptogenesis of electrical and GABAergic synapses of fast-spiking inhibitory neurons in the neocortex. *J Neurosci* **31:** 10767–10775. doi:10.1523/JNEUROSCI.6655-10.2011

Peerboom C, Wierenga CJ. 2021. The postnatal GABA shift: a developmental perspective. *Neurosci Biobehav Rev* **124:** 179–192. doi:10.1016/j.neubiorev.2021.01.024

Picardo MA, Guigue P, Bonifazi P, Batista-Brito R, Allene C, Ribas A, Fishell G, Baude A, Cossart R. 2011. Pioneer GABA cells comprise a subpopulation of hub neurons in the developing hippocampus. *Neuron* **71:** 695–709. doi:10.1016/j.neuron.2011.06.018

Piscopo DM, El-Danaf RN, Huberman AD, Niell CM. 2013. Diverse visual features encoded in mouse lateral geniculate nucleus. *J Neurosci* **33:** 4642–4656. doi:10.1523/JNEUROSCI.5187-12.2013

Pouchelon G, Dwivedi D, Bollmann Y, Agba CK, Xu Q, Mirow AMC, Kim S, Qiu Y, Sevier E, Ritola KD, et al. 2021. The organization and development of cortical interneuron presynaptic circuits are area specific. *Cell Rep* **37:** 109993. doi:10.1016/j.celrep.2021.109993

Powell N, Hein B, Kong D, Elpelt J, Mulholland H, Kaschube M, Smith G. 2022. Universality of modular correlated networks across the developing neocortex. *COSYNE 2022 Eposter*, Lisbon, Portugal. Available at https://www.worldwide.org/cosyne-22/universality-modular-correlated-networks-5a1134a0/#poster-details-anchor

Rahmati V, Kirmse K, Holthoff K, Schwabe L, Kiebel SJ. 2017. Developmental emergence of sparse coding: a dynamic systems approach. *Sci Rep* **7:** 13015. doi:10.1038/s41598-017-13468-z

Reyes A, Sakmann B. 1999. Developmental switch in the short-term modification of unitary EPSPs evoked in layer 2/3 and layer 5 pyramidal neurons of rat neocortex. *J Neurosci* **19:** 3827–3835. doi:10.1523/JNEUROSCI.19-10-03827.1999

Richter LM, Gjorgjieva J. 2017. Understanding neural circuit development through theory and models. *Curr Opin Neurobiol* **46:** 39–47. doi:10.1016/j.conb.2017.07.004

Rochefort NL, Garaschuk O, Milos R-I, Narushima M, Marandi N, Pichler B, Kovalchuk Y, Konnerth A. 2009. Sparsification of neuronal activity in the visual cortex at eye-opening. *Proc Natl Acad Sci* **106:** 15049–15054. doi:10 .1073/pnas.0907660106

Romagnoni A, Colonnese MT, Touboul JD, Gutkin BS. 2020. Progressive alignment of inhibitory and excitatory delay may drive a rapid developmental switch in cortical network dynamics. *J Neurophysiol* **123:** 1583–1599. doi:10 .1152/jn.00402.2019

Sadeh S, Clopath C. 2021. Inhibitory stabilization and cortical computation. *Nat Rev Neurosci* **22:** 21–37. doi:10 1038/ s41583-020-00390-z

Sanzeni A, Akitake B, Goldbach HC, Leedy CE, Brunel N, Histed MH. 2020. Inhibition stabilization is a widespread property of cortical networks. *eLife* **9:** e54875. doi:10 7554/ eLife.54875

Schuett S, Bonhoeffer T, Hübener M. 2002. Mapping retinotopic structure in mouse visual cortex with optical imaging. *J Neurosci* **22:** 6549–6559. doi:10.1523/JNEUROSCI .22-15-06549.2002

Seelke AMH, Blumberg MS. 2010. Developmental appearance and disappearance of cortical events and oscillations in infant rats. *Brain Res* **1324:** 34–42. doi:10.1016/j .brainres.2010.01.088

Shen J, Colonnese MT. 2016. Development of activity in the mouse visual cortex. *J Neurosci* **36:** 12259–12275. doi:10 .1523/JNEUROSCI.1903-16.2016

Siegel F, Heimel JA, Peters J, Lohmann C. 2012. Peripheral and central inputs shape network dynamics in the developing visual cortex in vivo. *Curr Biol* **22:** 253–258. doi:10 .1016/j.cub.2011.12.026

Smith GB, Hein B, Whitney DE, Fitzpatrick D, Kaschube M. 2018. Distributed network interactions and their emergence in developing neocortex. *Nat Neurosci* **21:** 1600– 1608. doi:10.1038/s41593-018-0247-5

Sollini J, Chapuis GA, Clopath C, Chadderton P. 2018. ON-OFF receptive fields in auditory cortex diverge during development and contribute to directional sweep selectivity. *Nat Commun* **9:** 2084. doi:10.1038/s41467-018-04548-3

Song S, Miller KD, Abbott LF. 2000. Competitive Hebbian learning through spike-timing-dependent synaptic plasticity. *Nat Neurosci* **3:** 919–926. doi:10.1038/78829

Stringer C, Pachitariu M, Steinmetz N, Reddy CB, Carandini M, Harris KD. 2019. Spontaneous behaviors drive multidimensional, brainwide activity. *Science* **364:** 255. doi:10 .1126/science.aav7893

Takahashi N, Kitamura K, Matsuo N, Mayford M, Kano M, Matsuki N, Ikegaya Y. 2012. Locally synchronized synaptic inputs. *Science* **335:** 353–356. doi:10.1126/science.121 0362

Tan HM, Wills TJ, Cacucci F. 2017. The development of spatial and memory circuits in the rat. *Wiley Interdiscip Rev Cogn Sci* **8:** 1424. doi:10.1002/wcs.1424

Teh KL, Sibille J, Gehr C, Kremkow J. 2023. Retinal waves align the concentric orientation map in mouse superior colliculus to the center of vision. *Sci Adv* **9:** eadf4240. doi:10.1126/sciadv.adf4240

Tezuka Y, Hagihara KM, Ohki K, Hirano T, Tagawa Y. 2022. Developmental stage-specific spontaneous activity contributes to callosal axon projections. *eLife* **11:** e72435. doi:10.7554/eLife.72435

Thompson A, Gribizis A, Chen C, Crair MC. 2017. Activity-dependent development of visual receptive fields *Curr*

Opin Neurobiol **42:** 136–143. doi:10.1016/j.conb.2016.12 .007

Tikidji-Hamburyan RA, Govindaiah G, Guido W, Colonnese MT. 2023. Synaptic and circuit mechanisms prevent detrimentally precise correlation in the developing mammalian visual system. *eLife* **12:** e84333. doi:10.7554/eLife .84333

Trägenap S, Whitney DE, Fitzpatrick D, Kaschube M. 2023. The nature-nurture transform underlying the emergence of reliable cortical representations. bioRxiv doi:10.1101/ 2022.11.14.516507

Tremblay R, Lee S, Rudy B. 2016. GABAergic interneurons in the neocortex: from cellular properties to circuits. *Neuron* **91:** 260–292. doi:10.1016/j.neuron.2016.06.033

Tuncdemir SN, Wamsley B, Stam FJ, Osakada F, Goulding M, Callaway EM, Rudy B, Fishell G. 2016. Early somatostatin interneuron connectivity mediates the maturation of deep layer cortical circuits. *Neuron* **89:** 521–535. doi:10.1016/j .neuron.2015.11.020

Turrigiano GG, Nelson SB. 2004. Homeostatic plasticity in the developing nervous system. *Nat Rev Neurosci* **5:** 97– 107. doi:10.1038/nrn1327

Valeeva G, Tressard T, Mukhtarov M, Baude A, Khazipov R. 2016. An optogenetic approach for investigation of excitatory and inhibitory network GABA actions in mice expressing Channelrhodopsin-2 in GABAergic neurons. *J Neurosci* **36:** 5961–5973. doi:10.1523/JNEUROSCI.3482- 15.2016

Vogels TP, Sprekeler H, Zenke F, Clopath C, Gerstner W. 2011. Inhibitory plasticity balances excitation and inhibition in sensory pathways and memory networks. *Science* **334:** 1569–1573. doi:10.1126/science.1211095

Wang HC, Bergles DE. 2015. Spontaneous activity in the developing auditory system. *Cell Tissue Res* **361:** 65–75. doi:10.1007/s00441-014-2007-5

Weber SN, Sprekeler H. 2018. Learning place cells, grid cells and invariances with excitatory and inhibitory plasticity. *eLife* **7:** e34560. doi:10.7554/eLife.34560

White LE, Coppola DM, Fitzpatrick D. 2001. The contribution of sensory experience to the maturation of orientation selectivity in ferret visual cortex. *Nature* **411:** 1049–1052. doi:10.1038/35082568

Wills TJ, Cacucci F. 2014. The development of the hippocampal neural representation of space. *Curr Opin Neurobiol* **24:** 111–119. doi:10.1016/j.conb.2013.09.006

Wills TJ, Cacucci F, Burgess N, O'Keefe J. 2010. Development of the hippocampal cognitive map in preweanling rats. *Science* **328:** 1573–1576. doi:10.1126/science.1188224

Wilmes KA, Sprekeler H, Schreiber S. 2016. Inhibition as a binary switch for excitatory plasticity in pyramidal neurons. *PLoS Comput Biol* **12:** e1004768. doi:10.1371/journal .pcbi.1004768

Wilson DE, Whitney DE, Scholl B, Fitzpatrick D. 2016. Orientation selectivity and the functional clustering of synaptic inputs in primary visual cortex. *Nat Neurosci* **19:** 1003– 1009. doi:10.1038/nn.4323

Winnubst J, Cheyne JE, Niculescu D, Lohmann C. 2015. Spontaneous activity drives local synaptic plasticity in vivo. *Neuron* **87:** 399–410. doi:10.1016/j.neuron.2015.06 .029

Cite this article as *Cold Spring Harb Perspect Biol* doi: 10.1101/cshperspect.a041511

Wong ROL, Meister M, Shatz CJ. 1993. Transient period of correlated bursting activity during development of the mammalian retina. *Neuron* **11:** 923–938. doi:10.1016/0896-6273(93)90122-8

Wosniack ME, Kirchner JH, Chao LY, Zabouri N, Lohmann C, Gjorgjieva J. 2021. Adaptation of spontaneous activity in the developing visual cortex. *eLife* **10:** e61619. doi:10.7554/eLife.61619

Wu YK, Miehl C, Gjorgjieva J. 2022. Regulation of circuit organization and function through inhibitory synaptic plasticity. *Trends Neurosci* **45:** 884–898. doi:10.1016/j.tins.2022.10.006

Xu X, Song L, Kringel R, Hanganu-Opatz IL. 2021. Developmental decrease of entorhinal-hippocampal communication in immune-challenged DISC1 knockdown mice. *Nat Commun* **12:** 6810. doi:10.1038/s41467-021-27114-w

Yang JW, Hanganu-Opatz IL, Sun JJ, Luhmann HJ. 2009. Three patterns of oscillatory activity differentially synchronize developing neocortical networks in vivo. *J Neurosci* **29:** 9011–9025. doi:10.1523/JNEUROSCI.5646-08.2009

Zhang Z, Jiao YY, Sun QQ. 2011. Developmental maturation of excitation and inhibition balance in principal neurons across four layers of somatosensory cortex. *Neuroscience* **174:** 10–25. doi:10.1016/j.neuroscience.2010.11.045

Zhang S, Xu M, Kamigaki T, Hoang Do JP, Chang WC, Jenvay S, Miyamichi K, Luo L, Dan Y. 2014. Long-range and local circuits for top-down modulation of visual cortex processing. *Science* **345:** 660–665. doi:10.1126/science.1254126

Mapping the Retina onto the Brain

Daniel Kerschensteiner[1,2,3] and Marla B. Feller[4,5]

[1]Department of Ophthalmology and Visual Sciences; [2]Department of Neuroscience; [3]Department of Biomedical Engineering, Washington University School of Medicine, St. Louis, Missouri 63110, USA

[4]Department of Molecular and Cell Biology; [5]Helen Wills Neuroscience Institute, University of California, Berkeley, Berkeley, California 94720, USA

Correspondence: mfeller@berkeley.edu; kerschensteinerd@wustl.edu

Vision begins in the retina, which extracts salient features from the environment and encodes them in the spike trains of retinal ganglion cells (RGCs), the output neurons of the eye. RGC axons innervate diverse brain areas (>50 in mice) to support perception, guide behavior, and mediate influences of light on physiology and internal states. In recent years, complete lists of RGC types (~45 in mice) have been compiled, detailed maps of their dendritic connections drawn, and their light responses surveyed at scale. We know less about the RGCs' axonal projection patterns, which map retinal information onto the brain. However, some organizing principles have emerged. Here, we review the strategies and mechanisms that govern developing RGC axons and organize their innervation of retinorecipient brain areas.

ORGANIZING PRINCIPLES, LESSER-KNOWN TARGETS, AND POSTSYNAPTIC CHOICES

Eye-Specific Territories

In mammals, retinal ganglion cell (RGC) axons from both eyes innervate each side of the brain. Contralaterally projecting RGCs (contra-RGCs) occupy the nasal retina, whereas ipsilaterally projecting RGCs (ipsi-RGCs) are found in the temporal retina, which views binocular visual space (Dräger and Olsen 1980). Across species, the decussation line between ipsi- and contra-RGCs moves temporally with more lateral eye positions (Walls 1942; Rodger et al. 1998). Depending on the species, different subsets of RGCs in the temporal retina contribute to the ipsilateral projections (mouse: 20%, cat: 70%, macaque: 100%) (Dräger and Olsen 1980; Illing and Wässle

1981; Fukuda et al. 1989; Johnson et al. 2021). Thus, during development, region- and cell-type-specific sets of RGC axons decide to cross or not to cross at the optic chiasm. In their targets, ipsi- and contra-RGC axons initially overlap but refine to occupy separate territories at maturity (Fig. 1; Godement et al. 1984).

Retinotopic Maps

Nearby RGCs innervate nearby neurons in their targets to establish topographic maps that maintain spatial information. Like eye-specific territories, retinotopic maps emerge from biased starting points by gradual refinement (Cang and Feldheim 2013). In addition to establishing order among neighboring RGCs, retinotopic maps align visual space across eye-specific territories (Dräger and Hubel 1976; Haustead et al. 2008). During devel-

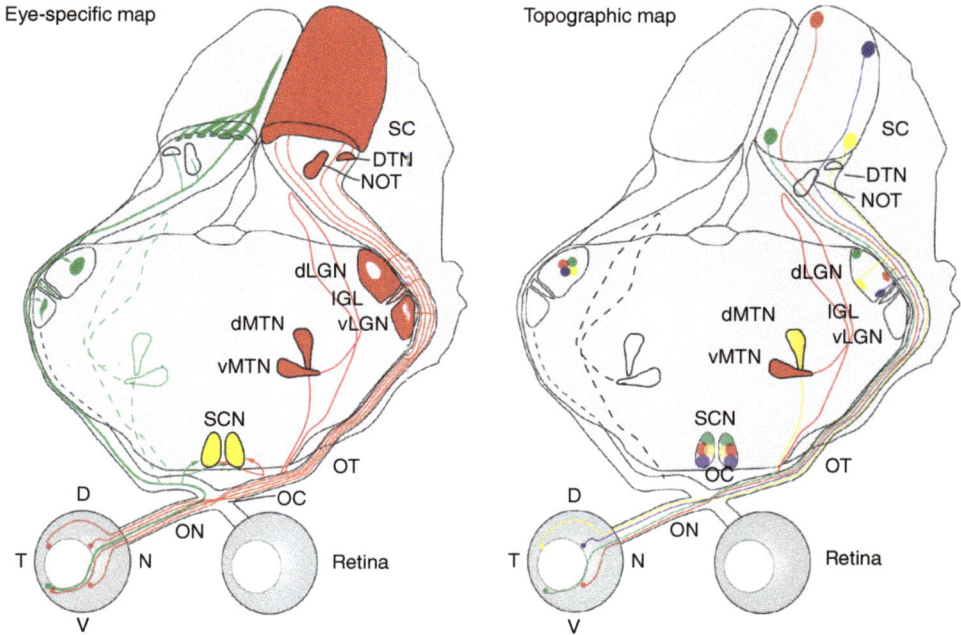

Figure 1. Organization of retinal projections into eye-specific territories and topographic maps in the mouse visual system. (ON) Optic nerve, (OC) optic chiasm, (OT) optic tract, (SCN) suprachiasmatic nucleus, (vLGN) ventrolateral geniculate nucleus, (dLGN) dorsolateral geniculate nucleus, (IGL) intergeniculate leaflet; (SC) superior colliculus, (MTN) medial terminal nucleus; (DTN) dorsal terminal nucleus, (NOT) nucleus of the optic tract, (D) dorsal, (N) nasal, (V) ventral, (T) temporal. (Image created from data in Assali et al. 2014.)

opment, retinorecipient targets also register ascending inputs from the retina with descending visual inputs (Rhoades and Chalupa 1978) and, in the superior colliculus (SC), other sensory modalities and motor outputs (Schiller and Stryker 1972; Wurtz and Goldberg 1972; Sparks et al. 1990; Knudsen and Brainard 1995; Triplett et al. 2012).

Type-Specific Targets and Layers

Neurons can be classified into types based on their morphology, gene expression pattern, and function (Zeng and Sanes 2017). Large-scale surveys of each modality identified similar numbers of RGC types (~45) in mice (Baden et al. 2016; Bae et al. 2018; Rheaume et al. 2018; Tran et al. 2019). Recent multimodal analyses revealed good correlation between the unimodal classification schemes and identified cross-modal correspondences (Goetz et al. 2022; Huang et al. 2022). Each of the mouse's ~45 RGC types sends its unique fea-

ture representation of the world to a specific subset of >50 retinorecipient brain areas (Morin and Studholme 2014; Martersteck et al. 2017; Kerschensteiner 2022). The same applies, with different numbers, to other species (e.g., macaques: ~18 RGC types and >20 retinorecipient areas) (Hendrickson et al. 1970; O'Brien et al. 2001; Dacey 2004; Peng et al. 2019). Most brain targets receive input from multiple RGC types, whose axons occupy type-specific layers in the target (Reese 1988; Huberman et al. 2009; Hong et al. 2011, 2019; Tien et al. 2022). In addition to choosing the right targets and correct layers, developing RGC axons replicate topographic maps in each type-specific layer and repeat layers across eye-specific territories (Cang and Feldheim 2013).

Lesser-Known Targets

The principles of RGC axon organization (i.e., eye-specific territories, retinotopic maps, and type-specific layers) were gleaned from studies of the two main retinorecipient targets in mammals: the dorsolateral geniculate nucleus (dLGN) of the thalamus and SC (Kerschensteiner and Guido 2017; Cang et al. 2018; Liang and Chen 2020). To what extent these principles and the mechanisms that control RGC axon organization are preserved across targets is unclear. In addition to dLGN and SC, this review summarizes recent insights into the RGC innervation of the lesser-known majority of targets.

Postsynaptic Choices

RGC axons lay out retinal information according to the principles described above. How this information is used (e.g., whether retinal representations are preserved downstream or combined to extract new features) depends on the dendritic sampling of these maps by postsynaptic target neurons. Morphological, transcriptomic, and functional surveys are beginning to define the diversity of postsynaptic neurons in retinorecipient targets (Krahe et al. 2011; Piscopo et al. 2013; Gale and Murphy 2014; Reinhard et al. 2019; Bakken et al. 2021; Liu et al. 2023b). We cover postsynaptic contributions to retina-to-brain mapping where they are known.

RETINAL WAVES

Retinal connectivity to the brain is established early in development. RGC axons reach their targets before birth and long before photoreceptors drive their responses (Wong 1999). Although guidance cues and molecular interactions are critical for this process, as RGCs innervate their targets, they are also spontaneously active. This spontaneous activity occurs in waves propagating across the developing retina (i.e., retinal waves). Retinal waves provide a robust source of depolarization and correlated activity to the retina and downstream visual areas, including the SC, dLGN, and primary visual cortex (V1) (Kirkby et al. 2014; Seabrook et al. 2017). Waves emerge in a retina with few synapses and a large ventricular zone populated by precursor cells (embryonic day E16 in mice) (Catsicas et al. 1998; Bansal et al. 2000) and per-

sist until eye opening (postnatal day P14 in mice) when all neurons have differentiated and synaptic circuits are nearly mature (Fisher 1979; Demas et al. 2003; Kerschensteiner 2020). Although there is a continuum of changes, retinal waves are typically classified into three stages (1–3), in which different circuit mechanisms generate waves with distinct activity patterns. Waves have mostly been studied in isolated retinas, but recent calcium imaging of RGC axons in the SC confirmed their existence in vivo and revealed similar spatiotemporal patterns (Ackman et al. 2012; Ackman and Crair 2014).

Retinal waves have been observed in many species, including primates (Wong 1999; Warland et al. 2006), but are best studied in mice. During stage 1 (E16-P1 in mice), retinal waves are characterized by two types of events: small nonpropagating events and large propagating ones (Bansal et al. 2000; Voufo et al. 2023). Both events are blocked by nicotinic acetylcholine receptor (nAChR) antagonists; the large propagating events are also blocked by gap junction antagonists (Voufo et al. 2023). Thus, stage 1 waves are mediated by a combination of electrical synapses, likely among specific RGC types, and volume transmission of ACh. During stage 2 (P2-P10 in mice), waves propagate across large distances, only limited by refractory periods induced by a previous wave (Ford et al. 2012). During this stage, all waves are blocked by nAChR antagonists and are mediated by volume transmission of ACh. The transition to stage 3 waves occurs as glutamatergic bipolar cell synapses in the inner retina mature (P11-P13 in mice) (Fisher 1979; Morgan et al. 2008; Akrouh and Kerschensteiner 2013; Kerschensteiner 2016). Stage 3 waves are unaffected by nAChR antagonists and are blocked by ionotropic glutamate receptor antagonists. Stage 3 waves have more complex propagation patterns and cover shorter distances than their predecessors (Kerschensteiner and Wong 2008; Gribizis et al. 2019; Ge et al. 2021). Interestingly, stage 3 waves can be triggered by photoreceptor activation, allowing light to enter through the closed eyelids to influence spontaneous activity patterns (Tiriac et al. 2018). During late stage 2 and early stage 3, waves exhibit a propagation bias, with more waves propagating in the nasal direction in retinal coordinates, mimicking the optic flow generated when mice move forward (Stafford et al. 2009; Elstrott and Feller 2010; Ge et al. 2021; Tiriac et al. 2022).

BINOCULAR VISION

Mammalian brains integrate information from both eyes to extract depth cues (i.e., stereopsis) and enhance visual sensitivity in the frontal visual field (Read 2021). To support binocular integration in the brain, most RGCs cross at the optic chiasm (i.e., contra-RGCs), but some innervate ipsilateral targets (i.e., ipsi-RGCs). This section reviews the development of contra- and ipsi-RGCs, the factors that guide their axons at the chiasm and shape their terminal arbors in retinorecipient targets.

Ipsi- and Contralaterally Projecting RGCs

Ipsi-RGCs are found in the temporal crescent of the mouse retina (Dräger and Olsen 1980). Retrograde tracing revealed that only a subset (~20%) of RGCs in the temporal retina innervate ipsilateral brain targets (Dräger and Olsen 1980), and a functional and morphological survey showed that ipsi-RGCs comprise nine of the ~45 RGC types in mice (Johnson et al. 2021).

The decision to cross or not at the optic chiasm is guided by molecular interactions between cues presented by radial glia and early-born neurons at the diencephalic midline and receptors on RGC axons (Mason and Slavi 2020). Thus, radial glia and midline neurons express ephrin-B2, which interacts with EphB1 receptors on ipsi-RGC axons, repelling them from the midline (Williams et al. 2003; Petros et al. 2010). Conversely, several pairs of midline ligands and axonal receptors (i.e., Semaphorin6D-PlexinA1, NrCAM-NrCAM, VEGF-A-Neuropilin 1) attract contra-RGC axons to the midline and promote their crossing (Williams et al. 2006; Erskine et al. 2011; Kuwajima et al. 2012). Repulsive signals from contra-RGCs, releasing Sonic hedgehog (Shh), to ipsi-RGCs, expressing the Shh-receptor Boc, further segregate their axons in the optic chiasm (Sánchez-Arrones et al. 2013; Peng et al. 2018).

Transcriptional programs differentiating ipsi- and contra-RGCs have been uncovered. The Zinc finger transcription factor Zic2 is restricted to ipsi-RGCs across species, including mice (Herrera et al. 2003). It determines the path of their axons by driving the expression of EphB1 receptors (García-Frigola et al. 2008; Lee et al. 2008) and the Shh-receptor Boc (Fabre et al. 2010; Sánchez-Arrones et al. 2013). Transcription factors associated with contra-RGCs include Islet2, Brn3a, and SoxC (Mason and Slavi 2020). Clear links from these transcription factors to axon guidance remain to be established.

Albinism, including in humans, is associated with a reduced ipsilateral RGC projection (Prieur and Rebsam 2017; Kruijt et al. 2018). Recent findings indicate that the down-regulation of CyclinD2 in the ciliary marginal zona (a neurogenic niche) of albino mice prolongs the G_1/S phase transition in the progenitor cell cycle, compromising the generation of ipsi-RGCs (Slavi et al. 2023).

Eye-Specific Organization in dLGN

In mice, ipsi-RGC axons innervate a dorsomedial patch of the dLGN that contra-RGC axons avoid (Rafols and Valverde 1973). During development, the position of the ipsilateral patch is defined by molecular cues and the segregation of ipsi- and contra-RGC axons is driven by activity-dependent mechanisms.

Ephrin-As and EphA receptors mediate repulsive interactions and are expressed in complementary gradients in the dLGN and the retina, respectively (Feldheim et al. 1998). Conversely, Ten_m3, a cell-adhesion molecule that mediates homophilic attraction, is expressed in matching gradients in the dLGN and the retina (Leamey et al. 2007). Disruption of either set of gradients disperses the termination zones of ipsi-RGC axons, whereas activity-dependent refinement drives their segregation from contra-RGCs, resulting in fragmented and displaced ipsi-RGC patches in dLGN (Huberman et al. 2005; Pfeiffenberger et al. 2005; Leamey et al. 2007).

The extracellular glycoprotein Nell2 localizes to the ipsilateral patch of dLGN and repels contra-RGC axons. Deletion of Nell2 results in an invasion of contra-RGCs, particularly in the caudal dLGN, and, after activity-dependent refinement, a similar fragmentation and dispersion of the ipsi-RGC projection to the gradient disruptions (Nakamoto et al. 2019).

The axons of ipsi- and contra-RGCs initially overlap and are gradually separated during development (between postnatal days 4 and 10 in mice) (Rossi et al. 2001; Muir-Robinson et al. 2002). Eye-specific segregation coincides with the period of stage 2 waves, and deleting the β2-subunit of nAChRs (i.e., β2-nAChR-KO mice), which significantly reduces stage 2 waves, prevents eye-specific segregation (Rossi et al. 2001; Muir-Robinson et al. 2002; Burbridge et al. 2014). Retinal waves correlate the activity of RGCs in each eye. Because stage 2 waves are short compared to their intervals and occur independently in the two eyes, wave activity can drive eye-specific segregation via correlation-based synaptic plasticity rules (e.g., burst-time-dependent plasticity) (Butts et al. 2007). Optogenetic stimulation demonstrated the importance of asynchronous activity in segregating ipsi- and contra-RGC axons in dLGN (Zhang et al. 2011). Activity-dependent synaptic plasticity promotes competition, and when ipsi-RGCs are silenced, they fail to expel contra-RGC axons from their dLGN territories (Penn et al. 1998; Koch and Ullian 2010; Koch et al. 2011).

Thus, retinal waves drive and maintain (Demas et al. 2006) the eye-specific segregation of RGC axons in territories whose position in dLGN is molecularly defined.

The small size of the ipsi-RGC axon patch compared to the thalamocortical (TC) neuron dendrite arbors means that ipsi- and contra-RGCs could converge onto TC neurons (Krahe et al. 2011). Indeed, monosynaptic rabies virus tracing found frequent binocular convergence (Rompani et al. 2017), and in vivo recordings identified TC neurons responding to visual stimulation through either eye (Howarth et al. 2014). However, TC neurons restrict strong RGC inputs to their proximal dendrites (Morgan et al. 2016), and optogenetic experiments revealed that when ipsi- and contra-RGCs converge, one eye's input is strongly dominant, and the other eye's input is arrested in synaptic development (Bauer et al. 2021). Thus, the choices made by the postsynaptic neurons largely maintain the functional separation established by the eye-specific segregation of RGC axons.

Eye-Specific Organization in SC

Ipsi- and contra-RGC inputs to the superficial SC (sSC) are laminarly separated. Ipsi-RGC axons stratify in the deepest layer of sSC (i.e., the stratum opticum [SO]), whereas contra-RGCs target the upper layers (i.e., stratum griseum superficiale [SGS] and stratum zonale [SZ]) (Godement et al. 1984; Hong et al. 2011). Unlike the gradual refinement in dLGN, ipsi-RGC axons are restricted to SO from the outset (García-Frigola and Herrera 2010). Su et al. (2021) showed that the extracellular glycoprotein nephronectin is limited to the SO and is required for its ipsi-RGC innervation. This positional cue is produced by wide-field neurons in SC, whose cell bodies reside in SO (Tsai et al. 2022). Ipsi-RGC axons express α8β1 integrin, which binds nephronectin and guides them to their laminar target (Su et al. 2021).

Although molecular cues establish eye-specific axon segregation in SC, retinal waves are required for its maintenance. Thus, ipsi-RGCs begin to stray into SGS and SZ in β2-nAChR-KO mice (Rossi et al. 2001; Burbridge et al. 2014) or when the activity of both eyes is optogenetically synchronized (Zhang et al. 2011).

The extent of binocular convergence in SC is still unclear. A recent characterization of light responses revealed complex interactions between stimuli presented through both eyes (Russell et al. 2022), which could be indirect. Moreover, how binocular convergence differs between different SC neuron types remains to be determined (Gale and Murphy 2014; Tsai et al. 2022; Liu et al. 2023b).

Eye-Specific Organization in Lesser-Known Targets

Most retinorecipient targets receive input from both eyes (Morin and Studholme 2014; Martersteck et al. 2017). The extent of ipsilateral input depends on the complement of RGC types innervating each target. For example, nuclei of the accessory optic system (i.e., the medial terminal nucleus [MTN], dorsal terminal nucleus [DTN], and nucleus of the optic tract [NOT]), which mediate the gaze-stabilizing optokinetic reflex, receive limited ipsilateral input because their retinal innervation is dominated by ON direction-selective (DS) RGCs, which are excluded from the ipsi-RGC set (Simpson 1984; Gauvain and Murphy 2015; Johnson et al. 2021). By contrast, the olivary pretectal nucleus (OPN), which mediates the pupillary light response (i.e., the pupil's constriction to increasing light levels), receives abundant input from M1 and M6 intrinsically photosensitive RGCs (ipRGCs), which are overrepresented in the ipsi-RGC set (Hattar et al. 2006; Levine and Schwartz 2020; Johnson et al. 2021). The M1 ipRGC population encodes global luminance in a distributed manner, and its ablation abolishes the pupillary light response (Chen et al. 2011; Milner and Do 2017; Liu et al. 2023a). Eye-specific axon segregation in OPN appears precise from the outset (McNeill et al. 2011); the mechanisms guiding this organization remain unknown.

Across species, the size of ipsilateral projections can be predicted from the eye positions: the more lateral the eyes, the smaller the ipsilateral projection (i.e., the Newton–Müller–Gudden law) (Walls 1942). The retinal projection to the suprachiasmatic nucleus (SCN), which entrains circadian rhythms to light, consistently breaks this law (Magnin et al. 1989). The SCN is predominantly innervated by M1 ipRGCs (Hattar et al. 2006; Baver et al. 2008; Güler et al. 2008). Single-cell labeling demonstrated that the axons of individual M1 ipRGCs frequently innervate both SCNs via collaterals branching at the chiasm or within the SCN (Fernandez et al. 2016). As a result, the mouse SCN receives approximately equal input from both eyes (Hattar et al. 2006; Fernandez et al. 2016). However, functional convergence appears limited, and 67% of cells respond exclusively to stimuli through one or the other eye (Walmsley and Brown 2015). Unlike dLGN and SC (Howarth et al. 2014; Russell et al. 2022), monocular responses in SCN are equally distributed between ipsi- and contralateral eyes (Walmsley and Brown 2015). The

mechanisms guiding the development of bilateral axonal projections and maintaining postsynaptic eye-specific segregation in SCN remain to be uncovered.

RETINOTOPY

Nearby RGCs innervate nearby neurons in retinorecipient targets to maintain spatial information. This section reviews regional specializations within the retina and the mechanisms that establish and refine retinotopic target innervation.

Regional Specializations of RGCs

Most RGCs are distributed across the retina in regular mosaics, in which the distances between same-type neighbors are relatively constant. This arrangement enables the retina to encode visual features uniformly across space. However, recent experiments have revealed striking variations in RGC density and receptive field properties across the mouse retina (i.e., regional specializations) (Rivlin-Etzion et al. 2018; El-Danaf and Huberman 2019; Heukamp et al. 2020).

The retinas of many species contain areas in which the RGC density is increased and dendritic and receptive fields are proportionally smaller, allowing for high-acuity vision (i.e., acute zones) (Baden et al. 2020). When all RGCs are labeled, no acute zones are visible in the mouse retina (Dräger and Olsen 1981). However, Bleckert et al. (2014) discovered that the density of sustained ONα and OFFα RGCs, which signal local light increments and decrements (i.e., luminance contrast), respectively, is elevated, and their dendritic and receptive fields are smaller in the temporal retina, forming a cell-type-specific binocular acute zone. An extreme asymmetry in distribution has also been observed in the non-image-forming pathway, where a subtype of M1 ipRGCs resides exclusively in the dorsal retina and projects to specific targets within the SCN, the supraoptic nucleus through which light regulates metabolism and social memories, and the zona incerta, a structure associated with novelty seeking and the integration of sensory and motivational cues for predation (Zhao et al. 2019; Monosov et al. 2022; Berry et al. 2023; Huang et al. 2023; Meng et al. 2023).

In addition, regional specializations in RGC function have been identified. Thus, the responses of transient Offα RGCs are more sustained in the dorsal than the ventral retina (Warwick et al. 2018). Transient OFFα RGCs in the ventral retina drive the innate defensive responses of mice to looming stimuli that signal approaching aerial predators in the sky (Münch et al. 2009; Kim et al. 2020; Wang et al. 2021). Furthermore, variations in the center-surround organization of receptive fields have been observed across multiple RGC subtypes that appear optimized to encode variations in spatiotemporal features of natural scenes across the visual field (Gupta et al. 2023). The developmental bases of these regional specializations in cell density and function have not yet been elucidated.

The preferred directions of DS RGCs vary with location in the retina, following a coordinate system defined by optic flow (Sabbah et al. 2017; Tiriac et al. 2022). Thus, the preferred directions of horizontal DS RGCs converge on a singularity in the temporal retina, from which optic flow emerges when mice move forward. Vertical DS RGC preferences converge on a ventral singularity. The mapping of horizontal but not vertical DS RGC motion preferences depends on retinal waves, which show a horizontal propagation bias aligned with optic flow fields during late stage 2 and early stage 3 (Stafford et al. 2009; Elstrott and Feller 2010; Sabbah et al. 2017; Ge et al. 2021; Tiriac et al. 2022).

Retinotopic Organization in dLGN

Contra-RGCs are organized retinotopically, with the nasal–temporal retinal axis mapping onto the ventrolateral–dorsomedial axis of the dLGN, and the dorsoventral retinal axis mapping onto the ventromedial–dorsolateral axis of the dLGN (Lund et al. 1974). Ipsilaterally projecting axons also follow a retinotopic organization with ventrotemporal RGCs projecting to the dorsal part of the ipsi-recipient region, which is shifted ventrally from the contra-RGC axons. Ephrin-A/EphA interactions guide nasotemporal mapping, with

high ephrin ligand expression in ventrolateral anterior dLGN (Pfeiffenberger et al. 2005). Mice lacking ephrin-As exhibit abnormal mapping along this axis, although the defects are less dramatic than in the SC (Pfeiffenberger et al. 2005). Also, in contrast to SC, there are no overshooting axons but increased branching in the correct retinotopic location (Dhande et al. 2011; Assali et al. 2014). Knocking out ephrin-A signaling and retinal waves leads to the most dramatic disruption along the horizontal visual field axis (Pfeiffenberger et al. 2006). Reconstructions of individual retinogeniculate axons showed that activity disruption prevents the refinement of their arbors (Dhande et al. 2011; Benjumeda et al. 2013).

After eye-opening, retinotopic connections between RGCs and TC neurons continue to refine, a process driven by visual experience (Hooks and Chen 2006, 2008).

Retinotopic Organization in SC

The mechanisms underlying the development of retinotopic maps have been most intensely studied in the contra-RGC projections to the SC (Triplett and Feldheim 2012; Cang and Feldheim 2013; Johnson and Triplett 2021; Tomar et al. 2023). In mice, contra-RGC axons reach the rostral end of the SC at E16 and grow across to the caudal end by P2. Over the first postnatal week, overshooting branches are retracted, and axons elaborate dense arbors in their final termination zone (Simon and O'Leary 1992; Yates et al. 2001). Distinct mechanisms appear to determine the size and location of the termination zone along the rostrocaudal axis versus the mediolateral axis, which represents the vertical visual field.

The retraction of overshooting branches and the location of the termination zone along the rostrocaudal axis are determined by contact-mediated repulsive interactions between ephrin-As and EphA receptors and retinal waves. Temporal RGCs with high receptor expression are strongly repelled from caudal SC, whereas nasal RGC axons with lower receptor expression are less sensitive to high ligand concentrations, and so remain. This model is supported by loss-of-function and gain-of-function studies (Nakamoto et al. 1996; Frisén et al. 1998; Feldheim et al.

2000; Cang et al. 2008). Single knockout mice have minor retinotopic defects, often manifesting as a normal projection with an ectopic termination zone, with more serious deficits associated with double and triple mutants, indicating it is the sum total of ephrin-A/EphA signaling, rather than specific ephrin-A/EphA receptor species that drive repulsion (Cheng et al. 1995; Triplett and Feldheim 2012). More recent studies based on conditional knockouts of ephrin-A5 in the retina indicate that SC-independent ephrin-A/EphA interactions between RGC axons also contribute to their topographic mapping (Suetterlin and Drescher 2014). A role for axon–axon interactions is also supported by a study in which the depletion of large percentages of RGCs led to the degradation of retinotopic maps (Maiorano and Hindges 2013).

Waves also contribute to the retinotopic mapping of RGC axons to SC. In β2-nAChR-KO mice or mice in which waves have been inhibited pharmacologically, axonal arbors are centered on the correct retinotopic locations but with expanded termination zones (McLaughlin et al. 2003; Burbridge et al. 2014). These expanded termination zones expand receptive fields in SC (Chandrasekaran et al. 2005; Mrsic-Flogel et al. 2005). Knocking out both β2-nAChRs and ephrin-As leads to a complete lack of refinement along the horizontal visual axis (Pfeiffenberger et al. 2006).

Refinement along the vertical visual axis (dorsoventral on the retina, mediolateral in the SC) follows a different developmental program. When entering the SC, RGC axons are partially organized along the mediolateral axis, branching off the primary axon in the correct termination zones (Plas et al. 2005). EphB/ephrin-B have counter gradients within the retina and SC. Loss-of-function and overexpression studies imply that targeting of these axonal branches depends on EphB/ephrinB interactions, although the details remain obscure (Hindges et al. 2002). Other molecular gradients have also been implicated, including Wnt signaling (Schmitt et al. 2006) and cell-adhesion molecules (Dai et al. 2013). Activity manipulations have limited effects along this axis, indicating that it is likely to be mediated primarily by guidance molecules.

Retinotopic Alignment of Feedforward and Feedback Connections in SC and dLGN

The SC and the dLGN receive feedback connections from V1 that are retinotopically organized and aligned with feedforward retinal inputs (Triplett et al. 2009; Wang and Burkhalter 2013). Descending cortical inputs appear to modulate the gain of SC neuron responses but not alter the feature preferences they inherit from the retina (Zhao et al. 2014; Liang et al. 2015; Cang et al. 2018). Following RGC innervation and refinement in the first postnatal week, axons from layer 5 neurons in V1 reach SC starting at P6 and refine their termination zones by P12 in alignment with the retinal input map. The topographic refinement of the cortical input is instructed by the retinal projection (Johnson and Triplett 2021) and depends on ephrin-A gradients (Savier et al. 2017) and activity (Triplett et al. 2009). Interestingly, the refinement of V1 projections but not retinal projections to SC depends on postsynaptic NMDA receptors (Johnson et al. 2023).

The dLGN receives input from layer 6 of V1, which is also aligned with the retinal input (Hasse and Briggs 2017). Layer 6 axons arrive in the dLGN around P0 but only innervate the nucleus after the retinal input has traversed the structure at P3. The cortical innervation of dLGN is accelerated in the absence of RGC input (Seabrook et al. 2013), a process modulated by neural activity (Moreno-Juan et al. 2023). Corticogeniculate feedback also regulates the experience-dependent refinement of retinogeniculate synapses after eye-opening (Thompson et al. 2016).

Retinotopic Organization in Lesser-Known Targets

In retinorecipient regions, other than dLGN and SC, the development of topographic organization has rarely been explored (Dhande et al. 2015). Inputs to the ventrolateral geniculate nucleus (vLGN), which may mediate a wide range of light influences on mood, learning, and behavior (Huang et al. 2019, 2021; Fratzl et al. 2021; Salay and Huberman 2021; Hu et al. 2022), arrive in the first postnatal week (Su et al. 2011), are organized retinotopically (Holcombe and Guillery 1984), and undergo eye-specific segregation (Monavarfeshani et al. 2017). The mechanisms that govern this organization remain unexplored.

In the SCN, which receives overlapping ipsilateral and contralateral inputs from M1 ipRGCs, there is some evidence for a coarse retinotopic organization (Fernandez et al. 2016). The limited retinotopic and eye-specific refinement in the SCN may partly be due to the lack of M1 ipRGC participation in retinal waves (Caval-Holme et al. 2022).

TYPE-SPECIFIC PATHWAYS

The retina extracts salient features from the environment and encodes them in the spike trains of diverse RGC types (∼45 in mice). RGCs innervate an equally diverse set of brain areas (>50 in mice) to guide actions, generate perception, and mediate influences of light on physiology (Morin and Studholme 2014; Martersteck et al. 2017). This section reviews how developing RGCs choose specific targets, how they organize their axons within, and how postsynaptic neurons select inputs from the diverse RGC offering.

Type-Specific RGC Projections

The number of targets varies widely between RGC types; some innervate one or two brain areas (Chen et al. 2011; Martersteck et al. 2017), whereas others innervate more than 10 (Hattar et al. 2006). Screens of transgenic mice revealed that single drivers, with few exceptions, label several RGC types (Siegert et al. 2009; Ivanova et al. 2010; Badea and Nathans 2011; Martersteck et al. 2017), and isolating individual RGC types requires intersectional strategies (Chen et al. 2011; Tien et al. 2022). A few studies have labeled single RGCs and shown that their axons innervate multiple targets via collaterals, forming complete retinotopic maps in each (Dhande et al. 2011; Fernandez et al. 2016).

Different developmental strategies for target innervation have been identified. ON–OFF DS RGC axons innervate the correct targets shortly after they arrive in the brain postnatally. Conversely, transient suppressed-by-contrast (tSbC) RGCs span the length of the optic tract by birth and remain poised there until they simultaneously innervate their four targets (i.e., dLGN, vLGN, SC, and NOT) around P3. Like ON–OFF DS RGCs, tSbC RGCs make no developmental errors in their target choices. The same applies to M1 ipRGCs (McNeill et al. 2011). By contrast, M6 ipRGCs have been reported to explore more targets during development than they keep at maturity. However, developmental changes in the labeling specificity of the Cdh3-GFP transgenic mice used to trace M6 ipRGC projections in this study indicate that this developmental deviation needs to be reevaluated (Osterhout et al. 2011, 2014; Quattrochi et al. 2019). Whether these different developmental strategies correlate with particular features of these RGC types or their targets remains to be determined.

Molecular mechanisms that guide RGC-type-specific target choices are discussed in the following sections.

RGC-Type-Specific Organization in dLGN

In primates, cats, and ferrets, dLGN neurons are divided into layers that get input from specific RGC types (Usrey and Alitto 2015; Liang and Chen 2020). Mice and rats lack clear separation in dLGN neurons, but RGC projections divide the dLGN into a dorsolateral shell and ventromedial core (Reese 1988). More than 30 mouse RGC types innervate the dLGN (Ellis et al. 2016; Román Rosón et al. 2019). The shell receives input from ON–OFF DS RGCs, orientation-selective RGCs, and F RGCs (Kim et al. 2008; Huberman et al. 2009; Kay et al. 2011; Rivlin-Etzion et al. 2011; Rousso et al. 2016; Nath and Schwartz 2017). The core is strongly innervated by α RGCs, Pix_{ON} RGCs, ipRGCs, and transient suppressed-by-contrast RGCs (Huberman et al. 2008; Ecker et al. 2010; Johnson et al. 2018; Tien et al. 2022). TC neurons in the dLGN shell and core innervate layers 1/2 and 4 of V1, respectively (Cruz-Martín et al. 2014), extending the parallel visual pathways from the retina. The shell and core contain complete retinotopic maps (Reese 1988). Eye-specific segregation is limited to the core (Liang and Chen 2020) because the

RGC types targeting the shell lack ipsilateral projections (Johnson et al. 2021). The mechanisms that guide RGC types to the dLGN shell or core remain unknown.

TC neurons initially receive weak input from many RGCs and consolidate and strengthen connectivity with synaptic partners across development (Chen and Regehr 2000; Liang and Chen 2020). Estimates of mature convergence (i.e., how many RGCs innervate each TC neuron) vary by technique and between target cells, but, on average, TC neurons receive input from ~10 RGCs (Chen and Regehr 2000; Hammer et al. 2015; Morgan et al. 2016; Litvina and Chen 2017; Rompani et al. 2017; Román Rosón et al. 2019).

The refinement of RGC–TC connections depends on spontaneous activity (Hooks and Chen 2006). In addition to correlating RGC activity in a distance-dependent manner, stage 3 waves desynchronize the activity of neighboring RGCs with opposite light responses (ON vs. OFF) (Kerschensteiner and Wong 2008; Akrouh and Kerschensteiner 2013). Given burst-time-dependent plasticity rules (Butts et al. 2007), the asynchronous activation of nearby ON and OFF RGCs is predicted to separate their innervation in retinotopically refined projections (Kerschensteiner and Wong 2008; Gjorgjieva et al. 2009). In ferrets, blocking stage 3 waves blocks ON/OFF segregation (Hahm et al. 1991; Cramer and Sur 1997), and mice with precocious stage 3 waves show excessive ON/OFF segregation (i.e., ON and OFF dLGN neurons, which normally intermingle, congregate separately) (Grubb et al. 2003).

At a fine scale (~6 μm), boutons of RGC axons with overlapping feature preferences (including ON vs. OFF responses and motion direction preferences) cluster to converge onto TC neurons (Liang et al. 2018). The formation of complex bouton clusters depends on Lrrtm1, a postsynaptic cell-adhesion molecule on TC neuron dendrites (Monavarfeshani et al. 2018). The organization of synaptic boutons within RGC axons is also shaped by visual experience and late dark rearing (i.e., from postnatal day 20 to 30) disperses clustered boutons and reduces their size (Hong et al. 2014).

Like TC neurons, local interneurons (LINs) in dLGN receive input from diverse RGC types (Morgan and Lichtman 2020; Muellner and Roska 2023). RGC convergence appears to be higher for LINs than TC neurons. Individual LINs select input from nonrandom sets of RGC types to preserve feature-selective responses, similar to TC neurons (Piscopo et al. 2013; Muellner and Roska 2023). The mechanisms that regulate RGC connectivity with dLGN LINs remain to be elucidated.

RGC-Type-Specific Organization in SC

More than 85% of mouse RGCs project to sSC (Ellis et al. 2016). Transgenic labeling of specific RGC populations and viral labeling of individual cells revealed that axons of different RGC types stratify at particular depths within the sSC (Huberman et al. 2008, 2009; Kim et al. 2008; Hong et al. 2011; Tien et al. 2022). Whereas some RGC axons target the correct depths from the outset (Kim et al. 2010; Tien et al. 2022), others establish mature patterns gradually (Huberman et al. 2008; Kim et al. 2010). The molecular or activity-dependent mechanisms that guide developing RGC axons to specific SC depths remain to be uncovered.

The sSC contains different neuron types that can be distinguished by their dendritic morphology, axonal projection patterns, visual functions, and gene expression (Gale and Murphy 2014; Tsai et al. 2022; Li and Meister 2023; Liu et al. 2023b). Synaptic physiology suggests that each SC neuron receives input from ~5 RGCs (Chandrasekaran et al. 2007). The observation that molecularly distinct SC neuron types differ in their functional responses, therefore, suggests that SC neurons select input from specific RGC types (Gale and Murphy 2014; Hoy et al. 2019; Liu et al. 2023b), a notion supported by retrograde tracing and transsynaptic sequencing experiments (Reinhard et al. 2019; Tsai et al. 2022). Connections between ipsilaterally projecting αRGCs and wide-field neurons in SC depend on α8β1 integrin expression of the RGC axons and nephronectin production by the target (Su et al. 2021; Tsai et al. 2022). The comprehensive transcriptomic data sets available on both sides should accelerate the discovery of molecular partners that match RGC and SC neuron types (Rheaume et al. 2018; Tran et al. 2019; Dimitrov et al. 2022; Tsai et al. 2022; Liu et al. 2023b).

Whether RGC-type-specific connectivity patterns vary across SC topography to form feature maps for motion direction and stimulus orientation and, if so, what drives the development of these maps are matters of ongoing debate and investigation (Ahmadlou and Heimel 2015; Feinberg and Meister 2015; de Malmazet et al. 2018; Li et al. 2020; Chen et al. 2021; Ge et al. 2021; Tiriac and Feller 2022; Tiriac et al. 2022).

RGC-Type-Specific Innervation of Lesser-Known Targets

M1 ipRGCs strongly innervate the vLGN and the intergeniculate leaflet (IGL) of the visual thalamus, a structure that combines photic and nonphotic information to regulate circadian behaviors (Morin et al. 2003; Hattar et al. 2006; Fernandez et al. 2020; Beier et al. 2021). This targeting requires the glycoprotein Reelin in the extracellular matrix of vLGN and IGL and the intracellular signaling molecule disabled-1 (Dab1) in M1 ipRGCs (Su et al. 2011). Dab1

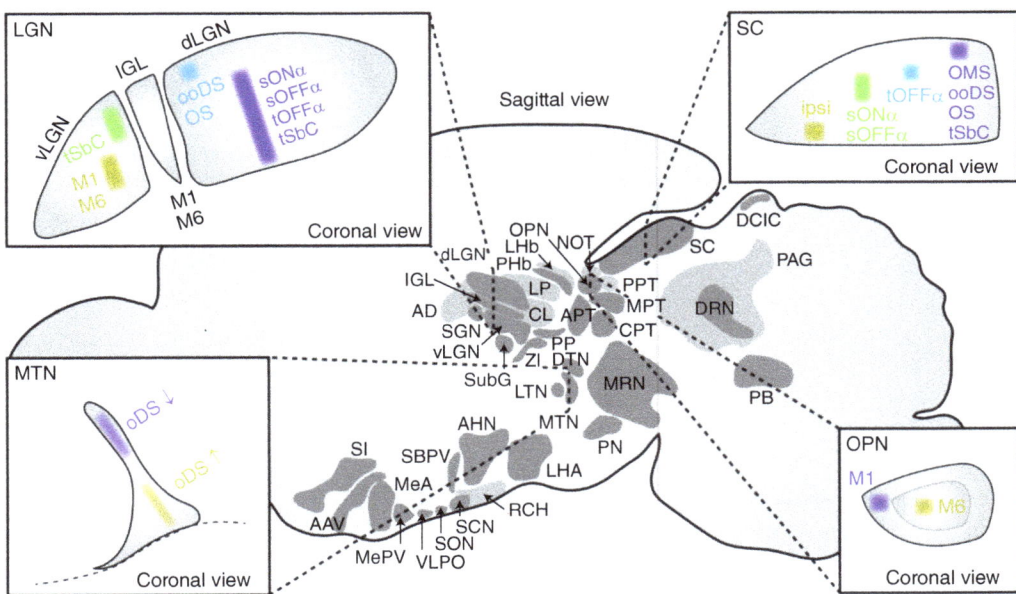

Figure 2. Retinorecipient brain targets and retinal ganglion cell (RGC)-type-specific lamination. Background illustration of the distribution of retinorecipient targets in a sagittal view of the mouse brain. (SI) Substantia innominata, (AAV) ventral anterior amygdaloid area, (MeA) anterior medial amygdala, (MePV) posteroventral medial amygdala, (VLPO) ventrolateral preoptic area, (SON) supraoptic nucleus, (SCN) suprachiasmatic nucleus, (RCH) retrochiasmatic area, (SBPV) subparaventricular zone, (AHN) anterior hypothalamic area, (LHN) lateral hypothalamic area, (PN) paranigral nucleus, (MRN) midbrain reticular nucleus, (MTN) medial terminal nucleus, (LTN) lateral terminal nucleus, (DTN) dorsal terminal nucleus, (ZI) zona incerta, (PP) peripeduncular nucleus, (SubG) subgeniculate nucleus, (vLGN) ventrolateral geniculate nucleus, (SGN) suprageniculate nucleus, (AD) anterodorsal thalamic nucleus, (IGL) intergeniculate leaflet, (CL) centrolateral thalamic nucleus, (dLGN) dorsolateral geniculate nucleus of the thalamus, (LP) lateral posterior nucleus of the thalamus, (APT) anterior pretectal nucleus, (CPT) commissural pretectal nucleus, (MPT) medial pretectal nucleus, (PPT) posterior pretectal nucleus, (PHb) perihabenular nucleus, (LHb) lateral habenula, (OPN) olivary pretectal nucleus, (NOT) nucleus of the optic tract, (SC) superior colliculus, (DCIC) dorsal cortex of the inferior colliculus, (PAG) periaqueductal gray, (DRN) dorsal raphe nucleus, (PB) parabrachial nucleus, (tSbC) transient suppressed-by-contrast. *Insets* in the foreground show the RGC-type-specific lamination of RGC axons in the LGN complex (*top left*), SC (*top right*), MTN (*bottom left*), and OPN (*bottom right*) in coronal views. (oDS) ON DS RGCs, (ooDS) ON-OFF DS RGCs.

mediates Reelin signals transmitted via VLDLR and ApoER2 receptors on the M1 ipRGC axons (Su et al. 2011).

The M6 ipRGC targeting of pretectal nuclei (i.e., OPN and the medial division of the posterior pretectal nucleus or mdPPN) depends on the matching expression of the homophilically interacting cell-adhesion molecule Cadherin-6 on the RGC axons and their postsynaptic partners (Osterhout et al. 2011). In the OPN, M6 ipRGCs target the core region (Quattrochi et al. 2019), whereas M1 ipRGCs innervate the OPN shell (Fig. 2; Chen et al. 2011). The function of this M6 ipRGC projection remains unknown.

Different ON DS RGC types innervate different nuclei of the accessory optic system (i.e., MTN, DTN, and NOT) to mediate horizontal and vertical gaze-stabilizing eye movements (Fig. 2; Simpson 1984). Interactions between the cell-adhesion molecule Contactin-4 on RGC axons and the amyloid precursor protein (APP) on target neurons drive the innervation of the NOT (Osterhout et al. 2015). Conversely, interactions between Semaphorin6A on RGC axons and PlexinA2/A4 on target neurons mediate ON DS RGC targeting of the MTN and DTN (Sun et al. 2015).

CONCLUDING REMARKS

The last 5 years represent significant progress toward a complete taxonomy of RGC types and retinorecipient brain areas in mice. This progress has opened the door for identifying the wide variety of molecular- and activity-based mechanisms that govern retina-to-brain mapping. Understanding the developmental mechanisms and sequence of retinal innervation will, in turn, advance our understanding of the adult organization of visual brain areas and how they mediate their diverse functions.

REFERENCES

Ackman JB, Crair MC. 2014. Role of emergent neural activity in visual map development. *Curr Opin Neurobiol* 24: 166–175. doi:10.1016/j.conb.2013.11.011

Ackman JB, Burbridge TJ, Crair MC. 2012. Retinal waves coordinate patterned activity throughout the developing visual system. *Nature* 490: 219–225. doi:10.1038/nature 11529

Ahmadlou M, Heimel JA. 2015. Preference for concentric orientations in the mouse superior colliculus. *Nat Commun* 6: 6773. doi:10.1038/ncomms7773

Akrouh A, Kerschensteiner D. 2013. Intersecting circuits generate precisely patterned retinal waves. *Neuron* 79: 322–334. doi:10.1016/j.neuron.2013.05.012

Assali A, Gaspar P, Rebsam A. 2014. Activity dependent mechanisms of visual map formation—from retinal waves to molecular regulators. *Semin Cell Dev Biol* 35: 136–146. doi:10.1016/j.semcdb.2014.08.008

Badea TC, Nathans J. 2011. Morphologies of mouse retinal ganglion cells expressing transcription factors Brn3a, Brn3b, and Brn3c: analysis of wild type and mutant cells using genetically-directed sparse labeling. *Vision Res* 51: 269–279. doi:10.1016/j.visres.2010.08.039

Baden T, Berens P, Franke K, Román Rosón M, Bethge M, Euler T. 2016. The functional diversity of retinal ganglion cells in the mouse. *Nature* 529: 345–350. doi:10.1038/nature16468

Baden T, Euler T, Berens P. 2020. Understanding the retinal basis of vision across species. *Nat Rev Neurosci* 21: 5–20. doi:10.1038/s41583-019-0242-1

Bae JA, Mu S, Kim JS, Turner NL, Tartavull I, Kemnitz N, Jordan CS, Norton AD, Silversmith WM, Prentki R, et al. 2018. Digital museum of retinal ganglion cells with dense anatomy and physiology. *Cell* 173: 1293–1306.e19. doi:10.1016/j.cell.2018.04.040

Bakken TE, van Velthoven CT, Menon V, Hodge RD, Yao Z, Nguyen TN, Graybuck LT, Horwitz GD, Bertagnolli D, Goldy J, et al. 2021. Single-cell and single-nucleus RNA-seq uncovers shared and distinct axes of variation in dorsal LGN neurons in mice, non-human primates, and humans. *eLife* 10: e64875. doi:10.7554/eLife.64875

Bansal A, Singer JH, Hwang BJ, Xu W, Beaudet A, Feller MB. 2000. Mice lacking specific nicotinic acetylcholine receptor subunits exhibit dramatically altered spontaneous activity patterns and reveal a limited role for retinal waves in forming ON and OFF circuits in the inner retina. *J Neurosci* 20: 7672–7681. doi:10.1523/JNEUROSCI.20-20-07672.2000

Bauer J, Weiler S, Fernholz MHP, Laubender D, Scheuss V, Hübener M, Bonhoeffer T, Rose T. 2021. Limited functional convergence of eye-specific inputs in the retinogeniculate pathway of the mouse. *Neuron* 109: 2457–2468. e12. doi:10.1016/j.neuron.2021.05.036

Baver SB, Pickard GE, Sollars PJ, Pickard GE. 2008. Two types of melanopsin retinal ganglion cell differentially innervate the hypothalamic suprachiasmatic nucleus and the olivary pretectal nucleus. *Eur J Neurosci* 27: 1763–1770. doi:10.1111/j.1460-9568.2008.06149.x

Beier C, Zhang Z, Yurgel M, Hattar S. 2021. Projections of ipRGCs and conventional RGCs to retinorecipient brain nuclei. *J Comp Neurol* 529: 1863–1875. doi:10.1002/cne.25061

Benjumeda I, Escalante A, Law C, Morales D, Chauvin G, Muça G, Coca Y, Márquez J, López-Bendito G, Kania A, et al. 2013. Uncoupling of EphA/ephrinA signaling and spontaneous activity in neural circuit wiring. *J Neurosci* 33: 18208–18218. doi:10.1523/JNEUROSCI.1931-13.2013

Cite this article as *Cold Spring Harb Perspect Biol* doi: 10.1101/cshperspect.a041512

Berry MH, Moldavan M, Garrett T, Meadows M, Cravetchi O, White E, Leffler J, von Gersdorff H, Wright KM, Allen CN, et al. 2023. A melanopsin ganglion cell subtype forms a dorsal retinal mosaic projecting to the supraoptic nucleus. *Nat Commun* **14:** 1492.

Bleckert A, Schwartz GW, Turner MH, Rieke F, Wong ROL. 2014. Visual space is represented by nonmatching topographies of distinct mouse retinal ganglion cell types. *Curr Biol* **24:** 310–315. doi:10.1016/j.cub.2013.12.020

Burbridge TJ, Xu HP, Ackman JB, Ge X, Zhang Y, Ye MJ, Zhou ZJ, Xu J, Contractor A, Crair MC. 2014. Visual circuit development requires patterned activity mediated by retinal acetylcholine receptors. *Neuron* **84:** 1049–1064. doi:10.1016/j.neuron.2014.10.051

Butts DA, Kanold PO, Shatz CJ. 2007. A burst-based "Hebbian" learning rule at retinogeniculate synapses links retinal waves to activity-dependent refinement. *PLoS Biol* **5:** e61. doi:10.1371/journal.pbio.0050061

Cang J, Feldheim DA. 2013. Developmental mechanisms of topographic map formation and alignment. *Annu Rev Neurosci* **36:** 51–77. doi:10.1146/annurev-neuro-062012-170341

Cang J, Wang L, Stryker MP, Feldheim DA. 2008. Roles of ephrin-as and structured activity in the development of functional maps in the superior colliculus. *J Neurosci* **28:** 11015–11023. doi:10.1523/JNEUROSCI.2478-08.2008

Cang J, Savier E, Barchini J, Liu X. 2018. Visual function, organization, and development of the mouse superior colliculus. *Annu Rev Vis Sci* **4:** 239–262. doi:10.1146/annurev-vision-091517-034142

Catsicas M, Bonness V, Becker D, Mobbs P. 1998. Spontaneous Ca²⁺ transients and their transmission in the developing chick retina. *Curr Biol* **8:** 283–286. doi:10.1016/S0960-9822(98)70110-1

Caval-Holme FS, Aranda ML, Chen AQ, Tiriac A, Zhang Y, Smith B, Birnbaumer L, Schmidt TM, Feller MB. 2022. The retinal basis of light aversion in neonatal mice. *J Neurosci* **42:** 4101–4115. doi:10.1523/JNEUROSCI.0151-22.2022

Chandrasekaran AR, Plas DT, Gonzalez E, Crair MC. 2005. Evidence for an instructive role of retinal activity in retinotopic map refinement in the superior colliculus of the mouse. *J Neurosci* **25:** 6929–6938. doi:10.1523/JNEUROSCI.1470-05.2005

Chandrasekaran AR, Shah RD, Crair MC. 2007. Developmental homeostasis of mouse retinocollicular synapses. *J Neurosci* **27:** 1746–1755. doi:10.1523/JNEUROSCI.4383-06.2007

Chen C, Regehr WG. 2000. Developmental remodeling of the retinogeniculate synapse. *Neuron* **28:** 955–966. doi:10.1016/S0896-6273(00)00166-5

Chen S-K, Badea TC, Hattar S. 2011. Photoentrainment and pupillary light reflex are mediated by distinct populations of ipRGCs. *Nature* **476:** 92–95. doi:10.1038/nature10206

Chen H, Savier EL, DePiero VJ, Cang J. 2021. Lack of evidence for stereotypical direction columns in the mouse superior colliculus. *J Neurosci* **41:** 461–473. doi:10.1523/JNEUROSCI.1155-20.2020

Cheng HJ, Nakamoto M, Bergemann AD, Flanagan JG. 1995. Complementary gradients in expression and binding of ELF-1 and Mek4 in development of the topographic retinotectal projection map. *Cell* **82:** 371–381. doi:10.1016/0092-8674(95)90426-3

Cramer KS, Sur M. 1997. Blockade of afferent impulse activity disrupts on/off sublamination in the ferret lateral geniculate nucleus. *Brain Res Dev Brain Res* **98:** 287–290. doi:10.1016/S0165-3806(96)00188-5

Cruz-Martín A, El-Danaf RN, Osakada F, Sriram B, Dhande OS, Nguyen PL, Callaway EM, Ghosh A, Huberman AD. 2014. A dedicated circuit links direction-selective retinal ganglion cells to the primary visual cortex. *Nature* **507:** 358–361. doi:10.1038/nature12989

Dacey DM. 2004. Origins of perception: retinal ganglion cell diversity and the creation of parallel visual pathways. *Cogn Neurosci* **3:** 281–301.

Dai J, Buhusi M, Demyanenko GP, Brennaman LH, Hruska M, Dalva MB, Maness PF. 2013. Neuron glia-related cell adhesion molecule (NrCAM) promotes topographic retinocollicular mapping. *PLoS ONE* **8:** e73000. doi:10.1371/journal.pone.0073000

de Malmazet D, Kühn NK, Farrow K. 2018. Retinotopic separation of nasal and temporal motion selectivity in the mouse superior colliculus. *Curr Biol* **28:** 2961–2969. e4. doi:10.1016/j.cub.2018.07.001

Demas J, Eglen SJ, Wong ROL. 2003. Developmental loss of synchronous spontaneous activity in the mouse retina is independent of visual experience. *J Neurosci* **23:** 2851–2860. doi:10.1523/JNEUROSCI.23-07-02851.2003

Demas J, Sagdullaev BT, Green E, Jaubert-Miazza L, McCall MA, Gregg RG, Wong ROL, Guido W. 2006. Failure to maintain eye-specific segregation in nob, a mutant with abnormally patterned retinal activity. *Neuron* **50:** 247–259. doi:10.1016/j.neuron.2006.03.033

Dhande OS, Hua EW, Guh E, Yeh J, Bhatt S, Zhang Y, Ruthazer ES, Feller MB, Crair MC. 2011. Development of single retinofugal axon arbors in normal and β2 knockout mice. *J Neurosci* **31:** 3384–3399. doi:10.1523/JNEUROSCI.4899-10.2011

Dhande OS, Stafford BK, Lim JHA, Huberman AD. 2015. Contributions of retinal ganglion cells to subcortical visual processing and behaviors. *Annu Rev Vis Sci* **1:** 291–328. doi:10.1146/annurev-vision-082114-035502

Dimitrov D, Türei D, Garrido-Rodriguez M, Burmedi PL, Nagai JS, Boys C, Ramirez Flores RO, Kim H, Szalai B, Costa IG, et al. 2022. Comparison of methods and resources for cell-cell communication inference from single-cell RNA-seq data. *Nat Commun* **13:** 3224. doi:10.1038/s41467-022-30755-0

Dräger UC, Hubel DH. 1976. Topography of visual and somatosensory projections to mouse superior colliculus. *J Neurophysiol* **39:** 91–101. doi:10.1152/jn.1976.39.1.91

Dräger UC, Olsen JF. 1980. Origins of crossed and uncrossed retinal projections in pigmented and albino mice. *J Comp Neurol* **191:** 383–412. doi:10.1002/cne.901910306

Dräger UC, Olsen JF. 1981. Ganglion cell distribution in the retina of the mouse. *Invest Ophthalmol Vis Sci* **20:** 285–293.

Ecker JL, Dumitrescu ON, Wong KY, Alam NM, Chen SK, LeGates T, Renna JM, Prusky GT, Berson DM, Hattar S. 2010. Melanopsin-expressing retinal ganglion-cell photoreceptors: cellular diversity and role in pattern vision. *Neuron* **67:** 49–60. doi:10.1016/j.neuron.2010.05.023

El-Danaf RN, Huberman AD. 2019. Sub-topographic maps for regionally enhanced analysis of visual space in the mouse retina. *J Comp Neurol* **527**: 259–269. doi:10.1002/cne.24457

Ellis EM, Gauvain G, Sivyer B, Murphy GJ. 2016. Shared and distinct retinal input to the mouse superior colliculus and dorsal lateral geniculate nucleus. *J Neurophysiol* **116**: 602–610. doi:10.1152/jn.00227.2016

Elstrott J, Feller MB. 2010. Direction-selective ganglion cells show symmetric participation in retinal waves during development. *J Neurosci* **30**: 11197–11201. doi:10.1523/JNEUROSCI.2302-10.2010

Erskine L, Reijntjes S, Pratt T, Denti L, Schwarz Q, Vieira JM, Alakakone B, Shewan D, Ruhrberg C. 2011. VEGF signaling through neuropilin 1 guides commissural axon crossing at the optic chiasm. *Neuron* **70**: 951–965. doi:10.1016/j.neuron.2011.02.052

Fabre PJ, Shimogori T, Charron F. 2010. Segregation of ipsilateral retinal ganglion cell axons at the optic chiasm requires the Shh receptor Boc. *J Neurosci* **30**: 266–275. doi:10.1523/JNEUROSCI.3778-09.2010

Feinberg EH, Meister M. 2015. Orientation columns in the mouse superior colliculus. *Nature* **519**: 229–232. doi:10.1038/nature14103

Feldheim DA, Vanderhaeghen P, Hansen MJ, Frisén J, Lu Q, Barbacid M, Flanagan JG. 1998. Topographic guidance labels in a sensory projection to the forebrain. *Neuron* **21**: 1303–1313. doi:10.1016/S0896-6273(00)80650-9

Feldheim DA, Kim YI, Bergemann AD, Frisén J, Barbacid M, Flanagan JG. 2000. Genetic analysis of ephrin-A2 and ephrin-A5 shows their requirement in multiple aspects of retinocollicular mapping. *Neuron* **25**: 563–574. doi:10.1016/S0896-6273(00)81060-0

Fernandez DC, Chang YT, Hattar S, Chen SK. 2016. Architecture of retinal projections to the central circadian pacemaker. *Proc Natl Acad Sci* **113**: 6047–6052. doi:10.1073/pnas.1523629113

Fernandez DC, Komal R, Langel J, Ma J, Duy PQ, Penzo MA, Zhao H, Hattar S. 2020. Retinal innervation tunes circuits that drive nonphotic entrainment to food. *Nature* **581**: 194–198. doi:10.1038/s41586-020-2204-1

Fisher LJ. 1979. Development of synaptic arrays in the inner plexiform layer of neonatal mouse retina. *J Comp Neurol* **187**: 359–372. doi:10.1002/cne.901870207

Ford KJ, Félix AL, Feller MB. 2012. Cellular mechanisms underlying spatiotemporal features of cholinergic retinal waves. *J Neurosci* **32**: 850–863. doi:10.1523/JNEUROSCI.5309-12.2012

Fratzl A, Koltchev AM, Vissers N, Tan YL, Marques-Smith A, Stempel AV, Branco T, Hofer SB. 2021. Flexible inhibitory control of visually evoked defensive behavior by the ventral lateral geniculate nucleus. *Neuron* **109**: 3810–3822.e9. doi:10.1016/j.neuron.2021.09.003

Frisén J, Yates PA, McLaughlin T, Friedman GC, O Leary DD, Barbacid M. 1998. Ephrin-A5 (AL-1/RAGS) is essential for proper retinal axon guidance and topographic mapping in the mammalian visual system. *Neuron* **20**: 235–243. doi:10.1016/S0896-6273(00)80452-3

Fukuda Y, Sawai H, Watanabe M, Wakakuwa K, Morigiwa K. 1989. Nasotemporal overlap of crossed and uncrossed retinal ganglion cell projections in the Japanese monkey (*Macaca fuscata*). *J Neurosci* **9**: 2353–2373. doi:10.1523/JNEUROSCI.09-07-02353.1989

Gale SD, Murphy GJ. 2014. Distinct representation and distribution of visual information by specific cell types in mouse superficial superior colliculus. *J Neurosci* **34**: 13458–13471. doi:10.1523/JNEUROSCI.2768-14.2014

García-Frigola C, Herrera E. 2010. Zic2 regulates the expression of Sert to modulate eye-specific refinement at the visual targets. *EMBO J* **29**: 3170–3183. doi:10.1038/emboj.2010.172

García-Frigola C, Carreres MI, Vegar C, Mason C, Herrera E. 2008. Zic2 promotes axonal divergence at the optic chiasm midline by EphB1-dependent and -independent mechanisms. *Development* **135**: 1833–1841. doi:10.1242/dev.020693

Gauvain G, Murphy GJ. 2015. Projection-specific characteristics of retinal input to the brain. *J Neurosci* **35**: 6575–6583. doi:10.1523/JNEUROSCI.4298-14.2015

Ge X, Zhang K, Gribizis A, Hamodi AS, Sabino AM, Crair MC. 2021. Retinal waves prime visual motion detection by simulating future optic flow. *Science* **373**: eabd0830. doi:10.1126/science.abd0830

Gjorgjieva J, Toyoizumi T, Eglen SJ. 2009. Burst-time-dependent plasticity robustly guides ON/OFF segregation in the lateral geniculate nucleus. *PLoS Comput Biol* **5**: e1000618. doi:10.1371/journal.pcbi.1000618

Godement P, Salaün J, Imbert M. 1984. Prenatal and postnatal development of retinogeniculate and retinocollicular projections in the mouse. *J Comp Neurol* **230**: 552–575. doi:10.1002/cne.902300406

Goetz J, Jessen ZF, Jacobi A, Mani A, Cooler S, Greer D, Kadri S, Segal J, Shekhar K, Sanes JR, et al. 2022. Unified classification of mouse retinal ganglion cells using function, morphology, and gene expression. *Cell Rep* **40**: 111040. doi:10.1016/j.celrep.2022.111040

Gribizis A, Ge X, Daigle TL, Ackman JB, Zeng H, Lee D, Crair MC. 2019. Visual cortex gains independence from peripheral drive before eye opening. *Neuron* **104**: 711–723.e3. doi:10.1016/j.neuron.2019.08.015

Grubb MS, Rossi FM, Changeux JP, Thompson ID. 2003. Abnormal functional organization in the dorsal lateral geniculate nucleus of mice lacking the β2 subunit of the nicotinic acetylcholine receptor. *Neuron* **40**: 1161–1172. doi:10.1016/S0896-6273(03)00789-X

Güler AD, Ecker JL, Lall GS, Haq S, Altimus CM, Liao H-W, Barnard AR, Cahill H, Badea TC, Zhao H, et al. 2008. Melanopsin cells are the principal conduits for rod-cone input to non-image-forming vision. *Nature* **453**: 102–105. doi:10.1038/nature06829

Gupta D, Młynarski W, Sumser A, Symonova O, Svatoň J, Joesch M. 2023. Panoramic visual statistics shape retina-wide organization of receptive fields. *Nat Neurosci* **26**: 606–614. doi:10.1038/s41593-023-01280-0

Hahm JO, Langdon RB, Sur M. 1991. Disruption of retinogeniculate afferent segregation by antagonists to NMDA receptors. *Nature* **351**: 568–570. doi:10.1038/351568a0

Hammer S, Monavarfeshani A, Lemon T, Su J, Fox MA. 2015. Multiple retinal axons converge onto relay cells in the adult mouse thalamus. *Cell Rep* **12**: 1575–1583. doi:10.1016/j.celrep.2015.08.003

Cite this article as *Cold Spring Harb Perspect Biol* doi: 10.1101/cshperspect.a041512

Hasse JM, Briggs F. 2017. A cross-species comparison of corticogeniculate structure and function. *Vis Neurosci* **34:** E016. doi:10.1017/S095252381700013X

Hattar S, Kumar M, Park A, Tong P, Tung J, Yau KW, Berson DM. 2006. Central projections of melanopsin-expressing retinal ganglion cells in the mouse. *J Comp Neurol* **497:** 326–349. doi:10.1002/cne.20970

Haustead DJ, Lukehurst SS, Clutton GT, Bartlett CA, Dunlop SA, Arrese CA, Sherrard RM, Rodger J. 2008. Functional topography and integration of the contralateral and ipsilateral retinocollicular projections of *ephrin-A*$^{-/-}$ mice. *J Neurosci* **28:** 7376–7386. doi:10.1523/JNEUROSCI.1135-08.2008

Hendrickson A, Wilson ME, Toyne MJ. 1970. The distribution of optic nerve fibers in *Macaca mulatta*. *Brain Res* **23:** 425–427. doi:10.1016/0006-8993(70)90068-5

Herrera E, Brown L, Aruga J, Rachel RA, Dolen G, Mikoshiba K, Brown S, Mason CA. 2003. Zic2 patterns binocular vision by specifying the uncrossed retinal projection. *Cell* **114:** 545–557. doi:10.1016/S0092-8674(03)00684-6

Heukamp AS, Warwick RA, Rivlin-Etzion M. 2020. Topographic variations in retinal encoding of visual space. *Annu Rev Vis Sci* **6:** 237–259. doi:10.1146/annurev-vision-121219-081831

Hindges R, McLaughlin T, Genoud N, Henkemeyer M, O'Leary DDM. 2002. Ephb forward signaling controls directional branch extension and arborization required for dorsal-ventral retinotopic mapping. *Neuron* **35:** 475–487. doi:10.1016/S0896-6273(02)00799-7

Holcombe V, Guillery RW. 1984. The organization of retinal maps within the dorsal and ventral lateral geniculate nuclei of the rabbit. *J Comp Neurol* **225:** 469–491. doi:10.1002/cne.902250402

Hong YK, Kim IJ, Sanes JR. 2011. Stereotyped axonal arbors of retinal ganglion cell subsets in the mouse superior colliculus. *J Comp Neurol* **519:** 1691–1711. doi:10.1002/cne.22595

Hong YK, Park S, Litvina EY, Morales J, Sanes JR, Chen C. 2014. Refinement of the retinogeniculate synapse by bouton clustering. *Neuron* **84:** 332–339. doi:10.1016/j.neuron.2014.08.059

Hong YK, Burr EF, Sanes JR, Chen C. 2019. Heterogeneity of retinogeniculate axon arbors. *Eur J Neurosci* **49:** 948–956. doi:10.1111/ejn.13986

Hooks BM, Chen C. 2006. Distinct roles for spontaneous and visual activity in remodeling of the retinogeniculate synapse. *Neuron* **52:** 281–291. doi:10.1016/j.neuron.2006.07.007

Hooks BM, Chen C. 2008. Vision triggers an experience-dependent sensitive period at the retinogeniculate synapse. *J Neurosci* **28:** 4807–4817. doi:10.1523/JNEUROSCI.4667-07.2008

Howarth M, Walmsley L, Brown TM. 2014. Binocular integration in the mouse lateral geniculate nuclei. *Curr Biol* **24:** 1241–1247. doi:10.1016/j.cub.2014.04.014

Hoy JL, Bishop HI, Niell CM. 2019. Defined cell types in superior colliculus make distinct contributions to prey capture behavior in the mouse. *Curr Biol* **29:** 4130–4138.e5. doi:10.1016/j.cub.2019.10.017

Hu Z, Mu Y, Huang L, Hu Y, Chen Z, Yang Y, Huang X, Fu Y, Xi Y, Lin S, et al. 2022. A visual circuit related to the periaqueductal gray area for the antinociceptive effects of bright light treatment. *Neuron* **110:** 1712–1727.e7. doi:10.1016/j.neuron.2022.02.009

Huang L, Xi Y, Peng Y, Yang Y, Huang X, Fu Y, Tao Q, Xiao J, Yuan T, An K, et al. 2019. A visual circuit related to habenula underlies the antidepressive effects of light therapy. *Neuron* **102:** 128–142.e8. doi:10.1016/j.neuron.2019.01.037

Huang X, Huang P, Huang L, Hu Z, Liu X, Shen J, Xi Y, Yang Y, Fu Y, Tao Q, et al. 2021. A visual circuit related to the nucleus reuniens for the spatial-memory-promoting effects of light treatment. *Neuron* **109:** 347–362.e7. doi:10.1016/j.neuron.2020.10.023

Huang W, Xu Q, Su J, Tang L, Hao ZZ, Xu C, Liu R, Shen Y, Sang X, Xu N, et al. 2022. Linking transcriptomes with morphological and functional phenotypes in mammalian retinal ganglion cells. *Cell Rep* **40:** 111322. doi:10.1016/j.celrep.2022.111322

Huang YF, Liao PY, Yu JH, Chen SK. 2023. Light disrupts social memory via a retina-to-supraoptic nucleus circuit. *EMBO Rep* **2023:** e56839. doi:10.15252/embr.202356839

Huberman AD, Murray KD, Warland DK, Feldheim DA, Chapman B. 2005. Ephrin-As mediate targeting of eye-specific projections to the lateral geniculate nucleus. *Nat Neurosci* **8:** 1013–1021. doi:10.1038/nn1505

Huberman AD, Manu M, Koch SM, Susman MW, Lutz AB, Ullian EM, Baccus SA, Barres BA. 2008. Architecture and activity-mediated refinement of axonal projections from a mosaic of genetically identified retinal ganglion cells. *Neuron* **59:** 425–438. doi:10.1016/j.neuron.2008.07.018

Huberman AD, Wei W, Elstrott J, Stafford BK, Feller MB, Barres BA. 2009. Genetic identification of an on-off direction-selective retinal ganglion cell subtype reveals a layer-specific subcortical map of posterior motion. *Neuron* **62:** 327–334. doi:10.1016/j.neuron.2009.04.014

Illing RB, Wässle H. 1981. The retinal projection to the thalamus in the cat: a quantitative investigation and a comparison with the retinotectal pathway. *J Comp Neurol* **202:** 265–285. doi:10.1002/cne.902020211

Ivanova E, Hwang GS, Pan ZH. 2010. Characterization of transgenic mouse lines expressing Cre recombinase in the retina. *Neuroscience* **165:** 233–243. doi:10.1016/j.neuroscience.2009.10.021

Johnson KO, Triplett JW. 2021. Wiring subcortical image-forming centers: topography, laminar targeting, and map alignment. *Curr Top Dev Biol* **142:** 283–317. doi:10.1016/bs.ctdb.2020.10.004

Johnson KP, Zhao L, Kerschensteiner D. 2018. A pixel-encoder retinal ganglion cell with spatially offset excitatory and inhibitory receptive fields. *Cell Rep* **22:** 1462–1472. doi:10.1016/j.celrep.2018.01.037

Johnson KP, Fitzpatrick MJ, Zhao L, Wang B, McCracken S, Williams PR, Kerschensteiner D. 2021. Cell-type-specific binocular vision guides predation in mice. *Neuron* **109:** 1527–1539.e4. doi:10.1016/j.neuron.2021.03.010

Johnson KO, Harel L, Triplett JW. 2023. Postsynaptic NMDA receptor expression is required for visual cortico-collicular projection refinement in the mouse superior colliculus. *J Neurosci* **43:** 1310–1320. doi:10.1523/JNEUROSCI.1473-22.2022

Kay JN, De la Huerta I, Kim I-J, Zhang Y, Yamagata M, Chu MW, Meister M, Sanes JR. 2011. Retinal ganglion cells with distinct directional preferences differ in molecular identity, structure, and central projections. *J Neurosci* 31: 7753–7762. doi:10.1523/JNEUROSCI.0907-11.2011

Kerschensteiner D. 2016. Glutamatergic retinal waves. *Front Neural Circuits* 10: 38. doi:10.3389/fncir.2016.00038

Kerschensteiner D. 2020. Mammalian retina development. In *The senses: a comprehensive reference*, 2nd ed (ed. Fritzsch B), pp. 234–251. Elsevier, Oxford.

Kerschensteiner D. 2022. Feature detection by retinal ganglion cells. *Annu Rev Vis Sci* 8: 135–169. doi:10.1146/annurev-vision-100419-112009

Kerschensteiner D, Guido W. 2017. Organization of the dorsal lateral geniculate nucleus in the mouse. *Vis Neurosci* 34: E008. doi:10.1017/S0952523817000062

Kerschensteiner D, Wong ROL. 2008. A precisely timed asynchronous pattern of ON and OFF retinal ganglion cell activity during propagation of retinal waves. *Neuron* 58: 851–858. doi:10.1016/j.neuron.2008.04.025

Kim IJ, Zhang Y, Yamagata M, Meister M, Sanes JR. 2008. Molecular identification of a retinal cell type that responds to upward motion. *Nature* 452: 478–482. doi:10.1038/nature06739

Kim I-J, Zhang Y, Meister M, Sanes JR. 2010. Laminar restriction of retinal ganglion cell dendrites and axons: subtype-specific developmental patterns revealed with transgenic markers. *J Neurosci* 30: 1452–1462. doi:10.1523/JNEUROSCI.4779-09.2010

Kim T, Shen N, Hsiang JC, Johnson KP, Kerschensteiner D. 2020. Dendritic and parallel processing of visual threats in the retina control defensive responses. *Sci Adv* 6: eabc9920. doi:10.1126/sciadv.abc9920

Kirkby LA, Sack GS, Firl A, Feller MB. 2014. Erratum to: A role for correlated spontaneous activity in the assembly of neural circuits [Neuron 80 (2013) 1129–1144]. *Neuron* 81: 218. doi:10.1016/j.neuron.2013.12.031

Knudsen EI, Brainard MS. 1995. Creating a unified representation of visual and auditory space in the brain. *Annu Rev Neurosci* 18: 19–43. doi:10.1146/annurev.ne.18.030195.000315

Koch SM, Ullian EM. 2010. Neuronal pentraxins mediate silent synapse conversion in the developing visual system. *J Neurosci* 30: 5404–5414. doi:10.1523/JNEUROSCI.4893-09.2010

Koch SM, Dela Cruz CG, Hnasko TS, Edwards RH, Huberman AD, Ullian EM. 2011. Pathway-specific genetic attenuation of glutamate release alters select features of competition-based visual circuit refinement. *Neuron* 71: 235–242. doi:10.1016/j.neuron.2011.05.045

Krahe TE, El-Danaf RN, Dilger EK, Henderson SC, Guido W. 2011. Morphologically distinct classes of relay cells exhibit regional preferences in the dorsal lateral geniculate nucleus of the mouse. *J Neurosci* 31: 17437–17448. doi:10.1523/jneurosci.4370-11.2011

Kruijt CC, de Wit GC, Bergen AA, Florijn RJ, Schalij-Delfos NE, van Genderen MM. 2018. The phenotypic spectrum of albinism. *Ophthalmology* 125: 1953–1960. doi:10.1016/j.ophtha.2018.08.003

Kuwajima T, Yoshida Y, Takegahara N, Petros TJ, Kumanogoh A, Jessell TM, Sakurai T, Mason C. 2012. Optic chiasm presentation of Semaphorin6D in the context of Plexin-A1 and Nr-CAM promotes retinal axon midline crossing. *Neuron* 74: 676–690. doi:10.1016/j.neuron.2012.03.025

Leamey CA, Merlin S, Lattouf P, Sawatari A, Zhou X, Demel N, Glendining KA, Oohashi T, Sur M, Fässler R. 2007. Ten_m3 regulates eye-specific patterning in the mammalian visual pathway and is required for binocular vision. *PLoS Biol* 5: e241. doi:10.1371/journal.pbio.0050241

Lee R, Petros TJ, Mason CA. 2008. Zic2 regulates retinal ganglion cell axon avoidance of ephrinB2 through inducing expression of the guidance receptor EphB1. *J Neurosci* 28: 5910–5919. doi:10.1523/JNEUROSCI.0632-08.2008

Levine JN, Schwartz GW. 2020. The olivary pretectal nucleus receives visual input of high spatial resolution. bioRxiv doi:10.1101/2020.06.23.168054

Li YT, Meister M. 2023. Functional cell types in the mouse superior colliculus. *eLife* 12: e82367. doi:10.7554/eLife.82367

Li YT, Turan Z, Meister M. 2020. Functional architecture of motion direction in the mouse superior colliculus. *Curr Biol* 30: 3304–3315.e4. doi:10.1016/j.cub.2020.06.023

Liang L, Chen C. 2020. Organization, function, and development of the mouse retinogeniculate synapse. *Annu Rev Vis Sci* 6: 261–285. doi:10.1146/annurev-vision-121219-081753

Liang F, Xiong XR, Zingg B, Ji XY, Zhang LI, Tao HW. 2015. Sensory cortical control of a visually induced arrest behavior via corticotectal projections. *Neuron* 86: 755–767. doi:10.1016/j.neuron.2015.03.048

Liang L, Fratzl A, Goldey G, Ramesh RN, Sugden AU, Morgan JL, Chen C, Andermann ML. 2018. A fine-scale functional logic to convergence from retina to thalamus. *Cell* 173: 1343–1355.e24. doi:10.1016/j.cell.2018.04.041

Litvina EY, Chen C. 2017. Functional convergence at the retinogeniculate synapse. *Neuron* 96: 330–338.e5. doi:10.1016/j.neuron.2017.09.037

Liu A, Milner ES, Peng YR, Blume HA, Brown MC, Bryman GS, Emanuel AJ, Morquette P, Viet NM, Sanes JR, et al. 2023a. Encoding of environmental illumination by primate melanopsin neurons. *Science* 379: 376–381. doi:10.1126/science.ade2024

Liu Y, Savier EL, DePiero VJ, Chen C, Schwalbe DC, Abraham-Fan RJ, Chen H, Campbell JN, Cang J. 2023b. Mapping visual functions onto molecular cell types in the mouse superior colliculus. *Neuron* 111: 1876–1886.e5. doi:10.1016/j.neuron.2023.03.036

Lund RD, Lund JS, Wise RP. 1974. The organization of the retinal projection to the dorsal lateral geniculate nucleus in pigmented and albino rats. *J Comp Neurol* 158: 383–403. doi:10.1002/cne.901580403

Magnin M, Cooper HM, Mick G. 1989. Retinohypothalamic pathway: a breach in the law of Newton-Müller-Gudden? *Brain Res* 488: 390–397. doi:10.1016/0006-8993(89)90737-3

Maiorano NA, Hindges R. 2013. Restricted perinatal retinal degeneration induces retina reshaping and correlated structural rearrangement of the retinotopic map. *Nat Commun* 4: 1938. doi:10.1038/ncomms2926

Martersteck EM, Hirokawa KE, Evarts M, Bernard A, Duan X, Li Y, Ng L, Oh SW, Ouellette B, Royall JJ, et al. 2017. Diverse central projection patterns of retinal ganglion cells. *Cell Rep* **18**: 2058–2072. doi:10.1016/j.celrep.2017.01.075

Mason C, Slavi N. 2020. Retinal ganglion cell axon wiring establishing the binocular circuit. *Annu Rev Vis Sci* **6**: 215–236. doi:10.1146/annurev-vision-091517-034306

McLaughlin T, Torborg CL, Feller MB, O'Leary DDM. 2003. Retinotopic map refinement requires spontaneous retinal waves during a brief critical period of development. *Neuron* **40**: 1147–1160. doi:10.1016/S0896-6273(03)00790-6

McNeill DS, Sheely CJ, Ecker JL, Badea TC, Morhardt D, Guido W, Hattar S. 2011. Development of melanopsin-based irradiance detecting circuitry. *Neural Dev* **6**: 8. doi:10.1186/1749-8104-6-8

Meng JJ, Shen JW, Li G, Ouyang CJ, Hu JX, Li ZS, Zhao H, Shi YM, Zhang M, Liu R, et al. 2023. Light modulates glucose metabolism by a retina-hypothalamus-brown adipose tissue axis. *Cell* **186**: 398–412.e17. doi:10.1016/j.cell.2022.12.024

Milner ES, Do MTH. 2017. A population representation of absolute light intensity in the mammalian retina. *Cell* **171**: 865–876.e16. doi:10.1016/j.cell.2017.09.005

Monavarfeshani A, Sabbagh U, Fox MA. 2017. Not a one-trick pony: diverse connectivity and functions of the rodent lateral geniculate complex. *Vis Neurosci* **34**: E012. doi:10.1017/S0952523817000098

Monavarfeshani A, Stanton G, Van Name J, Su K, Mills WA 3rd, Swilling K, Kerr A, Huebschman NA, Su J, Fox MA. 2018. LRRTM1 underlies synaptic convergence in visual thalamus. *eLife* **7**: e33498. doi:10.7554/eLife.33498

Monosov IE, Ogasawara T, Haber SN, Heimel JA, Ahmadlou M. 2022. The zona incerta in control of novelty seeking and investigation across species. *Curr Opin Neurobiol* **77**: 102650. doi:10.1016/j.conb.2022.102650

Moreno-Juan V, Aníbal-Martínez M, Herrero-Navarro Á, Valdeolmillos M, Martini FJ, López-Bendito G. 2023. Spontaneous thalamic activity modulates the cortical innervation of the primary visual nucleus of the thalamus. *Neuroscience* **508**: 87–97. doi:10.1016/j.neuroscience.2022.07.022

Morgan JL, Lichtman JW. 2020. An individual interneuron participates in many kinds of inhibition and innervates much of the mouse visual thalamus. *Neuron* **106**: 468–481.e2. doi:10.1016/j.neuron.2020.02.001

Morgan JL, Schubert T, Wong ROL. 2008. Developmental patterning of glutamatergic synapses onto retinal ganglion cells. *Neural Dev* **3**: 8. doi:10.1186/1749-8104-3-8

Morgan JL, Berger DR, Wetzel AW, Lichtman JW. 2016. The fuzzy logic of network connectivity in mouse visual thalamus. *Cell* **165**: 192–206. doi:10.1016/j.cell.2016.02.033

Morin LP, Studholme KM. 2014. Retinofugal projections in the mouse. *J Comp Neurol* **522**: 3733–3753. doi:10.1002/cne.23635

Morin LP, Blanchard JH, Provencio I. 2003. Retinal ganglion cell projections to the hamster suprachiasmatic nucleus, intergeniculate leaflet, and visual midbrain: bifurcation and melanopsin immunoreactivity. *J Comp Neurol* **465**: 401–416. doi:10.1002/cne.10881

Mrsic-Flogel TD, Hofer SB, Creutzfeldt C, Cloëz-Tayarani I, Changeux JP, Bonhoeffer T, Hübener M. 2005. Altered map of visual space in the superior colliculus of mice

lacking early retinal waves. *J Neurosci* **25**: 6921–6928. doi:10.1523/JNEUROSCI.1555-05.2005

Muellner FE, Roska B. 2023. Individual thalamic inhibitory interneurons are functionally specialized towards distinct visual features. bioRxiv doi:10.1101/2023.03.22.533751

Muir-Robinson G, Hwang BJ, Feller MB. 2002. Retinogeniculate axons undergo eye-specific segregation in the absence of eye-specific layers. *J Neurosci* **22**: 5259–5264. doi:10.1523/JNEUROSCI.22-13-05259.2002

Münch TA, da Silveira RA, Siegert S, Viney TJ, Awatramani GB, Roska B. 2009. Approach sensitivity in the retina processed by a multifunctional neural circuit. *Nat Neurosci* **12**: 1308–1316. doi:10.1038/nn.2389

Nakamoto M, Cheng HJ, Friedman GC, McLaughlin T, Hansen MJ, Yoon CH, O'Leary DD, Flanagan JG. 1996. Topographically specific effects of ELF-1 on retinal axon guidance in vitro and retinal axon mapping in vivo. *Cell* **86**: 755–766. doi:10.1016/S0092-8674(00)80150-6

Nakamoto C, Durward E, Horie M, Nakamoto M. 2019. Nell2 regulates the contralateral-versus-ipsilateral visual projection as a domain-specific positional cue. *Development* **146**: dev170704. doi:10.1242/dev.170704

Nath A, Schwartz GW. 2017. Electrical synapses convey orientation selectivity in the mouse retina. *Nat Commun* **8**: 2025. doi:10.1038/s41467-017-01980-9

O'Brien BJ, Abel PL, Olavarria JF. 2001. The retinal input to calbindin-D28k-defined subdivisions in macaque inferior pulvinar. *Neurosci Lett* **312**: 145–148. doi:10.1016/S0304-3940(01)02220-0

Osterhout JA, Josten N, Yamada J, Pan F, Wu S-W, Nguyen PL, Panagiotakos G, Inoue YU, Egusa SF, Volgyi B, et al. 2011. Cadherin-6 mediates axon-target matching in a non-image-forming visual circuit. *Neuron* **71**: 632–639. doi:10.1016/j.neuron.2011.07.006

Osterhout JA, El-Danaf RN, Nguyen PL, Huberman AD. 2014. Birthdate and outgrowth timing predict cellular mechanisms of axon target matching in the developing visual pathway. *Cell Rep* **8**: 1006–1017. doi:10.1016/j.celrep.2014.06.063

Osterhout JA, Stafford BK, Nguyen PL, Yoshihara Y, Huberman AD. 2015. Contactin-4 mediates axon-target specificity and functional development of the accessory optic system. *Neuron* **86**: 985–999. doi:10.1016/j.neuron.2015.04.005

Peng J, Fabre PJ, Dolique T, Swikert SM, Kermasson L, Shimogori T, Charron F. 2018. Sonic Hedgehog is a remotely produced cue that controls axon guidance trans-axonally at a midline choice point. *Neuron* **97**: 326–340.e4. doi:10.1016/j.neuron.2017.12.028

Peng YR, Shekhar K, Yan W, Herrmann D, Sappington A, Bryman GS, van Zyl T, Do MTH, Regev A, Sanes JR. 2019. Molecular classification and comparative taxonomics of foveal and peripheral cells in primate retina. *Cell* **176**: 1222–1237.e22. doi:10.1016/j.cell.2019.01.004

Penn AA, Riquelme PA, Feller MB, Shatz CJ. 1998. Competition in retinogeniculate patterning driven by spontaneous activity. *Science* **279**: 2108–2112. doi:10.1126/science.279.5359.2108

Petros TJ, Bryson JB, Mason C. 2010. Ephrin-B2 elicits differential growth cone collapse and axon retraction in retinal ganglion cells from distinct retinal regions. *Dev Neurobiol* **70**: 781–794. doi:10.1002/dneu.20821

Pfeiffenberger C, Cutforth T, Woods G, Yamada J, Rentería RC, Copenhagen DR, Flanagan JG, Feldheim DA. 2005. Ephrin-As and neural activity are required for eye-specific patterning during retinogeniculate mapping. *Nat Neurosci* **8:** 1022–1027. doi:10.1038/nn1508

Pfeiffenberger C, Yamada J, Feldheim DA. 2006. Ephrin-As and patterned retinal activity act together in the development of topographic maps in the primary visual system. *J Neurosci* **26:** 12873–12884. doi:10.1523/JNEUROSCI .3595-06.2006

Piscopo DM, El-Danaf RN, Huberman AD, Niell CM. 2013. Diverse visual features encoded in mouse lateral geniculate nucleus. *J Neurosci* **33:** 4642–4656. doi:10.1523/ JNEUROSCI.5187-12.2013

Plas DT, Lopez JE, Crair MC. 2005. Pretarget sorting of retinocollicular axons in the mouse. *J Comp Neurol* **491:** 305–319. doi:10.1002/cne.20694

Prieur DS, Rebsam A. 2017. Retinal axon guidance at the midline: chiasmatic misrouting and consequences. *Dev Neurobiol* **77:** 844–860. doi:10.1002/dneu.22473

Quattrochi LE, Stabio ME, Kim I, Ilardi MC, Fogerson PM, Leyrer ML, Berson DM. 2019. The M6 cell: A small-field bistratified photosensitive retinal ganglion cell. *J Comp Neurol* **527:** 297–311. doi:10.1002/cne.24556

Rafols JA, Valverde F. 1973. The structure of the dorsal lateral geniculate nucleus in the mouse. A golgi and electron microscopic study. *J Comp Neurol* **150:** 303–332. doi:10 .1002/cne.901500305

Read JCA. 2021. Binocular vision and stereopsis across the animal kingdom. *Annu Rev Vis Sci* **7:** 389–415. doi:10 .1146/annurev-vision-093109-113212

Reese BE. 1988. "Hidden lamination" in the dorsal lateral geniculate nucleus: the functional organization of this thalamic region in the rat. *Brain Res* **472:** 119–137. doi:10.1016/0165-0173(88)90017-3

Reinhard K, Li C, Do Q, Burke EG, Heynderickx S, Farrow K. 2019. A projection specific logic to sampling visual inputs in mouse superior colliculus. *eLife* **8:** e50697. doi:10 7554/ eLife.50697

Rheaume BA, Jereen A, Bolisetty M, Sajid MS, Yang Y, Renna K, Sun L, Robson P, Trakhtenberg EF. 2018. Single cell transcriptome profiling of retinal ganglion cells identifies cellular subtypes. *Nat Commun* **9:** 2759. doi:10 .1038/s41467-018-05134-3

Rhoades RW, Chalupa LM. 1978. Functional properties of the corticotectal projection in the golden hamster. *J Comp Neurol* **180:** 617–634. doi:10.1002/cne.901800312

Rivlin-Etzion M, Zhou K, Wei W, Elstrott J, Nguyen PL, Barres BA, Huberman AD, Feller MB. 2011. Transgenic mice reveal unexpected diversity of on-off direction-selective retinal ganglion cell subtypes and brain structures involved in motion processing. *J Neurosci* **31:** 8760–8769. doi:10.1523/JNEUROSCI.0564-11.2011

Rivlin-Etzion M, Grimes WN, Rieke F. 2018. Flexible neural hardware supports dynamic computations in retina. *Trends Neurosci* **41:** 224–237. doi:10.1016/j.tins.2018.01 .009

Rodger J, Dunlop SA, Beazley LD. 1998. The ipsilateral retinal projection in the fat-tailed dunnart, *Sminthopsis crassicaudata*. *Vis Neurosci* **15:** 677–684. doi:10 1017/ S095252389815407X

Román Rosón M, Bauer Y, Kotkat AH, Berens P, Euler T, Busse L. 2019. Mouse dLGN receives functional input from a diverse population of retinal ganglion cells with limited convergence. *Neuron* **102:** 462–476.e8. doi:10 .1016/j.neuron.2019.01.040

Rompani SB, Müllner FE, Wanner A, Zhang C, Roth CN, Yonehara K, Roska B. 2017. Different modes of visual integration in the lateral geniculate nucleus revealed by single-cell-initiated transsynaptic tracing. *Neuron* **93:** 767–776.e6. doi:10.1016/j.neuron.2017.01.028

Rossi FM, Pizzorusso T, Porciatti V, Marubio LM, Maffei L, Changeux JP. 2001. Requirement of the nicotinic acetylcholine receptor β2 subunit for the anatomical and functional development of the visual system. *Proc Natl Acad Sci* **98:** 6453–6458. doi:10.1073/pnas.101120998

Rousso DL, Qiao M, Kagan RD, Yamagata M, Palmiter RD, Sanes JR. 2016. Two pairs of ON and OFF retinal ganglion cells are defined by intersectional patterns of transcription factor expression. *Cell Rep* **15:** 1930–1944. doi:10 .1016/j.celrep.2016.04.069

Russell AL, Dixon KG, Triplett JW. 2022. Diverse modes of binocular interactions in the mouse superior colliculus. *J Neurophysiol* **127:** 913–927. doi:10.1152/jn.00526.2021

Sabbah S, Gemmer JA, Bhatia-Lin A, Manoff G, Castro G, Siegel JK, Jeffery N, Berson DM. 2017. A retinal code for motion along the gravitational and body axes. *Nature* **14:** 5267.

Salay LD, Huberman AD. 2021. Divergent outputs of the ventral lateral geniculate nucleus mediate visually evoked defensive behaviors. *Cell Rep* **37:** 109792. doi:10.1016/j .celrep.2021.109792

Sánchez-Arrones L, Nieto-Lopez F, Sánchez-Camacho C, Carreres MI, Herrera E, Okada A, Bovolenta P. 2013. Shh/Boc signaling is required for sustained generation of ipsilateral projecting ganglion cells in the mouse retina. *J Neurosci* **33:** 8596–8607. doi:10.1523/JNEUROSCI .2083-12.2013

Savier E, Eglen SJ, Bathélémy A, Perraut M, Pfrieger FW, Lemke G, Reber M. 2017. A molecular mechanism for the topographic alignment of convergent neural maps. *eLife* **6:** e20470. doi:10.7554/eLife.20470

Schiller PH, Stryker M. 1972. Single-unit recording and stimulation in superior colliculus of the alert rhesus monkey. *J Neurophysiol* **35:** 915–924. doi:10.1152/jn.1972.35.6 .915

Schmitt AM, Shi J, Wolf AM, Lu CC, King LA, Zou Y. 2006. Wnt–Ryk signalling mediates medial–lateral retinotectal topographic mapping. *Nature* **439:** 31–37. doi:10.1038/ nature04334

Seabrook TA, El-Danaf RN, Krahe TE, Fox MA, Guido W. 2013. Retinal input regulates the timing of corticogeniculate innervation. *J Neurosci* **33:** 10085–10097. doi:10 .1523/JNEUROSCI.5271-12.2013

Seabrook TA, Burbridge TJ, Crair MC, Huberman AD. 2017. Architecture, function, and assembly of the mouse visual system. *Annu Rev Neurosci* **40:** 499–538. doi:10.1146/an nurev-neuro-071714-033842

Siegert S, Scherf BG, Del Punta K, Didkovsky N, Heintz N, Roska B. 2009. Genetic address book for retinal cell types. *Nat Neurosci* **12:** 1197–1204. doi:10.1038/nn.2370

Simon DK, O'Leary DD. 1992. Development of topographic order in the mammalian retinocollicular projection. *J*

Neurosci **12:** 1212–1232. doi:10.1523/JNEUROSCI.12-04-01212.1992

Simpson JI. 1984. The accessory optic system. *Annu Rev Neurosci* **7:** 13–41. doi:10.1146/annurev.ne.07.030184.000305

Slavi N, Balasubramanian R, Lee MA, Liapin M, Oaks-Leaf R, Peregrin J, Potenski A, Troy CM, Ross ME, Herrera E, et al. 2023. CyclinD2-mediated regulation of neurogenic output from the retinal ciliary margin is perturbed in albinism. *Neuron* **111:** 49–64.e5. doi:10.1016/j.neuron.2022.10.025

Sparks DL, Lee C, Rohrer WH. 1990. Population coding of the direction, amplitude, and velocity of saccadic eye movements by neurons in the superior colliculus. *Cold Spring Harb Symp Quant Biol* **55:** 805–811. doi:10.1101/SQB.1990.055.01.075

Stafford BK, Sher A, Litke AM, Feldheim DA. 2009. Spatial-temporal patterns of retinal waves underlying activity-dependent refinement of retinofugal projections. *Neuron* **64:** 200–212. doi:10.1016/j.neuron.2009.09.021

Su J, Haner CV, Imbery TE, Brooks JM, Morhardt DR, Gorse K, Guido W, Fox MA. 2011. Reelin is required for class-specific retinogeniculate targeting. *J Neurosci* **31:** 575–586. doi:10.1523/JNEUROSCI.4227-10.2011

Su J, Sabbagh U, Liang Y, Olejníková L, Dixon KG, Russell AL, Chen J, Pan YA, Triplett JW, Fox MA. 2021. A cell-ECM mechanism for connecting the ipsilateral eye to the brain. *Proc Natl Acad Sci* **118:** e2104343118. doi:10.1073/pnas.2104343118

Suetterlin P, Drescher U. 2014. Target-independent ephrinA/EphA-mediated axon-axon repulsion as a novel element in retinocollicular mapping. *Neuron* **84:** 740–752. doi:10.1016/j.neuron.2014.09.023

Sun LO, Brady CM, Cahill H, Al-Khindi T, Sakuta H, Dhande OS, Noda M, Huberman AD, Nathans J, Kolodkin AL. 2015. Functional assembly of accessory optic system circuitry critical for compensatory eye movements. *Neuron* **86:** 971–984. doi:10.1016/j.neuron.2015.03.064

Thompson AD, Picard N, Min L, Fagiolini M, Chen C. 2016. Cortical feedback regulates feedforward retinogeniculate refinement. *Neuron* **91:** 1021–1033. doi:10.1016/j.neuron.2016.07.040

Tien NW, Vitale C, Badea TC, Kerschensteiner D. 2022. Layer-specific developmentally precise axon targeting of transient suppressed-by-contrast retinal ganglion cells (tSbC RGCs). *J Neurosci* **42:** 7213–7221. doi:10.1523/JNEUROSCI.2332-21.2022

Tiriac A, Feller MB. 2022. Roles of visually evoked and spontaneous activity in the development of retinal direction selectivity maps. *Trends Neurosci* **45:** 529–538. doi:10.1016/j.tins.2022.04.002

Tiriac A, Smith BE, Feller MB. 2018. Light prior to eye opening promotes retinal waves and eye-specific segregation. *Neuron* **100:** 1059–1065.e4. doi:10.1016/j.neuron.2018.10.011

Tiriac A, Bistrong K, Pitcher MN, Tworig JM, Feller MB. 2022. The influence of spontaneous and visual activity on the development of direction selectivity maps in mouse retina. *Cell Rep* **38:** 110225. doi:10.1016/j.celrep.2021.110225

Tomar M, Beros J, Meloni B, Rodger J. 2023. Interactions between guidance cues and neuronal activity: therapeutic insights from mouse models. *Int J Mol Sci* **24:** 6966. doi:10.3390/ijms24086966

Tran NM, Shekhar K, Whitney IE, Jacobi A, Benhar I, Hong G, Yan W, Adiconis X, Arnold ME, Lee JM, et al. 2019. Single-cell profiles of retinal ganglion cells differing in resilience to injury reveal neuroprotective genes. *Neuron* **104:** 1039–1055.e12. doi:10.1016/j.neuron.2019.11.006

Triplett JW, Feldheim DA. 2012. Eph and ephrin signaling in the formation of topographic maps. *Semin Cell Dev Biol* **23:** 7–15. doi:10.1016/j.semcdb.2011.10.026

Triplett JW, Owens MT, Yamada J, Lemke G, Cang J, Stryker MP, Feldheim DA. 2009. Retinal input instructs alignment of visual topographic maps. *Cell* **139:** 175–185. doi:10.1016/j.cell.2009.08.028

Triplett JW, Phan A, Yamada J, Feldheim DA. 2012. Alignment of multimodal sensory input in the superior colliculus through a gradient-matching mechanism. *J Neurosci* **32:** 5264–5271. doi:10.1523/JNEUROSCI.0240-12.2012

Tsai NY, Wang F, Toma K, Yin C, Takatoh J, Pai EL, Wu K, Matcham AC, Yin L, Dang EJ, et al. 2022. Trans-Seq maps a selective mammalian retinotectal synapse instructed by nephronectin. *Nat Neurosci* **25:** 659–674. doi:10.1038/s41593-022-01068-8

Usrey WM, Alitto HJ. 2015. Visual functions of the thalamus. *Annu Rev Vis Sci* **1:** 351–371. doi:10.1146/annurev-vision-082114-035920

Voufo C, Chen AQ, Smith BE, Yan R, Feller MB, Tiriac A. 2023. Circuit mechanisms underlying embryonic retinal waves. *eLife* **12:** e81983. doi:10.7554/eLife.81983

Walls GL. 1942. *The vertebrate eye and its adaptive radiation.* Cranbrook Institute of Science, Bloomfield Hills, MI.

Walmsley L, Brown TM. 2015. Eye-specific visual processing in the mouse suprachiasmatic nuclei. *J Physiol* **593:** 1731–1743. doi:10.1113/jphysiol.2014.288225

Wang Q, Burkhalter A. 2013. Stream-related preferences of inputs to the superior colliculus from areas of dorsal and ventral streams of mouse visual cortex. *J Neurosci* **33:** 1696–1705. doi:10.1523/JNEUROSCI.3067-12.2013

Wang F, Li E, De L, Wu Q, Zhang Y. 2021. OFF-transient α RGCs mediate looming triggered innate defensive response. *Curr Biol* **31:** 2263–2273.e3. doi:10.1016/j.cub.2021.03.025

Warland DK, Huberman AD, Chalupa LM. 2006. Dynamics of spontaneous activity in the fetal macaque retina during development of retinogeniculate pathways. *J Neurosci* **26:** 5190–5197. doi:10.1523/JNEUROSCI.0328-06.2006

Warwick RA, Kaushansky N, Sarid N, Golan A, Rivlin-Etzion M. 2018. Inhomogeneous encoding of the visual field in the mouse retina. *Curr Biol* **28:** 655–665.e3. doi:10.1016/j.cub.2018.01.016

Williams SE, Mann F, Erskine L, Sakurai T, Wei S, Rossi DJ, Gale NW, Holt CE, Mason CA, Henkemeyer M. 2003. Ephrin-B2 and EphB1 mediate retinal axon divergence at the optic chiasm. *Neuron* **39:** 919–935. doi:10.1016/j.neuron.2003.08.017

Williams SE, Grumet M, Colman DR, Henkemeyer M, Mason CA, Sakurai T. 2006. A role for Nr-CAM in the patterning of binocular visual pathways. *Neuron* **50:** 535–547. doi:10.1016/j.neuron.2006.03.037

Wong RO. 1999. Retinal waves and visual system development. *Annu Rev Neurosci* **22:** 29–47. doi:10.1146/annurev.neuro.22.1.29

Wurtz RH, Goldberg ME. 1972. The role of the superior colliculus in visually-evoked eye movements. *Bibl Ophthalmol* **82:** 149–158.

Yates PA, Roskies AL, McLaughlin T, O'Leary DD 2001. Topographic-specific axon branching controlled by ephrin-As is the critical event in retinotectal map development. *J Neurosci* **21:** 8548–8563. doi:10.1523/JNEUROSCI.21-21-08548.2001

Zeng H, Sanes JR. 2017. Neuronal cell-type classification: challenges, opportunities and the path forward. *Nat Rev Neurosci* **18:** 530–546. doi:10.1038/nrn.2017.85

Zhang J, Ackman JB, Xu HP, Crair MC. 2011. Visual map development depends on the temporal pattern of binocular activity in mice. *Nat Neurosci* **15:** 298–307. doi:10.1038/nn.3007

Zhao X, Liu M, Cang J. 2014. Visual cortex modulates the magnitude but not the selectivity of looming-evoked responses in the superior colliculus of awake mice. *Neuron* **84:** 202–213. doi:10.1016/j.neuron.2014.08.037

Zhao ZD, Chen Z, Xiang X, Hu M, Xie H, Jia X, Cai F, Cui Y, Chen Z, Qian L, et al. 2019. Zona incerta GABAergic neurons integrate prey-related sensory signals and induce an appetitive drive to promote hunting. *Nat Neurosci* **22:** 921–932. doi:10.1038/s41593-019-0404-5

Cite this article as *Cold Spring Harb Perspect Biol* doi: 10.1101/cshperspect.a041512

Interneuron Diversity: How Form Becomes Function

Natalia V. De Marco García[1] and Gord Fishell[2,3]

[1]Center for Neurogenetics, Brain and Mind Research Institute, Weill Cornell Medicine, New York, New York 10021, USA

[2]Harvard Medical School, Blavatnik Institute, Department of Neurobiology, Boston, Massachusetts 02115, USA

[3]Stanley Center for Psychiatric Research, Broad Institute of MIT and Harvard, Cambridge, Massachusetts 02142, USA

Correspondence: nad2018@med.cornell.edu; gordon_fishell@hms.harvard.edu

A persistent question in neuroscience is how early neuronal subtype identity is established during the development of neuronal circuits. Despite significant progress in the transcriptomic characterization of cortical interneurons, the mechanisms that control the acquisition of such identities as well as how they relate to function are not clearly understood. Accumulating evidence indicates that interneuron identity is achieved through the interplay of intrinsic genetic and activity-dependent programs. In this work, we focus on how progressive interactions between interneurons and pyramidal cells endow maturing interneurons with transient identities fundamental for their function during circuit assembly and how the elimination of transient connectivity triggers the consolidation of adult subtypes.

Attempts to understand the nervous system often begin with efforts to ascertain its modular components, the cell types. While this question permeates all parts of the central nervous system (CNS), nowhere is it more evident than in the bewildering diversity of cortical interneurons. But what defines a cell type? The problem in part derives from the question of how we measure neuronal diversity: by shape, through their intrinsic physiological properties, or perhaps by connectivity. With the advent of single-cell transcriptomics, this question is usually addressed at the level of gene expression. Although hugely informative, this can also be problematic. Gene expression per se is neither discrete nor categorical, but the clustering algorithms typically used to classify cells can mislead one to create new subtypes with each transcriptional cluster discovered. The question is further complicated by transient developmental states. For instance, induction of plasticity leads to fluctuations in activity-dependent transcription, increasing apparent genetic diversity. In a sensible compromise, the Allen Brain Institute has adopted their classification, which attempts to reconcile three major metrics of diversity: morphology, electrophysiology, and transcriptomics (MET). While a good starting point, further clarity to the question can be brought by adopting a developmental and functional lens, specifically through the analysis of progressive changes in developmental gene expression and connectivity. In this review, we focus on the impact of developmental events from their genesis on through to their establishment of mature connectivity. We begin with their origins in the ventral pallium and then trace their identities across time as they migrate first tangentially to the cortex and then radially within the cortical plate to find mature settling positions. Single-cell genomics has established that interneuron diversity is first evident, at the transcriptional level, upon them becoming postmitotic. Nonetheless, how this initial diversification relates to the maturation of subtype-specific identities has been debated. What makes this effort difficult is defining mature interneuron types on their developmental gene expression patterns, which are often shared across cell types, irrespective of their identities. Hence, when particular identities are established is inevitably in the eye of the beholder. Here, rather than debate the question of when subtypes emerge, we will focus on each of their developmental steps and how these impact their mature identities. By discussing the progressive interactions interneurons make to reach maturity, it is our hope to determine how this impacts their adult function. Moreover, as the same interneuron types appear to populate the entire cortex, a majority of studies have focused on the sensory regions, the primary visual and somatosensory cortices of mice in particular. As such, we will primarily focus our discussion on these areas.

EARLY PROGRAMMING OF MOLECULAR IDENTITY IN CARDINAL INTERNEURON GROUPS

The question of how interneurons achieve their mature identity is inherently a question of nature versus nurture. Their diversity arises through the unfolding of a combination of intrinsic genetic programs, coupled with extrinsic influences during development. Interneurons are unique in their broad migration from their site of origin to where they terminally differentiate, making the question of environmental influences particularly intriguing. Given their protracted develop-

opment, that there must be an environmental contribution seems obvious. Less clear is the degree to which intrinsic programs are embedded at birth versus shaped by environmental cues as they mature. In short, does the environment actively direct the emergence of subclass identity or simply passively allow for preselected subtypes to emerge? Most likely while some mix of these extremes occurs, the contribution of each is less certain. So rather than debate the specific role of intrinsic programs versus environment, we will describe each of the requisite steps for their differentiation and present the data as they support either hypothesis.

Inhibitory neurons are generated from common progenitor pools in the medial, caudal, and lateral ganglionic eminences (MGEs, CGEs, and LGEs, reviewed by Wamsley and Fishell 2017) as well as the preoptic area (POA) (Nery et al. 2001; Wichterle et al. 2001; Butt et al. 2005; Miyoshi et al. 2007; Gelman et al. 2009). Neuronal types generated from these embryonic structures can adopt strikingly different postmitotic identities (Mayer et al. 2018). Within specific lineages, CGE-derived siblings can give rise to vasoactive intestinal peptide (VIP) interneurons versus neurogliaform (NGF) cells, while MGE progenitors can adopt identities as distinct as GABA-ergic versus cholinergic precursor neurons (Bandler et al. 2022). In addition, all three of these eminences produce both interneuron and projection neuron subtypes (Mayer et al. 2018). Genetic analysis of precursors from the three eminences revealed that as early as embryonic day 13 (E13), newborn interneurons in the CGE and MGE express markers that to varying extents are predictive of their mature identity (e.g., somatostatin [Sst] expressed by MGE-derived interneurons; *Htr3A/AdarB2* expressed by CGE-derived interneurons), each of these genes being indicative of broad categories MGE or CGE classes, respectively (Fig. 1). Although this indicates that some of the molecular signatures characteristic of unique cell classes are established before interneurons reach the cortical plate (Mayer et al. 2018; Mi et al. 2018; Yu et al. 2021), the extent to which subtype identity is concurrently determined is debatable. By performing transcriptomic analysis of the broad

Cite this article as *Cold Spring Harb Perspect Biol* doi: 10.1101/cshperspect.a041513

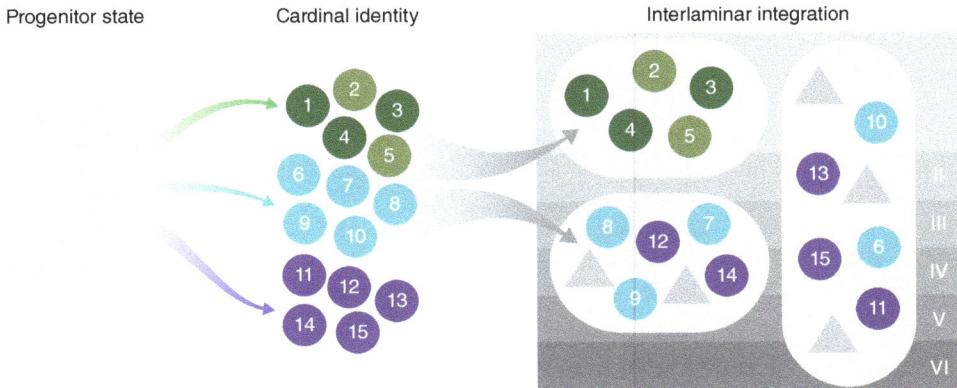

Figure 1. Tracking the progression of cortical interneurons as the transit from progenitors to mature cells within cortical circuits. Here, the progressive restriction of interneurons as they mature from progenitors to cardinally specified progenitors to mature cells integrated into cortical laminae is represented by their increased specification. When cells go from a progenitor state (purple) to a cardinal state (green, blue, or dark blue) they take on the properties of general classes of interneurons. The extent of cardinal specification is still unclear. While there certainly are more interneuron cardinal classes than the four cardinal groups (somatostatin [Sst], parvalbumin [PV], vasoactive intestinal peptide [VIP], and Lamp5) their level of specification has not reached the mature state of when they integrate into cortical circuits (interlaminar integration as shown on the *right*; background shows a schematic of cortical layer positioning). (Image designed by Julia Kuhl.)

types detected in the adult mouse brain, Mayer et al. (2018) mapped these types retrospectively in development. Although some transcriptomic signatures for the earliest born interneurons could be detected at embryonic stages (Mi et al. 2018; Shi et al. 2021), the separation of their various types was less certain before postnatal day (P)10 (Mayer et al. 2018). Indeed, given the large overlap in gene expression among developing interneurons, discerning particular subtypes during embryogenesis is challenging. For instance, the two largest classes of interneurons, Sst and parvalbumin (PV), show a 60% overlap in their gene expression up until P2, after which their genetic programs rapidly diverge (Allaway et al. 2021). While work by Mi et al. (2018) finds that further specificity may exist earlier, the sparse alignment of longitudinal gene expression across developmental times raises the question as to how precisely or faithfully specific subtype fates can be determined. Nonetheless, single-cell RNA-seq analysis suggests that at least some aspects of Sst interneuron diversity are established before these neurons reach the cortical plate (Mayer et al. 2018; Mi et al. 2018). Consistent

with this, a comparison of Sst interneurons embryonically provides compelling evidence that fate distinctions between dorsally migrating Martinotti cells versus subcortically migrating projection (CHODL) are evident before birth (Lim et al. 2018). Furthermore, recent evidence indicates that the myeloid translocation gene 8 (Mtg8), which is expressed in the MGE as cells exit mitosis, interacts with the LIM-HD transcription factor LHX6 to control the specification of Sst/NPY-expressing interneurons, suggesting early acquisition of neuronal identity (Asgarian et al. 2022). Altogether, these results indicate that interneuron subtype identity emerges slowly in development and is initiated by early specification events, and later driven by extrinsic environmental cues impinging on intrinsic gene regulatory networks (GRNs) in young postmitotic interneurons. That intrinsic identity of interneuron subclasses is at least partially fixed based on their site of origin is evident from heterotopic transplant experiments whereby their superclass identities remain fixed even when transplanted from one eminence to another (Butt et al. 2005). Moreover, the degree to which mature interneuron subtypes represent discrete cell fates versus environmental or circumstance-induced identity likely differs according to which superclass of interneurons are being explored. While Sst interneurons show clear subtype differences from early development, PV interneurons mature late and their differences in both gene expression and function across their classes may reflect their local position in the mature brain rather than discrete differences in their cell identity.

MOLECULAR AND ACTIVITY-DEPENDENT MECHANISMS FOR INTERNEURON MIGRATION

Dispersion of cortical interneurons from the germinal zones to their final position is achieved through complex migratory patterns, culminating with their allocation to select cortical laminae. This position reflects subtype identity and appears to be strongly influenced by neuronal activity. Upon undergoing protracted tangential migration, interneurons invade the cortex via the marginal zone (MZ) and subventricular zone (SVZ), and sort themselves into specific laminae (Miyoshi and Fishell 2011). In CGE-derived neurons, this in part is driven through activity-dependent migration, regulated by the engulfment and cell motility gene 1 (Elmo1) (De Marco García et al. 2011) and serotoninergic signaling (Murthy et al. 2014). Signals such as the chemokine Cxcl12 and its receptors, Cxcr4 and Ackr3 (formerly Cxcr7), may terminate migration by playing an important role in retaining interneurons within MZ (López-Bendito et al. 2008; Sánchez-Alcañiz et al. 2011). Expression of Cxcl12 in the MZ sequentially attracts MGE and CGE-derived interneurons. Down-regulation of Cxcr4 expression allows the cortical plate invasion of MGE interneurons. In contrast, CGE interneurons maintain Cxcr4 and Ackr3 expression and Cxcl12-mediated attraction results in the allocation of a substantial proportion of these neurons remaining in layer (L)1 (Venkataramanappa et al. 2022). Consistently, the developmental dysfunction of such signaling results in a ventral displacement of certain CGE subtypes (reelin-, Sp8-, calretinin-expressing interneurons) (Venkataramanappa et al. 2022). In addition to the

roles of activity in driving migration and Cxcl12-mediated attraction maintaining L1 populations, recent work shows that Cajal Retzius cells and superficial interneurons act to prevent ectopic pyramidal cell and interneuron invasion of upper cortical layers (Genescu et al. 2022; Vílchez-Acosta et al. 2022). Similarly, within the subplate, projection Sst populations are repulsed through fibronectin leucine-rich transmembrane FLRT2 and FLRT3 proteins (Fleitas et al. 2021). Hence, the migration of interneurons is balanced by coordinated mechanisms that act to direct, pull, or push various populations in accordance with their expected cortical positions.

Increasing evidence suggests that the positioning of interneurons is intimately dependent on their interactions with pyramidal neurons. Although the initial migratory patterns of some Sst populations are subtype specific (Lim et al. 2018), laminar positioning appears to be fully realized at late developmental stages, suggesting a role for the cortical environment in late diversification. Consistent with the hypothesis of interneurons being programmed in response to pyramidal neurons (Fig. 1), MGE interneurons are attracted to deep-layer pyramidal neurons (Lodato et al. 2011); CGE populations show a similar propensity for superficial populations (Wester et al. 2019). Specifically, neuregulin 3 (Nrg3) produced by pyramidal neurons attracts MGE interneurons through signaling via the receptor tyrosine–protein kinase ErbB4 (Bartolini et al. 2017). Shortly after interneurons reach their appropriate laminae, they develop characteristic morphologies that emerge due to the unfolding of activity-dependent programs. In particular, NGF interneurons in the somatosensory cortex rely on thalamic inputs acting via NR2B-containing NMDA receptors to achieve mature dendritic and axonal morphologies (De Marco García et al. 2015). Notably, the knockdown of such signaling in the hippocampus at later developmental stages appears to only minimally affect dendritic formation (Chittajallu et al. 2017), emphasizing how such interactions are time and/or context dependent. In juvenile mice, interneuron subtypes can be distinguished by the emergence of intrinsic electrophysiological properties (Butt et al. 2005; Miyoshi et al. 2007). Although some of these properties permit cardinal grouping as early as P5 (Anastasiades et al. 2016; Duan et al. 2020), they continue to mature through juvenile stages, suggesting modulation by the cortical environment. Cardinal specification refers to the partition of cortical interneuron identity into four nonoverlapping groups (PV, Sst, Lamp5, and VIP). All together, this evidence indicates that molecular identities define migratory routes for specific subtypes, and the final interneuron differentiation relies on interactions with circuit partners in a context-dependent manner.

A CIRCUIT MECHANISM FOR INTERNEURON CELL DEATH

Neuronal apoptosis sculpts emergent circuits to set an appropriate balance of excitatory and inhibitory neurons in the CNS. Although cortical interneurons are born in excess, cell death eliminates ~30%–40% of the initial population during postnatal development (Southwell et al. 2012; Denaxa et al. 2018; Priya et al. 2018; Wong et al. 2018; Luhmann 2021). Both MGE/ POA and CGE-derived interneurons undergo apoptosis initiated by intrinsic mechanisms and regulated by neuronal activity (Southwell et al. 2012; Priya et al. 2018; Wong et al. 2018, 2022; Duan et al. 2020). Experimental evidence indicates that apoptosis does occur stochastically, but survival is intrinsically linked to circuit and network integration. Immature MGE-derived interneurons and pyramidal cells participate in neuronal assemblies during the first postnatal week (Mòdol et al. 2020) and this participation protects interneurons from apoptosis (Duan et al. 2020). At the circuit level, pyramidal cell inputs onto maturing MGE interneurons are fundamental for survival (Wong et al. 2018), whereas GABA$_A$R-mediated feedback from MGE interneurons limits assembly size through the regulation of apoptosis (Duan et al. 2020; Mòdol et al. 2020). The presence of perisomatic innervation in these assemblies and the absence of a survival phenotype in mice in which Sst interneuron output is silenced during development (Duan et al. 2020; Mòdol et al. 2020) suggests that immature PV interneuron inputs onto

pyramidal cells may play a prominent role in this process. Interestingly, excessive density of PV interneurons due to transplantation of MGE-derived progenitors (Southwell et al. 2010, 2012) or deficit in the transcription factor Cux2 leads to a transient increase in their survival and long-lasting behavioral deficits (Magno et al. 2021). However, the circuit mechanism by which excessive immature interneurons lead to aberrant behavior remains unclear.

The importance of select inputs for the sculpture of emergent circuits appears to be a conserved feature across cell types and brain regions (Fig. 2). Local glutamatergic inputs are required for the survival of NGF and basket interneurons in the somatosensory cortex, whereas bipolar cells are depolarized by serotonin and systemic administration of serotonin reuptake inhibitors increases their survival (Wong et al. 2022). About 40% of 5HT3aR-expressing interneurons coexpress VIP, including bipolar and basket cells (Rudy et al. 2011; Prönneke et al. 2015). In contrast to the results reported by Wong et al. (2022), Priya et al. (2018) report no impact of excitability manipulations on VIP interneuron survival. This discrepancy could reflect differences in the ability of Kir2.1 to silence VIP cells under a chronic manipulation (Priya et al. 2018) and Gi DREADDS, which greatly reduced bipolar and basket cell survival (Wong et al. 2022). Interestingly, the circuit-specific control of interneuron survival is a conserved feature in the CNS. In the developing striatum, cortical inputs regulate the survival of PV interneurons, whereas medium spiny neuron inputs negatively impact the survival of ChAT$^+$ interneurons (Sreenivasan et al. 2022). Conversely, callosal projections in the visual cortex negatively regulate chandelier cell survival in the binocular visual cortex (Wang et al. 2021). At the molecular level, cortical interneuron survival is regulated by cell adhesion mediated by the γ protocadherin gene cluster (PCDHG, specifically isoforms Pcdhgc3, Pcdhgc4, and Pcdhgc5). Deletion of this cluster in GABAergic interneurons leads to enhanced apoptosis of MGE subtypes in both mutant mice and transplant models (Carriere et al. 2020; Mancia Leon et al. 2020). One possible interpretation of these results is

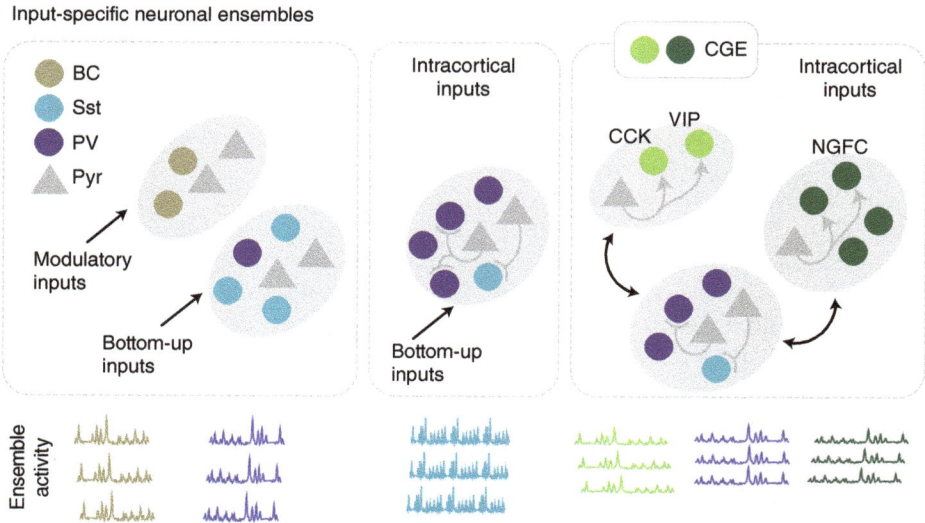

Figure 2. Cell-type-specific inputs regulate the emergence of synchronous activity in the developing brain. (*Top*) Neuronal ensembles composed of specific interneuron subtypes (shaded ovals) and afferent inputs fundamental for ensemble recruitment (arrows). (*Bottom*) Calcium events depicting similar synchronous activity among members of the same ensemble. (BC) Bipolar cell, (VIP) vasoactive intestinal peptide, (CCK) cholecystokinin, (NGFC) neurogliaform cell, (Sst) somatostatin, (CGE) caudal ganglionic eminence-derived interneurons, (PV) parvalbumin, (Pyr) pyramidal cell. (Image designed by Julia Kuhl.)

that PCDHG-mediated interaction between interneurons may facilitate assembly formation, protecting them from cell death. This model would reconcile intrinsic and activity-dependent mechanisms for interneuron survival.

MATURATION IN THE CORTEX DEPENDS ON TRANSIENT CONNECTIVITY, WHICH APPEARS TO CONTROL CORTICAL NETWORKS BECOMING DESYNCHRONIZED

The maturation of inhibitory cortical networks is an iterative process, whereby early scaffolding events are critical for the development of mature circuitry. In development, Sst interneurons regulate the frequency of spontaneous spindle bursts, restrict neuronal activation, and mediate facilitation upon sensory stimulation (Kastli et al. 2020; Leighton et al. 2021; Baruchin et al. 2022). At these early stages, Sst interneurons in deep layers are transiently recruited by the thalamus (Marques-Smith et al. 2016; Tuncdemir et al. 2016). These inputs promote the enlarge-

ment of GABAergic assemblies mostly composed of perisomatically innervating (prospective PV) interneurons at perinatal stages but restrict their activation later in development to initiate perisomatic inhibition (Mòdol et al. 2020). Coincidentally, neuronal activity in murine sensory cortices and the prefrontal cortex undergoes significant decorrelation from the first to the second postnatal weeks (Golshani et al. 2009; Rochefort et al. 2009; Chini et al. 2022; Babij et al. 2023). This decorrelation is thought to be a prerequisite for proper neuronal coding in the mature brain. Importantly, in the barrel cortex, interneuron dysfunction causes abnormal network correlation (Che et al. 2018; Duan et al. 2020; Babij et al. 2023). These events appear fundamental to ensure the proper emergence of functional topography in preparation for active exploration (Che et al. 2018; Duan et al. 2020; Murata and Colonnese 2020). Furthermore, recent work revealed that transient innervation of Sst interneurons coordinates synchronized rhythmic activity that ultimately becomes desynchronized as the developmental in-

Cite this article as *Cold Spring Harb Perspect Biol* doi: 10.1101/cshperspect.a041513

puts to Sst interneurons form the thalamus re-tract (Mòdol et al. 2024).

Developmental events critical for adult cir-cuitry are not solely occurring in deeper layers of cortex. During the first postnatal week, LI inter-neurons are transiently activated by thalamic in-puts (Che et al. 2018). These inputs are funda-mental for the sharpening of barrel fields and whisker-dependent sensory discrimination in adult mice (Che et al. 2018). Furthermore, recent work in the visual cortex has revealed a require-ment for the developmental input of first-order thalamus to LI NGF interneurons for their re-ceipt of proper maturation of top-down inputs from the anterior-cingulate cortex (Ibrahim et al. 2021). Thus, combined with the role of Sst and

PV interneurons in deep cortical layers, there is clear emerging evidence indicating that different interneuron populations are coordinated to sup-port the emergence of the cortical computations seen in the mature brain (Fig. 3). Hence, these findings emphasize the importance of inhibit-ory circuits in the assembly of mature cortical function.

THE IMPACT OF DEVELOPMENTAL DYSFUNCTION ON NEUROPSYCHIATRIC DISEASE

Notably, when the development of normal cor-tical inhibition fails to occur properly, it can re-sult in neuropsychiatric disease, implicating an

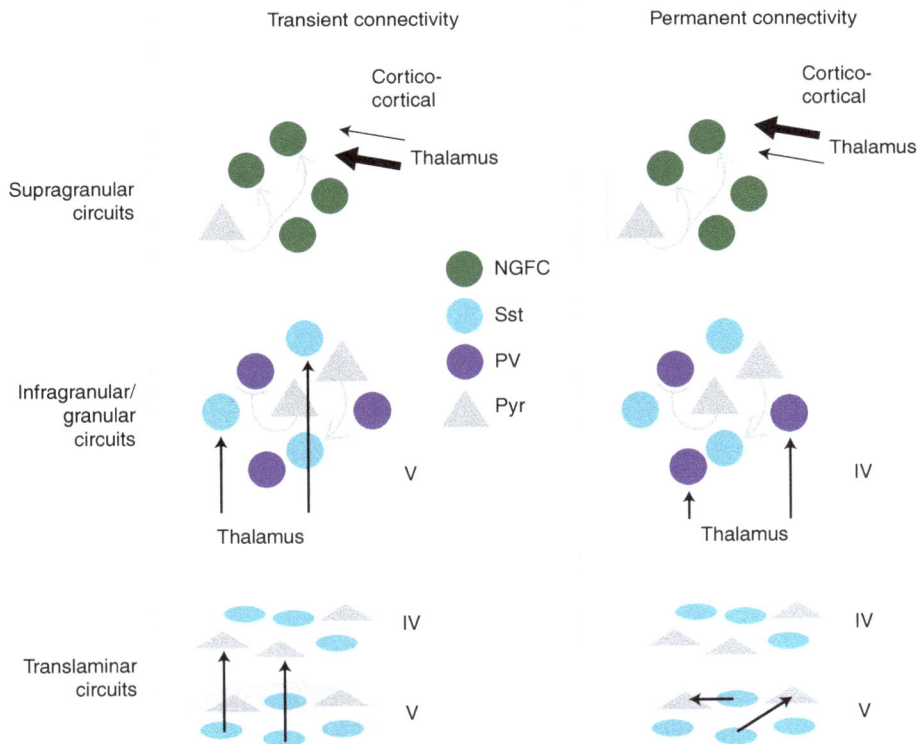

Figure 3. Developmental transition of interneuron afferent connectivity. (*Top*) The strength of afferent inputs onto layer 1 neurogliaform cell (NGFC) changes with development. (*Middle*) Thalamic inputs target somatostatin (Sst) interneurons in early development switching to parvalbumin (PV) interneurons at later stages. (*Bottom*) Sst interneurons transiently innervate LIV neurons. (Pyr) Pyramidal cell, (CGE) caudal ganglionic eminence-derived interneurons. (Image designed by Julia Kuhl.)

involvement of inhibitory cells in the etiology of developmental brain disorders. For instance, recent work shows that GABA$_A$R-mediated currents are required to suppress developmental pyramidal cell coactivation (Duan et al. 2020). Furthermore, the developmental deletion of the gene encoding for the β3 subunit of GABA$_A$R, a high-risk ASD gene, in forebrain pyramidal cells impairs the developmental decorrelation of cortical networks (Babij et al. 2023). Taken together, these observations emphasize how the emergence of mature cortical function is dependent on the role of developing interneurons in sculpting immature connections. The failure of this to occur, from this and other manipulations, causes tactile hypersensitivity and aberrant functional connectivity between cortical hemispheres (Babij et al. 2023). In agreement with these observations, dysregulation of PV activity at neonatal stages underlies tactile hypersensitivity in Fragile X mice (Kourdougli et al. 2023). Such defects not only impact sensory cortices but also affect associative areas. Collectively, desynchronization appears critically dependent on the late maturation of PV interneurons. Recent evidence indicates that developmental impairment in PV interneurons in the PFC leads to long-lasting cognitive impairments (Bitzenhofer et al. 2021; Canetta et al. 2022). Similarly, the loss of Cntnap2 or Cntnap4 in PV interneurons results in behavioral defects in mice and/or human (Karayannis et al. 2014; Vogt et al. 2018), highlighting the critical importance of perinatal stages for the maturation of these late-developing class of GABAergic interneurons.

COMPARTMENT SELECTIVITY OF INTERNEURON OUTPUT

During the later phases of maturation, interneurons establish connectivity patterns that are highly subtype specific. The selectivity of interneuron targeting to postsynaptic compartments relies on selective molecular programs. Specifically, synaptic organizers, which control synapse formation, localization, and maintenance, play a major role in this process. An extensive body of literature shows the importance of the developmental expression of ErbB4 for the formation of input and output synapses between pyramidal cells and PV interneurons (Fazzari et al. 2010; Del Pino et al. 2013; Batista-Brito et al. 2023). ErbB4 is expressed in these interneurons and allocated to both postsynaptic and presynaptic sites (Fazzari et al. 2010; Del Pino et al. 2013). ErbB4 is activated by Nrg3 present in pyramidal cell axons where it regulates excitatory synaptic formation onto recipient interneurons. This activation represses Tsc2, a mTORC1 regulator, driving the local translation of synaptic proteins (Bernard et al. 2022). In contrast, presynaptic ErbB4 is activated by postsynaptic Nrg1 present in the somatodendritic compartment of pyramidal cells and regulates inhibitory synaptic density (Exposito-Alonso et al. 2020). A similar degree of molecular specificity is observed in the targeting of cellular compartments. The synaptic organizers Cbln4 (a member of the C1q family), Lgi2 (leucine repeat-rich LGI family member), and Fgf13 control the development of dendritic, perisomatic, and axo-axonic inhibitory synapses (Favuzzi et al. 2019). Interestingly, the expression of these genes is dramatically up-regulated ∼P10, a stage of heightened synaptogenesis, suggesting that activity-dependent programs may trigger their expression. In part, these changes may engage activity-dependent alternative splicing as a mechanism to direct synaptogenesis (Wamsley and Fishell 2017; Wamsley et al. 2018; Ibrahim et al. 2023).

The selection of synaptic targets is followed by the remodeling of inhibitory synapses. Although intrinsic signals from neurons drive synaptic refinement, glial cells such as astrocytes and microglia largely execute it. Recently, we demonstrated that microglia are selectively recruited by GABAergic neurons to initiate inhibitory synapse remodeling (Favuzzi et al. 2021). Removal of microglial cells impairs synaptic elimination in PV interneurons, leading to an excess of inhibitory synapses. The normal elimination of PV synapses by microglia relies on the activation of metabotropic GABA$_B$ receptors on microglia and is impaired when the Gabbr1 subunit is removed from microglia. Importantly, excitatory synapses are not affected by these manipulations, indicating that different synapses recruit microglia through different signaling. Interestingly,

Sst synapses are also altered after microglia manipulations. However, the underlying mechanism seems to also involve a role for microglia-derived cytokines in regulating Sst axon arborization (Gesuita et al. 2022). Hence, subtype-specific cellular interactions between inhibitory neurons and nonneuronal cells drive the refinement of inhibitory circuits during postnatal development. Notably, microglia-mediated synaptic pruning not only affects their cellular targets but also their intracellular domain specificity.

In addition to synapse formation or elimination, is the functional diversification of interneuron subtypes constrained by synaptic targeting? Recent evidence points to a discrete diversification tailored to selective circuit function. Recent experimental evidence suggests that the diversity of the Sst cardinal class can be captured in nine subtypes in sensory cortices. These Sst subtypes show selectivity in laminar targeting, pyramidal cell innervation (intratelencephalic [IT] vs. pyramidal tract projecting [PT]), and dendritic compartment (basal, oblique, apical branch, and tuft) (Wu et al. 2023). Nonetheless, transcriptomic analysis indicates that 21 Sst types (13 MET) can in addition be identified (Gouwens et al. 2020). Perhaps this additional diversity may indicate that these nine subtypes can be further divided into a larger number of specific states. This among other things may indicate how they respond to local activity through modulation of gene expression, which could refine their synaptic connectivity and function substantially. Such activity-based changes likely impact the plasticity of all types of cortical interneurons. In support of this notion, experimental evidence indicates that mature interneurons may adapt their transcriptional profile in response to activity-dependent processes. For instance, even relatively late removal of the transcription factor Er81 from PV interneurons can regulate their function and likely impact their transcriptomic profile. Activity-dependent expression of this gene tunes intrinsic excitability in response to changes in pyramidal cell activity and learning (Dehorter et al. 2015). Notably, such observations may explain the present mismatch between interneuron types based on transcription versus other metrics, such as morphology or physiology. This demands that we explore whether these transcriptional differences represent how specific subtypes can alter their function based on activity-dependent circumstances. As such, the breadth of transcriptional diversity may indicate that interneuron types possess a level of plasticity that allows them to vary their gene expression in a state-dependent manner.

CONCLUDING REMARKS

By comparing interneurons in development and adults, it becomes clear that they undergo a series of changes as they mature. Interneurons are developmentally recruited during network assembly in a manner that differs from their function in the adult cortex. In this context, an interneuron's first job may involve a role in organizing cortical networks in a manner distinct from how they function in adult circuitry. For their developmental roles, cardinal identity may suffice to endow interneurons with the capacity to regulate cell death, as well as initiating translaminar and intralaminar control of cortical networks and only later partake in more specific regulation in cortical gating. For example, Sst and L1 interneurons appear to adapt their connectivity across development to allow them to shift signaling from the bottom up to a more specific role in synaptic inhibition. Notably, the differentiation of pyramidal cells is far from complete when interneurons first contact them. As the cortex matures, the targeting and refinement of IT axons continue into the second through third postnatal week of mouse development and shifts the balance from being driven by the thalamus to being more reliant on top-down feedback control (Fame et al. 2011). Indeed, ultrastructural electron microscopic analysis of the barrel cortex during development indicates that the synaptic specificity of inhibitory inputs to dendrites begins in the first postnatal week, followed by somatic and ultimately axonal specificity occurs sequentially (Gour et al. 2021). Together, this suggests that interneurons recalibrate their function from a more general role to ultimately align with more mature cortical activity. To achieve this transition, it remains possible that interneu-

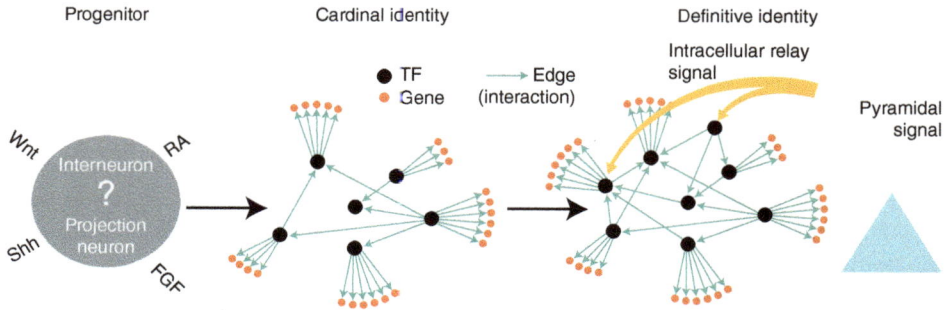

Figure 4. Interneuron specification is determined by a series of extrinsic and intrinsic influences. In this figure, progenitors become either interneurons or projection neurons. These choices are likely influenced by morphogenic factors including Wnts, FGF, RA, or SHH. These signals establish cardinal identities through the initiation of gene regulatory networks (GRNs), involving the expression of transcription factors (TFs), which regulate each other and target genes. In our model, these are intrinsically regulated in interneurons as they migrate into the cortex, and pyramidal neurons (blue triangles) provide further extrinsic signals that promote differentiation and survival of interneurons through their regulation of GRNs (blue arrow extrinsic influences, resulting in yellow arrows, intrinsic regulators of the GRN program). (Image designed by Julia Kuhl.)

ron subtypes are specialized from embryonic stages but adjust their function concomitant during the maturation of cortical circuits (Fig. 4). If so, this emphasizes how a natural balance between their cardinal programming, followed by their later postmigratory refinement, coordinates their contributions to the assembly of cortical circuits.

While the understanding of interneuron function in adult circuits is rapidly progressing, how the recruitment of different cardinal classes regulates circuit assembly, and the purpose of transitory innervation are not clearly understood. One possibility is that this connectivity may trigger the maturation of inhibition-stabilized networks seeding the emergence of spontaneous attractors (Rahmati et al. 2017; Kirmse and Zhang 2022). Our previous work supports this model, suggesting that cell death may facilitate the elimination of interneurons not integrated into stable networks (Duan et al. 2020; Mòdol et al. 2020). Interestingly, an attractor model for the acquisition of interneuron identity has been proposed by Fishell and Kepecs (2020). This model posits that transcription factors rather than specifying individual cardinal and subtype identities participate in GRNs and contribute to the generation of attractor states. It is conceivable that developmental circuit configu-

rations stabilize GRNs associated with attractor states. We propose that this stabilization may be achieved through interneuron integration into maturing circuits. At perinatal stages, transient thalamic and intracortical connectivity may consolidate cardinal identities, whereas the transition from transient to mature connectivity may stabilize the emergence of adult functional subtypes (Fig. 4). Interestingly, in both cases, the specification of interneurons appears to be part intrinsic character and part environmental influences, which in whole or part are dependent on their afferent and efferent synaptic drive. Intriguingly, the importance of excitatory influences on interneuron maturation is multifaceted (for review, see Wamsley and Fishell 2017), with thalamic inputs regulating their morphology earlier in development and pyramidal neurons affecting their later survival and maturation. Thus, functional interneuron diversity is tailored to facilitate the increasing demands of circuit configurations underlying cortical computations associated with the expansion of the behavioral repertoire from the pup to the adult.

ACKNOWLEDGMENTS

We are grateful to Marta Florio and Emilia Favuzzi for comments on the manuscript. Research in the Fishell laboratory is supported by grants from the National Institutes of Health (NIH) (R37MH071679, P01NS074972, UH3MH120096, R01NS081297, and UF1MH130701) and the Simons Foundation, while the De Marco laboratory is supported by grants from the NIH (NIMH R01MH110553, R01MH125006, and NINDS R01NS116137), and an Irma Hirschl/Monique Weill-Caulier Career Scientist Award.

REFERENCES

Allaway KC, Gabitto MI, Wapinski O, Saldi G, Wang CY, Bandler RC, Wu SJ, Bonneau R, Fishell G. 2021. Genetic and epigenetic coordination of cortical interneuron development. *Nature* 597: 693–697. doi:10.1038/s41586-021-03933-1

Anastasiades PG, Marques-Smith A, Lyngholm D, Lickiss T, Raffiq S, Kätzel D, Miesenböck G, Butt SJ. 2016. GABAergic interneurons form transient layer-specific circuits in early postnatal neocortex. *Nat Commun* 7: 10584. doi:10.1038/ncomms10584

Asgarian Z, Oliveira MG, Stryjewska A, Maragkos I, Rubin AN, Magno L, Pachnis V, Ghorbani M, Hiebert SW, Denaxa M, et al. 2022. MTG8 interacts with LHX6 to specify cortical interneuron subtype identity. *Nat Commun* 13: 5217. doi:10.1038/s41467-022-32898-6

Babij R, Ferrer C, Donatelle A, Wacks S, Buch AM, Niemeyer JE, Ma H, Duan ZRS, Fetcho RN, Che A, et al. 2023. Gabrb3 is required for the functional integration of pyramidal neuron subtypes in the somatosensory cortex. *Neuron* 111: 256–274.e10. doi:10.1016/j.neuron.2022.10.037

Bandler RC, Vitali I, Delgado RN, Ho MC, Dvoretskova E, Ibarra Molinas JS, Frazel PW, Mohammadkhani M, Machold R, Maedler S, et al. 2022. Single-cell delineation of lineage and genetic identity in the mouse brain. *Nature* 601: 404–409. doi:10.1038/s41586-021-04237-0

Bartolini G, Sánchez-Alcañiz JA, Osório C, Valiente M, García-Frigola C, Marín O. 2017. Neuregulin 3 mediates cortical plate invasion and laminar allocation of GABAergic interneurons. *Cell Rep* 18: 1157–1170. doi:10.1016/j.celrep.2016.12.089

Baruchin LJ, Ghezzi F, Kohl MM, Butt SJB. 2022. Contribution of interneuron subtype-specific GABAergic signaling to emergent sensory processing in mouse somatosensory whisker barrel cortex. *Cereb Cortex* 32: 2538–2554. doi:10.1093/cercor/bhab363

Batista-Brito R, Majumdar A, Nuño A, Ward C, Barnes C, Nikouei K, Vinck M, Cardin JA. 2023. Developmental loss of ErbB4 in PV interneurons disrupts state-dependent cortical circuit dynamics. *Mol Psychiatry* 28: 3133–3143. doi:10.1038/s41380-023-02066-3

Bernard C, Exposito-Alonso D, Selten M, Sanalidou S, Hanusz-Godoy A, Aguilera A, Hamid F, Oozeer F, Maeso P, Allison L, et al. 2022. Cortical wiring by synapse type-specific control of local protein synthesis. *Science* 378: eabm7466. doi:10.1126/science.abm7466

Bitzenhofer SH, Pöpplau JA, Chini M, Marquardt A, Hanganu-Opatz IL. 2021. A transient developmental increase

in prefrontal activity alters network maturation and causes cognitive dysfunction in adult mice. *Neuron* 109: 1350–1364.e6. doi:10.1016/j.neuron.2021.02.011

Butt SJ, Fuccillo M, Nery S, Noctor S, Kriegstein A, Corbin JG, Fishell G. 2005. The temporal and spatial origins of cortical interneurons predict their physiological subtype. *Neuron* 48: 591–604. doi:10.1016/j.neuron.2005.09.034

Canetta SE, Holt ES, Benoit LJ, Teboul E, Sahyoun GM, Ogden RT, Harris AZ, Kellendonk C. 2022. Mature parvalbumin interneuron function in prefrontal cortex requires activity during a postnatal sensitive period. *eLife* 11: e80324. doi:10.7554/eLife.80324

Carriere CH, Wang WX, Sing AD, Fekete A, Jones BE, Yee Y, Ellegood J, Maganti H, Awofala L, Marocha J, et al. 2020. The γ-protocadherins regulate the survival of GABAergic interneurons during developmental cell death. *J Neurosci* 40: 8652–8668. doi:10.1523/JNEUROSCI.1636-20.2020

Che A, Babij R, Iannone AF, Fetcho RN, Ferrer M, Liston C, Fishell G, De Marco García NV. 2018. Layer I interneurons sharpen sensory maps during neonatal development. *Neuron* 99: 98–116.e7. doi:10.1016/j.neuron.2018.06.002

Chini M, Pfeffer T, Hanganu-Opatz I. 2022. An increase of inhibition drives the developmental decorrelation of neural activity. *eLife* 11: e78811. doi:10.7554/eLife.78811

Chittajallu R, Wester JC, Craig MT, Barksdale E, Yuan XQ, Akgül G, Fang C, Collins D, Hunt S, Pelkey KA, et al. 2017. Afferent specific role of NMDA receptors for the circuit integration of hippocampal neurogliaform cells. *Nat Commun* 8: 152. doi:10.1038/s41467-017-00218-y

Dehorter N, Ciceri G, Bartolini G, Lim L, del Pino I, Marín O. 2015. Tuning of fast-spiking interneuron properties by an activity-dependent transcriptional switch. *Science* 349: 1216–1220. doi:10.1126/science.aab3415

Del Pino I, García-Frigola C, Dehorter N, Brotons-Mas JR, Alvarez-Salvado E, Martínez de Lagrán M, Ciceri G, Gabaldón MV, Moratal D, Dierssen M, et al. 2013. Erbb4 deletion from fast-spiking interneurons causes schizophrenia-like phenotypes. *Neuron* 79: 1152–1168. doi:10.1016/j.neuron.2013.07.010

De Marco García NV, Karayannis T, Fishell G. 2011. Neuronal activity is required for the development of specific cortical interneuron subtypes. *Nature* 472: 351–355. doi:10.1038/nature09865

De Marco García NV, Priya R, Tuncdemir SN, Fishell G, Karayannis T. 2015. Sensory inputs control the integration of neurogliaform interneurons into cortical circuits. *Nat Neurosci* 18: 393–401. doi:10.1038/nn.3946

Denaxa M, Neves G, Rabinowitz A, Kemlo S, Liodis P, Burrone J, Pachnis V. 2018. Modulation of apoptosis controls inhibitory interneuron number in the cortex. *Cell Rep* 22: 1710–1721. doi:10.1016/j.celrep.2018.01.064

Duan ZRS, Che A, Chu P, Mòdol L, Bollmann Y, Babij R, Fetcho RN, Otsuka T, Fuccillo MV, Liston C, et al. 2020. GABAergic restriction of network dynamics regulates interneuron survival in the developing cortex. *Neuron* 105: 75–92.e5. doi:10.1016/j.neuron.2019.10.008

Exposito-Alonso D, Osório C, Bernard C, Pascual-García S, Del Pino I, Marín O, Rico B. 2020. Subcellular sorting of neuregulins controls the assembly of excitatory-inhibitory cortical circuits. *eLife* 9: e57000. doi:10.7554/eLife.57000

Fame RM, MacDonald JL, Macklis JD. 2011. Development, specification, and diversity of callosal projection neurons. *Trends Neurosci* **34**: 41–50. doi:10.1016/j.tins.2010.10.002

Favuzzi E, Deogracias R, Marques-Smith A, Maeso P, Jezequel J, Exposito-Alonso D, Balia M, Kroon T, Hinojosa AJ, Maraver EF, et al. 2019. Distinct molecular programs regulate synapse specificity in cortical inhibitory circuits. *Science* **363**: 413–417. doi:10.1126/science.aau8977

Favuzzi E, Huang S, Saldi GA, Binan L, Ibrahim LA, Fernández-Otero M, Cao Y, Zeine A, Sefah A, Zheng K, et al. 2021. GABA-receptive microglia selectively sculpt developing inhibitory circuits. *Cell* **184**: 4048–4063.e32. doi:10.1016/j.cell.2021.06.018

Fazzari P, Paternain AV, Valiente M, Pla R, Luján R, Lloyd K, Lerma J, Marín O, Rico B. 2010. Control of cortical GABA circuitry development by Nrg1 and ErbB4 signalling. *Nature* **464**: 1376–1380. doi:10.1038/nature08928

Fishell G, Kepecs A. 2020. Interneuron types as attractors and controllers. *Annu Rev Neurosci* **43**: 1–30. doi:10.1146/annurev-neuro-070918-050421

Fleitas C, Marfull-Oromi P, Chauhan D, Del Toro D, Peguera B, Zammou B, Rocandio D, Klein R, Espinet C, Egea J. 2021. FLRT2 and FLRT3 cooperate in maintaining the tangential migratory streams of cortical interneurons during development. *J Neurosci* **41**: 7350–7362. doi:10.1523/JNEUROSCI.0380-20.2021

Gelman DM, Martini FJ, Nóbrega-Pereira S, Pierani A, Kessaris N, Marín O. 2009. The embryonic preoptic area is a novel source of cortical GABAergic interneurons. *J Neurosci* **29**: 9380–9389. doi:10.1523/JNEUROSCI.0504-09.2009

Genescu I, Aníbal-Martínez M, Kouskoff V, Chenouard N, Mailhes-Hamon C, Cartonnet H, Lokmane L, Rijli FM, López-Bendito G, Gambino F, et al. 2022. Dynamic interplay between thalamic activity and Cajal–Retzius cells regulates the wiring of cortical layer 1. *Cell Rep* **39**: 110667. doi:10.1016/j.celrep.2022.110667

Gesuita L, Cavaccini A, Argunsah AO, Favuzzi E, Ibrahim LA, Stachniak TJ, De Gennaro M, Utz S, Greter M, Karayannis T. 2022. Microglia contribute to the postnatal development of cortical somatostatin-positive inhibitory cells and to whisker-evoked cortical activity. *Cell Rep* **40**: 111209. doi:10.1016/j.celrep.2022.111209

Golshani P, Gonçalves JT, Khoshkhoo S, Mostany R, Smirnakis S, Portera-Cailliau C. 2009. Internally mediated developmental desynchronization of neocortical network activity. *J Neurosci* **29**: 10890–10899. doi:10.1523/JNEUROSCI.2012-09.2009

Gour A, Boergens KM, Heike N, Hua Y, Laserstein P, Song K, Helmstaedter M. 2021. Postnatal connectomic development of inhibition in mouse barrel cortex. *Science* **371**: eabb4534. doi:10.1126/science.abb4534

Gouwens NW, Sorensen SA, Baftizadeh F, Budzillo A, Lee BR, Jarsky T, Alfiler L, Baker K, Barkan E, Berry K, et al. 2020. Integrated morphoelectric and transcriptomic classification of cortical GABAergic cells. *Cell* **183**: 935–953.e19. doi:10.1016/j.cell.2020.09.057

Ibrahim LA, Huang S, Fernandez-Otero M, Sherer M, Qiu Y, Vemuri S, Xu Q, Machold R, Pouchelon G, Rudy B, et al. 2021. Bottom-up inputs are required for establishment of top-down connectivity onto cortical layer 1 neuroglia-

form cells. *Neuron* **109**: 3473–3485.e5. doi:10.1016/j.neuron.2021.08.004

Ibrahim LA, Wamsley B, Alghamdi N, Yusuf N, Sevier E, Hairston A, Sherer M, Jaglin XH, Xu Q, Guo L, et al. 2023. Nova proteins direct synaptic integration of somatostatin interneurons through activity-dependent alternative splicing. *eLife* **12**: e86842. doi:10.7554/eLife.86842

Karayannis T, Au E, Patel JC, Kruglikov I, Markx S, Delorme R, Héron D, Salomon D, Glessner J, Restituito S, et al. 2014. Cntnap4 differentially contributes to GABAergic and dopaminergic synaptic transmission. *Nature* **511**: 236–240. doi:10.1038/nature13248

Kastli R, Vighagen R, van der Bourg A, Argunsah AO, Iqbal A, Voigt FF, Kirschenbaum D, Aguzzi A, Helmchen F, Karayannis T. 2020. Developmental divergence of sensory stimulus representation in cortical interneurons. *Nat Commun* **11**: 5729. doi:10.1038/s41467-020-19427-z

Kirmse K, Zhang C. 2022. Principles of GABAergic signaling in developing cortical network dynamics. *Cell Rep* **38**: 110568. doi:10.1016/j.celrep.2022.110568

Kourdougli N, Suresh A, Liu B, Juarez P, Lin A, Chung DT, Graven Sams A, Gandal MJ, Martinez-Cerdeno V, Buonomano DV, et al. 2023. Improvement of sensory deficits in fragile X mice by increasing cortical interneuron activity after the critical period. *Neuron* **111**: 2863–2880.e6. doi:10.1016/j.neuron.2023.06.009

Leighton AH, Cheyne JE, Houwen GJ, Maldonado PP, De Winter F, Levelt CN, Lohmann C. 2021. Somatostatin interneurons restrict cell recruitment to retinally driven spontaneous activity in the developing cortex. *Cell Rep* **36**: 109316. doi:10.1016/j.celrep.2021.109316

Lim L, Pakan JMP, Selten MM, Marques-Smith A, Llorca A, Bae SE, Rochefort NL, Marín O. 2018. Optimization of interneuron function by direct coupling of cell migration and axonal targeting. *Nat Neurosci* **21**: 920–931. doi:10.1038/s41593-018-0162-9

Lodato S, Rouaux C, Quast KB, Jantrachotechatchawan C, Studer M, Hensch TK, Arlotta P. 2011. Excitatory projection neuron subtypes control the distribution of local inhibitory interneurons in the cerebral cortex. *Neuron* **69**: 763–779. doi:10.1016/j.neuron.2011.01.015

López-Bendito G, Sánchez-Alcañiz JA, Pla R, Borrell V, Picó E, Valdeolmillos M, Marín O. 2008. Chemokine signaling controls intracortical migration and final distribution of GABAergic interneurons. *J Neurosci* **28**: 1613–1624. doi:10.1523/JNEUROSCI.4651-07.2008

Luhmann HJ. 2021. Neurophysiology of the developing cerebral cortex: what we have learned and what we need to know. *Front Cell Neurosci* **15**: 814012. doi:10.3389/fncel.2021.814012

Magno L, Asgarian Z, Pendolino V, Velona T, Mackintosh A, Lee F, Stryjewska A, Zimmer C, Guillemot F, Farrant M, et al. 2021. Transient developmental imbalance of cortical interneuron subtypes presages long-term changes in behavior. *Cell Rep* **35**: 109249. doi:10.1016/j.celrep.2021.109249

Mancia Leon WR, Spatazza J, Rakela B, Chatterjee A, Pande V, Maniatis T, Hasenstaub AR, Stryker MP, Alvarez-Buylla A. 2020. Clustered γ-protocadherins regulate cortical interneuron programmed cell death. *eLife* **9**: e55374. doi:10.7554/eLife.55374

Cite this article as *Cold Spring Harb Perspect Biol* doi: 10.1101/cshperspect.a041513

Marques-Smith A, Lyngholm D, Kaufmann AK, Stacey JA, Hoerder-Suabedissen A, Becker EB, Wilson MC, Molnár Z, Butt SJ. 2016. A transient translaminar GABAergic interneuron circuit connects thalamocortical recipient layers in neonatal somatosensory cortex. *Neuron* **89:** 536–549. doi:10.1016/j.neuron.2016.01.015

Mayer C, Hafemeister C, Bandler RC, Machold R, Batista Brito R, Jaglin X, Allaway K, Butler A, Fishell G, Satija R. 2018. Developmental diversification of cortical inhibitory interneurons. *Nature* **555:** 457–462. doi:10.1038/nature25999

Mi D, Li Z, Lim L, Li M, Moissidis M, Yang Y, Gao T, Hu TX, Pratt T, Price DJ, et al. 2018. Early emergence of cortical interneuron diversity in the mouse embryo. *Science* **360:** 81–85. doi:10.1126/science.aar6821

Miyoshi G, Fishell G. 2011. GABAergic interneuron lineages selectively sort into specific cortical layers during early postnatal development. *Cereb Cortex* **21:** 845–852. doi: 10.1093/cercor/bhq155

Miyoshi G, Butt SJ, Takebayashi H, Fishell G. 2007. Physiologically distinct temporal cohorts of cortical interneurons arise from telencephalic *Olig2*-expressing precursors. *J Neurosci* **27:** 7786–7798. doi:10.1523/JNEUROSCI.1807-07.2007

Mòdol L, Bollmann Y, Tressard T, Baude A, Che A, Duan ZRS, Babij R, De Marco García NV, Cossart R. 2020. Assemblies of perisomatic GABAergic neurons in the developing barrel cortex. *Neuron* **105:** 93–105.e4. doi:10.1016/j.neuron.2019.10.007

Mòdol L, Moissidis M, Selten M, Oozeer F, Marín O. 2024. Somatostatin interneurons control the timing of developmental desynchronization in cortical networks. *Neuron* **112:** P2015–2030.E5. doi:10.1016/j.neuron.2024.03.014

Murata Y, Colonnese MT. 2020. GABAergic interneurons excite neonatal hippocampus in vivo. *Sci Adv* **6:** eaba1430. doi:10.1126/sciadv.aba1430

Murthy S, Niquille M, Hurni N, Limoni G, Frazer S, Chameau P, van Hooft JA, Vitalis T, Dayer A. 2014. Serotonin receptor 3A controls interneuron migration into the neocortex. *Nat Commun* **5:** 5524. doi:10.1038/ncomms6524

Nery S, Wichterle H, Fishell G. 2001. Sonic hedgehog contributes to oligodendrocyte specification in the mammalian forebrain. *Development* **128:** 527–540. doi:10.1242/dev.128.4.527

Priya R, Paredes MF, Karayannis T, Yusuf N, Liu X, Jaglin X, Graef I, Alvarez-Buylla A, Fishell G. 2018. Activity regulates cell death within cortical interneurons through a calcineurin-dependent mechanism. *Cell Rep* **22:** 1695–1709. doi:10.1016/j.celrep.2018.01.007

Prönneke A, Scheuer B, Wagener RJ, Möck M, Witte M, Staiger JF. 2015. Characterizing VIP neurons in the barrel cortex of VIPcre/tdTomato mice reveals layer-specific differences. *Cereb Cortex* **25:** 4854–4868. doi:10.1093/cercor/bhv202

Rahmati V, Kirmse K, Holthoff K, Schwabe L, Kiebel SJ. 2017. Developmental emergence of sparse coding: a dynamic systems approach. *Sci Rep* **7:** 13015. doi:10.1038/s41598-017-13468-z

Rochefort NL, Garaschuk O, Milos RI, Narushima M, Marandi N, Pichler B, Kovalchuk Y, Konnerth A. 2009. Sparsification of neuronal activity in the visual cortex at eye-opening. *Proc Natl Acad Sci* **106:** 15049–15054. doi:10.1073/pnas.0907660106

Rudy B, Fishell G, Lee S, Hjerling-Leffler J. 2011. Three groups of interneurons account for nearly 100% of neocortical GABAergic neurons. *Dev Neurobiol* **71:** 45–61. doi:10.1002/dneu.20853

Sánchez-Alcañiz JA, Haege S, Mueller W, Pla R, Mackay F, Schulz S, López-Bendito G, Stumm R, Marín O. 2011. Cxcr7 controls neuronal migration by regulating chemokine responsiveness. *Neuron* **69:** 77–90. doi:10.1016/j.neuron.2010.12.006

Shi Y, Wang M, Mi D, Lu T, Wang B, Dong H, Zhong S, Chen Y, Sun L, Zhou X, et al. 2021. Mouse and human share conserved transcriptional programs for interneuron development. *Science* **374:** eabj6641. doi:10.1126/science.abj6641

Southwell DG, Froemke RC, Alvarez-Buylla A, Stryker MP, Gandhi SP. 2010. Cortical plasticity induced by inhibitory neuron transplantation. *Science* **327:** 1145–1148. doi:10.1126/science.1183962

Southwell DG, Paredes MF, Galvao RP, Jones DL, Froemke RC, Sebe JY, Alfaro-Cervello C, Tang Y, Garcia-Verdugo JM, Rubenstein JL, et al. 2012. Intrinsically determined cell death of developing cortical interneurons. *Nature* **491:** 109–113. doi:10.1038/nature11523

Sreenivasan V, Serafeimidou-Pouliou E, Exposito-Alonso D, Bercsenyi K, Bernard C, Bae SE, Oozeer F, Hanusz-Godoy A, Edwards RH, Marín O. 2022. Input-specific control of interneuron numbers in nascent striatal networks. *Proc Natl Acad Sci* **119:** e2118430119. doi:10.1073/pnas.2118430119

Tuncdemir SN, Wamsley B, Stam FJ, Osakada F, Goulding M, Callaway EM, Rudy B, Fishell G. 2016. Early somatostatin interneuron connectivity mediates the maturation of deep layer cortical circuits. *Neuron* **89:** 521–535. doi:10.1016/j.neuron.2015.11.020

Venkataramanappa S, Saaber F, Abe P, Schütz D, Kumar PA, Stumm R. 2022. Cxcr4 and Ackr3 regulate allocation of caudal ganglionic eminence-derived interneurons to superficial cortical layers. *Cell Rep* **40:** 111157. doi:10.1016/j.celrep.2022.111157

Vílchez-Acosta A, Manso Y, Cárdenas A, Elias-Tersa A, Martínez-Losa M, Pascual M, Álvarez-Dolado M, Nairn AC, Borrell V, Soriano E. 2022. Specific contribution of Reelin expressed by Cajal–Retzius cells or GABAergic interneurons to cortical lamination. *Proc Natl Acad Sci* **119:** e2120079119. doi:10.1073/pnas.2120079119

Vogt D, Cho KKA, Shelton SM, Paul A, Huang ZJ, Sohal VS, Rubenstein JLR. 2018. Mouse *Cntnap2* and human *CNTNAP2* ASD alleles cell autonomously regulate PV+ cortical interneurons. *Cereb Cortex* **28:** 3868–3879. doi:10.1093/cercor/bhx248

Wamsley B, Fishell G. 2017. Genetic and activity-dependent mechanisms underlying interneuron diversity. *Nat Rev Neurosci* **18:** 299–309. doi:10.1038/nrn.2017.30

Wamsley B, Jaglin XH, Favuzzi E, Quattrocolo G, Nigro MJ, Yusuf N, Khodadadi-Jamayran A, Rudy B, Fishell G. 2018. Rbfox1 mediates cell-type-specific splicing in cortical interneurons. *Neuron* **100:** 846–859.e7. doi:10.1016/j.neuron.2018.09.026

Wang BS, Bernardez Sarria MS, An X, He M, Alam NM, Prusky GT, Crair MC, Huang ZJ. 2021. Retinal and callosal activity-dependent chandelier cell elimination shapes binocularity in primary visual cortex. *Neuron* **109:** 502–515.e7. doi:10.1016/j.neuron.2020.11.004

Wester JC, Mahadevan V, Rhodes CT, Calvigioni D, Venkatesh S, Maric D, Hunt S, Yuan X, Zhang Y, Petros TJ, et al. 2019. Neocortical projection neurons instruct inhibitory interneuron circuit development in a lineage-dependent manner. *Neuron* **102:** 960–975.e6. doi:10.1016/j.neuron.2019.03.036

Wichterle H, Turnbull DH, Nery S, Fishell G, Alvarez-Buylla A. 2001. In utero fate mapping reveals distinct migratory pathways and fates of neurons born in the mammalian basal forebrain. *Development* **128:** 3759–3771. doi:10.1242/dev.128.19.3759

Wong FK, Bercsenyi K, Sreenivasan V, Portalés A, Fernández-Otero M, Marín O. 2018. Pyramidal cell regulation of interneuron survival sculpts cortical networks. *Nature* **557:** 668–673. doi:10.1038/s41586-018-0139-6

Wong FK, Selten M, Rosés-Novella C, Sreenivasan V, Pallas-Bazarra N, Serafeimidou-Pouliou E, Hanusz-Godoy A, Oozeer F, Edwards R, Marín O. 2022. Serotonergic regulation of bipolar cell survival in the developing cerebral cortex. *Cell Rep* **40:** 111037. doi:10.1016/j.celrep.2022.111037

Wu SJ, Sevier E, Dwivedi D, Saldi GA, Hairston A, Yu S, Abbott L, Choi DH, Sherer M, Qiu Y, et al. 2023. Cortical somatostatin interneuron subtypes form cell-type-specific circuits. *Neuron* **111:** 2675–2692.e9. doi:10.1016/j.neuron.2023.05.032

Yu Y, Zeng Z, Xie D, Chen R, Sha Y, Huang S, Cai W, Chen W, Li W, Ke R, et al. 2021. Interneuron origin and molecular diversity in the human fetal brain. *Nat Neurosci* **24:** 1745–1756. doi:10.1038/s41593-021-00940-3

Cite this article as *Cold Spring Harb Perspect Biol* doi: 10.1101/cshperspect.a041513

Convergent Circuit Computation for Categorization in the Brains of Primates and Songbirds

Andreas Nieder

Animal Physiology Unit, Institute of Neurobiology, University of Tübingen, 72076 Tübingen, Germany

Correspondence: andreas.nieder@uni-tuebingen.de

Categorization is crucial for behavioral flexibility because it enables animals to group stimuli into meaningful classes that can easily be generalized to new circumstances. A most abstract quantitative category is set size, the number of elements in a set. This review explores how categorical number representations are realized by the operations of excitatory and inhibitory neurons in associative telencephalic microcircuits in primates and songbirds. Despite the independent evolution of the primate prefrontal cortex and the avian nidopallium caudolaterale, the neuronal computations of these associative pallial circuits show surprising correspondence. Comparing cellular functions in distantly related taxa can inform about the evolutionary principles of circuit computations for cognition in distinctly but convergently realized brain structures.

A key aspect to intelligent behavior is the ability to group objects and events into meaningful categories. Categorization enables humans and animals to group stimuli into behaviorally relevant classes that can easily be generalized to new circumstances to provide behavioral flexibility (Miller et al. 2003; Seger and Miller 2010; Mansouri et al. 2020). Oftentimes the qualitative appearance of stimuli allows us to group objects into perceptual categories. For instance, when seeing four-legged carnivores, we group them into the categories "cats" and "dogs" based on bodily features. For other categories, we use quantitative information as grouping criteria. A most abstract quantitative category is set size, the number of elements in a set (Nieder 2020). When assessing set size, the numerosity of a set, the sensory appearance of the elements is meaningless. For example, three flowers, three sounds, and three grasps all belong to the category "three." Numbers are particularly fascinating categories because they can and need to be processed based on rules in working memory by animals (Cantlon and Brannon 2005; Bongard and Nieder 2010; Vallentin et al. 2012) or during symbolic mental calculation in humans (Dehaene et al. 1999; Nieder 2004; Kutter et al. 2018, 2022, 2023).

Over the past years, studies in nonhuman primates (i.e., macaques) and later in corvid songbirds (i.e., crows) have identified cellular mechanisms that give rise to the representation of numerical categories in these animals (for reviews, see Nieder 2016a,b, 2021a,b). A key finding is that single neurons in the associative neocortex of monkeys (Nieder et al. 2002, 2006; Sawamura et al. 2002; Nieder and Miller 2004; Okuyama et al. 2015; Ramirez-Cardenas et al. 2016) and the associative telencephalon of crows

(Ditz and Nieder 2015; 2016; Wagener et al. 2018; Kirschhock et al. 2021; Kirschhock and Nieder 2022) responded selectively to a specific number of items. Such "number neurons" are tuned to preferred numerosities; they show a bell-shaped response curve as a function of the number of presented elements, with the preferred number at the center of this tuning curve. The sensory features of the elements have no effect on the neuronal activity of "number neurons," confirming that they represented abstract quantity categories (Nieder 2016a,b). This review demonstrates how neuronal number representations in the telencephalon provide a means to decipher exemplary microlevel circuit operations involved in cognitive functions.

Neural circuits implement the computations carried out by the brain (Dayan et al. 2011). In microcircuits, neuronal representations emerging from the activity of individual neuron types with specific patterns of synaptic connections are crucial (deCharms and Zador 2000). Such neuronal representations are typically depicted as tuning curves, the input–output (or "stimulus–response") functions of single neurons. The quality of a neuronal representation is mainly characterized by two features: the neuronal selectivity (the width of the tuning curve) to variations of stimulus parameters, and the neuronal discriminability (firing rate differences) between preferred and nonpreferred stimulus parameters. The crucial question addressed here is how high-quality neuronal representations to quantity categories are achieved by neurons forming microcircuits in the brain. Exploring this question in the telecephalic pallium of distantly related vertebrate groups—nonhuman primates and corvid songbirds (crows, ravens and jays)—can inform about the evolutionary principles of circuit computations for cognition in distinctly but convergently realized brain circuits (Fig. 1).

CIRCUITS AND COMPUTATIONS IN THE PRIMATE PREFRONTAL NEOCORTEX

Identifying Excitatory and Inhibitory Neurons

Improved perceptual discriminability of stimulus features is associated with more selective (i.e., narrower or sharper) tuning functions (Schoups et al. 2001; Yang and Maunsell 2004; Lee et al. 2012). Narrow numerosity tuning functions therefore allow a precise readout of the numerical values from neuronal responses. To sculpture tuning curves at the level of local microcircuits, inhibitory interneurons, which are outnumbered by excitatory neurons (mainly pyramidal projection neurons) by a ratio 4:1, play a crucial role (Markram et al. 2004; Wonders and Anderson 2006; Tremblay et al. 2016). By combining electrophysiological and anatomical methods, it has been shown that these two main cortical neuron types can be discriminated with sufficient precision in the primate prefrontal cortex (PFC) based on the waveforms of their action potentials (Fig. 2A–D; Merchant et al. 2012). Pyramidal cells tend to display relatively broad waveforms ("broad-spiking" neurons) and exhibit low firing rates ("regular-spiking" neurons) in extracellular PFC recordings of monkeys. In contrast, interneurons show narrow action potential waveforms ("narrow-spiking" neurons) and high firing rates ("fast-spiking" neurons) (Wilson et al. 1994; Rao et al. 1999; Constantinidis and Goldman-Rakic 2002; Johnston et al. 2009; Johnston and Everling 2009). In primate PFC, the broad waveforms of projection pyramidal neurons have been unequivocally identified based on antidromic stimulation (Johnston and Everling 2009). However, because these different classes of neurons are typically inferred from electrophysiological recordings in the absence of anatomical verification, they are addressed as "putative pyramidal cells" and "putative inhibitory interneurons."

We exploited these established differences between putative pyramidal cells and inhibitory interneurons while recording extracellularly in the PFC and interparietal sulcus (IPS) of macaques trained to discriminate numerical quantity (Diester and Nieder 2008). Using spike sorting techniques, we isolated different neurons based on their distinct action potential waveforms and assigned them based on statistics to the two (excitatory and inhibitory) neuron classes (Fig. 2A–D). This allowed us to gain insight into the functional roles these two classes of neurons assume in local microcircuits during number processing.

Cite this article as *Cold Spring Harb Perspect Biol* doi: 10.1101/cshperspect.a041526

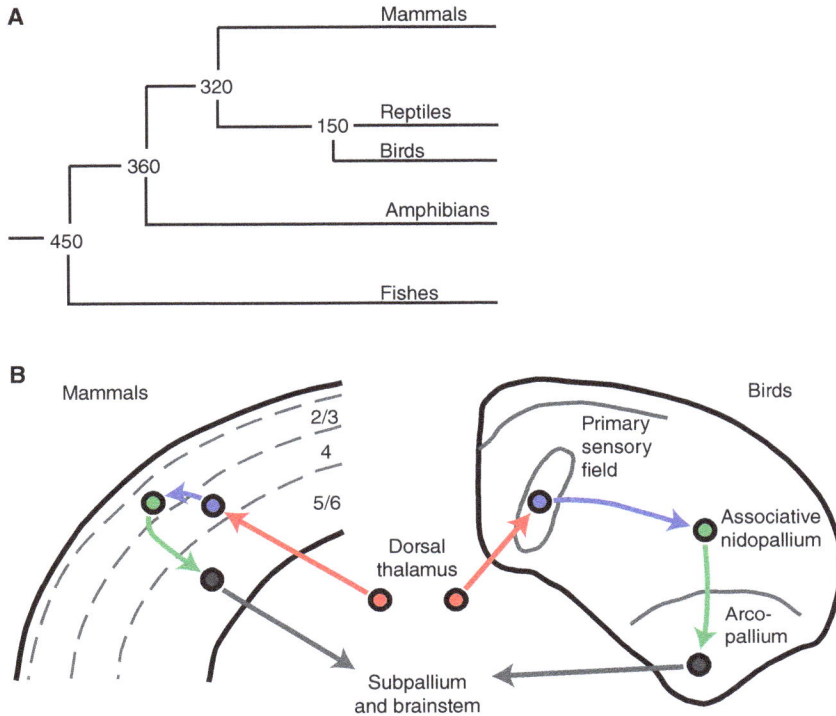

Figure 1. Phylogenetic modifications of major pallial circuits. (*A*) Phylogenetic relationship and divergence times (millions of years ago) of vertebrate taxa. Branch lengths are not proportional to time. (Panel *A* is based on data in Hedges 2002 and Striedter and Northcutt 2020.) (*B*) Cell-type homology hypothesis (Karten 1969) for input and output neurons of mammalian and avian pallium. (*Left*) In mammals, input from the dorsal thalamus (red) terminates on neurons in layer 4 (blue) of the neocortex. These neurons project to layers 2/3 neurons (green), which in turn connect with layer 5/6 pyramidal neurons (gray). These neurons send axons to subcortical regions. (*Right*) An equivalent circuit is present in the avian pallium. Here, input from the dorsal thalamus (red) terminates in primary sensory nuclei (such as the primary auditory field L) (blue). These neurons connect (likely via mediatory interneurons) to interneurons in associative nidopallial regions (green), which in turn project to output arcopallial regions (gray) that constitute the source of pallial output to subpallial regions. (Panel *B* created from data in Dugas-Ford et al. 2012, Dugas-Ford and Ragsdale 2015, and Striedter and Northcutt 2020.)

Response Characteristics of Putative Pyramidal Cells and Interneurons in PFC to Categories

In PFC, putative pyramidal cells and inhibitory interneurons were tuned to numerical categories according to the anatomically expected proportions (Diester and Nieder 2008). Compared to putative pyramidal cells, putative interneurons generally responded with higher firing rates and showed stronger stimulus-evoked responses. Both cell types showed interesting temporal and selectivity differences in numerosity tuning.

With respect to time, putative interneurons responded (by 46 msec) faster to visual stimulation and also discriminated numerical categories (by 51 msec) earlier than putative pyramidal cells. This faster time course could support a role of interneurons in feedforward inhibition of pyramidal cells. Faster response characteristics can be explained by significantly larger and faster excitatory postsynaptic potentials (EPSPs) of fast spiking interneurons compared to pyramidal cells (Povysheva et al. 2006), which in turn can cause lower thresholds for action potential gen-

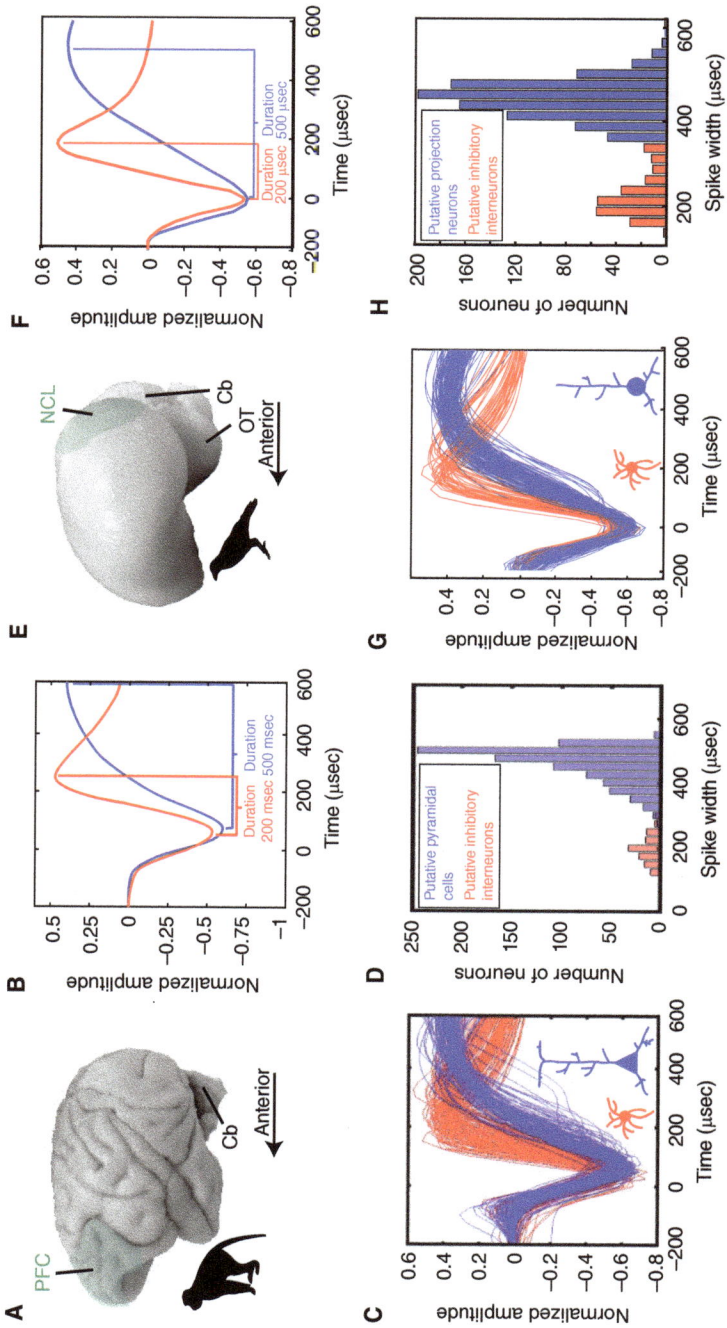

Figure 2. (*See following page for legend.*)

eration and thus shorter latencies for EPSP–spike coupling in interneurons.

In addition to temporal differences, the two cell types differed in neuronal selectivity to quantity categories. Putative inhibitory interneurons exhibited much broader tuning curves than putative pyramidal cells (Diester and Nieder 2008). This finding is interesting because inhibition by broadly tuned inhibitory interneurons could increase the selectivity of pyramidal cells by sharpening their tuning to preferred categories (Wang et al. 2004). As a consequence of sharper tuning, putative pyramidal cells were more strongly modulated by numerosity (i.e., higher ratio of firing rates elicited by the most- and least-preferred numerosity). However, putative inhibitory interneurons discriminated quantity categories more accurately, or reliably, compared to putative pyramidal cells (as measured by the area under the receiver operating characteristic curve) (Diester and Nieder 2008). The higher reliability of putative inhibitory interneurons could be explained by their lower membrane threshold rendering them more likely to fire in response to an input (Povysheva et al. 2006).

It is worth mentioning that the basic circuitries enabling categorical number representations seem to be hard-wired because numerosity tuning exists even in numerically naive animals that have never been trained to discriminate numerosity (Viswanathan and Nieder 2013; Wagener et al. 2018; Kobylkov et al. 2022a). At the same time, numerosity tuning of neurons can be shaped and sharpened through experience and attention (Viswanathan and Nieder 2015). More specifically, putative pyramidal cells show higher numerosity selectivity in discriminating monkeys, which suggests a preferential role of these projection neurons in active (explicit) numerosity processing. This was seen when we compared numerosity tuning selectivity of neurons in monkeys that performed an implicit numerosity task in which only the color of the dots in a set was behaviorally relevant with the situation in an explicit numerosity task in which the monkeys discriminated the number of dots in a set (Viswanathan and Nieder 2015). Indeed, when the same monkeys were retrained from an implicit numerosity task (discriminate only the color of dots) to perform an explicit numerosity task (discriminate the number of dots), PFC neurons became more selective to numerosity during active numerosity discrimination (Viswanathan and Nieder 2015). This improvement in numerosity coding was exclusively due to putative pyramidal neurons; putative inhibitory interneurons were unaffected by behavioral relevance. This effect was specific for the PFC as neither cell class in simultaneously recorded ventral intraparietal area (VIP) of the parietal lobe showed a corresponding effect (Viswanathan and Nieder 2015).

Beyond representing categories, putative pyramidal cells in PFC also seem to contribute to other explicit tasks, such as learning to memorize stimuli (Qi et al. 2011), judging a specific visual parameter (Hussar and Pasternak 2009),

Figure 2. Characterizing putative pyramidal cells and inhibitory interneurons from extracellular recordings in monkey prefrontal cortex (PFC) (*A–D*) and crow nidopallium caudolaterale (NCL) (*E–H*). (*A*) Lateral view of a macaque monkey brain depicting the dorsolateral PFC. (*B*) The width of an action potential waveform is defined as the time elapsed between the spike trough and the peak. Two mean waveforms show the spiking difference between putative inhibitory interneurons (red) and putative pyramidal cells (blue). (*C*) Normalized average waveforms of a random subset of extracellularly recorded PFC neurons aligned by their minimum. Waveforms of putative inhibitory neurons (narrow spiking neurons) are depicted in red; waveforms of putative pyramidal cells (broad spiking neurons) are depicted in blue. (*D*) Bimodal distribution of waveform widths shown in *C* indicating the two major cell-type classes in PFC. (*E*) Lateral view of a crow brain depicting the NCL. (*F*) Action potential waveform characterization for NCL neurons. Same layout as in *B*. (*G*) Normalized average waveforms of a random subset of extracellularly recorded NCL neurons aligned by their minimum. Same layout as in *C*. (*H*) Bimodal distribution of waveform widths shown in *G* indicating the two major cell-type classes in NCL. Same layout as in *D*. (Cb) Cerebellum, (OT) optic tectum. (Panels *B–D* are adapted from Diester and Nieder 2008 with permission from the Society for Neuroscience © 2008. *F–H* adapted from Ditz et al. 2022 under a Creative Commons Attribution 4.0 International License.)

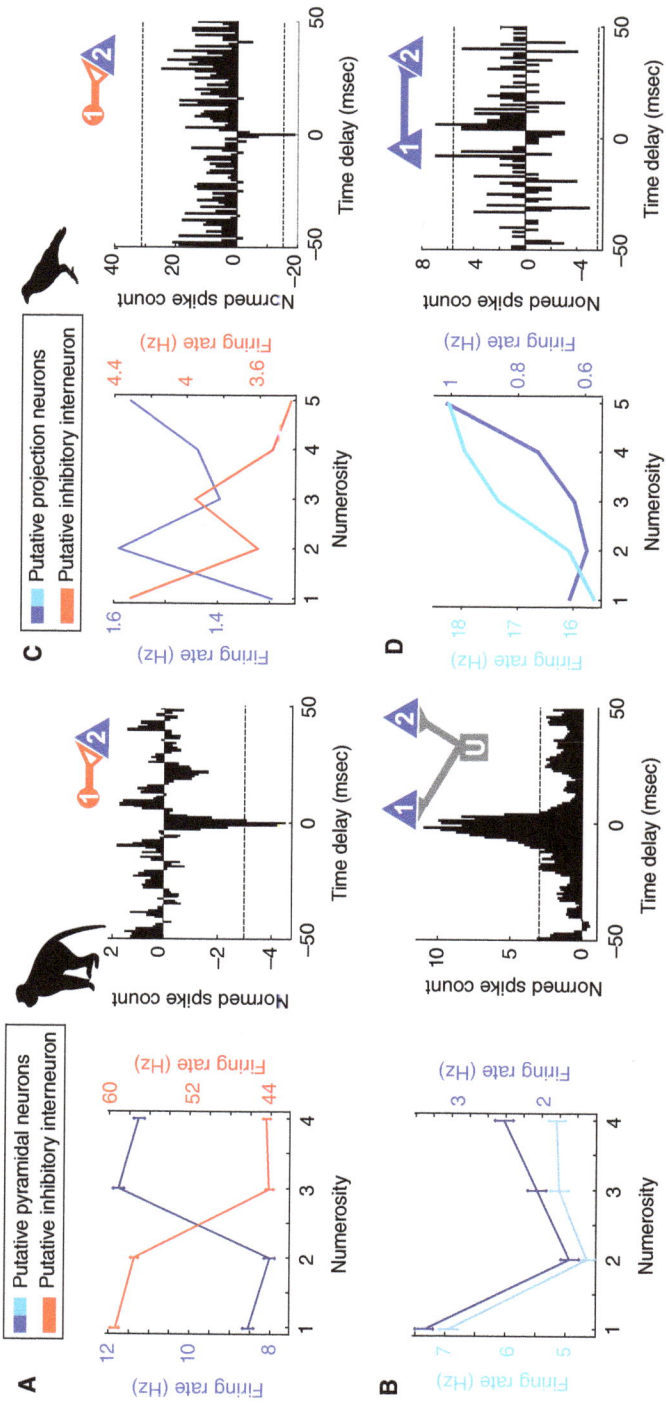

Figure 3. (*See following page for legend.*)

and making decisions (Ding and Gold 2012). Thus, neuronal coding improvements accompanying behavioral relevance seem to result from selectivity increases in putative pyramidal cells of the PFC. Given that PFC projection neurons project in a feedforward manner to premotor output structures, this enhanced coding could well be translated into behavioral precision (Gerbella et al. 2010; Borra et al. 2017; Battaglia-Mayer and Caminiti 2019). Alternatively, or in addition, feedback from PFC projection neurons to upstream areas could enhance cognitive control by influencing sensory areas (Gregoriou et al. 2014; Bichot et al. 2019).

Circuit Interactions between Functionally Coupled Category-Selective Pyramidal Cells and Interneurons

If adjacent putative inhibitory interneurons and pyramidal cells constitute elements of microcircuits interacting with inhibition and excitation, their category tuning profiles are expected to differ or even be inverted relative to each other. To test this, we investigated the tuning of single cells recorded simultaneously at the same electrode. Spike sorting (i.e., the assignment of different neurons based on their distinct action potential waveforms) enabled us to isolate more than one neuron that has been recorded at the same electrode tip and therefore in an immediate anatomical vicinity (Diester and Nieder 2008). Such juxtaposed neurons suggestive of microcircuit membership may interact more frequently than neurons recorded at different sites.

Indeed, inverted tuning to quantity categories between adjacent putative inhibitory interneurons and pyramidal cell pairs recorded at the same electrode tip in PFC was found (Diester and Nieder 2008). This inverse tuning between putative inhibitory interneurons and pyramidal cells occurred significantly more often compared to cell pairs consisting of neighboring putative pyramidal cells. This result, together with the finding that putative inhibitory interneurons showed shorter latencies, again suggests that inhibitory interneurons may exert feedforward inhibition on pyramidal cells. This notion also fits with the finding that putative inhibitory interneurons in PFC show larger receptive fields compared to pyramidal cells (Viswanathan and Nieder 2017a,b), potentially enabling inhibitory neurons to exert spatially broad lateral inhibition on pyramidal cells. This, in turn, may support PFC neurons' global and spatially released number representations beyond the neurons' classical receptive fields (Viswanathan and Nieder 2020).

More direct evidence for circuit interactions came from PFC neuron pairs for which functional connectivity could be established (Fig. 3A,B). One way of revealing functional coupling between neurons is the demonstration of temporally correlated discharges (Epping and Eggermont 1987; Barthó et al. 2004). Cell pairs consisting of

---

Figure 3. Tuning behavior of functionally connected pairs of numerosity-selective neurons recorded in monkey prefrontal cortex (PFC) (*A,B*) and crow nidopallium caudolaterale (NCL) (*C,D*). (*A*) (*Left* panel) Numerosity tuning curves of a pair of coupled PFC neurons consisting of tuned putative pyramidal cell and inhibitory interneuron. (*Right* panel) Corresponding spiking cross-correlogram (shift-corrected and baseline-subtracted; dotted lines indicate positive and negative significance thresholds, respectively) with a significant negative trough around time delay 0 msec indicating forward inhibition. *Inset* shows putative connectivity between inhibitory neuron (red) and excitatory neuron (blue). Open triangles indicate inhibitory synapse. (*B*) Same layout as in *A*, but for a pair of coupled putative pyramidal cells from monkey PFC. Both neurons exhibit equivalent numerosity tuning profiles while showing a positive peak around time delay 0 msec in the cross correlogram indicating shared excitation (from an unknown input neuron, *U*). *Inset* shows putative connectivity between the two excitatory neurons (blue). Closed triangles indicate excitatory synapses. (*C*) Numerosity tuning curves and cross correlograms of a crow NCL cell pair consisting of a putative projection neuron and an inhibitory interneuron with inverse tuning profiles that are functionally coupled. Same layout as in *A*. (*D*) Numerosity tuning curves and cross-correlograms of a crow NCL cell pair consisting of two coupled putative projection neurons Same layout as in *B*. *Inset* shows putative unidirectional excitatory input from one excitatory neuron to the other (blue). (Panels *A* and *B* are adapted from Diester and Nieder 2008 with permission from the Society for Neuroscience © 2008. *C* and *D* adapted from Ditz et al. 2022 under a Creative Commons Attribution 4.0 International License.)

putative pyramidal cells showed tuning to similar preferred numerosities and were mainly excited in synchrony (Fig. 3B); inhibitory effects between neighboring putative pyramidal cells were largely absent. These findings likely reflect shared excitatory input on putative pyramidal cells (Rao et al. 1999). In contrast, neuron pairs consisting of a putative inhibitory interneuron and a pyramidal cell were functionally connected via a negative correlation of temporal firing and showed inverted tuning relative to each other; in other words, if a putative inhibitory interneuron fired, the functionally connected pyramidal cell was significantly inhibited, and vice versa (Fig. 3A).

Our results during abstract categorization mirrored earlier findings showing that neighboring putative inhibitory interneurons and pyramidal cells in primate PFC exhibit opposite response properties in the spatial domain. In earlier studies, inverted spatial direction selectivity of nearby putative inhibitory interneurons and pyramidal cells was observed (Wilson et al. 1994; Constantinidis and Goldman-Rakic 2002; but see Rao et al. 1999 for contrasting results). Interestingly, blockade of GABAergic inhibition led to broadening of PFC neurons' spatial tuning profiles. This is in agreement with a role of inhibition in shaping neuronal tuning (Rao et al. 2000).

Mechanistically, these findings indicate that inhibitory interneurons systematically exert inhibition on numerosity-tuned pyramidal cells, a mechanism also proposed by recurrent network models for neuronal tuning in the service of spatial working memory (Compte et al. 2000). Within this framework, the preferred numerosities of the inhibitory interneurons represent the non-preferred numerosities at the flanks of a pyramidal cell's numerosity tuning function. The tuning curve shoulders of the pyramidal cell are lowered by this lateral inhibition, which thereby sharpens the pyramidal cell's tuning curve. Lateral inhibition may also explain the more precise tuning of neurons in the subitizing range below number 5 in the human medial temporal lobe (Kutter et al. 2023).

Interestingly, microcircuit operations are not identical across neocortical brain regions. A direct comparison of functional connectivity patterns of adjacent cortical cell types in PFC versus IPS resulted in marked differences (Viswanathan and Nieder 2017a,b). Relative to IPS, more inhibitory and temporally more precise connections exist in PFC. The functional connectivity patterns observed in IPS are more reminiscent of early visual areas, whereas prefrontal local circuits seem to have developed differently. Perhaps inhibition figures more prominently in PFC to suppress prepotent behavioral output and to shape behaviorally relevant representations.

A defining feature of the PFC is its support of working memory functions (Miller 2013; Miller et al. 2018). An extension of the mechanism depicted above may explain how numerosity selective activity could persist into delay periods to support tuning to categories during working memory. After a quantity category has been encoded during presentation of a display, an ensemble of pyramidal cells exhibits tuned sustained activity through recurrent excitations based on positive feedback loops of activity between excitatory pyramidal cells (Wang 1999, 2001). These tuned pyramidal cells activate local inhibitory interneurons, which, in turn, inhibit pyramidal cells with a different numerosity preference. Remarkably, such a reverberatory mechanism could persist after a stimulus has ceased into delay periods to support neuronal tuning to categories during working memory (Durstewitz et al. 2000).

Prefrontal Neuron Classes Are Differentially Affected by Dopamine Receptors

Neuronal microcircuits in PFC process cognitive information not in a static environment but are influenced by neuromodulators such as dopamine (Ott and Nieder 2019). Two main dopamine receptor families, D1R (subtypes D_1 and D_5) and D2R (subtypes D_2, D_3, and D_4), mediate dopamine's control of neuronal activity, and both are expressed in both excitatory pyramidal cells and inhibitory interneurons (Smiley et al. 1994; Mrzljak et al. 1996; Muly et al. 1998; Gao and Goldman-Rakic 2003; de Almeida and Mengod 2010; Santana and Artigas 2017). To test the causal impact of dopamine receptor activation

Cite this article as *Cold Spring Harb Perspect Biol* doi: 10.1101/cshperspect.a041526

on neuronal tuning, we used a combination of single-neuron recordings and simultaneous pharmacological stimulation of dopamine receptor families via micro-iontophoretic drug applications (Jacob et al. 2013; Stalter et al. 2020). The two dopamine receptor families had strong but differential impact on how the neurons represented the encoding and rule-based memorization of numerical categories (Ott et al. 2014; Ott and Nieder 2017).

In monkeys performing a delayed match-to-numerosity task with a distractor in the working memory period (Jacob and Nieder 2014), we found a cell-type-specific influence of D1Rs on the coding quality of the sample numerosity in the working memory period after a distracting numerosity had been shown: in putative pyramidal neurons, D1R inhibition improved neuronal coding quality after the interference, whereas D1R stimulation weakened it. However, the inverse pattern was observed in putative interneurons (Jacob et al. 2016). These results imply that dopaminergic neuromodulation of PFC circuits and their cell types via D1R regulates representations of behaviorally relevant categories that compete with task-irrelevant categories. The results emphasize the different roles of different cell types in working memory coding and highlight the need to consider cortical cell types.

CIRCUITS AND COMPUTATIONS IN THE SONGBIRD NIDOPALLIUM

Circuit Components in the Avian Pallium

Compared to mammals, the microcircuits in the avian telencephalon are significantly less well explored, despite some birds exhibiting sophisticated categorization capabilities (Soto and Wasserman 2014; Scarf et al. 2016; Huber and Aust 2017; Pusch et al. 2023). This lapse also has to do with a century-old misunderstanding concerning the building plan of the telencephala of sauropsids (i.e., reptiles and birds). The understanding of the avian telencephalon was revolutionized when new phylogenetic and developmental neuroanatomical studies convincingly demonstrated that the avian pallium, despite its structural independencies that emerged in evolutionary time, is homologous to that of mammals and similarly dominates the telencephalon (Jarvis et al. 2005). Unravelling the functional-level convergence between neuronal operations in nonhuman primates and corvid songbirds will permit a deeper understanding of evolutionarily superior circuit computations that are realized in independently evolved brains.

Birds are vertebrates that possess an alternative telencephalic layout compared to mammals. Most remarkable, the avian telencephalon lacks a layered cerebral cortex. Since birds and mammals diverged from a last common stem-amniote 320 million years ago (Fig. 1A; Hedges 2002), they evolved a profoundly distinct cellular arrangement as integrative structures from different territories of the embryonic pallium. Of the four original pallial territories (ventral, lateral, dorsal, and medial), birds evolved rather nuclear integration centers out of the ventral pallium, whereas the major integration center of mammals, the six-layered neocortex, develops from the embryonic dorsal pallium (Jarvis et al. 2005; Puelles 2017; Cárdenas and Borrell 2020; Striedter and Northcutt 2020; Nieder 2021b).

Despite these differences in structural origins, the avian ventral pallium (of which the dorsal ventricular ridge [DVR] emerges) and the mammalian neocortex converged on comparable circuit operations (Fig. 1B). The layered neocortex exhibits three circuit building blocks (Shepherd 2009). First, input neurons that receive sensory information relayed by the thalamus enter the thalamo-recipient. Second, intracortical neurons process this information locally in projecting layers 2 and 3. Third, output neurons in the subcortically projecting layers 5 and 6 project to subcortical motor control centers. Remarkably, these circuit building blocks can all be identified in the avian pallium. Columnar organizations and layering may even occur in sensory pallial areas of the bird brain (Wang et al. 2010; Ahumada-Galleguillos et al. 2015; Stacho et al. 2020; Fernández et al. 2021). Beyond sensory pallial areas, however, these circuit components are arranged along a sequence of nuclei rather than layers.

The three major circuit components of the neocortex can be retraced in the bird pallium

(Fig. 1B; Karten 1969, 2013; Reiner 2013). Recent molecular studies show that markers of mammalian cortical layer 4 neurons are expressed by neurons in the major thalamo-recipient nuclei of chicken telencephalon (Dugas-Ford et al. 2012; Dugas-Ford and Ragsdale 2015). Moreover, neurons of the avian mesopallium, an associative avian brain region, have the connectivity of cortical layers 2/3 neurons and express several genes that uniquely mark layers 2/3 in mammals (Suzuki et al. 2012). In mammals, neurons of cortical layers 2/3 receive input from layer 4 thalamo-recipient neurons and project to layer 5 output neurons (Shepherd 2009; Harris and Shepherd 2015). Indeed, different layer 5 markers are expressed in avian pallial structures, such as the arcopallium, that provide telencephalic output projections to the brainstem. From an evolutionary point of view, the three main types of pallial neurons (thalamo-recipient, intrapallially projecting, and extra-telencephalically projecting neurons) likely represent three ancient neuron types found in the common ancestor stem amniote pallium. From these neuron types, comparable circuits seem to be built in their descendants (mammals and birds, respectively), independent of whether they were arranged in a nuclear or laminar fashion (Reiner 2013).

On a more fine-grained level, microcircuits both in the mammalian neocortex and the avian pallium are composed of glutamatergic excitatory projection neurons and GABA (γ-aminobutyric acid)-ergic local inhibitory interneurons (Spool et al. 2021). These major cell types were molecularly identified in the avian brain using promoter-specific viral optogenetics (Spool et al. 2021). These authors explored the molecular phenotypes of mammalian neocortical excitatory (via calmodulin-dependent kinase α [CaMKIIα] promoters) and inhibitory neurons (via glutamate decarboxylase 1 [GAD1] promoters) (Lee et al. 2012; Pfeffer et al. 2013) in the nidopallium of zebra finches. They found that promoter-driven molecular cell identity segregated nidopallial neurons with distinct physiological properties, such as the known spike waveform differences, into excitatory and inhibitory cell types as it does in mammalian pallium (Spool et al. 2021).

Importantly, however, avian pallial circuits engage genetically separate classes of excitatory and inhibitory neurons that are not present in the mammalian neocortex. Colquitt et al. (2021) showed that excitatory (glutamatergic) neurons have transcription factor profiles similar to the mammalian ventral pallium but not to the neocortex, which develops from the dorsal pallium (Colquitt et al. 2021). In addition, and consistent with the assumption that avian nido-/arcopallial regions are of ventral pallial origins, the most abundant inhibitory (GABA-releasing) neuron type in these avian brain regions resembles inhibitory neurons in mammalian ventral pallial derivatives but is absent from the neocortex (Colquitt et al. 2021). Thus, the songbird nidopallium and the mammalian neocortex contain transcriptionally relatively similar neurons, which, however, have distinct developmental origins.

The Avian Pallial Association Area Nidopallium Caudolaterale

In birds, a highly associative pallial brain area termed "nidopallium caudolaterale (NCL)" was identified as an essential cognitive brain area giving rise to complex avian cognition (Güntürkün 2005; Nieder 2017a,b; Nieder et al. 2020). The NCL is certainly not homologous to the mammalian PFC but it is often called a functional equivalent of the PFC. Recordings in behaving crows showed NCL neurons that represent stimulus association (Moll and Nieder 2015; Veit et al. 2015), encode working memory information (Veit et al. 2014; Rinnert et al. 2019), signify motor plans (Rinnert and Nieder 2021), and are engaged in abstract magnitude categorization (Wagener and Nieder 2023). Like monkey PFC and IPS neurons, NCL neurons in crows are also tuned to numerosity (Ditz and Nieder 2015, 2016, 2020; Wagener et al. 2018; Kirschhock et al. 2021; Kirschhock and Nieder 2022). Thus, both primates and corvids convergently evolved pallial structures that generate executive processes.

The similarities between the NCL and the PFC even extend to the neuromodulator dopamine. In fact, the anatomical delineation of the NCL in the avian telencephalon is based on the criterion of a strong dopaminergic innervation

Cite this article as *Cold Spring Harb Perspect Biol* doi: 10.1101/cshperspect.a041526

arising from the midbrain (Divac et al. 1985; Waldmann and Güntürkün 1993; Wynne and Güntürkün 1995; von Eugen et al. 2020; Kersten et al. 2022; Kobylkov et al. 2022b). Similar to the primate PFC, the avian NCL also contains D1Rs and D2Rs that play a functional role in cognition (Dietl and Palacios 1988; Durstewitz et al. 1998). Dopamine concentration in pigeon NCL (measured via microdialysis) increased during a working memory task compared to the same task without a memory delay (Karakuyu et al. 2007). Stimulation of D1R in NCL (and striatum) improved performance on working memory tasks (Herold et al. 2008), whereas blockade of D1R in NCL of pigeons resulted in severe disruptions of discrimination reversal learning (Diekamp et al. 2000). Currently, the influence of the neurotransmitter dopamine on single neurons and microcircuits in the avian pallium is unknown. However, in light of these findings in cognitive tasks and given that D1R manipulation in the auditory nidopallium changes the songbirds' behavioral song preference (Barr et al. 2021), dopamine most likely has an impact on the categorical tuning properties of neurons.

Contributions of Inhibitory and Excitatory Neurons in the Avian Nidopallium Caudolaterale in Abstract Categorization

How tuning functions to abstract categories are shaped by neuronal circuit computations in the avian pallium remained elusive until recently. Applying the same methods as for extracellular PFC recordings, we demonstrated that local microcircuits containing excitatory projection neurons and inhibitory interneurons play a key role in sculpting tuning functions (Fig. 2E,F; Ditz et al. 2022). As crucial circuit components, these major neuron types exist in song-related pallial brain areas of songbirds (Spiro et al. 1999; Calabrese and Woolley 2015; Kosche et al. 2015; Yanagihara and Yazaki-Sugiyama 2016; Bottjer et al. 2019). As outlined above, these major cell type classes and their physiological characteristics were molecularly identified in the bird brain (Spool et al. 2021), supporting earlier studies reporting network computations via "putative" excitatory and inhibitory cell types segregated based on extracellular waveform separation.

In the NCL of crows discriminating numerosities, recorded neurons can also be classified into broad-spiking putative projection neurons and narrow-spiking putative inhibitory interneurons based on their action potential profile (Ditz et al. 2022). These avian neuron classes show the same characteristics as those in the primate PFC. First, the cellular proportion of roughly 80% projection neurons and 20% interneurons corresponded to cell counts reported in the avian (Yanagihara and Yazaki-Sugiyama 2016; Bottjer et al. 2019; Ditz et al. 2022) and mammal brains (Markram et al. 2004). Second, putative NCL interneurons showed higher firing rates compared to putative projection neurons, confirming earlier findings in the auditory/song system (Schneider and Woolley 2013; Yanagihara and Yazaki-Sugiyama 2016). Third, stimulus-evoked responses were greater in putative NCL interneurons compared to projection neurons. Fourth, putative NCL interneurons responded significantly faster to the onset of stimuli (Ditz et al. 2022).

Because the distinct physiological properties of putative inhibitory interneurons and projection neurons were suggestive of them having different functions in neuronal circuits, we probed these NCL neuron types' contributions in quantity categorization. To that aim, we used the same delayed-match-to-numerosity task in crows as previously described for monkeys. Surprisingly, the results were virtually identical to those earlier obtained for the PFC of primates (Diester and Nieder 2008). Of those NCL neurons that were tuned to preferred numerosities, putative projection neurons show a sharper and more selective tuning than putative inhibitory interneurons (Ditz et al. 2022). In addition, neighboring putative projection neurons recorded at the same electrode tip exhibited comparable numerosity preferences and tuning profiles. In contrast, nearby putative inhibitory interneurons and projection neurons tended to show inverse tuning relative to one another.

The commonalities between PFC and NCL neurons also extended to functionally connected neuron pairs (Ditz et al. 2022). NCL neuron of temporally coupled pairs consisting of nearby putative inhibitory interneurons and projection neurons tended to inhibit each other and showed inverse numerosity tuning profiles (Fig. 3C). This is the circuit operation required if interneurons sharpen tuning curves of projection neurons. In contrast, adjoining and functionally connected putative projection neurons showed similar numerosity tuning and usually became excited in synchrony (Fig. 3D). The combination of spike timing and category tuning properties in functionally connected putative inhibitory interneurons and projection neurons implies that these cell classes assume distinct roles in microcircuits of the crow NCL. Remarkably, the circuit operations witnessed in the crow NCL showed surprising correspondence with those observed in the primate PFC, despite their independent evolution.

CONCLUDING REMARKS

When evaluating the findings described in this article, two evolutionary scenarios may explain the apparent commonalities in circuit operations between corvids and primates. One scenario posits that the respective pallial microcircuits have been conserved over 640 million years of parallel evolution in birds and mammals (320 Mio years of evolution for each taxon since the last common ancestor) to maintain their functionalities in the avian nidopallium and the mammalian neocortex. Alternatively, the deciphered microcircuit mechanisms together with their building blocks, the excitatory projection neurons and the inhibitory interneurons, have been re-invented independently in the avian and mammalian lineages to serve similar computational functions in shaping and sharpening quantity tuning.

The latter hypothesis is supported by recent single-cell and spatial transcriptomics used in all tetrapod classes (amphibians, reptiles, birds, and mammals) to investigate cell-type evolution at the brain scale, and even prior to the phylogenetic reptile/avian–mammal separation (Tosches et al. 2018; Colquitt et al. 2021; Hain et al. 2022; Lust et al. 2022; Woych et al. 2022). The evidence from these combined studies pictures a complex scenario and suggests that the telencephalon is not composed of phylogenetically old and new regions but rather consists of a mosaic of conserved and new cell types (Faltine-Gonzalez and Kebschull 2022). A consistent finding is that the classes of inhibitory interneurons are largely conserved in the telencephalon of different species of tetrapods. In contrast, excitatory neurons in the telencephalon are much less conserved, arguing for the evolution of new excitatory cell types (Tosches et al. 2018; Hain et al. 2022; Lust et al. 2022; Woych et al. 2022). In a species of salamander (Amphibia), molecular and structural commonalities suggest that amphibian ventral pallium neurons are homologous to parts of the reptile DVR (a ventral pallium derivative) that is complemented by reptile-specific novelties. However, the amphibian dorsal pallium lacks molecular and cellular characteristics of excitatory cell types in the mammalian neocortex, a dorsal pallium derivative (Woych et al. 2022). This suggests that excitatory pyramidal neuron types in the mammalian six-layered neocortex are evolutionary novelties in mammals. Even though glutamatergic neurons in the avian DVR show greatest transcriptional similarities to neocortical projection neurons, these similarities are not restricted to specific layers but are found across neocortical layers (Colquitt et al. 2021). This finding is difficult to reconcile with the prediction of the cell-type homology hypothesis (nucleus-to-layer model) that glutamatergic neurons defined by processing level (i.e., thalamo-recipient, intra-pallially projecting, and extra-telencephalically projecting neurons) in specific nuclei of the bird pallium should be localized in specific neocortical layers that operate at the equivalent processing level in mammals. Rather, these insights suggest that the transcriptional similarities are at the level of function (i.e., projections and connections) within three major circuit components of the ventral pallium/neocortex (Fig. 1B). These cell-type differences across the vertebrate lineage together with the different regional origins of the reptilian/avian DVR and the mammalian neocortex corroborate the hypothesis that the functional similarities of the reptilian/avian DVR and neocortex are the result of convergence instead of homology. Notably, however, evolution seems to have re-

Cite this article as *Cold Spring Harb Perspect Biol* doi: 10.1101/cshperspect.a041526

invented the major circuit components based on partial diversified and specialized neuron types in either or both lineages.

How exactly the emergence of old and new pallial territories and the intermingling of old (inhibitory interneurons) and new (primarily excitatory) neuron types can be explained is an open question. One scenario posits that new and old cell types may segregate into evolutionary newer and older pallial subregions that evolved potentially by duplication and divergence of sets of cell types (Kebschull et al. 2020). Alternatively, well-conserved telencephalic interneurons intermingle by (ontogenetic and phylogenetic) cell migration with newly arising excitatory cells in divergent regions of the pallium (Tosches et al. 2018; Colquitt et al. 2021). Whatever the precise mechanisms, based upon general circuitries inherited from common ancestry, the microscale networks in associative pallial areas reveal that both birds and mammals evolved similar neuronal and computational principles in parallel and partly independently based on convergent evolutionary forces (Nieder 2021c). It stands to reason that the microcircuit operations enabled by excitatory projection neurons and inhibitory interneurons constitute a superior solution to a common computation problem when representing abstract categories.

ACKNOWLEDGMENTS

This work was supported by DFG grants NI 618/10-1, NI 618/11-1 (SPP 2205), and NI 618/13-1 (FOR 5159) to A.N.

REFERENCES

Ahumada-Galleguillos P, Fernández M, Marin GJ, Letelier JC, Mpodozis J. 2015. Anatomical organization of the visual dorsal ventricular ridge in the chick (*Gallus gallus*): layers and columns in the avian pallium. *J Comp Neurol* **523**: 2618–2636. doi:10.1002/cne.23808

Barr HJ, Wall EM, Woolley SC. 2021. Dopamine in the songbird auditory cortex shapes auditory preference. *Curr Biol* **31**: 4547–4559.e5. doi:10.1016/j.cub.2021.08.005

Barthó P, Hirase H, Monconduit L, Zugaro M, Harris KD, Buzsáki G. 2004. Characterization of neocortical principal cells and interneurons by network interactions and extracellular features. *J Neurophysiol* **92**: 600–608. doi:10.1152/jn.01170.2003

Battaglia-Mayer A, Caminiti R. 2019. Corticocortical systems underlying high-order motor control. *J Neurosci* **39**: 4404–4421. doi:10.1523/JNEUROSCI.2094-18.2019

Bichot NP, Xu R, Ghadooshahy A, Williams ML, Desimone R. 2019. The role of prefrontal cortex in the control of feature attention in area V4. *Nat Commun* **10**: 5727. doi:10.1038/s41467-019-13761-7

Bongard S, Nieder A. 2010. Basic mathematical rules are encoded by primate prefrontal cortex neurons. *Proc Natl Acad Sci* **107**: 2277–2282. doi:10.1073/pnas.0909180107

Borra E, Gerbella M, Rozzi S, Luppino G. 2017. The macaque lateral grasping network: a neural substrate for generating purposeful hand actions. *Neurosci Biobehav Rev* **75**: 65–90. doi:10.1016/j.neubiorev.2017.01.017

Bottjer SW, Ronald AA, Kaye T. 2019. Response properties of single neurons in higher level auditory cortex of adult songbirds. *J Neurophysiol* **121**: 218–237. doi:10.1152/jn.00751.2018

Calabrese A, Woolley SM. 2015. Coding principles of the canonical cortical microcircuit in the avian brain. *Proc Natl Acad Sci* **112**: 3517–3522. doi:10.1073/pnas.1408545112

Cantlon JF, Brannon EM. 2005. Semantic congruity affects numerical judgments similarly in monkeys and humans. *Proc Natl Acad Sci* **102**: 16507–16511. doi:10.1073/pnas.0506463102

Cárdenas A, Borrell V. 2020. Molecular and cellular evolution of corticogenesis in amniotes. *Cell Mol Life Sci* **77**: 1435–1460. doi:10.1007/s00018-019-03315-x

Colquitt BM, Merullo DP, Konopka G, Roberts TF, Brainard MS. 2021. Cellular transcriptomics reveals evolutionary identities of songbird vocal circuits. *Science* **371**: eabd9704. doi:10.1126/science.abd9704

Compte A, Brunel N, Goldman-Rakic PS, Wang XJ. 2000. Synaptic mechanisms and network dynamics underlying spatial working memory in a cortical network model. *Cereb Cortex* **10**: 910–923. doi:10.1093/cercor/10.9.910

Constantinidis C, Goldman-Rakic PS. 2002. Correlated discharges among putative pyramidal neurons and interneurons in the primate prefrontal cortex. *J Neurophysiol* **88**: 3487–3497.

Dayan P, Feller M, Feldman D. 2011. Networks, circuits and computation. *Curr Opin Neurobiol* **21**: 661–663. doi:10.1016/j.conb.2011.07.003

de Almeida J, Mengod G. 2010. D2 and D4 dopamine receptor mRNA distribution in pyramidal neurons and GABAergic subpopulations in monkey prefrontal cortex: implications for schizophrenia treatment. *Neuroscience* **170**: 1133–1139. doi:10.1016/j.neuroscience.2010.08.025

deCharms RC, Zador A. 2000. Neural representation and the cortical code. *Annu Rev Neurosci* **23**: 613–647. doi:10.1146/annurev.neuro.23.1.613

Dehaene S, Spelke E, Pinel P, Stanescu R, Tsivkin S. 1999. Sources of mathematical thinking: behavioral and brain-imaging evidence. *Science* **284**: 970–974. doi:10.1126/science.284.5416.970

Diekamp B, Kalt T, Ruhm A, Koch M, Güntürkün O. 2000. Impairment in a discrimination reversal task after D1 receptor blockade in the pigeon "prefrontal cortex." *Behav Neurosci* **114**: 1145–1155. doi:10.1037/0735-7044.114.6.1145

Diester I, Nieder A. 2008. Complementary contributions of prefrontal neuron classes in abstract numerical categorization. *J Neurosci* **28**: 7737–7747. doi:10.1523/JNEURO SCI.1347-08.2008

Dietl MM, Palacios JM. 1988. Neurotransmitter receptors in the avian brain. I: Dopamine receptors. *Brain Res* **439**: 354–359.

Ding L, Gold JI. 2012. Neural correlates of perceptual decision making before, during, and after decision commitment in monkey frontal eye field. *Cereb Cortex* **22**: 1052–1067. doi:10.1093/cercor/bhr178

Ditz HM, Nieder A. 2015. Neurons selective to the number of visual items in the corvid songbird endbrain. *Proc Natl Acad Sci* **112**: 7827–7832. doi:10.1073/pnas.1504245112

Ditz HM, Nieder A. 2016. Sensory and working memory representations of small and large numerosities in the crow endbrain. *J Neurosci* **36**: 12044–12052. doi:10 .1523/JNEUROSCI.1521-16.2016

Ditz HM, Nieder A. 2020. Format-dependent and format-independent representation of sequential and simultaneous numerosity in the crow endbrain. *Nat Commun* **11**: 686. doi:10.1038/s41467-020-14519-2

Ditz HM, Fechner J, Nieder A. 2022. Cell-type specific pallial circuits shape categorical tuning responses in the crow telencephalon. *Commun Biol* **5**: 269. doi:10.1038/s42003-022-03208-z

Divac I, Mogensen J, Björklund A. 1985. The prefrontal "cortex" in the pigeon. Biochemical evidence. *Brain Res* **332**: 365–368. doi:10.1016/0006-8993(85)90606-7

Dugas-Ford J, Ragsdale CW. 2015. Levels of homology and the problem of neocortex. *Annu Rev Neurosci* **38**: 351–368. doi:10.1146/annurev-neuro-071714-033911

Dugas-Ford J, Rowell JJ, Ragsdale CW. 2012. Cell-type homologies and the origins of the neocortex. *Proc Natl Acad Sci* **109**: 16974–16979. doi:10.1073/pnas.1204773109

Durstewitz D, Kröner S, Hemmings HC Jr, Güntürkün O. 1998. The dopaminergic innervation of the pigeon telencephalon: distribution of DARPP-32 and co-occurrence with glutamate decarboxylase and tyrosine hydroxylase. *Neuroscience* **83**: 763–779. doi:10.1016/S0306-4522(97) 00450-8

Durstewitz D, Seamans JK, Sejnowski TJ. 2000. Neurocomputational models of working memory. *Nat Neurosci* **3** (Suppl.): 1184–1191. doi:10.1038/81460

Epping WJ, Eggermont JJ. 1987. Coherent neural activity in the auditory midbrain of the grassfrog. *J Neurophysiol* **57**: 1464–1483. doi:10.1152/jn.1987.57.5.1464

Faltine-Gonzalez DZ, Kebschull JM. 2022. A mosaic of new and old cell types. *Science* **377**: 1043–1044. doi:10.1126/ science.add9465

Fernández M, Reyes-Pinto R, Norambuena C, Sentis E, Mpodozis J. 2021. A canonical interlaminar circuit in the sensory dorsal ventricular ridge of birds: the anatomical organization of the trigeminal pallium. *J Comp Neurol* **529**: 3410–3428. doi:10.1002/cne.25201

Gao WJ, Goldman-Rakic PS. 2003. Selective modulation of excitatory and inhibitory microcircuits by dopamine. *Proc Natl Acad Sci* **100**: 2836–2841. doi:10.1073/pnas .262796399

Gerbella M, Belmalih A, Borra E, Rozzi S, Luppino G. 2010. Cortical connections of the macaque caudal ventrolateral prefrontal areas 45A and 45B. *Cereb Cortex* **20**: 141–168. doi:10.1093/cercor/bhp087

Gregoriou GG, Rossi AF, Ungerleider LG, Desimone R. 2014. Lesions of prefrontal cortex reduce attentional modulation of neuronal responses and synchrony in V4. *Nat Neurosci* **17**: 1003–1011. doi:10.1038/nn.3742

Güntürkün O. 2005. The avian "prefrontal cortex" and cognition. *Curr Opin Neurobiol* **15**: 686–693. doi:10.1016/j .conb.2005.10.003

Hain D, Gallego-Flores T, Klinkmann M, Macias A, Ciirdaeva E, Arends A, Thum C, Tushev G, Kretschmer F, Tosches MA, et al. 2022. Molecular diversity and evolution of neuron types in the amniote brain. *Science* **377**: eabp8202. doi:10.1126/science.abp8202

Harris KD, Shepherd GM. 2015. The neocortical circuit: themes and variations. *Nat Neurosci* **18**: 170–181. doi: 10.1038/nn.3917

Hedges SB. 2002. The origin and evolution of model organisms. *Nat Rev Genet* **3**: 838–849. doi:10.1038/nrg929

Herold C, Diekamp B, Güntürkün O. 2008. Stimulation of dopamine D1 receptors in the avian fronto-striatal system adjusts daily cognitive fluctuations. *Behav Brain Res* **194**: 223–229. doi:10.1016/j.bbr.2008.07.017

Huber L, Aust U. 2017. Mechanisms of perceptual categorization in birds. In *Avian cognition* (ed. ten Cate C, Healy S), pp. 208–228. Cambridge University Press, Cambridge.

Hussar CR, Pasternak T. 2009. Flexibility of sensory representations in prefrontal cortex depends on cell type. *Neuron* **64**: 730–743. doi:10.1016/j.neuron.2009.11.018

Jacob SN, Nieder A. 2014. Complementary roles for primate frontal and parietal cortex in guarding working memory from distractor stimuli. *Neuron* **83**: 226–237. doi:10.1016/ j.neuron.2014.05.009

Jacob SN, Ott T, Nieder A. 2013. Dopamine regulates two classes of primate prefrontal neurons that represent sensory signals. *J Neurosci* **33**: 13724–13734. doi:10.1523/ JNEUROSCI.0210-13.2013

Jacob SN, Stalter M, Nieder A. 2016. Cell-type-specific modulation of targets and distractors by dopamine D1 receptors in primate prefrontal cortex. *Nat Commun* **7**: 13218. doi:10.1038/ncomms13218

Jarvis ED, Güntürkün O, Bruce L, Csillag A, Karten H, Kuenzel W, Medina L, Paxinos G, Perkel DJ, Shimizu T, et al. 2005. Avian brains and a new understanding of vertebrate brain evolution. *Nat Rev Neurosci* **6**: 151–159. doi:10.1038/nrn1606

Johnston K, Everling S. 2009. Task-relevant output signals are sent from monkey dorsolateral prefrontal cortex to the superior colliculus during a visuospatial working memory task. *J Cogn Neurosci* **21**: 1023–1038. doi:10.1162/jocn .2009.21067

Johnston K, DeSouza JF, Everling S. 2009. Monkey prefrontal cortical pyramidal and putative interneurons exhibit differential patterns of activity between prosaccade and antisaccade tasks. *J Neurosci* **29**: 5516–5524. doi:10.1523/ JNEUROSCI.5953-08.2009

Karakuyu D, Herold C, Güntürkün O, Diekamp B. 2007. Differential increase of extracellular dopamine and serotonin in the "prefrontal cortex" and striatum of pigeons during working memory. *Eur J Neurosci* **26**: 2293–2302. doi:10.1111/j.1460-9568.2007.05840.x

Cite this article as *Cold Spring Harb Perspect Biol* doi: 10.1101/cshperspect.a041526

Karten HJ. 1969. The organization of the avian telencephalon and some speculations on the phylogeny of the amniote telencephalon. *Ann NY Acad Sci* **167:** 164–179. doi:10.1111/j.1749-6632.1969.tb20442.x

Karten HJ. 2013. Neocortical evolution: neuronal circuits arise independently of lamination. *Curr Biol* **23:** R12–R15. doi:10.1016/j.cub.2012.11.013

Kebschull JM, Richman EB, Ringach N, Friedmann D, Albarran E, Kolluru SS, Jones RC, Allen WE, Wang Y, Cho SW, et al. 2020. Cerebellar nuclei evolved by repeatedly duplicating a conserved cell-type set. *Science* **370:** eabd5059. doi:10.1126/science.abd5059

Kersten Y, Friedrich-Müller B, Nieder A. 2022. A brain atlas of the carrion crow (*Corvus corone*). *J Comp Neurol* **530:** 3011–3038. doi:10.1002/cne.25392

Kirschhock ME, Nieder A. 2022. Number selective sensorimotor neurons in the crow translate perceived numerosity into number of actions. *Nat Commun* **13:** 6913. doi:10.1038/s41467-022-34457-5

Kirschhock ME, Ditz HM, Nieder A. 2021. Behavioral and neuronal representation of numerosity zero in the crow. *J Neurosci* **41:** 4889–4896. doi:10.1523/JNEUROSCI.0090-21.2021

Kobylkov D, Mayer U, Zanon M, Vallortigara G. 2022a. Number neurons in the nidopallium of young domestic chicks. *Proc Natl Acad Sci* **119:** e2201039119. doi:10.1073/pnas.2201039119

Kobylkov D, Musielak I, Haase K, Rook N, von Eugen K, Dedek K, Güntürkün O, Mouritsen H, Heyers D. 2022b. Morphology of the "prefrontal" nidopallium caudolaterale in the long-distance night-migratory Eurasian blackcap (*Sylvia atricapilla*). *Neurosci Lett* **789:** 136869. doi:10.1016/j.neulet.2022.136869

Kosche G, Vallentin D, Long MA. 2015. Interplay of inhibition and excitation shapes a premotor neural sequence. *J Neurosci* **35:** 1217–1227. doi:10.1523/JNEUROSCI.4346-14.2015

Kutter EF, Bostroem J, Elger CE, Mormann F, Nieder A. 2018. Single neurons in the human brain encode numbers. *Neuron* **100:** 753–761.e4. doi:10.1016/j.neuron.2018.08.036

Kutter EF, Boström J, Elger CE, Nieder A, Mormann F. 2022. Neuronal codes for arithmetic rule processing in the human brain. *Curr Biol* **32:** 1275–1284.e4. doi:10.1016/j.cub.2022.01.054

Kutter EF, Dehnen G, Borger V, Surges R, Mormann F, Nieder A. 2023. Distinct neuronal representation of small and large numbers in the human medial temporal lobe. *Nat Hum Behav* doi:10.1038/S41562-023-01709-5

Lee SH, Kwan AC, Zhang S, Phoumthipphavong, V, Flannery, JG, Masmanidis, SC, Taniguchi, H, Huang ZJ, Zhang F, Boyden ES, et al. 2012. Activation of specific interneurons improves V1 feature selectivity and visual perception. *Nature* **488:** 379–383. doi:10.1038/nature11312

Lust K, Maynard A, Gomes T, Fleck JS, Camp JG, Tanaka EM, Treutlein B. 2022. Single-cell analyses of axolotl telencephalon organization, neurogenesis, and regeneration. *Science* **377:** eabp9262. doi:10.1126/science.abp9262

Mansouri FA, Freedman DJ, Buckley MJ. 2020. Emergence of abstract rules in the primate brain. *Nat Rev Neurosci* **21:** 595–610. doi:10.1038/s41583-020-0364-5

Markram H, Toledo-Rodriguez M, Wang Y, Gupta A, Silberberg G, Wu C. 2004. Interneurons of the neocortical inhibitory system. *Nat Rev Neurosci* **5:** 793–807. doi:10.1038/nrn1519

Merchant H, de Lafuente V, Peña-Ortega F, Larriva-Sahd J. 2012. Functional impact of interneuronal inhibition in the cerebral cortex of behaving animals. *Prog Neurobiol* **99:** 163–178. doi:10.1016/j.pneurobio.2012.08.005

Miller EK. 2013. The "working" of working memory. *Dialogues Clin Neurosci* **15:** 411–418. doi:10.31887/DCNS.2013.15.4/emiller

Miller EK, Nieder A, Freedman DJ, Wallis JD. 2003. Neural correlates of categories and concepts. *Curr Opin Neurobiol* **13:** 198–203. doi:10.1016/S0959-4388(03)00037-0

Miller EK, Lundqvist M, Bastos AM. 2018. Working memory 2.0. *Neuron* **100:** 463–475. doi:10.1016/j.neuron.2018.09.023

Moll FW, Nieder A. 2015. Cross-modal associative mnemonic signals in crow endbrain neurons. *Curr Biol* **25:** 2196–2201. doi:10.1016/j.cub.2015.07.013

Mrzljak L, Bergson C, Pappy M, Huff R, Levenson R, Goldman-Rakic PS. 1996. Localization of dopamine D4 receptors in GABAergic neurons of the primate brain. *Nature* **381:** 245–248. doi:10.1038/381245a0

Muly EC 3rd, Szigeti K, Goldman-Rakic PS. 1998. D_1 receptor in interneurons of macaque prefrontal cortex: distribution and subcellular localization. *J Neurosci* **18:** 10553–10565. doi:10.1523/JNEUROSCI.18-24-10553.1998

Nieder A. 2004. The number domain—can we count on parietal cortex? *Neuron* **44:** 407–409. doi:10.1016/j.neuron.2004.10.020

Nieder A. 2016a. Representing something out of nothing: the dawning of zero. *Trends Cogn Sci* **20:** 830–842. doi:10.1016/j.tics.2016.08.008

Nieder A. 2016b. The neuronal code for number. *Nat Rev Neurosci* **17:** 366–382. doi:10.1038/nrn.2016.40

Nieder A. 2017a. Inside the corvid brain—probing the physiology of cognition in crows. *Curr Opin Behav Sci* **16:** 8–14. doi:10.1016/j.cobeha.2017.02.005

Nieder A. 2017b. Evolution of cognitive and neural solutions enabling numerosity judgements: lessons from primates and corvids. *Philos Trans R Soc Lond B Biol Sci* **373:** 20160514. doi:10.1098/rstb.2016.0514

Nieder A. 2020. The adaptive value of numerical competence. *Trends Ecol Evol* **35:** 605–617. doi:10.1016/j.tree.2020.02.009

Nieder A. 2021a. Neuroethology of number sense across the animal kingdom. *J Exp Biol* **224:** jeb218289. doi:10.1242/jeb.218289

Nieder A. 2021b. The evolutionary history of brains for numbers. *Trends Cogn Sci* **25:** 608–621. doi:10.1016/j.tics.2021.03.012

Nieder A. 2021c. Consciousness without cortex. *Curr Opin Neurobiol* **71:** 69–76. doi:10.1016/j.conb.2021.09.010

Nieder A, Miller EK. 2004. A parieto-frontal network for visual numerical information in the monkey. *Proc Natl Acad Sci* **101:** 7457–7462. doi:10.1073/pnas.0402239101

Nieder A, Freedman DJ, Miller EK. 2002. Representation of the quantity of visual items in the primate prefron-

tal cortex. *Science* **297**: 1708–1711. doi:10.1126/science
.1072493

Nieder A, Diester I, Tudusciuc O. 2006. Temporal and spatial enumeration processes in the primate parietal cortex. *Science* **313**: 1431–1435. doi:10.1126/science.1130308

Nieder A, Wagener L, Rinnert P. 2020. A neural correlate of sensory consciousness in a corvid bird. *Science* **369**: 1626–1629. doi:10.1126/science.abb1447

Okuyama S, Kuki T, Mushiake H. 2015. Representation of the numerosity "zero" in the parietal cortex of the monkey. *Sci Rep* **5**: 10059. doi:10.1038/srep10059

Ott T, Nieder A. 2017. Dopamine D2 receptors enhance population dynamics in primate prefrontal working memory circuits. *Cereb Cortex* **27**: 4423–4435.

Ott T, Nieder A. 2019. Dopamine and cognitive control in prefrontal cortex. *Trends Cogn Sci* **23**: 213–234. doi:10
.1016/j.tics.2018.12.006

Ott T, Jacob SN, Nieder A. 2014. Dopamine receptors differentially enhance rule coding in primate prefrontal cortex neurons. *Neuron* **84**: 1317–1328. doi:10.1016/j.neuron
.2014.11.012

Pfeffer CK, Xue M, He M, Huang ZJ, Scanziani M. 2013. Inhibition of inhibition in visual cortex: the logic of connections between molecularly distinct interneurons. *Nat Neurosci* **16**: 1068–1076. doi:10.1038/nn.3446

Povysheva NV, Gonzalez-Burgos G, Zaitsev AV, Kröner S, Barrioneuvo G, Lewis DA, Krimer LS. 2006. Properties of excitatory synaptic responses in fast-spiking interneurons and pyramidal cells from monkey and rat prefrontal cortex. *Cereb Cortex* **16**: 541–552. doi:10.1093/cercor/bhj002

Puelles L. 2017. Comments on the updated tetrapartite pallium model in the mouse and chick, featuring a homologous claustro-insular complex. *Brain Behav Evol* **90**: 171–189. doi:10.1159/000479782

Pusch R, Clark W, Rose J, Güntürkün O. 2023. Visual categories and concepts in the avian brain. *Anim Cogn* **26**: 153–173. doi:10.1007/s10071-022-01711-8

Qi XL, Meyer T, Stanford TR, Constantinidis C. 2011. Changes in prefrontal neuronal activity after learning to perform a spatial working memory task. *Cereb Cortex* **21**: 2722–2732. doi:10.1093/cercor/bhr058

Ramirez-Cardenas A, Moskaleva M, Nieder A. 2016. Neuronal representation of numerosity zero in the primate parieto-frontal number network. *Curr Biol* **26**: 1285–1294. doi:10.1016/j.cub.2016.03.052

Rao SG, Williams GV, Goldman-Rakic PS. 1999. Isodirectional tuning of adjacent interneurons and pyramidal cells during working memory: evidence for microcolumnar organization in PFC. *J Neurophysiol* **81**: 1903–1916. doi:10.1152/jn.1999.81.4.1903

Rao SG, Williams GV, Goldman-Rakic PS. 2000. Destruction and creation of spatial tuning by disinhibition: GABA$_A$ blockade of prefrontal cortical neurons engaged by working memory. *J Neurosci* **20**: 485–494. doi:10.1523/JNEUROSCI.20-01-00485.2000

Reiner A. 2013. You are who you talk with—a commentary on Dugas-Ford et al. PNAS, 2012. *Brain Behav Evol* **81**: 146–149. doi:10.1159/000348281

Rinnert P, Nieder A. 2021. Neural code of motor planning and execution during goal-directed movements in crows.

J Neurosci **41**: 4060–4072. doi:10.1523/JNEUROSCI.07
39-20.2021

Rinnert P, Kirschhock ME, Nieder A. 2019. Neuronal correlates of spatial working memory in the endbrain of crows. *Curr Biol* **29**: 2616–2624.e4. doi:10.1016/j.cub
.2019.06.060

Santana N, Artigas F. 2017. Laminar and cellular distribution of monoamine receptors in rat medial prefrontal cortex. *Front Neuroanat* **11**: 87. doi:10.3389/fnana.2017.00087

Sawamura H, Shima K, Tanji J. 2002. Numerical representation for action in the parietal cortex of the monkey. *Nature* **415**: 918–922. doi:10.1038/415918a

Scarf D, Boy K, Uber Reinert A, Devine J, Güntürkün O, Colombo M. 2016. Orthographic processing in pigeons (*Columba livia*). *Proc Natl Acad Sci* **113**: 11272–11276. doi:10.1073/pnas.1607870113

Schneider DM, Woolley SM. 2013. Sparse and background-invariant coding of vocalizations in auditory scenes. *Neuron* **79**: 141–152. doi:10.1016/j.neuron.2013.04.038

Schoups A, Vogels R, Qian N, Orban G. 2001. Practising orientation identification improves orientation coding in V1 neurons. *Nature* **412**: 549–553. doi:10.1038/3508
7601

Seger CA, Miller EK. 2010. Category learning in the brain. *Annu Rev Neurosci* **33**: 203–219. doi:10.1146/annurev
.neuro.051508.135546

Shepherd GM. 2009. Intracortical cartography in an agranular area. *Front Neurosci* **3**: 337–343. doi:10.3389/neuro.01
.030.2009

Smiley JF, Levey AI, Ciliax BJ, Goldman-Rakic PS. 1994. D1 dopamine receptor immunoreactivity in human and monkey cerebral cortex: predominant and extrasynaptic localization in dendritic spines. *Proc Natl Acad Sci* **91**: 5720–5724. doi:10.1073/pnas.91.12.5720

Soto FA, Wasserman EA. 2014. Mechanisms of object recognition: what we have learned from pigeons. *Front Neural Circuits* **8**: 122.

Spiro JE, Dalva MB, Mooney R. 1999. Long-range inhibition within the zebra finch song nucleus RA can coordinate the firing of multiple projection neurons. *J Neurophysiol* **81**: 3007–3020. doi:10.1152/jn.1999.81.6.3007

Spool JA, Macedo-Lima M, Scarpa G, Morohashi Y, Yazaki-Sugiyama Y, Remage-Healey L. 2021. Genetically identified neurons in avian auditory pallium mirror core principles of their mammalian counterparts. *Curr Biol* **31**: 2831–2843.e6. doi:10.1016/j.cub.2021.04.039

Stacho M, Herold C, Rook N, Wagner H, Axer M, Amunts K, Güntürkün O. 2020. A cortex-like canonical circuit in the avian forebrain. *Science* **369**: eabc5534. doi:10.1126/science.abc5534

Stalter M, Westendorff S, Nieder A. 2020. Dopamine gates visual signals in monkey prefrontal cortex neurons. *Cell Rep* **30**: 164–172.e4. doi:10.1016/j.celrep.2019.11.082

Striedter GF, Northcutt RG. 2020. *Brains through time.* Oxford University Press, Oxford.

Suzuki IK, Kawasaki T, Gojobori T, Hirata T. 2012. The temporal sequence of the mammalian neocortical neurogenetic program drives mediolateral pattern in the chick pallium. *Dev Cell* **22**: 863–870. doi:10.1016/j.devcel.2012
.01.004

Cite this article as *Cold Spring Harb Perspect Biol* doi: 10.1101/cshperspect.a041526

Tosches MA, Yamawaki TM, Naumann RK, Jacobi AA, Tushev G, Laurent G. 2018. Evolution of pallium, hippocampus, and cortical cell types revealed by single-cell transcriptomics in reptiles. *Science* **360**: 881–888. doi:10 .1126/science.aar4237

Tremblay R, Lee S, Rudy B. 2016. GABAergic interneurons in the neocortex: from cellular properties to circuits. *Neuron* **91**: 260–292. doi:10.1016/j.neuron.2016.06.033

Vallentin D, Bongard S, Nieder A. 2012. Numerical rule coding in the prefrontal, premotor, and posterior parietal cortices of macaques. *J Neurosci* **32**: 6621–6630. doi:10 .1523/JNEUROSCI.5071-11.2012

Veit L, Hartmann K, Nieder A. 2014. Neuronal correlates of visual working memory in the corvid endbrain. *J Neurosci* **34**: 7778–7786. doi:10.1523/JNEUROSCI.0612-14.2014

Veit L, Pidpruzhnykova G, Nieder A. 2015. Associative learning rapidly establishes neuronal representations of upcoming behavioral choices in crows. *Proc Natl Acad Sci* **112**: 15208–15213. doi:10.1073/pnas.1509760112

Viswanathan P, Nieder A. 2013. Neuronal correlates of a visual "sense of number" in primate parietal and prefrontal cortices. *Proc Natl Acad Sci* **110**: 11187–11192. doi:10 .1073/pnas.1308141110

Viswanathan P, Nieder A. 2015. Differential impact of behavioral relevance on quantity coding in primate frontal and parietal neurons. *Curr Biol* **25**: 1259–1269. doi:10 .1016/j.cub.2015.03.025

Viswanathan P, Nieder A. 2017a. Visual receptive field heterogeneity and functional connectivity of adjacent neurons in primate frontoparietal association cortices. *J Neurosci* **37**: 8919–8928. doi:10.1523/JNEUROSCI.0829-17 .2017

Viswanathan P, Nieder A. 2017b. Comparison of visual receptive fields in the dorsolateral prefrontal cortex and ventral intraparietal area in macaques. *Eur J Neurosci* **46**: 2702–2712. doi:10.1111/ejn.13740

Viswanathan P, Nieder A. 2020. Spatial neuronal integration supports a global representation of visual numerosity in primate association cortices. *J Cogn Neurosci* **32**: 1184–1197. doi:10.1162/jocn_a_01548

von Eugen K, Tabrik S, Güntürkün O, Ströckens F. 2020. A comparative analysis of the dopaminergic innervation of the executive caudal nidopallium in pigeon, chicken, zebra finch, and carrion crow. *J Comp Neurol* **528**: 2929–2955. doi:10.1002/cne.24878

Wagener L, Nieder A. 2023. Categorical representation of abstract spatial magnitudes in the executive telencephalon of crows. *Curr Biol* **33**: 2151–2162.e5. doi:10.1016/j.cub .2023.04.013

Wagener L, Loconsole M, Ditz HM, Nieder A. 2018. Neurons in the endbrain of numerically naive crows spontaneously encode visual numerosity. *Curr Biol* **28**: 1090–1094.e4. doi:10.1016/j.cub.2018.02.023

Waldmann C, Güntürkün O. 1993. The dopaminergic innervation of the pigeon caudolateral forebrain: immunocytochemical evidence for a "prefrontal cortex" in birds? *Brain Res* **600**: 225–234. doi:10.1016/0006-8993(93) 91377-5

Wang XJ. 1999. Synaptic basis of cortical persistent activity: the importance of NMDA receptors to working memory. *J Neurosci* **19**: 9587–9603. doi:10.1523/JNEUROSCI.19-21-09587.1999

Wang XJ. 2001. Synaptic reverberation underlying mnemonic persistent activity. *Trends Neurosci* **24**: 455–463. doi:10.1016/S0166-2236(00)01868-3

Wang XJ, Tegner J, Constantinidis C, Goldman-Rakic PS. 2004. Division of labor among distinct subtypes of inhibitory neurons in a cortical microcircuit of working memory. *Proc Natl Acad Sci* **101**: 1368–1373.

Wang Y, Brzozowska-Prechtl A, Karten HJ. 2010. Laminar and columnar auditory cortex in avian brain. *Proc Natl Acad Sci* **107**: 12676–12681. doi:10.1073/pnas.10066 45107

Wilson FA, O'Scalaidhe SP, Goldman-Rakic PS. 1994. Functional synergism between putative γ-aminobutyrate-containing neurons and pyramidal neurons in prefrontal cortex. *Proc Natl Acad Sci* **91**: 4009–4013.

Wonders CP, Anderson SA. 2006. The origin and specification of cortical interneurons. *Nat Rev Neurosci* **7**: 687–696. doi:10.1038/nrn1954

Woych J, Ortega Gurrola A, Deryckere A, Jaeger ECB, Gumnit E, Merello G, Gu J, Joven Araus A, Leigh ND, Yun M, et al. 2022. Cell-type profiling in salamanders identifies innovations in vertebrate forebrain evolution. *Science* **377**: eabp9186. doi:10.1126/science.abp9186

Wynne B, Güntürkün O. 1995. Dopaminergic innervation of the telencephalon of the pigeon (*Columba livia*): a study with antibodies against tyrosine hydroxylase and dopamine. *J Comp Neurol* **357**: 446–464.

Yanagihara S, Yazaki-Sugiyama Y. 2016. Auditory experience-dependent cortical circuit shaping for memory formation in bird song learning. *Nat Commun* **7**: 11946. doi:10.1038/ncomms11946

Yang T, Maunsell JH. 2004. The effect of perceptual learning on neuronal responses in monkey visual area V4. *J Neurosci* **24**: 1617–1626. doi:10.1523/JNEUROSCI.4442-03 .2004

Index